EVERYTHING FITS

by

Brendt Vitale

Oracle Origin Books

First Published by Oracle Origin Books 06/30/2025

www.howeverythingfits.com

ISBN: 979-8-218-68088-6

Library of Congress Control Number: 2025910193

Printed in the United States of America
Seattle Washington

CONTENTS

PART TWO: F I T S

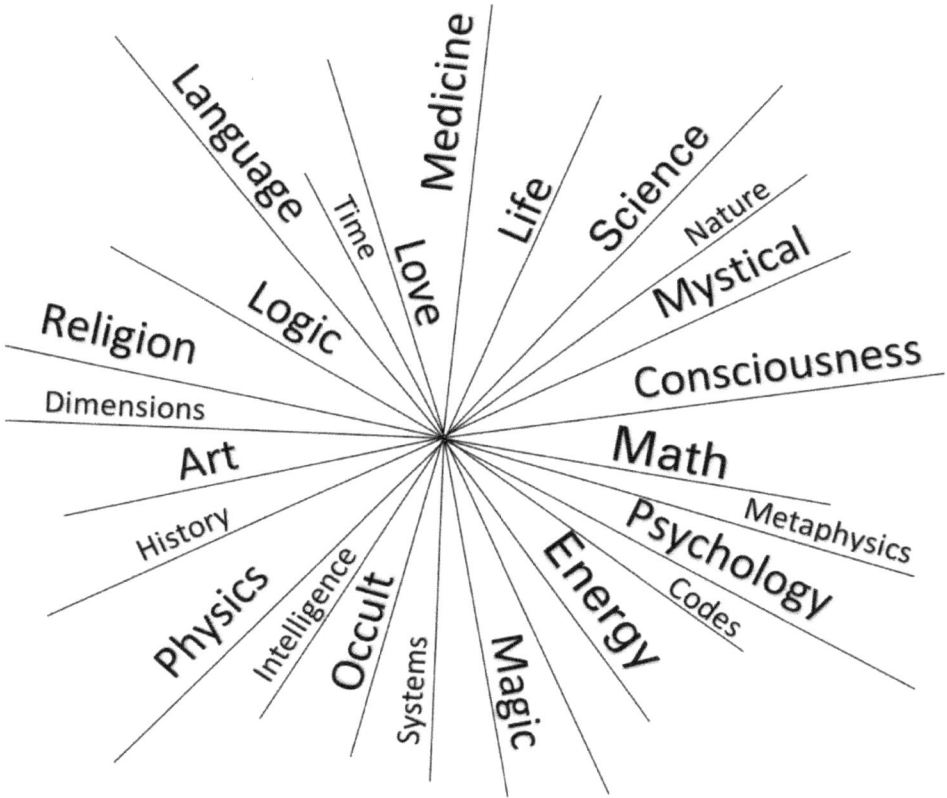

INTRODUCTION

When I wrote "The Levels Theory" 40 years ago the idea of a Theory of Everything resonated deeply within, stirring up a degree of excitement that captivates any young mind naturally curious to begin with. For me it was an ever-present yearning to understand how the world works at every level, along with an instinctive feeling that everything is interconnected in some harmonious dynamic way. I suppose this quest was inevitable given my relentless curiosity and early tendency of taking things apart to see how they worked. My father reminded me that at the age of five I completely disassembled a Singer sewing machine with nothing more than a screwdriver and hammer – back when they were built solid and compact. Of course, this insatiable curiosity had a downside since opening up toys with mechanical and electrical components usually never worked again. I can still remember the cool smell of transistor circuitry inside a radio…and the intricate electric components inside the expensive model train engine I ruined – costly at the time but cheap in the long run considering the path it led me on.

You are now on that same path – a seeker looking for answers to life's mysteries and hidden patterns. Your interest in reading this subject reveals a similar sense of curiosity and wonder about how the world works. Over the years I've maintained a peripheral interest in this subject, waiting for some science-oriented genius or a philosophical guru to present a truly comprehensive Theory of Everything; one that covers the entire spectrum of categories, actually worthy of using the term "everything". So far no one's done it, or even come close. I cringe when I hear the physics community repeatedly tout their "theory of everything" as they pursue the elusive goal of unifying four fundamental forces. Sorry folks but physics is just one field of science among many others and doesn't even come close to addressing "everything". Other philosophers and science emissaries have occasionally offered grand theories or complex models but unfortunately, they always come down to limited perspectives of specialized fields. In fact, it's the very self-limiting nature of being an "expert" in a field that's prevented anyone from developing a comprehensive model that truly captures everything.

This book *IS* about everything. Gulp. Yeah, that's right, *everything*, including science and non-science, the tangible and intangible, metaphysics, religion, the arts, abstract and hidden realities, life and non-life, time and space, human reality, and much, much more. The trick is to define what *everything* actually includes, organize it in a practical manner, and present it at the right level – not too general and not too mired in the weeds. What's needed is a perfect balance of scope that captures universal categories, elements, qualities and the right depth of detail and analysis, all while avoiding technical jargon and confusing esoteric language.

This book is also about one thing; *Unity*. We're going to reveal how everything is connected as one single comprehensive whole. That's really what the theory or model is all about. We'll boldly reveal how everything and one thing are simply two sides of the same coin, symbiotically connected and metaphorically identical.

Finally, this book is also about one more thing – YOU. Everything and Unity only appear separately and differently because of YOU. Reality is created and experienced by your perception of the world *out there* based on your middle position in the center of it all. Human reality is entirely based on each individual occupying a center reference point which everything else is connected to. YOU are literally a separating filter at the center of the universe, which is a perfect starting point for making sense of it all.

What follows is essentially a two-part presentation mirroring the book's title: Part 1 "Everything and Part 2 "Fits". In part one we'll define what *everything* actually is, what it covers and how it relates to us. We'll identify broad categories and construct a model that shows how they collectively relate to each other. This model will comprehensively contain everything there is, spanning all levels of perspective across universal fields of interest. Most importantly, it will make a lot more sense when we insert YOU right at the center of it. Everything isn't just *out there*, it's also *in here* and all around each of us as you'll soon see.

Part two is the grand synthesis putting it all together. It's where we'll uncover unities in every key field that reveal the interconnectedness of everything. We'll show universal themes, recurring patterns and connecting spectrums present everywhere. Laws of nature and hidden codes will confirm things are not as separate as they appear. Unities in every field of science will reveal a single connected hierarchy of mankind's collective accumulated knowledge. We'll do the same with life itself, showing how all living things are connected along a single spectrum connected with everything else above and beyond. Our grand model of everything wouldn't be complete without addressing unities in physics, where we'll reveal the universal connecting infrastructure of physical reality and how it applies to everything. Finally, we'll identify the connecting universal patterns that tie it all together, at every level, on every scale.

Producing a grand theory of everything isn't just about the final product, which in this case is a practical model that interconnects everything. It's also about the process and journey getting there. In this quest to reverse engineer reality and figure out how it all works, I discovered the secret pattern that ties everything together. It applies at all levels and is present in every field of interest. If you don't see it in the earlier chapters, you'll definitely catch on in the later parts of the book. Once you know the pattern and look around, you'll see it everywhere hiding in plain sight. That's an Aha moment waiting for you just up ahead…and once you get it, you'll clearly understand how everything fits.

PART ONE: *EVERYTHING*

Chapter 1   What's Everything?

The problem with theories of everything is they never actually cover *everything*. Philosophers, scientists and various scholars through history have attempted to explain big picture reality. Some simple, others elegant but limited, while few actually qualified as a genuine *theory of everything* (TOE). What they all have in common is a lack of comprehensive scope. *No one* has ever constructed a TOE that captures or covers everything in reality…ever. What they have done is produce models or concepts that are heavily skewed toward a single discipline, unwittingly excluding the vast majority of other areas and fields. So far, the most common attempts have leaned toward science, especially physics, along with fields related to consciousness and metaphysical perennial philosophies. A few also tried to base their models entirely on mathematics.

This makes sense when you consider most of the individuals we look upon as intellectual giants are experts in a particular field – specialists who focus on one or a few areas. Dealing with the subject of everything requires a generalist with a strategic perspective, purposefully limiting the specific. What's needed here is an intellectual jack of all trades, or in the classic spirit, a renaissance man or woman – someone with broad knowledge of diverse fields and an appreciation of interdisciplinary subjects. Since modern society rewards specialists over generalists, there's a built-in disincentive to developing a broad scope *TOE*.

DEFINING EVERYTHING

Where should we start? There's so much to cover and it's such a broad scope. We really have to begin with general areas. We'll have to include all people, places, things, ideas, man-made objects, and all mental constructs. We must capture everything we see, feel and experience in our world and beyond. We must consider spatial and temporal limits – here, everywhere and everywhen. We must span vast orders of magnitude, grasping big numbers beyond comprehension, the big and small, from subatomic particles to galaxies and superclusters, or the universe itself.

The first step is to assume the right mindset; a big picture, strategic perspective with a sense of detachment. A cosmic eye that perceives all things individually, collectively and wholistically. On this higher level of perception, we'll cross scales of big and small, few and many, near and far – appreciating the full spectrum of dimensions and degrees in between. We'll balance breadth and depth, taking in parts and wholes, accommodating every partial fraction, segment and fractal aspect while maintaining clarity of the bigger picture.

The next step is to identify major categories that break down and summarize the many into the few, manageable groups. We don't want to be too specific or too general, but just the right balance of sets. A good start would be referencing the Dewey decimal system used in libraries to organize content. It breaks down into 10 major categories which are then further refined within each one:

Dewey Decimal System

General	001-099	Encyclopedias, curiosities, wonders, unexplained mysteries
Philosophy	100-199	The self, psychology, logic, ethics, metaphysics,
Religion	200-299	Christianity, Judaism, Buddhism, Hinduism, Mythology
Social Science	300-399	Customs, cultures, laws, political science, education, etiquette
Languages	400-499	Dictionaries, speech, linguistics, sign language, foreign
Natural Science	500-599	Mathematics, physics, astronomy, chemistry, biology, earth
Applied Science	600-699	Engineering, manufacturing, medical, agriculture, management
Arts & Recreation	700-799	Drawing, painting, graphics, handicrafts, music, games, sports
Literature	800-899	Classics, Latin, English, European, rhetoric, poetry, plays
Geography & History	900-999	Landforms, travel, atlases, exploration, countries, genealogy
Biography	92-920	Single person, multiple people

These groups are further divided by specific topics in each area. For example:

Geography and History (900's)

910 Geography, Travel	*Further Category Break Down*
920 Genealogy, Biography	990 Pacific Ocean Islands
930 Ancient history	993 New Zealand and Melanesia
940 Europe	994 Australia
950 Asia	995 New Guinea
960 Africa	996 Polynesia
970 North America	997 Atlantic Ocean
980 South America	998 Arctic Region
990 Pacific Ocean Islands	999 Antarctic Regions

This is a fairly practical system for organizing a broad array of information. It also reflects the type of information that is most prominent in general culture. What it doesn't do is differentiate the quality of that information or relative importance of each category. For instance, geography fills most of the 900's group but is largely redundant material with variations by location. Compare that with the 700's group covering arts and recreation where you'll find a rich diversity of types, forms and applications. So, some categories vary by degree while other do in kind. It's the difference between quantity and quality of information. Not all data is equal and there's a heck of a lot out there to sift through. Maintaining quality of data is a real challenge in today's society where information overload makes it difficult to distinguish what's important and what's just noise. But again, this strategic grouping of sets is a good place to start. To capture *everything,* we can build on the

Dewey decimal system categories and add a few more. Our Taxonomy must cover all bases and leave nothing out. Consider the following comprehensive list:

CATEGORIES OF EVERYTHING

Outer and Inner worlds – Micro & Macro levels, Infinite scales
Tangible and Intangible things – Quantifiable and unquantifiable
Hidden things – Invisible, Abstraction, Beyond senses
Relational properties – Patterns, Forms, Dimensions, Laws, Codes, Mathematics
Nature – Environment, Animal kingdom, Weather & Climate, Geography and Earth
Societal realm – Government, Economy, Culture, History, Language, Politics, Institution
Sciences – Hard & Soft science, Earth & Life sciences, Formal, Social & Applied science
Non-sciences – Religion, Pseudoscience & Occult, Mythology, Parascience, Paranormal,
                Metaphysics, Magical & Mystical, Fiction & Fantasy
Human reality – Consciousness, Behavior, States of being, Experience, Development,
                Relationships, Expressions, Personality
Dynamics of change – Processes, Cycles, Waves, Transformation, Transcendence,
                Dialectics, Quantum mechanics
Living and non-living things – Organic Life, Pseudo Life, Machines, Systems,
                Artificial Intelligence, Evolution

I challenge you to think of something, anything, that's not covered somewhere within these categories. This will be our starting basis for developing, refining and completing a comprehensive *Theory of Everything* (TOE). First, we'll go over these categories in a little more detail, then address what's needed to put it all together in a connected working model.

SCIENCES. What better place to start than fields where we study things. There's a specific science for just about every area of interest you can think of. Just organizing the various types of sciences is a task in itself. To keep it simple, there are 2 broad types: Hard and Soft science. The first includes natural and formal, the latter (soft science) refers to the humanities.

Hard Science consists of physics, chemistry, biology and astronomy. A subset of this area is *physical science*, relating to the natural world: geology, botany, anthropology, meteorology, zoology, paleontology, materials science. *Earth science* is a further subset with geology, oceanology, geography, paleontology, environmental science, geophysics, hydrology, ecology, climatology, tectonics, atmospheric science and minerology (includes some duplicates)

Life Sciences focus more in the area of organic life and include biology, animals, plants, human life, genetics, immunology, microbiology, molecular biology, zoology, botany, anatomy, and toxicology.

Formal Sciences analyze abstract structures, including mathematics, logic, statistics, computer science, information theory, artificial intelligence, systems science and theoretical linguistics.

Soft Science transitions to areas that involve human interaction, so they aren't as precise nor objective as their hard science counterparts. They're referred to as the humanities and include psychology, history, sociology, political science and communication studies. Social sciences are a subset expanding on these to include anthropology, archaeology, linguistics, economics, education, ethics, criminology, public health, gerontology, law, plus cognitive science, library science, behavioral science, information science, developmental humanities (human culture/experience), philosophy, literature, art, dance, theater, and music.

Applied Sciences are those fields that deal with real-world applications of research and development, solving current problems and making systems more efficient. They include *Engineering* (mechanical, biomedical, civil, physics, chemical, industrial, electrical, bio), *Sciences* (Aeronautical, forensic, food, health, environmental, computer, materials), plus applied mathematics, biochemistry, microbiology, architecture and criminology.

These broad categories of science are just a generalized overview and don't represent the entire comprehensive list. Many fields within science have branches and subsets that further break down into areas of focus. Take the field of physics for instance where there are at least 24 separate branches related to the field:

Mechanics, aerodynamics, fluid mechanics, thermodynamics, astronomy, astrophysics, planetary science, stellar astronomy, astrobiology, Optics, quantum physics, particle physics, atomic physics, nuclear physics, molecular physics, biophysics, neuro physics, chemical physics, geo physics, atmospheric physics, computational physics, mathematical physics, materials physics, applied physics.

Sciences are obviously an important category to include in any grand theory that connects the dots of everything. But it's still just one piece of the puzzle and shouldn't diminish our attention to other areas of equal importance. There's so much more left to examine. We'll now move beyond the limits of a *TOE* based purely on science and look at areas considered *non-science*.

NON-SCIENCES: Many fields of study resemble science with their methodical, structured approaches, but they can't produce exact results in repeatable tests so they cannot be considered science. Referred to as pseudoscience, these various disciplines may use math, rules and logical methods but they aren't precise or 100% consistent. They are overtly subjective and routinely favor intuition over actual evidence. To be fair though the same could be said about some areas of recognized science. Consider the inaccuracies of forecasting from meteorologists and economists. One could make a case that they qualify, at least in part as pseudoscience since their results are generally no better than amateur stock pickers who often just break even with the market as a whole.

Pseudoscience and the Occult. Like the humanities, these disciplines examine human behavior which is subject to the unpredictable nature of individual people. Obviously, it's difficult to forecast the future to a certainty, but with experience, insight and orderly methodology, specialists can make likely estimations. Unlike their scientific counterparts, fields depending on human response are limited to probabilities. Like the "experts" forecasting weather or the economy, practitioners here offer similar degrees of likelihood, fully acknowledging limited precision.

The occult is associated with mystical practices covering a broad spectrum of disciplines, varying by degree of structure. Some are systematic and methodical while others are loosely applied. They include astrology, tarot, palm reading, numerology, graphology, I-Ching, and Feng Shui. These practices generally rely on unquantifiable insight and gut feeling. Supporting their case is made more difficult when skeptics group all variations of pseudoscience into one basket, equating those following well established rules and methods with others that are borderline quackery. Unscientific practices that deal with human health are looked down upon by medical professionals who consider them as fringe. This includes faith healing, homeopathy, aromatherapy, healing crystals, acupuncture, biorhythms, magnet therapy, reiki, qigong, phrenology, hypnotherapy, and psychoanalysis. While pseudo sciences may be practical and even effective on a case-by-case basis, they are not provable.

Paranormal fields examine phenomena that are beyond the realm of normal scientific understanding. With little empirical evidence, practitioners rely on anecdotal evidence and personal testimony. Areas of interest include extrasensory perception (ESP), telepathy, parapsychology, paranormal phenomena, ghost hunting, and ufology. Because paranormal phenomena are not quantifiable or measurable by scientific standards, the field is generally considered fringe.

Religion is not only non-science; it's literally the opposite of science. Its fundamentally faith based and oriented toward the inner spiritual world, not the physical outer one. Accordingly, there's a built-in aspect of indirectness in how it's practiced. Religions vary greatly but possess common attributes, including rituals, sermons, sacred texts, sanctified places, holy persons and prophets. Traditional practices may involve festivals, feasts, initiations, sacrifice, prayer, dance, singing, baptisms, matrimonial and funeral services. Where science examines empirical evidence, religion pursues spirituality, soul and the divine. Religion spans the entire globe and all of human history, most notably with Christianity, Judaism, Islam, Confucianism, Buddhism, Hinduism, Sikhism, Taoism, and many others.

Mythology is a forerunner of religion, sharing much in common with it. Both are systems relating to the sacred divine source directing the affairs of mankind. Myths are pre-science, folklore stories operating along supernatural lines to explain life in the world, attributing cause and effect to gods and supernatural figures. Stories and parables describing human behavior and events are linked to other worldly influences. Legends and fairy tales serve a similar purpose, using stories of supernatural influences from ancient times that offer insight into the current affairs of man in the world. Religion and mythology overlap when it comes to using stories of folklore to address both the unexplainable and the difficult to explain. Virtually all religions have a creation myth to explain how the world was formed by an all-powerful deity.

Fiction and Fantasy relate to an imaginary realm where anything one can conjure up may exist, similar to mythical folklore narratives. Fantasy is often just an exaggerated version of current reality; where laws of physics may be violated while your own limitations are conveniently removed. There's a purposeful suspension of reality for temporary enjoyment in the moment, though there are some who may get lost in the practice and lose complete touch of what's real and what isn't.

Metaphysics is a cross between philosophy and religion, bringing a higher-level, deep-dive perspective into the nature of *being* and examining universal truths. It's a foray into existence, purpose, knowing, causality, and God. Metaphysics draws from religion by pursuing the meaning of life, while leaning philosophically into the essences of life and the nature of consciousness.

Outer and Inner Worlds exist at the extremes of our senses and comprehension limits. They represent everything at the macro and micro level of reality. Outer space and inner space are boundless realms with seemingly infinite scales of length, size, distance, time, numbers – each having extreme upper and lower limits. Realms on opposite sides of the spectrum possess different phenomena, different rules, and different behaviors. It's as if there's dual universes, one in here and one out there. There's quantum level weirdness at the micro level and relativistic oddity at the macro level. Inner space contains trillions of cells while outer space contains trillions of stars. And our immense universe may be but one tiny part of a much bigger multiverse.

Tangible substance includes most of the stuff in our physical world, made up of "solid" matter with atomic structure. There are 118 elements (natural and man-made) comprised of semi-tangible particles that form the variety of compounds and mixtures in our environment. All this material stuff possesses specific physical attributes that vary by degree. Qualities such as hardness, texture, volume, mixture, flexibility, shape, color, weight, density, size, etc. are what differentiates elements, compounds and mixtures.

Intangible phenomena are immaterial in essence and possess significantly different attributes than their physical-material compliments do. Non-physical items include energy, spirit, thoughts, feelings, intuitions, dimensions and anything not of this world. It relates to the difference between brain and mind, heart and soul, being and becoming. Consciousness itself is a key aspect separating man from lesser life forms, yet it is completely intangible. Ironically, physics which is the king of "hard science" is fundamentally dependent on intangible phenomena, including space, time, light, dimensions and numerous abstract realities.

Unquantifiable things are a subset of the intangible group, where phenomena are difficult to define as they have serpentine, ambiguous natures. Simple examples would be feelings, ideas, imagination, love, and the divine.

Nature is the worldly realm that includes all organic life, the entire animal kingdom, the total environment, ecosystems, weather, a wide variety of features on earth's surface: land, sea, and air. Most importantly, it involves the complete interactive network of connections between every element present in the natural environment.

Hidden Reality covers a wide variety of things including the invisible, intangible, abstract, mysteries, unknown worlds, as well as non-things and Nothing (*NO thing*). Electromagnetic energy waves which are invisible and essential to life literally permeate our environment and are present wherever there is light of any kind, visible or not. Other phenomena are hidden in time, such as relics and artifacts, buried or left behind over millions of years never to be seen again. On the earth there are astronomical numbers of hidden microbes never seen – over 5 million-trillion-trillion. That's 5×10 to the 30th power. Hidden realms include the unexplored depths of earth's oceans and the far reaches of the Universe that can never be seen or known.

Relational Properties are the connecting links that tie things together, creating context and meaning. They include patterns, form, order, hierarchies, spectrums, levels and scales. These integrative principles are expressed in a variety of ways:

branching possibility trees, cause and effect sequences, mathematic correlations, relativity, codes, laws, symmetry, dimensions, and connecting links in networks. Relationships provide context to everything, establishing meaning and significance in the things related – a key component in our model unifying *everything*.

Living & Nonliving. What we define as life is not a black and white, objective reality but rather a somewhat ambiguous spectrum varying by degrees. On one end of the scale are organic forms, such as plants, animals, and man, while the other end includes things possessing a pseudo life, including machines, systems, networks, and AI. Where life begins and ends is relative, depending on arbitrary conditions and one's level of perspective. The essence of life is a key component required in any model that attempts to unify *everything*.

Dynamics of Change. The process nature of reality is *the* primary component of everything, serving as a foundational aspect for constructing our comprehensive *TOE*. Ironically, the only constant in the universe is change, and it manifests in a variety of expressions and forms. The dynamics of change include cycles and waves, quantum mechanics, relative motion, dialectics, geometrics, transcendence, transformation, and emergence. Change drives reality through processes driven by force and applied with directionality resulting in creative novelty.

Human Reality should be a fairly obvious component of a *TOE* because as the Greeks observed, man is the measure of all things. Taking it one step further, YOU are the measure of all things, and any model unifying everything must connect the dots back to YOU. Human reality covers a lot of ground, including universal human experiences, human nature, behavior, personality, consciousness, dreams, relationships, health, love, creative spirit, states of being, the mind, objective and subjective perception, and the meaning of life. Human beings possess multiple dimensions, including body, mind, soul and spirit. Man's inner nature is expressed outwardly in a variety of actions, expressions and creations. The Human experience is comprehensively captured in books, TV, movies, songs, and plays, providing snapshots of how man's needs and desires lead to engagement in pursuits of self-interest in a variety of arenas. Examining human reality reveals a spectrum of personality types, character traits, capabilities, and universal qualities.

Society is the collective level of human reality where all man-made things, ideas, systems and accumulated knowledge create a perplexing world of opportunities and threats. It's an enormous external arena presenting ongoing challenges, tests, competitions and rewards at every turn, always changing and moving in unpredictable directions, sweeping up all individuals, including you, along the way. Society is a complex integrated network of culture, economics, politics, language, technology and values. Individuals must navigate its intricate web of rules, regulations, hierarchies of power, unspoken customs, ambiguous messaging, information overload and ever-changing cultural tastes.

History is the cumulative record of societal progression – a comprehensive chronicle of all events, periods, experiences (both individual and collective), and social evolution. It provides a track record of developments in every field of human endeavor, including science, technology, military, agriculture, entertainment, arts, music, transportation, pop culture, apparel, housing, economics, medicine, education, etc. History serves as a documentation of how the world works, covering the rise and fall of nations and empires, movements and changes in populations, boom and bust cycles of economies, waves of social unrest and revolutions, perpetual states of conflict among bordering countries, and the recurring instances of large-scale catastrophe and renewal.

SUMMARY OF EVERYTHING

Everything breaks down into the following broad categories: Sciences and non-science, religion and metaphysics, tangible & intangible, inner and outer realms, living and non-living things, nature and the world environment, human reality, society, history, hidden reality, unquantifiable phenomenon, relational properties and the dynamics of change and processes. Everything that exists or can be imagined falls somewhere within these general areas. This is just the starting point of constructing a model of everything. No grand scheme can be considered comprehensive and complete without addressing ALL of these areas. It's a tall order but a necessary one. The tricky part is taking that broad, diverse and exhaustive compilation of stuff and organizing it into a cohesive, connected, relatable model – one that's simple, elegant and accurate.

From here the goal is to comprehensively unify ALL reality. To connect ALL the dots, reveal omnipresent unifying patterns and uncover universal truths. This requires a strategic perspective of expanded awareness, pushing the limits of all boundaries and ensuring maximum inclusiveness. To comprehend the big picture, we must embrace a detached perspective, see on all levels, go beyond the familiar, embrace both knowns & unknowns, and open up to higher insights.

We'll cover and capture both Micro and Macro, Inner and Outer, Spatial and Temporal, Here and Everywhere. We'll include reality beyond perception: The Millennium and the Nanosecond, a lightening flash and a droplet splash. We'll span vast orders of magnitude – Powers of 10 that produce enormously big numbers relating to extremes on both ends of the scale; atoms and galaxies, cells and populations, microbes and stars in the cosmos.

THEORY OF EVERYTHING. As we proceed with constructing our comprehensive grand model, we should first acknowledge previous attempts at this colossal endeavor. The quest for an explanation of how the world works began many centuries ago, long before modern science and the immense accumulation of knowledge at our disposal today. Ancient Egyptians, Greeks, Romans, Babylonians and Chinese philosophers offered theories revealing the essence of life, the structure of reality and the primary elements comprising all things. Ironically, some of those early models were not far off the mark, at least on a metaphorical level.

Greeks made transformational advances in knowledge, building the foundation for a grand scheme describing reality, introducing essences of life as four primary elements: *Earth, Water, Air, Fire*. Impressively, they introduced the notion of atoms as the building blocks of material reality. That was in the 5th century BC. Later Greeks developed theories of life based on math and music, emphasizing geometry and harmony of ratios. Their cumulative efforts provided many of the key ingredients needed to piece together a big picture of reality.

Beyond the Greeks other ancient civilizations let their imaginations run wild, creating mythological stories that introduced fictious gods to explain every mystery. Conveniently these gods all possessed human traits and qualities. Their stories provided subtle clues to the underlying essences hidden within nature. Universal experiences of man surfaced through those narratives along with common themes and enduring truths.

Rudimentary pseudoscience gradually became the accepted standard practice, commingling philosophy and religion. The result was a mushy blend of rationale offering mystical explanations for anything unknown or unknowable at the time. Intellectuals of the day leaned heavily on intuition and imaginative perspectives. What science calls chemistry today was preceded by ancient alchemy, an early pseudoscience combining objective data with subjective mysticism. Though the methods were crude and inconsistent much of what early man developed was again, metaphorically close to truth.

Scientific thought and disciplined methodology eventually emerged as the pathway to discovering truth. Collective wisdom expanded geometrically as science found its way into every field of human interest. Lingering questions and curiosities instilled further motivation for continued exploration, fueling the pursuit of knowledge in areas previously unknown. But these advances in scientific progress reinforced the tendency to examine reality in a reductionist manner; the whole is the sum of its parts. Missing in the process was any appreciation for synergy, where novel aspects emerge, adding an extra unquantifiable quality to the whole. Also absent was any incentive for strategic, holistic analysis. The science revolution elevated and celebrated the specialist, embracing those who could hone in on a single area of interest with total commitment.

Today we sit atop a mountain of knowledge, standing on the shoulders of great pioneers and passionate scientists, many of whom dedicated entire lives in a single, narrow field. We also find ourselves drowning in information in the age of high-tech mass communication where data and content are available 24-7. Yet we still lack meaningful answers to the same questions our ancestors struggled with: What is the essence of life? Why are we here? Is there an explanation for things we don't seem to understand? How does everything connect with everything else? In this modern age of information and high-speed global internet access how is it that we still lack a big picture model that could answer all those questions? A TOE or model that unifies everything is certainly more possible now that during any other time in human history, so where is it? Well, the lack of a comprehensive theory of everything isn't because no one has tried because many certainly have. There's a distinguished trail of partial models left by intelligent pioneers who's only flaw was directing their efforts in a specialized manner;, with too much Micro and not enough Macro:

PARTIAL TOES – *From well-constructed models to misguided, weak efforts:*

Modern Era	*Ancient Times*
Systems theory, Chaos Theory, Universe as a computer	Platonic Solids
Physics – Unified forces, *Star Wars - The Force	Sacred Geometry
Perennial Philosophers – Consciousness, great chain of being	Music of Spheres
Walter Russell – Electric polarized universe	4 Divine Elements
Alfred Whitehead – Dynamic Being (*Process and Reality*)	Myths & Gods, Pantheism
P.D. Ouspensky – Physics and mysticism	
David Bohm – Holographic Universe	
Arthur Young – Reflexive Universe	
Prigogine – Nonlinear bifurcating dissipative structures	
Ken Wilber – 4 Quadrant model of consciousness	
Chris Langan – Universe as self-writing information system	
Thomas Campbell – Universe as a virtual reality simulation	

*Einstein spent the latter half of his career pursuing a physics TOE and failed
**Stephen Hawking concluded that a physics TOE was not obtainable

TOE REQUIREMENTS:

Unifying *everything* covers a great deal of ground, requiring a comprehensive organization of content which must be cohesive, consistent, meaningful and understandable. Building upon the broad categories identified earlier, the task at hand is first grouping fields by similarities and differences, then uncovering patterns that connect each and every one. Along the way we'll reveal common aspects that were hidden in seemingly unrelated things. Given the enormous range of content by type and scale, the challenge will be finding just the right balance of depth and breadth; painting the big picture while avoiding too many particulars or getting lost in the weeds.

A genuine Theory of Everything should not only quantitatively cover a lot of ground; it should also meet a high qualitative level of content and presentation.

The following criteria set a much higher standard basis to judge the merits of such an all-encompassing undertaking, where a truly *comprehensive theory of everything* must include all of the following:

* Define and capture *everything*
* Identify the essential elements within *everything*
* Reveal how the world works – mechanisms, processes, systems, etc.
* Identify universal truths, laws and operating principles
* Reverse engineer reality – reveal its fundamental infrastructure
* Demonstrate how seemingly random parts are actually unified wholes
* Bridge gaps and expose the misleading appearance of separateness
* Produce a simple model relating how everything connects
* Reveal the deep source of connectedness and unmask hidden unity
* Uncover universal patterns by connecting the dots

Answer Fundamental Questions:

What is the essence of life?
What are the *universal laws* of life and reality?
What are the *hidden codes* connecting everything?
What is *time* and how does it actually work?
What are the *secret patterns* in the universe?
What is the foundational structure of physical reality
How do human beings connect with everything?
How do YOU connect with everything?

In short, a *TOE* must break the codes of reality. This process will involve detective work on both the micro and macro levels, pattern recognition, and reverse engineering to reveal the highest source connecting all there is. The final product should include a graphic roadmap unifying every aspect of reality. It must bridge diverse fields and link disparate disciplines. It must accommodate opposites, ambiguities and paradoxes. It must link the most remote, unquantifiable phenomenon with the ordinary, the knowns with the unknowns. It must produce an elegant model that captures everything with simplicity and accuracy. In the end we'll discover that *everything* and *unity* are simply two sides of same coin.

$$\infty \equiv 1$$

2   The Essences of Life

What is the fundamental nature of reality? It's an important question we must address up front to construct a model of *everything*. Now that we've defined "everything", we'll next examine the root level, primal stuff from which it's all derived. In short, we must define what the essence of life is. To begin with there isn't just one, there are several. Also, what we perceive as "life" is ambiguous and somewhat difficult to simplify. At the most basic level life or living is generally experience based. *Essence* derives from the Latin root "Essere" meaning "to be". For Aristotle the essence of life was to serve and do good, emphasizing a humanistic flavor. Modern philosophy has expanded the scope of that term to take into account all of reality, not just the perception and experience aspects of it.

Western thought blossomed with ancient Greeks, whose great philosophers saw essences as element based. Four primary elements were viewed as the foundation of reality: *Earth, Air, Water, Fire*. Over time, each was given special emphasis and considered *the* primary essence. Both Empedocles and Aristotle combined the four, concluding they were the building block ingredients of everything. Subsequent analysis led to Anaxagoras focusing on the Mind as primary. Plato elaborated further by establishing a Mind/Body dualism (materialism-idealism). Later Greeks reduced the essence of life to substance, first with Leucippus developing a theory of Atomism (Atoms and the Void) and Democritus concluding that all matter is made up of tiny atoms. He came up with that theory in 460 BC!

Early Western analysis of The Essence of Life

Thales	– *Water*
Anaximander	– *Air*
Heraclitus	– *Fire*
Empedocles	– *Earth, Water, Air, Fire*
Aristotle	– *Earth, Water, Air, Fire, Aether*
Anaxagoras	– *Mind*
Plato	– *Mind/Matter dualism* (idealism/materialism)
Democritus	– *Atoms*

Eastern philosophy took the opposite approach, emphasizing a holistic, synergistic perspective. Where Western analysis focused on reductionism – parts explain the whole, Eastern thought favored wholes to explain the parts. They appreciated the relative, indeterminate aspects of reality and concluded an ongoing cyclic unfoldment created everything. In the East the essence of reality was the experience of an enlightened soul, a state of being attained through proper living. No parts or elements, just an ever-flowing plenum surrounding and penetrating everything.

The TAO (Lao Tzu) establishes the underlying origin of order in the universe. This nameless essence called Tao begot 1, 1 begot 2, 2 begot 3, and 3 begot everything (infinity). In other words, all reality starts with 1 unified whole which separates into 2, then 3, then everything else. The key principle in Eastern philosophy is the holistic essence. Ultimately a comprehensive theory of everything must incorporate and unify both Western and Eastern perspectives.

SUBSTANCE

Ancient Greeks did their deep dive on the basic ingredients of reality, concluding that it all begins with materiality, the "stuff" of life. What are the building blocks, units, parts, elements? What's the bottom-line foundation of all substance? Modern science has since revealed that materiality is composed of matter and energy in various combinations, including mixtures, compounds, objects, parts, and raw materials. In addition to these physical ingredients, one could include organic components and their biological parts. Ironically the "Stuff" science dissected is itself already the original ingredient. *Substance* itself is a primary essence of reality, regardless of the many forms it may come in. The atoms and elements described by Greeks were all just subsets of the same category; *Substance*. We need not narrow the search to any particulars for the essence is already there in front of us. Simply put, an essential ingredient of reality is the aspect of substance.

Anything serving as part of something greater than itself takes on the quality of substance, like a building block supporting a larger whole. Units, parts, pieces, bits, lots, lumps, scraps, slices, fragments and elements are all variations of stuff – fillers that possess attributes of material substance. Cells and organs in your body are the stuff of human beings, yet that's just one level in a much greater hierarchy. YOU may be the whole of those parts but you're also just a single unit in a population of many people. YOU and others are the stuff that comprises a town, city, state or nation – now an attribute of a piece, part, or unit of a greater whole.

Materiality is a primary quality of physical reality present everywhere, imbedded in everything and manifested in countless variations. Matter and energy assume nearly infinite forms, each with unique attributes. The elements of the periodic table set the foundation enabling variation in all substances. Categories include metals (alkali, alkaline earth, transition) nonmetals, halogens, Nobel gases and rare earth. Original elements form mixtures and compounds to further expand the unique qualities of different substances. Physical traits differ in a wide variety of ways: shape, size, texture, hardness, stickiness, brittle, elastic, malleable, etc. Substances also differ by state: solid, liquid and gas. Note how minor changes in atomic composition result in great differences in elemental attributes, an emergent effect influenced by our next primary essence.

Material Attributes

Malleable – flexibility	Elasticity – keep original shape	Temperature – hot/cold
Texture - smooth/rough	Stickiness – adhesiveness	Density – compactness
Sheen – shinny, dull	Mass – volume of matter	Magnetic – attract or repel
Shape – uneven/symmetric	Resilience – resist wear	Conductive – heat, electricity
Hardness – resist breakage	Size – dimensional magnitude	Corrosive – dissolve substances

Energy is immaterial pseudo-substance, present in all things, imbedded in all matter and substances.

ORDER

Next, we look at the **structure** of substance – how it comes in many types and varieties…and just what is it that differentiates types of things or stuff? This brings us to the next essence or ingredient of reality; *ORDER*. Out of a relatively small number of ingredients we get an almost infinite assortment of things, types, expressions, and manifestations, solely based on the way things are uniquely ordered. Variety is largely taking the few and mixing it, arranging it, molding, shaping, manipulating and just setting it in a different form. More importantly, a new reality emerges from a previous original state.

Form is present everywhere in everything. It defines the things we perceive. Form creates order that serves as the glue shaping our external reality. Your mind's primary function is to establish order out of chaos, both outside and within. Form isn't limited to static structures but includes relationships between anything and everything. It connects all the moving parts that comprise a system. Form can range from the very simple to the highly complex. It's expressed as the integrative glue emerging in a network of interdependent elements that work together. Order is always present, though it's often hidden within the dynamic equilibrium between opposite forces in balance, such as a balloon held together by inner pressure pushing outward against the outside air pressure pushing inward.

Order Hidden in Dynamic Equilibrium

Moon Orbit	-Centrifugal motion vs centripetal gravity
Tug of War	-Both sides equal pull vs pull
Arm Wrestle	-Both sides equal push vs push
Floating on Water	-Gravity pull vs Buoyancy push

*Once 1 side overpowers the other, hidden static order becomes visible locomotive order

Order is a fundamental aspect of reality, much less obvious or visible than substance. Our external environment may appear as an arena of random events however it's actually a highly integrated, largely predictable series of outcomes. Everything we see, perceive and experience is an effect with a correlating cause. Life is a nonstop series of causes and effects, continued infinitum. There is order imbedded in every situation or outcome whether we're aware of it or not.

We cope and adapt to our dynamic interactive environment by using models and mental constructs. Maps, graphs, charts and visual aids provide graphic structure for connecting the dots to clarify abstract things – a practical short cut to comprehending reality. Anytime we pick up a pen and write a list we're creating a mental construct of order. Same with drawing pictures, setting boundaries or visualizing outcomes. Anything representational is again a mental construct capturing order and form.

Order equals stability and enables identity. More importantly, it provides specific value to the parts that are ordered. Things are put in a sequence for a reason. Ordered stuff becomes more than the original random ingredients. Consider the Periodic Table where every element is comprised of similar atoms, each with identical ingredients – Protons, Neutrons, and Electrons. But *every* element is unique despite similar composition. What's different? *Form* and *Order*. Just by adding a few more identical particles, an atom can change from being a gas to a metal. The very same particle composition changes dramatically just by adding a few more protons and electrons. So, the "stuff" that makes up our entire physical material reality (Atoms) is composed of identical particles but structured in a variety of forms. Form-order-structure thus emerge as our 2nd key essence of reality. I wholeheartedly agree with Nobel prize winning physicist Erwin Schrodinger who once concluded "Form, not substance is the ultimate concept".

4 States of Matter Revealed in Water – (*Order changes qualities of a single substance*)
      Plasma   – little to no order
      Steam   – loose order maximizes movement with minimal integrity
      Liquid   – balanced order allows both integrity and movement
      Ice   – rigid order, stable, fixed position – loss of movement

   Each state change of water produces dramatic changes in its key attributes

PROCESS

Since *Stuff* and *Order* seem static by nature, how do we account for the dynamic moving aspects of life? Everything around us appears to be in a perpetual state of change and fluid motion. Anything we consider to be alive is defined by its motion or activity. Life moves forward, energy is transferred, work gets done. Cycles of growth and decay press on while rhythms of opposing forces clash to create new circumstances, opportunities and threats. In this dynamic environment change is the only constant. And therein lies our next distinguished essence: *PROCESS*. Reality is a continual flux of interdependent actions with various cause-effect relationships progressing through time. During all the ongoing transitional flow, order and form are continually reset to create novel outcomes.

    Life at its very essence is a perpetual state of dynamic change. It's a process of ebb and flow where different variables steer and influence the way things and events emerge. Change can be simple or transformational. From simple numerical addition/subtraction to more significant change in composition (change in ratios) to the most significant change in kind (different stuff altogether). The continual ongoing process of change varies by both degree and level, measured by relative aspects of time. Change can occur slowly or quickly. It can be temporary or long-lasting. It may assume a recurring rhythm or cyclic pattern. In each case the change process results in novelty. Cyclic change is often a spiral pattern where one circle is completed and a new one begins, different from the preceding one. In contrast to orderly cyclic change there are also sporadic, less predictable forms of change. Some involve mixed intervals of slow change followed by sharp, big changes, followed again by gradual change. Evolution follows such a pattern where eons pass with very little change only to be suddenly interrupted by a major transitional event (punctuated evolution). There are also chaotic processes appearing disorderly on the surface while possessing hidden orderly internal patterns. This includes fluid dynamics, population shifts, cosmic events and aggregate dice rolls.

Change is generally the result of two opposing forces meeting and combining to produce a third "new" force. Hegel called this process "dialectic", where there's an original Thesis accompanied by an opposing Antithesis, which then clash to produce a Synthesis. A related dynamic is the Tipping Point, where gradual change builds until a threshold is reached resulting in large-scale change. An example would be a dam breaking or an avalanche. When change occurs through slow, long-term accumulation, it's hard to notice. Glaciers move and canyons form ever so slowly but over time the difference is immense.

Multifaceted Aspects of Change

o Tempo – speed of change, time period dependent
o Degree – amount of change, proportional aspect
o Degree vs Transform – Partial vs Complete, Change in Type, new quality emerges
o Modes of Change – Smooth, uneven, continuous, punctuated, acceleration, chaotic
o Dialectics – Force A interacts with Force B, creating a new Force C.
o Cyclic – Regularity, recurring conditions

PURPOSE

We've identified Substance, Order and Process as key ingredients of reality. They continuously interact together, shaping our dynamic environment and the way we experience everything. But underneath the veneer of this perpetual action lies a driving force, one that must be present to account for the WHY? If in fact our universe is orderly, symmetrical and integrated as a whole then why this particular set of outcomes? Why are things going the way they are? It all starts with intention. There must be a first cause, an initial spark…a beginning motive. Despite apparent randomness here and there, everything has a directionality about it. Something always starts the action. Let's call this primal beginning force; *PURPOSE*.

In a random, unorganized universe there would be no purpose. Ours is an orderly connected network of directed actions, so intentionality is a necessary ingredient. At the human level it manifests as motivation; what makes a person tick. The human mind may provide necessary order but it's the soul that impels purpose. It emerges within us as desire linked to our emotional state of being. Thoughts are essential to planning and goal setting, but emotions are the driving force to achieving them. Human action is ultimately driven by passion, which we metaphorically associate with the heart. Purpose is expressed in many forms: Military-*mission*, Crime-*motive*, Politics-*power*, Business-*profit*, Sports-*victory*.

Reality is comprised of substance and order working codependently in a directed manner. It's a creative process that begins, continues and re-establishes novelty though intentional forces. Purpose generates dynamic forces that influence preferred outcomes, acting as a choice agent that leads to selective results. This orderly process links first causes with secondary effects in a manner implying direction and Purpose, establishing our 4th key essence of reality. This attribute is present everywhere, and as Dr. Wayne Dyer observed "Every aspect of nature, without exception, has intention built into it ".

Purpose Built into the Universe

o Cosmic forces of gravitational attraction and radiation
o Physical forces of atomic particle attraction, fusion, magnetism and pressure
o Chemical forces of molecular bonding and reactions
o Biological forces of instinct, survival, reproduction
o Human needs, wants, desires, general motivation

RELATION

Our final and most encompassing essence to account for is *RELATIONSHIP*. Each of the prior 4 root qualities are subsets of a greater principle. Anything and everything are what they are only in relation to something else. There is nothing that exists or could exist by itself. Even nothing becomes a something once it's consciously conceptualized. Contrast and meaning emerge only after a comparison exists. Everything *out there* depends on what's *in here*.

The relational essence implies a minimum of 3 things: a first thing (1), something else (2), and the relationship between them (3). This is a fundamental aspect of reality. The TAO follows this same principle: In the beginning there was a single thing – a monad or ONE. That one begat 2, that 2 begat 3, and 3 begat everything else (infinity). In our world and the greater universe, a boundless number of things, phenomena and activities are interwoven relationally in ways we don't comprehend. Not only is a thing related to another thing but everything else. Reality is largely an interconnected web of direct and indirect relationships, changing moment to moment, providing meaning to an infinite number of individual elements which by themselves mean nothing.

Symmetry patterns throughout nature provide subtle clues. They inform us of a connected, orderly universe that must abide by invariant laws and principles. Fractals exhibit super-symmetry, producing endless self-similar forms. Patterns are a language that tells us something about the things they relate. Some are obvious, others subtle or hidden. Whether these patterns are symmetrical or not they serve as a cypher rich with information linking connected parts of a whole.

Patterns are the very essence of the relational principle. Numbers are a basic form of pattern. Spatial and temporal relationships involve more complex forms. We start with "0", the symbol for nothing. It can be a pattern representing the smallest possible quantity. It can represent a point in time. It can imply a variety of different, meaningful things, but it's a pattern in itself, deriving meaning based solely on context. Next, there's "1", the symbol of the monad. It too can represent a low quantity. But it also can represent a beginning point, a completed whole, a separated part or a unity. Again, context is everything. 2 is the symbol of duality and polarity. It can represent opposition and conflict or express balance and harmony. It's the primary symbol of relationship, emerging whenever there are at least 2 elements of anything. 3 is the symbol of trinity, the first actual form of structure. It too can represent a whole or part. It's also the first relational arrangement establishing a recognizable pattern. 1 or 2 dots, lines or parts are just random elements until a 3rd element is added. Same is true with musical beats.

Relational Pattern Expressed in Foundational Numbers

0 – Nothing, emptiness, circle, center, seed, womb
1 – *Monad*, Oneness, Unity, beginning, identity, *point*
2 – *Dyad*, Twoness, duality, polarity, separation, division, tension, *line* joins/separates
3 – *Triad,* Threeness, minimum structure, form, balance, relationship, *plane*
4 – *Quad*, square, earth, matter, stability, constraint, *volume*
∞ – *Infinity*, Transcends numbers, unquantifiable

* 0 – 4 are foundational numbers producing all relationships and structures.
** 0 – ∞ (zero/infinity) nothing - everything, transcendent sides of the same coin.

The relational essence at the most fundamental level is symbolized by numbers. The first 3 numbers plus 0 set the foundation for all other number patterns. It's no coincidence that the first 3 are all prime numbers, meaning they can't be divided.

From 4 to ∞ (infinity) there are variations, repetitions and combinations that represent everything. These basic symbols are both patterns in themselves and representations of patterns manifested in the real world. Relationships among elements create more meaning than the elements themselves. It begins initially with the pattern of twoness, where we have elements in comparison, contrast and difference. This basic pattern in reality establishes degrees of separation. 2 elements can be close or far, related or unrelated and slightly or greatly different in a variety of ways. The difference between any 2 things is always by degree and can be represented by a scale. The relational essence therefore is largely the measure of difference between things. Scales of difference always have a maximum and minimum degree. 0 would represent no difference and 180 would be the maximum or opposite. *Polarity* is the relation of opposites. This *Twoness* quality of pattern is omnipresent, expressed in countless manifestations of things equal and opposite. Everything in the Universe is part of and subject to a polar spectrum.

Omnipresent Polarities

Yin-Yang, positive-negative, growth-decay, good-bad, big-small, fast-slow, male-female, light-dark, odd-even, stable-chaos, short term-long term, inside-outside, freedom-constraint, healthy-sick, rest-motion, near-far, abundant-scarce

*Every "thing" or concept experienced by human beings has polar attributes

Scales and spectrums representing differences among things are present everywhere we look, both quantitively and qualitatively. While both are generally interchangeable, scales are quantitative based whereas spectrums can include qualitative, subjective values that are perspective based. Consider the difference between the size of things versus the desirability of things. Or the number of plums versus the taste of them. This is an important conceptual point because not all differences are created equal in our subjective reality.

Levels and hierarchies are scales implying differences in relative disposition. Generally, things seen at a higher level are perceived to be more important, especially where power or authority is involved, though less applicable when the scale simply involves quantities. But even when differences are just limited to numeric elements, we often associate greater respect with larger amounts versus smaller ones. It's a completely subjective perspective.

Some scales and levels fit into a special category called *Holons,* where each element along a spectrum is both a part and a whole. An example of this would be letters on this page. Each letter is part of a sentence. Each sentence is part of a paragraph. Each paragraph is part of a chapter, and so on. Every element in this sequence is simultaneously both a whole in itself and part of a greater whole. Holons are an essential relationship pattern built into the infrastructure of reality and provide a key insight into the way everything fits.

Differences and relationships create variety. Just a few basic elements can be mixed, combined and distributed in multiple ways where the result is boundless diversity. Rearranging stuff can create a completely new identity without changing the original contents. Structures can be strengthened or weakened by just moving parts around a little bit. Simply changing the number of elements in a group can result in a different quality of that group. Consider the difference between a few angry customers and a protest mob outside a store. A simple quantitative difference can produce a qualitative variation – more is different.

Transcendence is a special category of relational difference. It's a combination of polarity and qualitative difference, going beyond a simple change in type to a change in kind. It's transformational. Again, not all differences are equal, especially quantitative versus qualitative. Here it's a complete change in the nature of the thing. Consider the difference between water and ice. Both are comprised of the same substance yet when frozen the very nature of water changes dramatically into something with completely different attributes. Transcendent relationships are significant because they link elements that seem disconnected on the surface yet remain closely related in a different dimension.

Examples: *Five Primary Essences of Life, Nature and Physical Reality*

SUBSTANCE	ORDER	PROCESS	PURPOSE	RELATION
Parts	Form	Action	Intention	Difference
Raw materials	Structure	Change	Desire/need	Dimension
Ingredients	Regulation	Motion	Direction	Space/Time
Solids-liquids-gases	Balance	Development	Attraction	Context
Business: Product	*Management*	*Production*	*Profit*	*Market*

DUAL NATURE OF THE ESSENCES

Defining qualities present in everything is tricky when things exhibit multiple aspects, differences of expression from the dimension of time. Our experience of here and now is very different from what happens over longer periods of time. This dynamic makes tracking changes more elusive since we have multiple perspectives to choose from. Measuring things, actions or processes normally involves either a snapshot or a graph – here/now versus a stretch of time. How we define or experience anything varies depending on these two expressions.

Each essence of life exhibits a here and now aspect along with an alternate variation expressed over time – one static, the other fluid. Beginning with *Substance* we can look at material parts as rigid objects or the ambiguous leftover residue from stuff that breaks down over time. Water has a static form (ice) and fluid states (water or steam). Matter and energy are similarly two sides of the same coin; one a structured form, the other flowing in constant motion.

Order manifests as form or structure in the here and now while expressing itself as force over time. Order is about force. When static it exhibits a constraining presence, manifested as control, regulation and balance. Over time, order emerges as motive force, setting things in motion to producing action. Dynamic equilibrium is a common expression; opposing forces settle into a temporary balance. Many things and systems, including YOU are in this state of being most of the time.

Purpose is the originating source generating the force present in order. In its static mode, purpose is simply intention, which generates needs, wants and desires. With time, intention emerges as directed action, steering the forces that create order and make things happen.

Process is already a fluid essence which makes its dual nature more intriguing. Its basic state is change, however there are both stable and fluid forms of change. Constant motion is a stable form of change; a first level version of transition. Acceleration is a 2nd level variation, a more fluid nature of change. And higher yet is accelerating acceleration, a 3rd level variant.

Relation manifests both spatially and temporally. Its structured aspect is expressed in space while its fluid nature is expressed in time. While spatial relationships between objects are measured by dimensions of distance, temporal relationships between events are measured by intervals of time. Both express unifying patterns of symmetry as well; spatially there are mirror-like symmetrical forms and temporally there are recurring cycles in time. Symmetrical unifying patterns emerging in different relational dimensions.

Dual aspects expressed in each primary essence

ESSENCE	STATIC	*FLUID*
Substance	Matter	*Energy*
Order	Structure	*Sequence*
Purpose	Intention	*Direction*
Process	Constant	Acceleration
Relation	Space	*Time*

CONCLUSION. Identifying primary essences of reality is the first step towards unifying everything. Sharing common attributes reinforces linkages through unifying qualities. At the physical level all things exhibit materiality and are shaped by form. They're guided and directed by intentionally, which subjects them to continual processes. They acquire identity or disposition through relationships with other things. These primary essences are present everywhere, are interrelated with everything, interacting in a completely integrative manner.

These shared common attributes will serve as a foundation for constructing our universal model. Like the border pieces of a jigsaw puzzle we can use these 5 fundamental parameters to set the general framework and fill in the rest from there. Since primary essences are universal, we can expect they'll apply at every level. Now we just need to identify levels of reality in a user-friendly, easy to understand manner. We'll greatly simplify this task by placing man at the center of the puzzle, right in the middle of everything. The rest will literally fall into place. And when we place man in the middle, we're really putting YOU in the middle, at the very center of it all. The following chapter will connect everything, including the primary essences identified here and relate them right back to YOU

The Essences of Life expressed in YOU

	STATIC	*FLUID*	EXPRESSION
Substance	Body	*Senses*	Body Language
Order	Brain	*Mind*	Thoughts
Purpose	Soul	*Spirit*	Feelings
Process	Homeostasis	*Metabolism*	Health
Relation	Identity	*Personality*	Behavior

3 Theory of Everything (*TOE*) Model

The challenge of capturing *everything* in a single graphic model should be obvious; how do you portray the most comprehensive subject in a meaningful, clear manner without getting so detailed that it becomes chaotically cluttered. Less is more. We can keep it simple with big picture, representative categories covering a broad spectrum without needing to list every element. Since everything is interconnected and interrelated, we just need to establish a central starting point and branch out from there. And that ground zero point is YOU.

Our *Theory of Everything* model is man (human) centered. This greatly simplifies our orientation to connecting everything from a central perspective. Everything traces back to us, occupying a median, nodal point along a series of seemingly infinite scales and spectrums. YOU in the center are connected to everything above and below, inside and out, here and there, before and after, including everything visible and hidden. Although this implies a circular relationship with nested spheres, it is much easier to graphicly portray the connections in a grid format.

We start out with YOU in the middle and determine what's above and below:

ABOVE YOU		BELOW YOU
Groups		Natural world
Organizations	YOU	Physical substance
Society	(*in the middle*)	Lower life forms – Plants, animals
Nations		Raw materials

These categories can be summarized as *Society, Man* and *Nature*. They emerge as nodal points representing 3 different levels of being: Societal, Biological (*Living*), Physical. Each category-level is connected on a continuous vertical spectrum with elements gradually transitioning as you go higher up. The bottom level begins with physical substance. Particles, atoms, molecules and mixtures form the lower base, followed by organic compounds and living creatures as complexity increases up the spectrum. Plants, animals and man extend above this level and merge into the middle, biological category. Then man clusters into groups with families, clans, neighborhoods, towns, cities, states and nations. This extension up the spectrum produces greater complexity, culminating with modern sophisticated societies at the top.

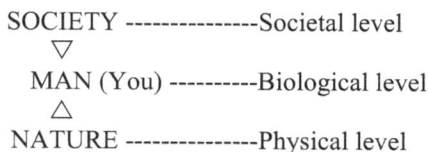

SOCIETY ---------------Societal level

▽

MAN (You) ----------Biological level

△

NATURE ---------------Physical level

Next, we can link these vertical levels horizontally by applying the essences identified in the preceding chapter. Each level contains an essence of Substance,

Order and Purpose. These qualities are built into physical reality as portrayed along the bottom of the matrix. Going up into higher levels of complexity these qualities transition into different forms. At the lower physical level these essences manifest as Matter, Forces, and Quantum choice. Transitioning upward to the biological level where man resides, we see these essences as Body, Mind and Soul. Extending all the way up toward the top these same essences manifest at the Societal level as Economy, Government and Culture. So, the matrix consists of 3 primary levels with 3 primary essences of life, expressed at each stage.

Society	*Economy*	*Government*	*Culture*	(Social level)	
Man	*Body*	*Mind*	*Soul*	(Biological level)	
Nature	*Matter*	*Forces*	*Quantum*	(Physical level)	
	Substance	**Order**	**Purpose**		

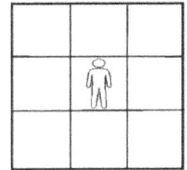

Quantum mechanics exhibits intentionality where 1 outcome is chosen out of infinite possibilities

If we stopped here, we would be committing the same oversight that so many other TOE models are limited by. The problem is we're missing the dynamic aspect of life. This static 2-dimensional graphic is missing a key essence – *Process*. Life is about action and change; purpose in motion. This quality of process is transcendent to the other 3 essences and thus must be represented by a different dimension; one that's perpendicular to the rest. We can represent this essence using dotted or dashed lines at 90-degree angles to the entire matrix. This will also capture the dual nature of each essence at every level.

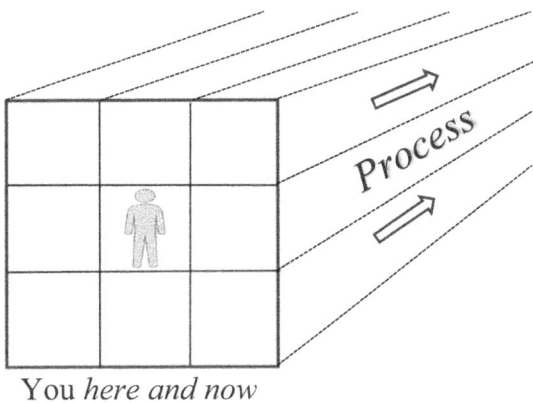

You *here and now*

Everything is connected in time, expressed in action, change and process dynamics. Static models fail to capture this essential aspect connecting all levels and essences.

The physical level now more accurately portrays each essence as a duality: Static Matter and flowing *Energy*, static Fields and flowing *Current*, singular *Quantum action* and it's flowing, ambiguous *Cloud of Potential*. Next, at the biological level we supplement a *Body* with active *senses*, a *Mind* with dynamic *thinking*, and a Soul with fluid *feelings*. At the higher societal level, we see an Economy with fixed *Goods* and fluid *Services*, fixed Government *Agencies* driven by dynamic *Politics*, and fixed Cultural *Values* supplemented by changing *Fads*. Every level within the spectrum contains its own unique manifestation of the essences expressed in dual form; one static, the other fluid.

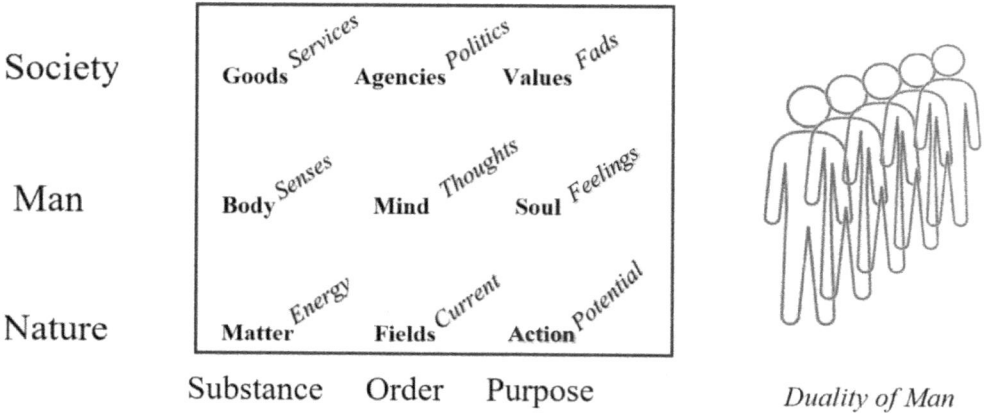

Society — Goods *Services* Agencies *Politics* Values *Fads*

Man — Body *Senses* Mind *Thoughts* Soul *Feelings*

Nature — Matter *Energy* Fields *Current* Action *Potential*

Substance   Order   Purpose

Duality of Man

Duality of ESSENCES

Sublevels – Our 3D TOE matrix is intentionally oversimplified, presenting a big picture, generalized model without getting too cluttered or bogged down with every variation and manifestation that exists somewhere within the framework. Every spectrum or scale represents a comprehensive scope of possibilities that could not practically be displayed in a single graphic or meaningful way. Just realize that between each primary level there are multiple other levels, each with their own unique expressions of the same basic essences.

TOE Model contains every level and scale

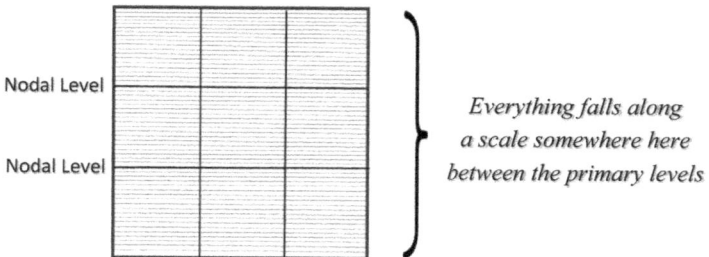

Nodal Level

Nodal Level

*Everything falls along
a scale somewhere here
between the primary levels*

Simplified 3 Tier Matrix masks the actual multi-level reality within for practicality

Each level along the vertical axis is linked to those just above and below, varying in degree only. Even the primary levels are just nodal points where the transition upward gradually becomes so significant that it results in a transformational change (Physical to organic/biological to societal). This can be likened to states of water: solid (ice) warms up until it melts into a liquid, and further warming turns it into steam. Levels within the matrix transition in an identical manner, increasing in complexity going upward until a tipping point emerges as a nodal stage with new, transformational qualities.

TOE Model: Primary and Sublevels between

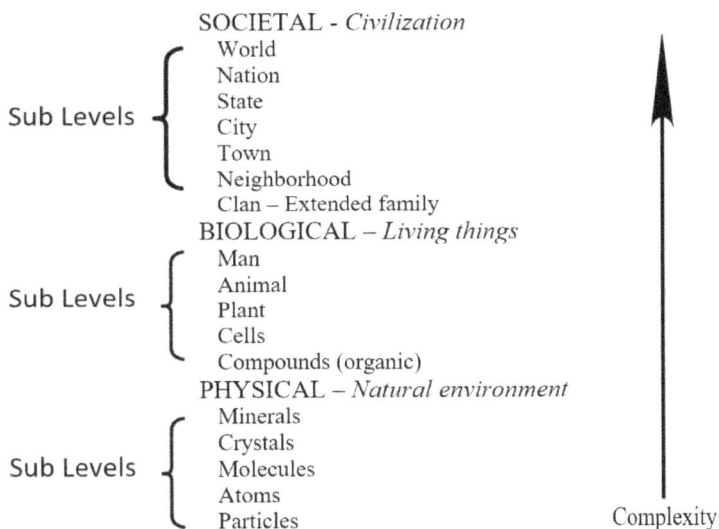

SOCIETAL - *Civilization*

Sub Levels
- World
- Nation
- State
- City
- Town
- Neighborhood
- Clan – Extended family

BIOLOGICAL – *Living things*

Sub Levels
- Man
- Animal
- Plant
- Cells
- Compounds (organic)

PHYSICAL – *Natural environment*

Sub Levels
- Minerals
- Crystals
- Molecules
- Atoms
- Particles

Complexity

*Primary levels are simply Nodal points on a continuous connected spectrum

Model Aspects – We can extract a lot from just the basic framework connecting everything on a comprehensive basis. General patterns emerge as a byproduct of its integrative connectedness. We can deduce that things increase in their degree of complexity as we go vertically higher up each level. We can also see the material aspect of things diminish as we go from the left side to the right. Even diagonally we can see the lower left corner grounded in pure science while going diagonally up-right there's a transition to non-science.

Universal Conflicts Correspond to the Primary Levels

CONFLICT	LEVEL	NODAL POINT
Man vs Man	Social level	Society
Man vs Himself	Biological Level	Man (*You*)
Man vs Nature	Physical Level	Nature

On the 3rd dimensional dotted line (process) we can track the flow of time. This axis accounts for the dual nature of everything in the matrix. All things in life express duality, difficult to track in 2 dimensional models. But understand that the dotted line actually goes in both directions, infinitely, representing the past and future. The 2-D structure of the matrix represents here and now, flowing toward us moment by moment. We can see how time is expressed differently at each primary level: *Physical time* traced as periods of evolution, *Biological lifetimes* captured as biographies, and *Societal time* recorded as history. Each expression of time is parallel to the others, just differing in scope.

Process Nature of Life Expressed at the Primary Levels

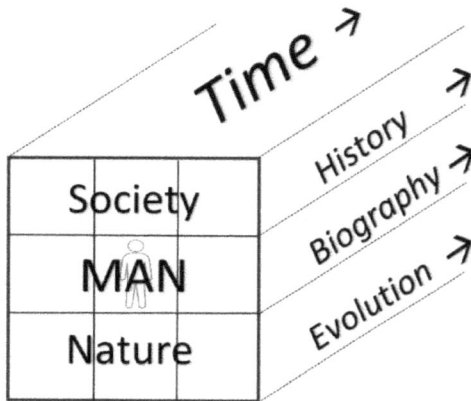

Time connects every level in a parallel linked process

Also understand that the *process* dotted line is more fundamental than the solid lines representing a static snapshot of physical reality. All the action takes place along the process dimension. This is where life happens – growth, motion, change, music, spirit, transformation. But the static grid is where we naturally focus our attention, embracing temporary forms and facades.

2D Grid	3D Process Dimension
Static	Transcendent (90-degree shift)
Digital	*Analog*
Being	*Doing*
Temporary Form	*Action and transformation*

* 2D Grid – *Resonant forms, illusions of stable tangible stuff, temporary patterns*
* 3D Dual – *Action line where life happens, creatively manifesting everything*

The first 4 primary essences are all present and connected within the TOE model. The 5th Essence, *Relation*, is represented by the model as a whole and everything interconnected within it. The relational essence is what links and connects everything, spatially and temporally. It's the reason we can build the entire model around man with YOU in the middle (center). Every relational aspect of reality begins with YOU and extends in every direction: Above/Below, Inside/Outside, Before/After (diagonal). Man is a microcosm of universe connected to everything; a self-contained system with an inside and outside. There's a microcosm within and a macrocosm without – both whole and part of an even greater whole. The matrix is both our map and MOE (Model of Everything). Its interconnected relationships link YOU to everything else, setting the foundation for building a comprehensive TOE (*Theory of Everything*), which comes later.

Big Picture Categories Connected to Man (*YOU*)

We captured the comprehensive nature of everything in the opening chapter, including a lengthy list of sciences, non-science, and a wide variety of other categories. All of it fits somewhere within the matrix model, which simplifies organizing content without getting too far into the weeds. This facilitates the next step; examining general categories of *everything* and connecting the dots. We can do this in a practical manner by simply relating it all back to man in the middle, with YOU occupying the center point.

  As a microcosm of the universe, YOU are both part and whole. Being a whole, self-contained system means there's an inside and an outside. The inside is an extensive world of its own, which we can categorize as human reality. It contains great depth with layers upon layers of attributes, nuances and fascinating realities. The outside is equally vast, with the natural environment, a world of challenges and pitfalls, and higher society with its share of threats and opportunities. This external realm is actually a combination of relationships between YOU and everything above plus everything below – the outside is both the levels below and above man (Nature and Society). It not only includes differences in complexity but differences in relationships, including size (bigger and smaller), speed (slower and faster) and other scales that include elements above and below our middle position.

  Occupying the center point of the matrix represents the here and now of reality. We must also include the process axis, perpendicular to the grid that accounts for the passage of time. This reveals the Before/After aspect of temporal reality as it relates to YOU in the present time:

Man (You) in the middle – Inside/Outside, Above/Below, Before/After

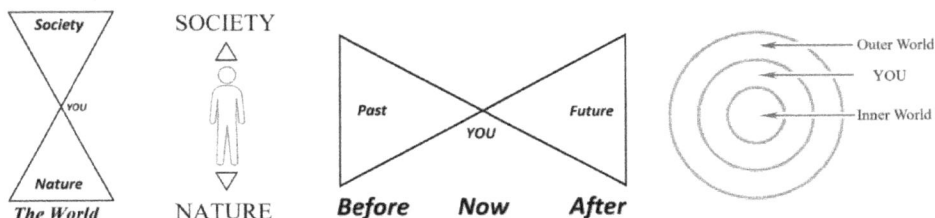

Human centered reality has its obvious advantages since it enables a privileged perspective of reality from an inside vantage point right in the middle of it all. There is however a distinct disadvantage that must be accounted for. Our awareness of reality is locally biased. We see a much clearer picture of stuff right in front of us in the here and now. We're also constrained by senses defaulting to things closer to us, limiting anything on the periphery. This creates a supplemental category of reality: things that are hidden and beyond. In our middle position along any scale, we are either less aware or unaware of anything on the extreme ends of each scale. This includes the very small or large, the very slow or fast, the very far in distance and time, and anything that is hidden or obscure from our limited, fixed middle position.

Hidden Reality – Very Large and Small, Very Fast and Slow, Far Past and Future

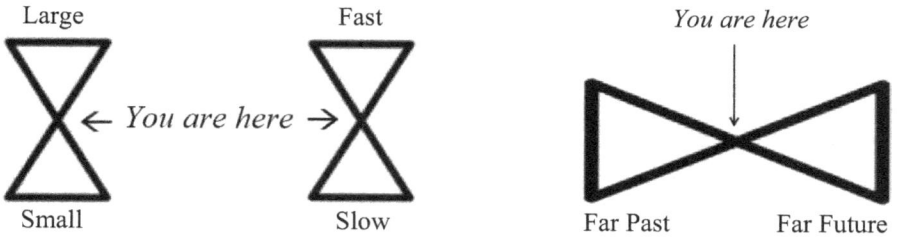

Going forward, our analysis of *everything* will be based on the following broad categories: *The World*, *Human reality*, *Temporal reality* and *Hidden reality*. These major areas comprehensively cover our experience of life from a middle position. The following chapters will extensively capture each of these general categories in depth and then connect them to YOU. They fill in the guts of what's inside the TOE model, serving as the basis for mapping out the content of everything as it fits into a relative position to everything else.

Chapter 4 The World   Above & below man (Nature, environment, society)
Chapter 5 Human Reality   Inside Man (Psychology, experience and behavior)
Chapter 6 Temporal Reality   Before & after man (History, past/future, time dynamics)
Chapter 7 Hidden Reality   Beyond man (Invisible reality, perception limitations)

Man (YOU) are always in the middle of Everything

4   THE WORLD

YOU are at the center of everything. All reality begins with you and extends to everything around you. From a relative perspective, those things around you are either above or below, lesser or greater. Man is in the middle of it all with nature and the physical world below, society and civilization above. Our physical realm contains things both very small and very large (atoms and galaxies) but it's a lower level of complexity beneath man. Think of this physical realm as the world, with a lower case "w", as it relates to nature of the earth and the external environment. Above this is the social realm where massive complex systems of mankind integrate economic, governmental and cultural domains. When we combine these two realms together as one unified external arena, we get *The World*. This super-realm is what we all find ourselves at the center of – no one is exempt.

The world (lowercase "w") includes the earth, nature, weather and the animal kingdom. It consists of the physical realm of raw stuff and the biological realm of organic life. Above this level is the social realm where man clusters into groups, organizes complex systems and creates entire civilizations. Adding the physical realm of outer space completes the elements that now comprise *The World*. It's our reality consisting of man in the middle and everything out there, above and below.

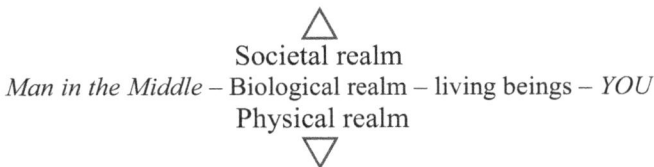

$$\triangle$$
Societal realm
Man in the Middle – Biological realm – living beings – *YOU*
Physical realm
$$\triangledown$$

Comparing the classical "world" with *The World*

Classical "world" is just Nature	*The World* is the comprehensive external environment
o Earth (Land, Sea, Sky)	o Nature (earth, weather, animals – the classic "world"
o Weather/Climate	o Society and civilization
o Animal Kingdom	o Cosmos (nature above and beyond the earth)

NATURE

What we call nature consists of 3 basic parts: The Earth, Weather/Climate, and Organic Life. It has both a microcosm and macrocosm: Physical elements including compounds of earth vs. outer space and the greater cosmos. It includes 3 general environments of earth: *Land* – geology, *Sea* – oceanography, and *Air* – Climatology/Meteorology. There's a vast spectrum of organic life; the animal kingdom and its integrated web of predation and parasitic relationships. Survival is a daily challenge, driven by a food cycle where energy is transferred between the physical realm and the biological.

The Earth – We generally focus on the planet's surface, rarely thinking about the 99% hidden beneath. We also don't pay much attention to oceans which cover the majority of the earth's surface. Earth's subterranean depths consist of inner layers making up the core. Below the thin crust under the surface is a thick mantle, then both a liquid outer core and solid inner core. Temperature increases the deeper you go, right down to the white-hot solid iron core, which is as hot or hotter than the surface of the sun! Convection currents of molten iron in the deep outer core produce earth's magnetic poles. Ongoing motion means the poles shift over time, periodically making a complete North-South reversal. But the real action that impacts man happens on or near the surface.

The planet's top layer is bustling with activity including tectonic plate shifts, magma pressure displacement, erosion, eruptions, organic growth, decay and rotting, floods, falling trees, rock movements, avalanches, earthquakes, sediment layering, etc. Further action involves water flowing and channeling: lakes, ponds, rivers, falls, marshes, swamps, brooks; all interacting with air and land activity. Ongoing surface dynamics are interlinked with the entire biosphere, where organic life symbiotically reciprocates with the environment. Organic activity is driven by a food cycle where energy from the sun and nourishment from the earth facilitate a comprehensive process of life feeding on life. Minerals, plants, vegetation, trees, brush, grass, plants, and all the edibles they produce serve as links in a circle of sustainment, beginning with the sun, and ending with a living being. It's all driven by solar power, with the sun regulating a water circulation cycle, then fueling organic energy via photosynthesis. Earth provides the raw material where life is fertilized, formed, energized, developed, and eventually decomposes back to whence it came. Man is essentially a parasite of the earth, living off its land, water, air, plant life and other organic life forms.

Weather and Climate. What happens in the skies above significantly impacts the world environment; weather in the short term, climate in the long term. Air, land and sea dynamically interact with the earth's surface, creating conditions that both sustain life and imperil it. Energy in constant motion transforms the landscape, alters the atmosphere and produces a perpetually changing environment.

Weather patterns are primarily the result of temperature and pressure interactions that regulate the amount of moisture in the air and create emergent, chaotic flowing currents. Variation in temperature-pressure will increase or decrease the capacity of the atmosphere to hold water, resulting in either dryness or humidity, and ultimately saturation in the form of rain. It's an integrated process linked by clouds, which provides a simple, visible gauge of what's going on. Of the ten types of clouds, there are four major ones that foretell the weather – Cirrus, Stratus, Cumulus, and Nimbus. Water is omnipresent in the air and its density determines the disposition of clouds, leading to variations of emergent liquid forms, including humidity, rain, hail, sleet and snow. The condensation-evaporation cycle of water is driven by solar energy (heat) and pressure (unequal heating), which also manifests as directional winds.

Weather systems involve several distinct variables that continuously interact together: winds blowing west to east, south to north, then pressure, density, and temperature. Computers are fairly successful at tracking these comprehensive interacting variables, to generate fairly accurate forecasts.

Storms emerge when regional temperature-pressure differences collide. Surface heat causes air to rise while cooler atmospheric layers make air sink. Earth's exterior is an overactive theater of High-Low pressure fronts meeting warm-cold air interactions, energy transformations and force-counterforce conflicts, combined to produce various weather events. Storms mix the moisture in the air with strong winds to produce a wide range of surface chaos including tornados, hurricanes, hail storms, and blizzards. The only difference between rain, hail and snow is the temperature-pressure dynamic shaping the water that permeates all our air.

When conditions are just right and a major storm emerges, it can take on a life of its own, behaving like a living being. Weather in general mimics the moods of sentient life, especially man, with temperamental expressions that changes day to day. Sunny, cloudy, hot and cold, damp and rainy, all correspond to human emotions, including happy, sad, excited, ambivalent, moody and depressed. Throw in the anger of a violent storm or the playful joy of a mild snow fall, and it starts to resemble human drama. Organic life in all of its forms connect to the earth and its weather as one interacting whole. That's why they can be described metaphorically with the same connecting language - a nervous breakdown and an earthquake or tornado are symbolically equivalent.

Topography serves as a regulating buffer between the earth and its ongoing environmental activity. Land features filter the flow of winds and channels the path of rainwater. Earth's surface temperature mutually interacts with the atmosphere to influence weather dynamics. Land, air and sea integratively link all climate and weather activity.

Indifferent Natural World – Despite the wonder and beauty of nature in all of its alluring splendor, it doesn't really care about you, nor does the earth. Regardless of your elevated status as the most advanced evolved creature living on the planet, it just doesn't care about you or your well-being. It will swallow you whole and spit you out without remorse. People are routinely killed by nature; today, tomorrow and throughout history. Land, sea and sky all readily contribute to human death. Earthquakes will crush you, avalanches will bury you and sink holes will suck you in without warning. Oceans will strand you, lakes will drown you and rivers will sweep you into a rocky blunt-force demise. Sailors tell a cautionary tale of an ocean described as a "cold hearted bitch who couldn't care less whether I live or die". And that lovely blue sky can suddenly transform into a chaotic dark whirlwind and throw you around like a cheap ragdoll. Our dear friend, the earth may appear benevolent while offering life-sustaining resources for survival but it will turn on a dime and end your fragile life if you aren't paying attention. Tens of thousands of skeletons buried around the world serve as lasting trophies of nature's indifferent temperamental wrath.

The Indifferent Malevolent Earth

Lethal Land	Wicked Waters	Sinister Skies
Earthquake	Whirlpool	Hurricane
Avalanche	Tidal wave	Tornado
Mudslide	Monsoon	Sand storm
Quick sand	Waterfall	Hail storm
Molten lava	Tsunami	Sunburn
Falling trees	Rocky rapids	Volcanic ash
Sink hole	Flash flood	Arid scorching desert

Animal Kingdom — The biological level covers a broad spectrum of life including animals, plants, and microorganisms. Earth's biomass is mostly non-human. Despite a world population of around 8 billion humans, mankind represents less than 1 ten thousandth of the total biomass. We may stand at the top of the hierarchy but like any pyramid structure, the lower parts take up all the space. Animals occupy every level of the biosphere (land-sea-air), with millions of species covering every space and filling every nook and cranny around the planet.

Man's linkage with animals is both direct and hierarchal. Human dominance of the animal kingdom is complete with extensive "domestication", which is just a euphemism for exploitation. A large quantity and variety of animals are captured daily and used for food, work, clothing and medicine. Not all animals acquired are eaten or put to work – some are taken in and treated as cohabitants in the home. Some even provide a measure of psychological support, to the point that a person is codependent on them. But no matter how advanced humans have become, we'll always retain our linkage to the animal kingdom. That heritage includes raw instincts, a strong survival sense, basic physical needs, and a biological blueprint where well over 90% of our genes are identical with other animals. Though man may be figuratively *above* the animal kingdom, we will always be *of* it.

Organic Life Forms Permeating the Earth's Surface

Animals	**Plants**	**Microorganisms**
(You are here)		
Mammals	Seed plants	Bacteria
Birds	Flowering	Protozoa
Reptiles	Ferns	Algae
Fish	Mosses	Fungus, mold, yeast
Insects	Grass, shrubs	Virus
Crustaceans	Trees	
Arachnids	Cactus	
Worms	Ivy	
Mollusks	Herbs	
Sponges	Climbers, creepers	

*Organic Life is the mid-point of our reality (raw physical world below & higher social realm above)

Nature's indifference to man applies equally to the animal kingdom, where any number of species in the wild will eagerly do harm to you, often without provocation. Threats on land include wolves, hyenas, hawks, spiders, snakes, scorpions, bees, lions and tigers, rhinos, gorillas, bears, and so many more. Death can also come from the sky and sea as well, with attacks from hawks, sharks, eels, stingrays and piranhas to name a few. Life in the wild is an ongoing battle for survival, 24 hours a day, and don't think for a moment that being a highly evolved, superior creature ruling the planet makes you exempt from any of it.

Fortunately for man earth has a benevolent side to it that is life sustaining and necessary to our good health and well-being. It provides all the food and water anyone could ever need along with plentiful fresh air to breathe. Abundant supplies of material resources are available for building shelter, manufacturing products of convenience and crafting a wide variety of creature comforts. Earth and nature are really all things to all occasions, neither good or bad, just the world we're born into. Both are far greater than any of us, regardless of our privileged position. And though civilized man had mastered the local environment, it doesn't take much to be reminded of how insignificant we are, both individually and collectively. All it takes is mother nature periodically erupting as the powerful bitch she's capable of becoming. Such moments are rare but humbling, and often deadly.

The earth's land, sea and air elements, combined with climate-weather and the diverse organic biosphere, make up the lower-level realm of *the world*. This broad physical arena is a navigational challenge, offering both threats and necessary resources for survival. It's the primal source man emerged from and eventually rose above, though perpetually tethered to its enduring constraints. Above this realm is the societal level mankind created, developed and ensconced into a new, broader reality – an evolutionary leap beyond nature. The result is a multilayered framework with the natural world below and a societal world above, leaving man right in the middle of everything, which now comprises a comprehensive macro arena, emerging as *The World*.

THE WORLD

SOCIETY – Civilization ⟵ Economics
Government
Culture

MAN   (*You are here*)

NATURE – Classic world ⟵ Climate/Weather
Earth – *Land, Sea, Sky*
Animal Kingdom

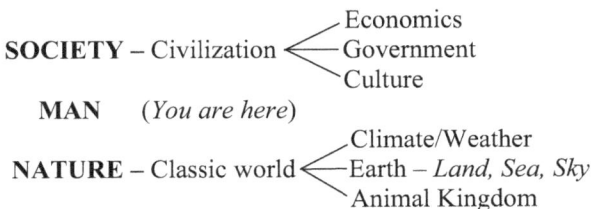

MAN IN THE MIDDLE

You are at the center of The World, and the universe for that matter. Everything around you is either above, below, inside or outside, near or far. As biological beings we've evolved above the physical realm and created a higher, more complex realm above (society). For each of us, living in The World requires navigating threats from nature below and laws, constraints and complexities from society above. We're all born at the highest rung of the animal kingdom and the lowest rung of civilization. We enter society just by being a part of a family – the lowest collective entity of the social level. Here we're both an individual being and a member of a collective group. This relationship continues as we grow and progress up the hierarchy. With time we transition through a series of larger collective memberships; neighborhood, town, city, state, nation, world.

Perceptions, Experiences & Characters Navigating through The World

Child – Trains, Trucks, Buildings. *Bullies, Teachers, Coach, Parents*
Young Adult – 1st Car, 1st Job, 1st Intimate relationship. *Boss, Traffic Cop, Romantic partner,*
Adult – Home maintenance, Salaried Job, Spouse. Office politics. *Kids, Bus. partner, Lawyer, IRS*
Senior – Hospital, Isolation, Medine, Reminiscing, Ret. Home, Probate. *Doctor, clinic manager*

At our anchored position in the middle, we naturally focus on interaction and relationships with others close to us, oriented towards siblings, parents, and friends. We can even include domesticated animals we call pets as part of this domain. It's the bottom rung of the social ladder that will gradually expand upwards with age and experience. A family and their home are like a single cell, which in the aggregate serves as the collective substance that societies are made of – the building blocks of civilization.

HIERARCHY of THE WORLD

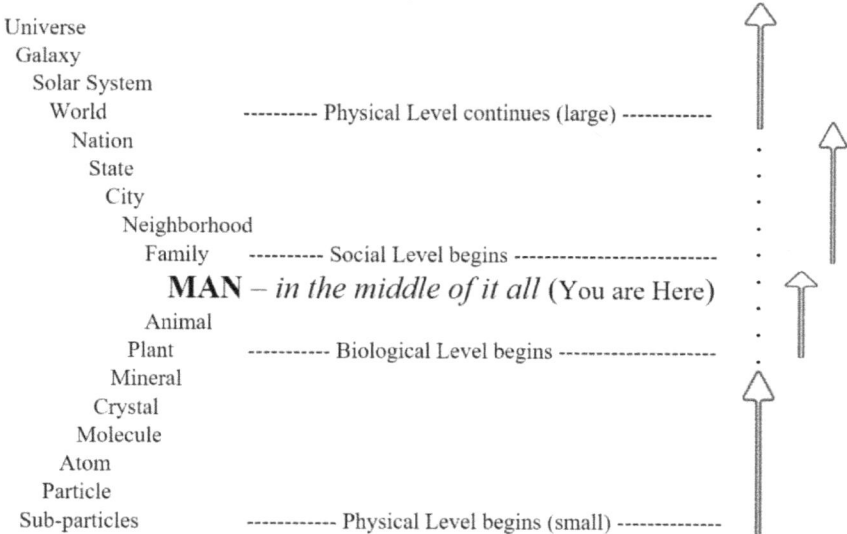

```
Universe
  Galaxy
   Solar System
    World          ---------- Physical Level continues (large) ------------
     Nation
      State
       City
        Neighborhood
         Family    ---------- Social Level begins --------------------------
    MAN – in the middle of it all (You are Here)
        Animal
       Plant       ----------- Biological Level begins --------------------
      Mineral
     Crystal
    Molecule
   Atom
  Particle
 Sub-particles     ------------ Physical Level begins (small) --------------
```

SOCIETY

When individuals cluster into groups, share ideas and engage in joint pursuits, a higher order emerges with upward progress. Collective groups create synergy where each member benefits from the efforts of others. Civilization brings added value to the lives of individuals who by themselves could never produce the variety and abundance of products and services that an organized group can. Membership in society requires acceptance of an implied social contract – you must sacrifice some of your individual freedom and submit to rules and guidelines from higher authority in return for services you could not otherwise receive. A healthy society maintains the right balance of individuals pursuing self-interest while abiding by collective restrictions set by higher authority. Revolt and revolution emerge when higher authority goes too far in constraining or abusing individual freedom.

Early civilization established primitive social organization: Hunting-gathering, horticultural, pastoral, and agricultural. They developed a rudimentary system of order with select individuals bestowed with authority over the group. Tribes and villages emerged, followed by larger, more structured organizations with greater populations and territories. Initial class divisions included kings, government officials, nobles, priests, scribes, soldiers, artisans and merchants, peasants and slaves. Eventually cities and states formed, bringing with them more comprehensive organization with integrated functions and services. Modern societies still contain distinct class divisions, primarily based on income levels; upper, middle, and lower class. Variations exist within each level, including working class, working poor, poverty level and elites. In recent history, civilization evolved into complex advanced industrial and post-industrial systems.

Society is comprised of multiple interdependent elements, including a common language, customs, traditions, religion, art, literature, and shared values. Collective unity is reinforced through a governing body, an economic system, justice system, recreation, education, and unique demographics. Society maintains order through these venues and features but really gets its cohesiveness through shared interests and values of the members. Degrees of unity fluctuate as segments of division arise over time. Issues of contention become polarizing and may result in agitated subgroups. Tolerance of differences is necessary to maintain stability. Dissatisfaction and perceived injustice may lead to civil unrest, sowing the seeds of rebellion. Stability is also challenged by dysfunctional elements: crime, vandalism, gangs, vandalism, cults, assault, harassment, intolerance, and feuding. Fringe members further destabilize as some individuals refuse to "fit in": living off the grid, non-conformists, outcasts, criminals, addicts and homeless.

Advanced civilizations are distinguished by their degree of organization and general order. Authority from above provides hierarchical control to maintain necessary stability. Order is enhanced by other sources including religious clergy, cultural leaders and local authorities, reinforced by legitimacy and trust. Other elements supplementing order are symbols, badges, official titles and uniforms. Society's infrastructure is a combination of vertical hierarchies and lateral networks, becoming what is commonly referred to as *The Establishment*. It's comprised of social institutions at the local, state and national levels, forming an interlinked web of authoritative hubs and communication centers. Within this matrix you'll find individuals with different degrees of power. They're recognizable by rank, position, wealth, title, clout, celebrity, reputation, or social status. There's a built-in pecking order placing certain individuals above others, with added degrees of authority, control, power or privilege. It's an implied trade-off agreement citizens accept as part of the social contract.

Modern societies have evolved into complex supersystems that provide just about every possible need the public may have. Societal services in the modern era include transportation, sanitation, housing, healthcare, police, fire, utilities, communication, maintenance, etc. Specialization and exchange are the force multipliers creating synergy that makes it all work. The whole becomes greater than the parts. And those parts include you and me. No matter how large or complex societies get, they're still people powered. Each of us has a role to play. Some are leaders, facilitators, producers, inventors, servers, creators, protectors and teachers. But even just a simple, ordinary citizen has a certain responsibility or obligation as part of the social contract. A healthy society requires citizen participation, expressed by the following 3 V's: be vocal, volunteer, and vote. Other acts of participation include military service, paying taxes, following laws and abiding the rules of good etiquette.

Collective dynamics influence the way societies work as individuals engage and interact with other individuals in unique situations. Desires and personal agendas clash, adjust, cooperate or compete, producing a public arena of opportunities and challenges. Peer pressure and unwritten rules play a role in steering individual behaviors in preferred directions. Rewards and punishments do the same, offering pain or pleasure. Social networks offer venues where people can engage in exchanges of ideas, goods or services, while providing an outlet for individual expression to a wide audience. Popular initiatives come and go as the zeitgeist changes over time. What's acceptable to the group today may become uncool tomorrow. The key takeaway of collective social dynamics is how unexpected conditions emerge when individuals cluster together into groups. Behavior is altered, preferences change and the whole takes on an unpredictable life of its own.

Societies operate like complex integrated living entities, with a body, mind and soul. They require physical sustainment, energy, order and a sense of purpose. It's manifested at the social level as economics, government and culture. The body of a society is its economy, where raw resources are processed and converted into products of sustainment. The mind function emerges as government, which controls, regulates and directs the whole from above. Culture is the spiritual aspect of societies, where purpose and a sense of identity can be found. All three of these primary elements combine to form the personality of a society.

Economy. Societies are maintained by production and sustainment systems generating both goods and services. Economics always involves choosing between finite resources, requiring some form of management to maximize efficient and effective results. Other considerations leading to hard choices include which products to make, how they will be made, who will make them, and who will get them. How these questions are answered reveals what kind of economic system a society actually has. It really comes down to how much central planning directs the economy versus letting people engage freely in open markets.

Early systems relied on uncoordinated trade and bartering, without use of monetary instruments. Feudalism was a semi-organized system of trade where kings and wealthy barons traded land to peasants, offering military security in exchange for their toiling the lands; those who did the work gave back a portion of the proceeds as a form of taxation. As societies grew in size and sophistication, it became necessary to facilitate exchange with coinage and paper money.

Economic systems have evolved into regulated frameworks of interconnected elements operating like a choreographed symphony, without needing a conductor since natural market forces drive the action. This *invisible hand*, as described by Adam Smith hundreds of years ago, emerges when supply and demand interact in open markets that set prices, something economists call equilibrium. Similar invisible forces influence production and distribution of products and services that corelate with the 3 traditional economic elements (land, capital, labor), plus energy, knowledge and physical resources. The result is a dynamic interdependent system of moving parts that sustain a society and power its engines of production.

Modern economies have evolved into complex structures with lots of networked hubs and moving parts. Societies are now sustained by a multitude of diverse yet connected pieces, including businesses, factories, warehouses, wholesale outlets, retail stores and distribution centers. These elements are themselves sustained via networks of transportation, communication, fueling, shipping, and raw materials. Production of end products is further supported by subsystems leveraging productivity through machines, assembly lines, utilities, and some form of management. Supplemental support systems include designing, refining, manufacturing, packaging and marketing. Finally, the integrated whole must be operated by labor and sponsored by finance – requiring banks, lending, credit and various financial instruments.

Economic Dynamics. The economic environment is comprised of multiple interconnected interacting elements. Economists face the ongoing challenge of balancing all the interrelated aspects where changing one affects every other one, making economic forecasting imprecise, on par with meteorologists trying to predict weather. Multiple variables interacting in a changing macro environment create a moving target that's difficult to consistently track and manage.

Other factors impacting the economic environment include social, legal, technology, political, environmental, demographics, risk and uncertainty. The most direct impact comes from government, where laws and policies create constraints and incentives that alter both business and consumer behavior. Markets are the arenas where all these macro and micro level aspects connect – the hub where forces of economic activity clash. It's where supply and demand meet, establishing equilibrium points expressed in prices of goods and services.

Macroeconomics refers to the big picture, dealing with aggregates and wholes while microeconomics focuses on individual parts, including human behavior. Both levels interact and follow a natural business cycle proceeding in four sequential phases: expansion, peak, contraction, recovery. Governments try to manipulate this natural process through reactive policies ranging from minor tweaks to major over-reactions, often causing more harm than help. Classical, Keynesian and Supply-Side approaches are competing philosophies intended to positively impact economic growth and employment. Fiscal policy involves taxation and government spending while monetary policy focuses on manipulating the money supply. These policies influence inflation, employment, price stability and GDP. Mismanagement of either can result in excessive debt, inflation, recession, or worse – depression.

Economic Factors   *The Ever-changing Economic Environment*

Interest rate – cost of money	MICRO Level	MACRO Level
Exchange rate – cost of import/exports	Competitors	Unemployment
Tax rate – cost of income/purchases	Demand	Inflation
Wages – cost of labor	Market size	Interest Rates
Labor – production resource	Supply	GDP
Supply and Demand	Suppliers	Exchange rates
Inflation – diminished value of money	Distribution chain	Consumer confidence

Money serves as a conduit for economic activity, similar to blood in a body. While providing a practical medium of exchange, currency itself has no intrinsic value. A paper dollar by itself is worthless, however it does represent an arbitrary value set by government, and it requires our collective perception of legitimacy to work. If the public ever loses faith in the system, it can all collapse. Managing the money supply involves establishing and regulating a banking system, beginning with the federal reserve and extending through national, regional and local banks. Monetary policy then impacts the economy similar to the way blood circulation regulates body functions; it affects interest rates, exchange rates, investment demand and inflation.

The modern era introduced two primary approaches for operating a large-scale economy: capitalism and socialism. They differ by degree of government intervention and dependence. Most systems use elements of both. The US system is capitalist centric but has seen increasing degrees of socialism injected over time, resulting in greater government intervention, regulation, taxation and added constraints. Socialism sounds fair and feels good but has floundered economically throughout history, largely because it overlooks human nature. Behavior is incentive based as people will rationally pursue their own self-interest. Capitalism openly embraces this self-evident fact and thrives when individuals are allowed to engage in free market exchanges with minimal government intervention. The profit motive is a powerful incentive that outperforms artificial government directives and restrictions. Top-down meddling policies rarely anticipate unintended consequences that occur when the public is forced to do what someone above wants rather than individual choice. Socialism's worst hybrid form is communism – a feel-good political ideology more than an actual economic model. Despite promising fairness and equal sharing, it always produces government managed socialism with extreme emphasis on authoritarian control. A small faction of elites tends to make all the decisions while retaining great power at the expense of everyone else. The best economic systems are those which maximize scarce resources, produce growth and prosperity, incentivize entrepreneurial spirit, and like a rising tide, lifts all boats. With these metrics in mind, history provides plenty of evidence of consistent under performance by socialism when directly compared to capitalist alternatives.

Socialism vs Capitalism
   East vs West Germany
   East vs West Europe
   USSR vs US
   North Korea vs South Korea
   Venezuela vs Venezuela
   (Present)   *(Past)*

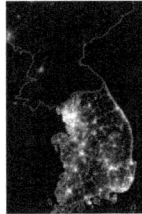

← North Korea

← South Korea

An economy is just an aggregate whole made up of lots of small, individual activities between people of all types and persuasions, engaged in making things, providing services and exchanging possessions. Like cells in a body, each person contributes to the whole by simply doing their own thing, performing some function and adding a measure of input into the greater collective system. Similarly, each small business and independent retailer acts like an organ in our metaphoric body, serving as hubs for microeconomic activity supporting the whole. Larger components include wholesalers, factories, distribution centers and corporations.

Goods and services cover just about every conceivable item or task one could imagine. If someone wants something bad enough, somebody else will produce it, legally or illegally. Businesses make profits by providing things somebody wants. Factories and assembly lines churn out products continually to satiate the collective demands of desirous consumers. Success lies in identifying wants or needs that aren't being fulfilled. Success also emerges where new desires are created by slick marketing and persuasive salesmanship. Opportunities abound wherever demand goes unfulfilled: black markets, independent dealers, flippers, scavengers, schemers and individual opportunists. Modern economies may be powered by corporate entities and business-to-business networks, but they still contain vestigial remnants of early systems, including farming and bartering, along with elements from preceding economic periods (Agrarian, industrial, informational). It's a comprehensive system of interrelated parts that co-evolve over time.

Government. Groups that operate without an organizing principle fall into chaos. The bigger the group the greater the need for order and control. Leadership is necessary to keep things on track. A single individual may suffice in a small group but for larger ones a defined structure becomes necessary. For societies and nations, government is the central organizing force holding everything together. Hierarchies arise as demands for maintaining control increase with broader spans, creating multiple levels and layers of distributed power. Positions, titles, uniforms badges and licenses serve as tools of legitimacy and authority. Branches and agencies naturally emerge to address specific areas of concern in a focused manner. Formal written documents establish legal grounding, providing guidelines for implementing policy. Recurring meetings ensure directives are carried out and provide tracking-oversight of new developments as they occur.

Emergent group order began in early history when humans first clustered into small factions with shared interests. In the beginning, bands, tribes and clans grew in size and organization, requiring greater order and control. Physical security was essential. Villages became towns, then cities. Early states formed civilized societies, established rudimentary government hierarchies, developed unique cultures and claimed lands. With expansion and complexity came city-states with significant swaths of territory and influence. Inevitably, societies butted up against each other necessitating governmental resolution to minimize conflicts. Nations became sovereign states: a permanent population, defined territory and independent government.

Modern states arrived several hundred years ago, accelerated by the Peace of Westphalia treaties recognizing sovereign states in the mid-17th century, followed by periods of nationalism that spread during the late 18th and 19th centuries. It took two world wars in the 20th century to expose the unhealthy aspects of extreme nationalism, setting the stage for a geopolitical transition towards globalization near the end of that same century. Ironically, after just two decades into the 21st century it appears nationalism is making an unexpected comeback.

Governments are basically structures of power and control, defined by who is in charge and how they run the state. A variety of types have been tried throughout history, varying between degrees of concentrated power in the hands of a few to democratic versions with power shared by many. Some emphasize rule by a single authoritarian leader while others rely more on consensus. Most employ a combination of hierarchy and networking distributions of power. Judging which type is better or worse depends upon what you value most: rights of individual people, needs of the larger whole, efficiency, effectiveness, continuity, or responsiveness to change.

Forms of Government (most are a mix, very few pure forms)

Democracy – majority rule
Oligarchy – minority rule
Republic – citizens elect representatives
Authoritarian (Dictator, fascism, totalitarianism) – top-down control
Monarchy – King's divine right to rule based on royal family lineage
Marxism – total government run socialism
Communism – classless society in theory, Marxist in actual practice
Aristocracy/plutocracy – controlled by wealth and social position
Theocracy – controlled by religious leaders

Government Components. Governments are set up and guided by founding documents that specify and limit powers, usually via branches and agencies. The US Constitution serves this purpose, separating powers between 3 branches: executive, legislative and judicial. It's supplemented by a Preamble and

Declaration of Independence setting a tone of *power to the people*, plus a bill of rights protecting the citizenry and their individual states.

Each of the three primary branches are present at each of the major levels of government. National, State and local each contain an elected executive serving as the leader at their level: President, governor and mayor. Each level also contains a legislative body that makes and passes laws: US Congress, state assembly, and city council. Finally, each level has an associated court handling judicial matters at their particular level (federal, state and municipal). Not to confuse the matter but there are also sublevels at each primary level (Federal court has district trial courts, circuit courts for appeals and the Supreme Court as the final level of appeal).

The US *executive branch* starts with the Whitehouse with a president and vice president, including 15 departments and cabinet members, plus numerous commissions, administrative bodies, boards, counsels, agencies and authoritative foundations. Directives and executive orders are made in a top-down manner. Policy initiatives are pushed down and out to garner public and congressional support. Speeches and appearances set the agenda with narratives constructed to shape public opinion and gain political advantage. This process is carried out at each level, with governors and mayors operating in a very similar manner.

Legislative Process. Any citizen can take an idea and solicit it to their respective representatives, who may then sponsor it as a bill. First it gets assigned to a committee for study. If approved it gets scheduled for either a vote, further debate or amendment. If passed it goes onto the Senate where the process is repeated. Those that pass both houses of congress (federal or state level) go to the executive branch for final approval (president or governor), where the bill can become a law.

At the local level there's a similar process where citizens forward issues to their local rep, who then gets it on the council agenda. Goals and objectives are defined, alternatives are discussed, and budget implications are considered. Localities can involve a combination of city, town and county governments that overlap, making the landscape seem a little confusing. Local councils and county commissions set policies which are then implemented by a mayor, county executive or city manager. Every community has a unique, local flavor which makes for some interesting differences between each. Despite the popular focus on national level politics, citizens' daily lives are often impacted much more by the ongoing changes in local ordinances, referendums, new initiatives and state statutes.

Modern government representatives have large bureaucratic staffs that do all the real work while they spend most of their energy speaking, persuading and engaging in back-room dealing. Halfway through their elected term politicians routinely spend more energy campaigning for reelection than legislating. Representatives and executive managers are themselves affected by two major things: elections and budgets. Anyone holding an elective office in government at any level spends a significant portion of their time and energy focused on getting elected and staying elected. Campaigns for 2, 4 and 6 year terms seem to be running continuously, not to mention supplemental special elections in between.

Periodically the outcomes of elections result in significantly different directions in government. Some are watershed events that lead to sweeping changes across the board. Equally impactful are the levies collected by government to spend, first on "essential" services (needs), then on discretionary items (wants). The process gets especially political whenever nonessential spending is desired, requiring a case to be made by persuading or coercing citizens into paying more for them.

Judicial system. Governments maintain order through regulation, rule of law, and securing *justice*; attained through a system based on fair process, transparency, impartiality, and the opportunity to defend one's actions. It's enforced by the branch of government dedicated to *interpreting* laws, how they should be applied and whether they conform to the founding principles, which in the US case is the constitution. A structure of courts is arranged in a hierarchy of levels, including local, municipal, circuit, federal and Supreme court. These courts are the arenas where disputes are resolved and individuals who violate the law are sentenced for punishment. Lawyers represent defendants, swearing an oath to the court and their clients, whose confessions must remain completely confidential. Judges are appointed, not elected, to minimize political influence and maximize impartiality. Many are appointed for life, not specific terms, for the very same reason.

Most judicial systems follow *common law*, where the precedents of previous court rulings are used to make current decisions. *Rule of law* is a principle of fairness where all citizens and institutions are subject and accountable to the same laws, including ruling elites in positions of power. Justice is best served when individual citizens have a right to a fair and speedy trial before a competent judge and a jury of their own peers. Natural rights are clearly defined in the US Declaration of Independence as life, liberty and the pursuit of happiness. A justice system's sacred role is to ensure that's preserved.

*Magna Carta of 1215 introduced the radical concept that the King and government are not above the law, but subject to it with its citizens – influenced US government founders.

Legal cases are either civil or criminal. *Trial courts* address both criminal cases (felonies, misdemeanors, traffic tickets) along with civil cases (probate, juvenile, family law, other civil matters) and appeals of civil cases. *Civil trials* address disputes between citizens or organizations seeking remedy after being wronged. *Criminal trials* are between a citizen and the state, where an individual is judged to be either innocent or guilty. Parties who feel their trial was conducted unfairly may appeal the outcome in a higher-level court. Despite the general fairness of the system, justice is never guaranteed. All it takes is a weak lawyer, biased judge, false evidence or a feeble jury.

Role of Government. Collective groups require many of the same basic needs that individuals do: safety, security, survival and maintenance. The challenge is determining what people can and can't do for themselves. Over time, governments tend to usurp more and more of the things that are naturally provided in the private sector. Much of the political discourse at all levels of government comes down to this matter of where to draw the line.

Essential Government Tasks:

Security	Police, Military
Safety	Business regulation, Code enforcement
Utilities	Water, Electric, Gas, Sewer
Health	Hospitals, Ambulatory service
Maintenance	Waste disposal, snow removal
Transportation	Vehicle registration, road repairs
Education	Schools, buses
Courts	Civil conflicts, Criminal punishment

Beyond these categories are supplemental tasks governments engage in that are certainly not "essential". Even those deemed essential are often stretched beyond actual need. Private businesses can provide much of the same services, depending on the scale and scope. The Department of Defense routinely contracts private services that support military needs. An example of overreach in an area considered "essential" is the notion of welfare. How much should a central government get involved with an individual's personal health and well-being? At what point is it fair to take money from a healthy person to subsidize a sick person? Is it fair to take money from a hard-working, productive person and give it to someone who doesn't work? These and many other conflicts arise as governments grow and expand in their reach – a central theme in political discourse.

POLITICS. How people get what they want involves a political process, where the dynamics of opposing viewpoints and competing desires interact, resulting in a variety of possible outcomes: one side is happier, both sides are happy, neither side is happy, or somewhere in between all three. Politics is really about conflict resolution, culminating with an allocation of resources – *who gets what and how*. Where government is a noun, politics is a verb (the process side of governing).

Political Process. Vying for power produces focused efforts between opposing sides, leading to extensive campaigns to achieve a certain end. The process can include staging elections, consensus building, generating support, decision making and policy formulation. It involves a wide variety of activities, tactics, approaches and responses, including negotiation, education, persuasion, coercion, lying, and compromise. Political dynamics produce an interactive series of action-reaction, measure and countermeasure. Intensity and scope vary, ranging from a simple verbal disagreement to a violent fist fight; even nations going to war. Whenever two or more people get together it's likely some form of politics is at play. With larger populations there's a greater degree of organized politicking where representatives are chosen to pursue the will of groups. At city, state and national levels there are dynamics requiring greater sophistication and coordinated efforts to achieve desired outcomes. Higher levels bring greater amounts of power which also invites misguided, disingenuous behavior and incentivized corruption.

Political activities are carried out at 3 basic levels: Macro, Middle, Micro. *Macro-politics* involves issues at the highest levels of government as well as geopolitical concerns crossing national borders. The international global stage includes multinational corporations, non-governmental organizations (NGOs), and international organizations (UN, ICJ, IMF, ILO, IBE, WHO, UNESCO, etc.). *Mid-level politics* are the most familiar activities, involving political campaigns, party efforts at the state and national level, plus general elections at any level. *Micropolitics* are the actions carried out by individuals, mostly at the local level in an effort to inform, agitate, destabilize and draw attention to a cause. Tactics include boycotts, demonstrations, petitions, strikes, picketing, heckling, and general civil disobedience. The most basic action any individual can engage in to achieve their pursued ends is simply voting.

Politics in pop culture tends to focus on the executive branch, despite the fact that most of the action really takes place in the legislative arena. Law making requires ginning up support with various factions who don't always share the same values. Covert backroom deals are made while overt public debate runs its course. Compromise is routinely required. As political parties become more polarized the legislative process gets uglier and less accommodating. Modern societies tend towards internal polarization; major issues generate equal and opposite views. Sensitive topics create significant tension between opposing sides. Some differences follow party lines while others cross those same lines. Where compromise fails, agitation and extremism grow, testing the resilience of the entire political system. Balance and reason are difficult to achieve once tribalism becomes the driving force. Though courts and judges are expected to be apolitical and have been through most of history, it's gradually changing as polarization and partisan media incentivize the politicization of the judicial system. Decorum and fairness are giving way to ruthless tactics and win at all cost actions. Thousands of years ago Plato encouraged harmonious compromise in matters of conflicting interests, urging resolutions allowing each side to be placated, never at the expense of the other. Today it seems all bets are off.

Power. The ability to get what you want and impose your will on others is the essence of politics. It's all about obtaining power, directing it and using it to achieve goals. It's the capacity to control others, situations and influence outcomes.

Types of Power

Formal	Bestowed by the group
Legitimate	Position or title
Expert	Knowledge based
Referent	Leading by example
Coercive	Use of force or threats
Reward	Use of benefits
Informational	Exclusive access to info
Connections	Influence through networking

Political tension surfaces when naturally changing conditions are not appropriately addressed or responded to in a timely manner. The ongoing process of societal adjustment can create moments of dissatisfaction among citizens who perceive things are not going their way. Lack of response by government representatives can lead to unrest, uprising, rebellion, rioting, civil disobedience, or in extreme cases, revolution or civil war. Political dysfunction also occurs when privileged power bestowed by a group is used for illegitimate private gain. The higher someone goes up the hierarchy of power, the greater the temptation for abuse and corruption. Typical forms include cronyism, nepotism, bribery, pork barreling, slush funds, kickbacks and exclusive perks. Sometimes the level of corruption is so pervasive within a government culture that the system qualifies as a kleptocracy, meaning *ruled by thieves*.

Political Spectrums. Perspectives on most issues come down to differences along a scale of freedom vs control, stability vs change, or cooperation vs opposition. One spectrum is based on how much power is acquiesced to a central government. This scale has extremes between total control and anarchy. Socialists favor more government control, libertarians less. The major political parties mirror these positions; Democrats favoring government socialism and Republicans leaning to individual freedom. These parties and their positions on a spectrum are synonymous with *liberal* and *conservative* respectively. The scale is also based on what degree of social change is preferred. Most tend to be either conservative or liberal. The former prefers keeping things as they are while the latter pushes for change. Liberals focus on grievances needing further attention, conservatives tweak progress already achieved. For liberals the cup is always half empty, for conservatives it's half full. When taken to extreme, liberals will exaggerate problems to obtain support, conservatives will dig in and resist major change.

Government Control Spectrum	*Conflict Resolution Spectrum*
100% ◄-------------------► 0%	LOVE ◄-------------------► WAR
Authoritarian Anarchy	Embrace Fight
Fascism Libertarian	Cooperate Oppose
Socialism Capitalism	Negotiate
Liberal Conservative	

Democrat ------ Republican

Labels and definition. Prior to today's political climate, classic Liberalism in its purest form embraced liberty, rights of the individual, equality before the law and consent of the governed. High priority was placed on free speech, freedom of the press and freedom of religion. Ironically, these rights are no longer as important; neoliberalism has morphed into authoritarianism, elevating government power over individual freedom. Classic Liberalism was closer to Libertarianism but now neoliberalism is labeled as *Progressive*, moving closer to autocracy. Republicans shifted as well, from once embracing big business and corporate wealth to now championing small business and the middle class. Conservativism favors restraint of government, lower taxation, deregulation, and traditional values. Culturally there's emphasis on family, religion, and pride of national heritage. When it comes to the individual, liberals link your troubles to others while conservatives link them to yourself. A Moderate is anyone in the middle, neither liberal or conservative, yet capable of siding with either side. There are also moderate versions of both liberals and conservatives who don't blindly follow the party line.

Conflict resolution is another related spectrum reflecting the very essence of politics. Whether it's individuals, groups or nations as whole, there are and always will be differences of opinion, values and desire. How these differences work out falls somewhere along a scale between full cooperation and total war. The middle ground is anchored in negotiation, which is simply the act of mutually resolving differences with give and take. Outcomes generally depend on what prevails – rationality or passion.

Confrontation	Vs	Conflict Resolution		
Violent		Peacefull		
Verbal threat		Talk	--------	Avoid
Psychological attack		Agree to disagree	--------	Diffuse
Physical violence		Mediate solution	--------	Confront

Government as a necessary evil. Control over a large group requires authority and power, which in the wrong hands can lead to serious trouble. Despite good intentions, people often let power go to their heads, falling into traps of misuse and abuse. Absolute power corrupts absolutely, and the road to hell is paved with good intentions. The founders of America's government knew this too well and with great skill and purposefulness constructed a roadmap for governing constraint, separating powers with checks and balances. Still, a government will only be as good as those who run it, regardless of how many built-in restraints are provided.

History has shown that governments tend to take on a life of their own, gradually progressing into a self-serving entity. Higher levels become more distant from the citizens below. The whole conglomerate grows from within, adding layers of overlapping bureaucracy like a virus feeding off itself. Complexity progresses, new branches are formed, and special agencies are established to meet some specific concern. Eventually, efficiency diminishes, even to the point where different branches or agencies engage in redundant or conflicting efforts with others. Government funding is rarely performance based nor incentive driven.

Top-down directives tend to lose touch with the needs of common folks. There's always someone sitting in a high-level meeting proposing a great idea with the best of intentions that sounds good in isolation but doesn't have any basis in street reality. Even when ideas are tested and backed by research, implementation may be short sighted, causing more problems than they solve. One of the biggest drawbacks to top-down government policy making is failing to anticipate unintended consequences, which typically occur when imposing new restrictions on people or when providing arbitrary penalties to redirect human behavior.

Government Control Dynamic: *Healthy balance difficult to maintain*

Freedom --------------⌃-------------- Control
                    (Healthy balance)

→ → → → → → → → → → → → → → → → → → → →
Natural tendency to expand control, limit freedom

Sensible limits vs. individual restrictions
Basic order vs. overbearing regulations
Free to choose vs. forced directives
Personal savings vs. higher taxes & fees

Governments gradually grow and ossify, becoming slow, inefficient and cumbersome over time. They routinely react rather than pro-act. Responses in the *here and now* often linger well beyond their usefulness, when they're no longer needed and actually make things worse. Once an added restriction or new tax is imposed, it rarely goes away. Yet there's really no way around the need to hand over great power to a chosen few, entrusted to manage your large, complex society. But it must be accompanied by competent oversight, transparency and an ongoing vigilance to keep the beast reigned in. Conservatives understand this better than liberals. When it comes to government as a necessary evil, liberals embrace the necessary while conservatives emphasize the evil. This inconvenient tendency is present in all governing bodies regardless of size or scope. When you hold a hammer in your hand everything looks like a nail. Consider the heavy hand that even small groups or official agencies wield over citizens in bodies that are intended to help, such as a homeowner association, code compliance department, the DNR, IRS, and any official who utters the infamous "we're from the government and we're here to help". Governments may be formed by the people and for the people, but over time they inevitably become self-serving entities of their own, and for their own.

CULTURE. Every group has a unique identity; a personality derived from a combination of attributes setting it apart. Societies are just very large groups whose personality is called culture. No two are exactly alike, differing in aspects and elements of composition. Culture is a collective whole of numerous individuals linked together by shared commonality, most importantly its language. This isn't limited to just verbal and written communication but extends into customs, codes, etiquette, rituals and traditions.

The group personality of a society is expressed in a variety of modes with collective traits. Its cultural spirit manifests in styles of dress, art, cuisine, accepted behavior, music, and a wide range of popular activities. The whole takes on a collective consciousness generated by its individuals but then reflected back on each participant. Everyone contributes, consciously or unconsciously, sharing similar values and abiding by established norms. A composite societal personality forms out of shared interests, common language and group consensus.

Components of Culture. In 1945 George Murdoch identified 67 *Universals of Culture*; all the things that contribute to the whole. His comprehensive list included sports, cooking, courtship, dancing, arts, education, ethics, botany, cosmology, games, gestures, hair styles, housing, kinship, language, medicine, religion, customs, plus many more. Any one of these categories can easily be broken down into further variations on a theme. Consider arts where there are 4 major types: Visual (architecture, ceramics, drawing, painting, photography, film, sculpting), Literary (drama, poetry, prose), Performing (dance, music, theater, comedy, magic), Culinary (cooking, baking, winemaking). Similarly, we have Entertainment (television, movies, music, sports, games), Special interests & hobbies (crafts, collecting, etc.), Education/School system (foundational, graduate, specialty, trade), Religion (theology, spiritual guidance, mentors, ritual, dogma, worship, holidays).

Modern society's culture is dominated by various forms of media, especially electronic. The first wave involved print media where mass communication in the form of books, newspapers, magazines, journals and newsletters connected citizens on a large scale. Subsequent tech innovations lead to networks of linked electronic media sources, including radio, TV, movies, videos and cell phones that make sharing ideas to a mass audience a daily experience. The final wave emerged with the introduction of a global internet, linking everything from home and office, individuals and groups, smart phones and home computers. It created a socially connected culture of sharing ideas and events anywhere within a society in real time. Modern culture was dominated by television during most of the last century but is now sharing the stage with social media on smart phones. Both serve as the most influential sources of pop culture, shaping values of the masses and directing attention to what's percolating in the zeitgeist.

Culture as the Collective Expression of Human Spirit. The outside is always connected to the inside just as parts form the whole. Large groups are simply a mix of unique individuals, each with their own traits, behaviors and desires. When you add up all the hearts and minds of individual people you get a composite whole with a distinct group identity. Culture is formed in this way as creative energies of the many get channeled into the mix. It's a collective process of spirit filtering into the physical world. We see this everywhere we look as people participate in a wide variety of activities in diverse types of arenas. Creative energy is expressed on either physical, mental or emotional levels, or a combination of all three. Examples include sports and hard work (physical), accounting and education (mental), romance and music (emotional). Most have aspects of all three in different degrees.

Just about everything human beings do contributes in some small part to the greater culture. Where an economy is driven by individuals pursuing self-interest, culture is steered by individuals expressing their inner desires. When work is finished the rest of the day is open for everything else. Sports, games, music, dance, art and crafts, or any number of hobbies; options are endless. Modern man is privileged with a healthy balance of work vs. leisure time, and societies benefit with richer culture. This wider range of activities and options is matched by a greater variety of arenas to participate in. Higher demand for leisure activity requires expanded public spaces: parks, town squares, riverwalks, shopping malls, atriums, designated sightseeing areas, etc. Culture is the melting pot of society where diverse creative spirit is expressed in a multitude of daily expressions.

Cultural Dynamics. Societies are enduring entities with long term attributes complimented by short term desires. This dual nature plays out in culture as established traditions that persist over time, flavored by the ebb and flow of changing moods, tastes, styles and preferences in the masses. Fixed cultural institutions provide continuity to society while punctuated events like holidays, festivals, celebrations, and performances express the current ambience. Language provides continuity to the social order. Though fairly structured as a long-term fixture, it's modified gradually and continually as new terms are introduced, and old sayings fade away. Similarly, household products may be standard items in daily life, but they too tend to become obsolete in time. Everything in society goes in and out of favor in waves, mirroring the fickleness of citizens who change their minds and desires after a while. Cultural trends can impact specific areas or wider swaths, bringing change to a variety of areas in harmonic unison. Styles shift in cycles, especially in the arts (music, painting, literature, comedy, movies, etc.) as tastes oscillate among the masses.

Pop culture is the flowing, change aspect complimenting the permanent, more stable social institutions. It's the surface layer that reflects what people and groups are feeling in the here and now. Fads that go in and out of favor are merely products of collective inner spirit manifesting externally. Pop culture is a composite shared experience of everyone participating in the social arena. Memes and viral creations emerge with unpredictability, taking on a life of their own at the collective level. The entire entertainment field follows a similar pattern where popularity hits on an almost arbitrary basis, influencing the success or failure of movies, television programs, celebrities, and artists in general. But popularity is a fickle phenomenon; today's hero can suddenly become tomorrow's villain, or a complete afterthought.

Modern culture is impacted by a variety of social networks with influencers driving current narratives. A single popular voice can move masses when the content resonates and timing is right. Electronic media empowers the few to achieve great influence over the many. Moreover, the widespread integration of television, movies, video clips and streaming programs creates an extensive web mixing real and fictional narratives that create new imaginative realities. Popular characters from movies and television often become more influential than real ones. Famous quotes and memorable deeds from fictional characters get imbedded into the web of society and become real aspects of the cultural personality.

Large groups are usually comprised of both individuals and small subgroups. The cultural equivalent would be subcultures, which are like mini societies with their own unique blend of attributes. Subcultures come in all sizes, shapes and types, including traditional, urban, rural, night life, avant-garde, youth, political, ethnic, and even counterculture. Different regions of the country have their own cultural identity, including the South, Midwest, Northeast and the coasts. Sports has become a multi-billion-dollar subculture with its own celebrity athletes, team mascots, logos, merchandise, promotional events and fans (short for fanatic). Special interest groups of all kinds pervade society, many with their own rituals, preoccupations, passions and uncommon celebrities – persons elevated to a higher status just because. These include the corporate world, military, inner city, universities, generational cohorts, theater and entertainment, Hip Hop, religious cults, Hollywood, government bureaucracy, "stoners", fitness enthusiasts, country, and so many more. Social media is a facilitating breeding ground for these and all other subcultures seeking their own special space.

Cultural disfunction naturally surfaces as societies have both good and bad people. Healthy groups are founded upon shared values, common interests and general unity. These are offset when individuals and subgroups splinter from the whole, engaging in activities not aligned with the rest. Symptoms of disfunction appear as crime, laziness, discrimination (race, age, sex, class), tribalism, low productivity, low quality, unhappiness, unfriendliness, corruption, violation of rights, lack of freedom, lack of respect, abusiveness, poor communication, and domestic violence. The common denominator is divisiveness within. A whole cannot endure when it gets too divided or beset by destructive splinter groups.

A healthy culture produces a spirit of cooperation and collaboration, exemplified by peace, individual rights, and quality of life. Strong, enduring societies exude a distinct sense of unified community – common values and shared interests. Modern societies seem to only experience peak unification following tragedies or large-scale crises, only to fall back to a state of diminished unity as time passes. Religion has historically served as both unifier and stabilizer within cultures though it's also been a major source of conflict between competing cultures. Its impact seems to have waned in western societies which have become increasingly secular.

Culture is the heart and soul of a society, a composite personality expressing the human spirit collectively with symbiotic connections between each individual and the whole. Its long-term enduring identity of shared values, customs and codes is passed on generation after generation. Short term surface changes ripple through each generation, shaped by ongoing ebb and flow dynamics of a restless citizenry. Language unifies culture in numerous ways and forms; verbal and nonverbal, unwritten codes, signs, symbols, customs, rituals, and norms of behavior. Modern culture is electronically linked in a comprehensive network of connected media making every individual a virtual participant in real time. The composite whole with all its subcultures and special interest groups forms a unique societal personality, greater than the sum of its parts.

SOCIETY's PRIMARY ELEMENTS – Economic, Political and Cultural.
The big 3 operate both independently and interdependently within a society in the same way a human being's components of body, mind and soul interact. Each performs a specific function supporting the whole, affecting and being affected by the others. Each has both fixed and variable aspects: goods/services, laws/politics, values/fads. Each operate collectively with emergent properties that take on a life

of their own, often moving in unpredictable directions. Together they combine to form a societal identity – a wholistic personality.

ECONOMY	GOVERNMENT	CULTURE
Body	Mind	Soul
Sustainment	Order	Purpose
Business	Legislation	Values
Work	Regulation	Creative expression

Healthy Society	US	China	Denmark

Most societies are skewed. The West generally emphasizes economics, East slants toward culture.

Economy as a Living System

ECONOMY	BODY
Resources	Food
Fuel	Carbs, sugars
Labor	Muscle
Management	Brain
Money	Blood
Inflation	High blood pressure
Transportation	Lymph system
Communication	Nervous system
Equilibrium	Homeostasis
Factory	Organ
Ponzi scheme	Cancer
Debt	Fat

A business is a microcosm of the economy

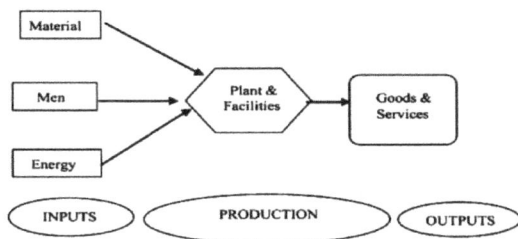

THE WORLD MATRIX.

The gradual evolution of civilized man created a super complex multifaceted system greater than anyone can comprehend or master. It's evolved into a world-wide superstructure with a life of its own, moving in unpredictable directions while steering, shifting, ebbing and flowing, occasionally unleashing great surprise events, but normally just chugging along taking everyone with it. Sure, we can tweak and steer parts of it locally but the greater whole is a behemoth difficult to budge. What we're left with is a global societal system with rules, constraints, competing forces, incentives, temporary opportunities, long-term threats and arbitrary rulings. It's a grand labyrinth of twisting pathways with pitfalls, yet there are rewards for those who skillfully adapt, break the code and exploit advantages.

This chapter is essentially a Red-Pill revelation exposing how *The World* works without judgement or prejudice, just clarity. Acknowledging the true nature of the World Matrix is an essential step for achieving mastery of it, especially when that very success is defined by The World itself. We're all born into this arena and must spend the rest of our lives navigating through it as we pursue our goals, desires and interests. Unfortunately, there's no simple roadmap to do it, just a few occasional self-help guides and mentors along the way. It's a daunting task that some manage better than others.

How The World Works. Everyone is subject to the grand framework of constraints in the World Matrix, working through a maze of unpredictable turns, twists, threats and opportunities. Every new day opens unknown potentials influenced by a variety of competing forces from all sides. Headwinds signal a wrong direction while tailwinds provide an assist, offering temporary reassurance that the present course is ok, for now. Constraints are everywhere, limiting choices and narrowing options. A ubiquitous system of control measures directs preferred behavior, including, laws, financial costs, peer pressure, penalties, time limits, legal requirements, family ties and contractual obligations. This comprehensive regulating framework manifests as police departments, courthouses, prisons, the IRS, military bases, and government institutions. To make matters more confusing, rules change over time and are applied differently in various settings. What works here and now may not work later or somewhere else.

The World Matrix is a comprehensive superstructure anchored in hierarchies, connected in frameworks of order and power. It functions like a semi-automated machine with multiple operators working both independently and in coordination. Those operators are people like you and me but in positions of power with official titles and degrees of authority. They set rules and regulations pushed downward for you and others to follow. Authority is maintained with official badges, titles, ID's, qualifications, certifications, licenses, special caps and uniforms. Hierarchies are pervasive in government, business, corporate structure, and even the cultural realm (church, entertainment world, education, etc.) all mirrored by big cities, skyscrapers and downtown financial districts. Power and authority are connected to certain classes, privileged elites and exclusive groups. Status is directly linked to one's position on a particular hierarchy. No matter how high up the totem pole you get, there's always someone above you. Age and experience factor into it as most committees, organizations, and boards are run by older people (good and bad aspects). One positive counterbalance to this hierarchical structure is the post-modern emergence of networks which incentivize collaboration and decentralization.

The World Matrix is bigger than anyone, no matter how rich, famous or powerful. It will swallow you whole on short notice without warning. It will bring sudden reversals of fortune. Great leaders can be taken down, overthrown or assassinated while sitting on *top of the world.* Power and glory are temporary and fleeting. Dominant industry titans are often swept aside by innovative competitors.

Famous celebrities may ruin their reputations overnight from a slip of the tongue, a compromising video clip or an inconvenient photo exposing unacceptable private behavior. All sorts of unexpected reversals can claim the comfortable status of anyone riding high, including a heart attack, car accident, embezzlement, non-curable disease, or any number of negligent actions committed during a momentary lapse of judgement. Anyone who is not ordinarily humble will soon be.

Recurring Events in your experience of The World
> 1st Occurrence is ordinary happenstance
> 2nd Occurrence is likely a coincidence
> 3rd Occurrence indicates directed intentional action.

News, Weather and Sports. The matrix system provides free daily updates and reminders of the arbitrary nature of living in *The World*. Every day there's a new headline, a major event, an accident, a murder, new political development or unresolved conflict. Much of the content is skewed towards negativity with an occasional feel-good story inserted just to round it out. Weather headlines remind us that we still have to navigate constraints of nature to succeed in *The World*. And sports are metaphorical dramas mirroring the ongoing competitive challenges we all have to face every day as we work through the matrix. Our attachment and identification with the local team draws us into a different realm where *we* feel *their* thrill of victory or agony of defeat. It's no coincidence that sports are a multi-billion-dollar industry built into the very fabric of society, playing a thematic role in the greater culture.

Arenas and Realms. The World is a stage, full of various arenas where people compete, share or exchange with each other. Sports offer the most obvious set of arenas but there are plenty of other types present everywhere in The World Matrix, including the work place, courtrooms, stadiums, fields, rings, stages, platforms, airwaves, game rooms, public squares, battlefields, pools, clubs, forums, and any place where two or more individuals or groups compete or engage with one another. Even the humble abode where you live and relax often becomes a place where competition may surface. Modes where you compete include work, sports, family, courting, career, play, and the most elementary opponent of all – yourself.

Life is a competition, and The World is the arena – a *super-arena* of arenas. While competing is the dominant dynamic, cooperation is its counterpart, also present in a variety of arenas. Engagements may be win-lose or win-win; both equally effective approaches to getting what you want. Places of cooperative exchange include businesses selling goods or services, venues for entertainment, schools, churches, and artistic outlets. Ironically, many of these "cooperative" outlets involve a significant degree of competition within themselves and between other similar venues.

Arenas in The World include repetitive frameworks appearing everywhere in a variety of related spaces. Every town, municipality and modern city share common features across the national landscape. Every big city has shopping malls, banks, bars, restaurants, schools, gas stations, fast food, a post office, library, police and fire stations, government centers, etc. It's a recurring template with minor variations repeated wherever you go. Even street names are duplicated hundreds if not thousands of times in different cities. Presidents' names are most common along with main, center, canal, broadway, park, elm, pine and numbered streets.

Realms are equally ubiquitous as arenas, serving as general domains of particular conditions, rules, traditions, etiquette, language and unspoken requirements. Every place we go, every club we join and every organization we belong to has a unique subculture favoring certain behaviors and actions while disapproving others. Some realms are naturally comfortable to us while others seem like foreign lands. Unfortunately, our goals and desires often require us to navigate realms we're not comfortable in and some we may even hate: corporate management, government bureaucracy, military, inner city, service industry, gated community, retail sales, jury duty, politics, senior nursing, country club, etc.

The World is a collective product of individuals competing and cooperating to get what they want. It's the same dynamic present in economics where supply and demand reflect clashing desires. Society as a whole is driven by individuals pursuing self-interest: Money exchange facilitates it, courts mediate it, governments regulate it, and politicians manipulate it. Consider the arena of driving in traffic, where your car represents you and the destination represents your desires. Roads are regulated by government, restricted with traffic lights, turn lanes, multiple signs, toll booths, construction zones and detours. Other cars can get in the way or just slow you down, just variable obstacles that must be managed. Unexpected back-ups and accidents are always lurking in the mix. Weather is a factor, occasionally slowing things down, sometimes causing a complete halt or even stranding you. It's a journey with required pitstops, refueling and maintenance, attractions, distractions, and unpredictable pitfalls…a perfect metaphor for the ongoing daily vigilance of pressing forward though *The World*.

Everything is already out there somewhere if you know where to look. Unimaginable wealth and abject poverty are not that far away. Untapped resources, niches, strange places to explore, new and different approaches to everything can be found close by. Somethings just require more effort to find with convoluted routes, extra twists and turns, and a labyrinth of uncertain pathways. The internet has added a whole new dimension to the grand matrix with even more pathways, options and possibilities – both positive and negative.

The World is a vast spectrum of all things good and bad, much of it never experienced by us directly. Most of us have no idea how good and bad each of the extremes can be. From a life of luxury and riches where maids and servants take care of your wants and needs to seedy drug houses where every day is a struggle to get the next fix. Most people have no clue what great wealth and opulence exists out there, nor do they have any grasp how miserable life in prison or solitary confinement is. Whatever you think life is like out there for other people, just remember it can be much better or much worse than you could ever imagine.

World Population scaled to 100 People

1 Australia	66 No access to affordable surgery
5 North America	26 No access to safe clean water
10 Europe	14 No access to toilets
16 Africa	10 Undernourished
60 Asia	10 No access to electricity
65 *Adults* (age 15-64)	34 Not finish high school
15 *English Speaking*	14 Illiterate

The World *out there* is a collective projection of the world *in here*, inside each of us individually. All our hopes and dreams, quirks, fears, pathologies and general character traits inside us are manifested collectively onto the world stage where we interact and experience it all. Ironically, we're actually confronting ourselves – our own internal creation. The results are often shocking on both ends of the scale, with inspirational highs and embarrassing lows. Everything we're capable of is projected out where it gets mixed with all sorts of positives, negatives and neutrals. Our very human nature is what creates the nonsensical absurdity of life; one that comedians gleefully exploit. Any one person is insignificant in the grand scheme but when you add up every soul, living and passed, and draw them into the collective mix, a complex matrix is created transcending everyone and everything. This super entity here and now includes legacy influences of past generations. Tribal tendencies, ancestral instincts and psychic baggage are all part of the mix.

World of Characters. Pathways through The World are populated with a diverse collection of personality types. Odd characters are present every way you turn, differing by language, dress, appearance, mannerisms, class, style and various quirks. It's a wide range of types representing the best and worst of man. A few outliers are larger than life, leaving a lasting impression, sometimes complicating the path ahead. To progress forward you may have to work around individuals who are pushy, deceptive, exploitive, know-it-alls, contrarians, negative nellies, downers, intimidators and energy vampires.

People we engage as we continue our journey are either enablers or resisters; they either help or hinder your forward progress. They can serve as gates, detours or catalysts. Progress depends on your ability to adjust to all the different types you come across. Key gatekeepers include your parents, bosses, close friends, and teachers-mentors-coaches. Each has a unique personality and style. Some you may not gel with but you can certainly adapt to them and just get along. There's no simple how-to manual, so you just have to learn as you go.

Money and power. Money is the life blood of an economically driven society; the symbolic measure of power. It's a derivative measure of freedom, influence, and importance. Money is a ticket of entry and access to exclusive things. It carries with it a coded message expressing individual and collective desires, values, and general interests. We can learn a lot about an individual by what they spend money on, what they'll do to get it, and how they'll respond when it's taken from away. Just ask yourself these related questions: What do you do with your money? How do you treat it? How do you look at others who have more or less than you? Money is a subtle code providing insight to personal character, revealed by how you get it, your efforts to get it, and how you use it or spend it. The same applies collectively: how do nations, states, or governments get it, share it or spend it?

Money and the pursuit of it drives the business world and plays a key role in politics. Cultural realms are not above it as movies and literature are dominated by themes of greed and ambition. Society is inundated by omnipresent marketing, advertising and selling since business has one primary goal; to get your money. The same is true for socialist politicians, scam artists and greedy individuals. There's a fine line between a willingness to purchase something in mutual exchange versus being persuaded, coerced or tricked into giving your money to someone else. Then there's a middle sweet spot where money brings happiness: too little (poor) or too much (super rich) carries a burdened nuance that diminishes peace or contentment.

Navigating the World Matrix. Every successful journey requires flexibility and adjustment. Life is a perpetual ongoing adjustment to the outside world, beginning right after birth and continuing through every stage afterwards. We start off with an innocent, vulnerable perspective in a relatively safe cocoon and then day by day react to the outside pushing back from every imaginable angle. We experience both pain and pleasure, resistance and assistance, threats and opportunities. Our behavior gets shaped by trial and error as we find out what's acceptable and what won't be tolerated. We learn as we go that certain things are preferred while others are not. It's a challenging journey for those who are stubborn, inflexible and press forward without regard to consequences. For some it takes physical pain to finally get the message that a course correction is needed.

Succeeding in The World is all about making good choices; ones that are approved by said World. It's best to know where you want to go and then embrace adjustments along the way. Progress is tied to your ability to roll with the punches, go with the flow and get back on your feet when knocked down. It also helps to choose the right paths and then stay on them when distracting forks in the road tempt you to go off on a tangent.

Success or survival in The World requires code breaking. Every arena and realm we navigate through is regulated by its own set of rules, policies, procedures, parameters and language. Some of these are obvious and stated up front while others are ambiguous, vague, confusing or concealed. This is where you must decipher the code to succeed, either through patience and persistence or assistance from someone who knows the code. Consider the workplace where you first have to discern the syntax of an effective resume to get an interview, then figure out what to say during the interview, and once hired learning do's and don'ts, reading your boss, coworkers, and subordinates, and picking up the workplace lingo.

Every journey includes assistance and resistance. Doors open and close along the route. No path is ever straight and level – there are always turns, twists, curves, rough patches, detours and dead ends. When goals are set high and your experience is limited, pathways will become treacherously ambiguous, like a clouded labyrinth of multiple branches with unknown outcomes. The way ahead may present any number of gates, walls, traps, pitfalls and challenging tests. Even simple, seemingly direct routes toward goals can be subject to unexpected blockages and distractions that will either slow the process down or end it. Sometimes the goal is simply unrealistic or one that isn't good for you – a misguided waste of time.

Anyone who's lived long enough recognizes the arbitrary nature of The World Matrix. Some people just fall into money, some inherit millions while others work hard all their lives struggling just to make ends meet, and some lose everything from unpredictable events, accidents or extreme weather disasters. One person can have their house demolished by a tornado while their neighbor's house is left untouched. Some people are struck by rare, debilitating diseases while others can smoke, drink, do drugs and live beyond 80. It doesn't matter if you're a good, likeable person who follows the rules and treats others well. It doesn't matter if you're a highly religious person and attend church every Sunday. Living in *The World* means every day contains an element of *anything is possible*, good or bad, and what's fair has nothing to do with it.

The World is certainly not based on fairness and many of man's futile attempts to impose fairness often backfire. Unfortunately, all men are not created equal. There's a big difference between equal rights, equal opportunity, equal ability or equal effort. No two people are the same. Any coerced effort to make them "equal" is fraught with peril. Ironically, the perceived unfair nature of inequality among people is equally matched by the bland indifference of a society where everyone is the same. Our differences are what makes life interesting, challenging and worthy of self-improvement. Consider how the coerced fairness from misguided governments giving out "welfare" checks created an entire generation of dependent people; drones incentivized to stop pursuing self-reliance. And the politicians behind it all probably had good intentions.

Relativity of Fairness or "Luck"

1st person found a 5-dollar bill on the ground and was able to buy food for the day
2nd person got a loan for a used car valued at $5000 with no money down
3rd person bought a foreclosed house worth $500,000 for only $200,000
4th person invested in apartment complex at $1,000,000, sold it a year later for $2,000,000
5th person bought a pro sports team for 50 million and years later it's worth half a billion

Cooperation, Competition, Conflicts and Challenges. Achieving goals involves a combination of helpful assistance from supporters while overcoming resistors. Typical sources of resistance can include random obstacles, distractions, competition and purposeful opposition. Success is more likely when you embrace assistors and avoid resistors, which requires an honest assessment of who you associate with. Sometimes a long-term roadblock can be instantly bypassed by meeting the one right person with inside information.

Assistance	*Resistance*
Partner	Competitor
Teammate	Naysayer
Mentor	Backstabber
Fans	Enemies
Sponsor	Embezzler

Conflicts and challenges come in many forms and directions, frequently unexpected. While you're moving along making decent progress you can be sidetracked by any number of headwinds that slow things down. Disputes, illness, miscommunication, theft, loss, accidents, false accusations, lack of resources and deceptive partners can stop your progress dead in its tracks. Most journeys involve a fair share of headwinds and personal tests, but the more you overcome the more satisfying final achievement becomes. Success without challenges leaves a degree of emptiness and sense of hollow victory.

Some people think they can beat the system by taking shortcuts or discovering a secret hack that bypasses the normal procedure. There are both good and bad versions of this. Certainly, we can get ahead by taking advantage of inside information or exploiting a temporary opportunity. But trying to skip stages or phases of an established process too often can backfire or result in a less-than-optimal outcome. Short cuts are sometimes pursued out of laziness, animus or apathy, and can descend into corrupt efforts such as scams, fraud, theft and general corruption. Beating the system is occasionally accomplished but it's definitely the exception. Those who do it with bad intentions unwittingly become a further extension of the system itself. And yes, the system does seem to possess a karmic quality where no good deed goes unpunished.

Navigating Relationship Dynamics. No one achieves success in The World without the help of others. In addition to working around a world of characters we also need to operate in a variety of relationships that require adjustments. First, it's family (parent/child, brother/sister, older/younger, cousins, nieces/nephews), then it's school (teacher/student, coach/player, friends), then work (boss/subordinate, coworkers), and onto a wide variety of others. We soon learn there are roles and expectations put upon us by others in diverse situations. More importantly, we learn through trial and error about relationship dynamics where our actions lead to counteractions and counter responses. It's a process of discovery where we find out that others aren't always who they appear to be and our expectations are sometimes way off base. Life lessons abound as we find out the hard way that relationships require constant adjustment. Feel good relationships can mask underlying control dramas, false friends, biases, stereotypes, and hidden agendas.

Codes of Success. Getting ahead in The World is easier once you figure out what works in various areas of the grand system. Techniques, hacks, best practices, and winning formulas are embedded in the matrix and can be discovered if you know where to look. This inside information can be acquired through dedicated learning, persistent inquiry, mentors, connections or insatiable curiosity. Once you're on the right path these inside tips are easier to uncover. Sometimes the only difference between two identical seekers pursuing the same goal is that one deciphered the success code while the other couldn't.

Business success starts with identifying an unfulfilled need and then filling it. Success can be achieved by finding a better way to do something, expanding upon it and developing an exclusive niche. It can happen either intentionally or by accident. Long-term success is sustained through product or service improvement. It's imperative to avoid complacency, adapt to industry innovations and respond to competitors in a timely manner.

Political success is achieved through gaining power and influence. Progressing upward requires talent, intelligence and communication skills, or from sheer persistence, plain luck or exploitation of unethical methods. Both well intentioned souls and ruthless tyrants can achieve the same goals. The upward climb towards greater power is beset with tests and traps, coming with a price that may be too costly for the uninitiated to bear. Political success requires being nimble during crises and possessing the ability to spin bad news into good. A lot of it simply comes down to a good sense of timing; knowing what to say and when to say it.

The ***Cultural sphere*** has its success codes as well, though less structured and more subtle. Talent and passion are essential in this realm however luck and timing also play a key role. Entertainers and performers rarely reach the highest levels without lots of practice, preparation and dedication. One's sheer persistence and hard work can overcome lack of natural ability. But even talent and effort may not be enough without the right connections and timing. It's hard enough to breakthrough and achieve celebrity status, but it's even harder to maintain it, especially since tastes change and fickle public interest always moves onto the next popular thing.

Moving up in The World Hierarchy. Upward progress generally involves achieving the 3 P's: Promotion, Prestige, Power. Success is a vertical perception in a system supported by hierarchies. Moving up rungs or levels sustains the machine. Higher positions are associated with greater importance, influence, and respect. Power is a universal catch all for higher levels and it's often interchangeable with money. Both can influence the other. Power is also enhanced through connections, tenure, persona and ambition. Success is defined differently both within societies and between them. Economically it's defined by material things and wealth. Politically it's about power and control. Culturally it's about creativity and popularity. In the West there's greater emphasis on economic and political status while in the East it's more about culture and personal well-being.

Elements for Rising to the Top:

Cunning – especially to stay on top
Communication skills
Intelligence
Luck
Personality
Persistence
Positive attitude
Physical energy
Work ethic
Good timing
Connections
Raw talent

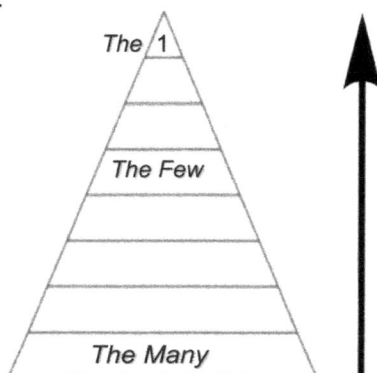

The 1
The Few
The Many

* Each rung up is a filtering stage or ceiling for those lacking a particular element

The World Microcosm in a City. Aspects of The World Matrix are captured and exemplified in the framework of any modern big city. You can see it in the hustle and bustle of everyday life in the inner city, the downtown mecca, street vendors, pedestrian malls, garbage trucks, traffic cop whistles, city smells, flashing signs, boarded up windows and yellow cabs parked with the engine running. The scene is normally hectic and sometimes frantic, yet things get done and life goes on. Every business, store, food truck and coffee shop has a unique story to tell with diverse customers of all backgrounds and interesting tales of particular experiences. Nightlife brings a transition in the action with a mix of odd characters, seedy night clubs, dark alleys, illegal activities and questionable transactions. This microcosm of The World illustrates the orderly chaos of uncoordinated, complex interactions that go on when diverse individuals of all types, pursuing self-interest, collectively comingle and somehow get most of what they want, then repeat variations of that same drama day after day after day.

Evolution of The World. In the beginning man was challenged by a world limited in nature where the primary threat to survival came from weather, elements of earth, and other creatures. Finding shelter and gathering food was front and center. Abrupt changes in weather was often life threatening. Mythic gods directed one's fate, expressed in dramatic events *out there*, always uncertain on any given day. Over time man progressed into mankind, developing collective organizations that stabilized and enrich daily life. Mastery of the planet surface catapulted civilized man far above the animal kingdom, exploiting a new privileged position via bridging, tunneling, damming, road networks, farms, canals and locks, airstrips and seaports. Soon cities, states and nations evolved with concentrations of bustling activity centers further entrenching man as the master of the planetary environment. With further progress came opportunities for individual gain, but it required acquiescing to preferred behaviors incentivized by the greater order. Rewards and punishments followed, derived from rules, regulations and directives from higher authority. Society then became the preeminent self-serving yet semi-pilotless force in The World while nature receded to a lower, peripheral nuisance.

Summary of The World. The human experience is about pursuing happiness within a worldwide system of constraints, navigated through a variety of ever-changing external parameters. There's no simple roadmap since everyone's desires, needs and goals are different. The machine doesn't care about your feelings or good intentions, just your compliance and conformity. No one is exempt from The World Matrix, but some adapt better than others. Rewards and punishments incentivize compliance. Success comes to those who comply and abide. It may seem harsh and unfair, but we must remember that the matrix is a product of our own making. Everything *out there* came from within us, *in here*. The World is a macrocosm of man – a collective expression of every individual. Man and The World are unified in a symbiotic relationship, complimentary sides of the same coin.

The World is a dynamic environment of constant change, adaption, emergent threats and opportunities. It contains both Hardware and Software – goods and services, institutions and political processes, as well as core values and temporary fads. Man operates in The World as both a free individual and subservient part of the whole, acting as a functioning subroutine in a greater system of systems. We're all parts of many wholes spanning several levels and layers – members of a greater community of nested communities (family-relatives, coworkers, organizations, affiliations, and subcultures).

The World is the collective aggregate of individuals projecting outward the sum total of our desires, preferences, fears, personalities, strengths and weaknesses, and the baggage of our past heritage. Its interlinked components include nature, the earth's environment, biosphere, animal kingdom, and greater society above. It all co-evolves and works together as a collective integrated whole; one that is locally driven but collectively takes on an unpredictable life of its own.

Conclusion. *The World* is an intricate encoded system we all must interpret and decipher, largely by simple trial and error as we navigate through the labyrinth of possible pathways, options, roadblocks and curveballs. Life lessons capture our experiences learning the hard way; the subtle nuances and counterintuitive nature of how The World works. It's a life-long process of adjustment, adaption and ongoing reality checks. Those who break the codes can unlock doors to remove roadblocks. Once you pragmatically accept the objective, arbitrary nature of the grand matrix, achieving success becomes a little easier. You are all born *into* The World and must abide *in* it but certainly don't have to be *of* it.

Spectrum of *The World*

The Earth (Air, Land, Sea) – Climate/Weather – Animal Kingdom – Man – Society

Benevolent-Malevolent Duality of The World

Amiable	Hostile
Society	
Opportunities	Threats
Friends	Enemies
Generosity	Hoarding
Abundance	Scarcity
Knowledge	Ignorance
Security	Instability, Chaos
Fortune	Loss
Heaven, angels	Hell, demons
Weather	
Sunshine	Overcast
Light rain	Monsoon
Cool breeze	Hurricane
Sunny 70's (Goldilocks temp)	Boiling desert heat / freezing temps
Animal Kingdom	
Domesticated pets	Wild kingdom
Dutiful Sled dogs	Vicious Wolves
Kitten	Tiger
Butterfly	Wasp
Dolphin	Shark
Man	
Generous	Thief
Leader	Rebel
Lover	Hater
Creative	Destructive
Successful	Failure
Productive	Lazy
Free	Slave
Nurture	Kill
Pleasure	Pain
The Earth	
Fertile land	Desert
Fresh water lake	Salt water ocean
Natural shelter	Tunnel collapse
Clean Well water	Chemical spill stream
Fresh fruit	Poison berries
Solid ground	Earthquake

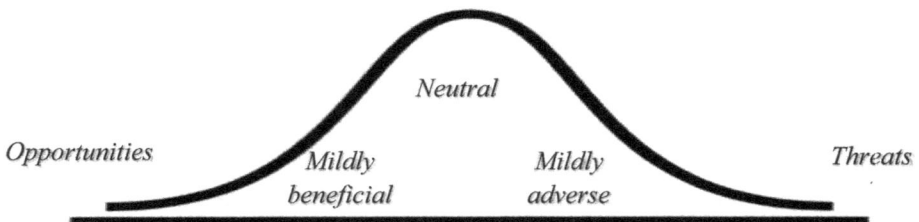

Opportunities   *Neutral*   *Threats*

Mildly beneficial   *Mildly adverse*

* The Bell Curve applies as most human experiences falls within the middle area, extremes on either side are rare.

Secret Cypher of The World. The benevolent-malevolent spectrum on the previous page captures the potential threats and opportunities we *may* experience at any given point in time, depending on the arbitrary nature of *The World*. Everybody faces challenges navigating through this grand two-sided matrix, with positive and negative forces dynamically operating in the background. It's an unpredictable environment; a super arena of arenas where literally anything *can* happen at any time. However, YOU can change the outcomes in ways that are favorable to YOU. The symbiotic dual connection between inner man and outer world creates a two-way channel of influence where what YOU experience *can* be self-directed. It's a mode of synchronicity where magic happens, from the inside-out. The secret code of life is that your inner state of being can draw either end of The World Spectrum your way, positive or negative, success or failure. Wonderful opportunities can open up by simply changing your mindset; through visualization, intention and self-determination. The power of the outside world is linked to YOU. All you have to do is change within. Inside man and outside world are always connected. Simply change yourself and you change *The World*.

Comedy-Tragedy duality of experience in The World

5   HUMAN REALITY

Human beings are fascinating creatures navigating a diverse world – one that originally created them but now is co-created by them. Composed of the same cosmic ingredients of stars, man is a physical microcosm of the world *out there*. But the connection goes much deeper and broader as systems and processes operating within us are derivative copies of similar mechanisms at work on a grand scale throughout the universe. At every level within man are operating principles mirroring universal patterns everywhere in the cosmos. Human beings are physically and spiritually sculpted by the same forces that produced the world, the sun and everything else. And the same wonderous mystique and grandeur of the heavens above are equally matched by the breathtaking complexity within YOU.

PSYCHOLOGY AND BEHAVIOR

Human nature is a product of internal dynamics continually at work inside that are then expressed outside. Behavior is the resultant by-product of multiple inner aspects, including desires, needs and perceptions jointly interacting to influence outward actions. Humans are a super complex species topping the entire animal kingdom, benefiting from greater intelligence with the capacity to think abstractly, conjure ideas, create concepts and imagine possibilities. Together these capabilities produced an entire immaterial realm where thoughts are collectively exchanged, recorded and passed onto successive generation, transcending time and space.

To understand man is to delve into the dynamics of with human beings where body, mind and spirit interact to generate formation of the *self* and its unique personality. Individuals develop a sense of identity while working through changing states of being, where daily reality is experienced in a fluid, malleable fashion. This process involves continual learning, adapting and developing. Internal motivations attract certain types of experiences, which in turn shape an individual. Relationships are formed, lessons are learned and paths are chosen. The human journey is marked by recurring themes, revealed in universal experiences occurring consistently through the history of mankind.

A deep dive into humanity begins with psychology and internal human dynamics; the interaction between body, mind, emotions and behavior. Everything originates inside where needs, wants, desires and motivations translate into outward expressed action. Clouding the process is a deep, shadow realm anchored beneath; an unconscious level, periodically leaking out, surfacing in unexpected expressions. The root word psyche actually means soul, not mind.

What makes one tick is a complex relation of various aspects and modes interacting in unison. Man in the middle is a process center where inner forces clash with exterior demands resulting in external behavior. A holistic perspective is needed to make sense of it all yet there must first be an appreciation of the multiple aspects at play. A short list of these would include thoughts, feelings, sensations, perceptions, age, gender, culture, size, climate, race, heredity, etc. These and many other aspects play a role in man's response to experiences in The World.

Human reality is thus a comprehensive interconnected process of inner aspects interacting with external constraints, producing patterns of expression and the emergence of identity, ego and personality. You are co-created through a developmental process as you adjust to challenges in The World and in doing so, you mature as an individual. This evolutionary growth takes place in both gradual and layered stages of progress, resulting in a well-rounded human being.

Man transcends the entire animal kingdom through evolutionary advances in mental capacity and consciousness. Our triune brain developed in successive stages: Hindbrain, midbrain and forebrain. A larger forebrain and its cerebrum portion are the real game changers, providing us with exceptional capacity for reasoning, judgment, learning and problem solving. A human brain is also binary, with left/right hemispheres specializing in dual aspects of reality: Thinking/feeling, logic/artistic, linear/non-linear, details/wholes. Human intellectual capacity also enables linguistic mastery where thoughts and concepts can be exchanged using symbols, opening up an entire transcendent realm of abstract ideas. This has completely transformed The World out there by creating a social information collective; an emergent realm of interacting ideas, concepts, and knowledge recorded and shared by all humans in a cumulative progressing dominion. This relationship is two-way, meaning any change mankind projects onto the world reflects right back onto participating individuals, affecting your inner being.

Psychology is focused on the human mind since most internal elements directly integrate with mental processes. While sense and perception are primary windows to reality, those filters are subject to biases and inaccuracies. The 5 classic senses (sight, hearing, smell, taste, touch) and many others, are largely dependent on interpretation. No two individuals ever experience the outside world the same way – too many variables unique to each of us influence the process. Even memory is inexact where two individuals experiencing the same event often recall it differently. Perception, storage and recall are further diminished by limitations between short and long term memory; both get worse with time, age and distance.

More important than the mind itself is the dynamic interaction between thoughts, emotions and sensations. Aspects of body, mind and soul interplay continuously in different degrees and combinations. Where the mind provides order to a chaotic world, the heart intermediates to bridge the two. Emotions serve as a language of the soul, responding to experiences in a way that provides richer meaning to the logic-oriented mind. Either one can be taken to extremes if not balanced by the other, resulting in either a cold, black and white existence or a chaotic mess of unrestrained feelings. Body-mind-soul working in balance and in sync are essential to health – physically, mentally, emotionally and spiritually.

Mental — Emotional — Physical — Healthy Balance

Mental — Emotional — Physical

Emotional — Physical — Mental

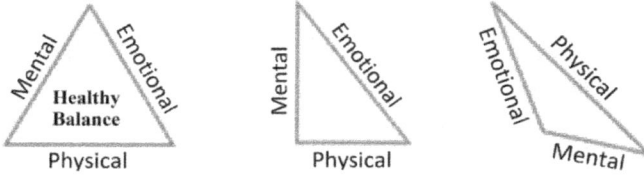

Obsessive attachments and inherited limitations lead to unhealth imbalances

A young, newly formed human being gradually establishes a concept of self, separate from the world *out there*. There's a sudden transition from oneness and wholeness to me, myself and I… and everything else. Then self-awareness leads to identity, and separation from The World. This creates a perception of isolation and detachment, which is limiting but necessary for survival and self-preservation. We all learn to cope, adjust and make sense of things in the exterior environment. Over time we begin to identify with things out there. Inner needs, wants and desires soon translate into attachments and fixations in The World, altering your identity. Who you are really comes down to what you focus your attention on, where you spend your energy, and what you think about most of the time. Self and its associated identity become a fluid essence that even the individual inside doesn't fully grasp. It's developed and reinforced by ongoing habits. Repetitive actions, obsessive thoughts and attachments create resonant patterns that persist and ossify.

The *Self*-concept is a necessary self-imposed limitation. It's a biased perception of isolated being, creating a localized reality of here and now, separate from everything else. With maturity and experience, an individual transcends the self and re-integrates with everything out there. The trick is to maintain a proper balance between maintaining a healthy self-identity while losing the perception of separateness or disconnectedness from The World. Ideally the inside syncs with the outside. It's ok to fully participate in The World, but it requires ongoing vigilance to avoid all the tempting trappings and attachments lurking everywhere.

The fundamental question everyone faces is "who am I?". Most answer with things they identify with such as titles, roles, occupation, personal history or other mental constructs. A common reductionist version of the question is "what do you do for a living?", implying that your occupation is what defines you. A more honest approach considers where you generally spend your energy and time. We could borrow the perspective "you are what you eat" and supplement it with "you are what you think about" or "you are what you feel strongly about". There's usually a big difference between who we think we are and who we actually are. It's a discrepancy exposed during times of crisis and personal challenge, where unacknowledged shortcomings suddenly become obvious.

Self and identity were defined by Freud as a product of 3 competing modes: ID, Ego and Super-Ego, representing raw instincts, pragmatic rationality and morality respectively. This dynamic involves the Ego working through and balancing low natured impulsive tendencies with higher moral values. In this model the Ego is a necessary limiting mechanism enabling an individual to cope within the boundaries of civilized society. However, that same safety valve tends to overcompensate and becomes a hinderance to development. Ego by its nature rationalizes, separates and defaults toward defensiveness. Self-preservation, security and pleasure come first. Short cuts will be taken to meet immediate needs, even if they're self-defeating in the long term. It reinforces separation and isolation, fosters perceived threats and attracts control dramas. All of this hinders individual maturity and development.

Ego at its worst separates us from everything, even our true selves. Where love unites, ego divides. Instead of pursuing real fulfillment, it falls for material possessions and immediate gratification. Attachments to things in The World override appreciation for spiritual attainment and joyful inner being. When left unchecked Ego makes the fundamental flaw of seeking what it needs out there, completely missing that it was right here inside all along. Most of the unhappiness people experience is the unwitting result of self-imposed mental constructs, self-created dramas and unnecessary ego attachments. Despite all these negative connotations, understand that a healthy ego is necessary to achieve success in The World, whether it's sports, business or competition in general. The key is to maintain balance, manage it, and then transcend it so you can integrate with everything *out there* without getting caught up in the trappings of it all. The choice is simple; operate as an ego driven, selfish oriented person or a soul centered, spirit driven being. But do realize that while ego dissolution is essential for spiritual growth, the transition can be disorienting and unsettling.

Behavior Dynamics. Why do human beings act and do the things they do? Obviously, inner motivations involve a multitude of competing internal aspects which form a state of being that gets externally projected – a combination of one's thoughts, feelings and psychic energy. Attachments and obsessions are derivative products of what motivates us. Our reality is really determined by what we channel our inner being into *out there*. Whether it's positive productive work, creative innovations or negative attachments and draining fixations, the residue of our outward behavior is a mere reflection of what's inside us.

The clash between inner desires and exterior worldly constraints drives human behavior. It's an ongoing process of adjustment, coping, trial and error. It's the source of all drama where a soul incarnate must mediate the intricacies of a bodily vessel with a piloting mind journeying through a world full of limitations, twists and pitfalls. Lots of moving parts and changing variables complicate the process, making human behavior the fascinating phenomenon it is. Issues emerge all along the way. Some internal, some external, and many others occur in the clash between both. Peace and harmony are often temporary since continual adjustments require ongoing vigilance. Failure to adjust leads to abnormal behavior and pathology.

Normal behavior is defined by the outside world. Conformity is expected. This includes anything within established limits, subject to change at any time. Of course, what's "normal" is entirely subjective but those who make the rules decide. But the world can be a harsh, threatening environment for the uninitiated, creating internal pain and discomfort when adjustment is insufficient. Abnormal behavior begins with imbalances of degree; channeling too much or too little energy into something. This includes obsessions, fixations, addiction and isolation, which are misuses of energy via blockage, leakage and stagnation. Unhealthy attachments encumber an individual, becoming a constant energy drag and burden. Mental and emotional dwelling produce similar results. Creative energy is stifled, and forward progress is halted.

Abnormal behavior at its worst goes beyond degree and becomes transformative in its diminishing effects. Sever pathology occurs when a delicate inner being can't adjust to demands of The World. Disorders and defense mechanisms can override the individual, making daily life an ongoing threatening conflict. Fear and isolation prevail. Challenging disorders include anxiety, mood, personality, dissociation, the physical and the psychotic. These disorders lead to a variety of debilitating states of being: Phobia, obsessive-compulsive, depression-mania, paranoia, narcissism, anti-social, schizophrenia, and multiple personalities. The difference between a healthy well-adjusted individual and one who is ill or plagued by disorders can be subtle by degree or dramatic and transformative. Balance within and without determines which will prevail. Normal, acceptable behavior surfaces when an individual properly balances the competing forces between progress and regression, integration and separation. Balance is attained and maintained through a healthy self-identity anchored in a centered state of being

Human Behavior Pitfalls (Destructive actions and habits)

Substance abuse, addiction	Herd mentality
Financial malfeasance, fraud	Self-destructive, self-sabotage
Career or education neglect	Lack of responsibility
Gambling, excessive risk-taking	Criminal acts
Toxic relationships	Poor decision making
Reckless, impulsive	Loss aversion

Human Behavior includes a wide range of actions and responses expressed in an even wider variety of situations and circumstances. Behaviors can be overt or covert, voluntary or involuntary, or simply knee-jerk reactions. Much of it can be considered a form of nonverbal communication since it conveys a lot about what's going on inside. In addition to body language, posture, gestures, facial expressions and touching, a person often communicates what their motives are simply by what they do and how they do it. Even your personality can be considered part of this same behavior-communication dynamic where what's inside is projected outside, providing a measurable signature carrying a subtle message.

Human Nature. Individual and collective human behavior is consistent over time and somewhat predictable. History provides recurring themes of different people behaving in different situations, revealed in literature, movies, songs and oral stories. Shakespeare famously highlighted the fatal flaws human beings exhibit regardless of social position. Likewise, character traits are documented in ancient religious books (Bible, Quran, Tao Te Ching, Vedas, Bhagavad Gita), parables, myths and sacred texts where common themes of human behavior transcend culture and time period. The universal drive to pursue self-interest leaves an unmistakable trail of repetitious actions – common, recurring patterns consistently played out over recorded history. Humanity has demonstrated the tendency to do the same things over and over – getting the same results. This pattern may not meet the classic definition of insanity, but it confirms the consistency of human nature. Despite conditions on the world stage changing greatly over time, the acts and behaviors of the actors doesn't change much at all. Names and faces are different, but the events and outcomes are quite similar. That's simply because human beings haven't changed much. While each one of us is certainly unique, our needs, desires and goals are similar on the inside and we face most of the same challenges on the outside.

Where we are different is in degrees of character. What makes some people fascinating is how they take a specific trait or approach to extremes, both in the popular culture and local less knowns. Famous examples are celebrities, pro athletes, gifted entertainers, experts in a field and those who are simply uniquely talented. But understand these extreme individuals lie on both ends of a spectrum, including good and bad, famous and infamous. We find these rare individuals in every field of endeavor. Many remain obscure at the local level, out of the general public square yet just as special in their own way as those who acquire fame.

Negative	VS	Positive	Famous	VS	Obscure	
Criminal		Saint	Celebrity		Local Hero	YOU
Phy Disability		Pro Athlete	Sports MVP		College star	Are Here
Evil Genius		Scientist	Pop Rock band		Garage band	
				Extreme		Extreme

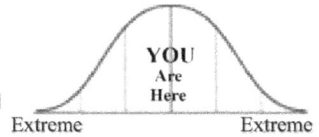

Extreme opposites are more alike than you in the middle; a fine line between genius and insanity

Human nature is consistent over time yet there's often diversity in the mix, revealed in countless examples of unique individuals. While popular culture tends to focus on those at the extreme ends of the spectrum, the simple truth is that the vast majority of us fall within a stable, boring center area. The great lesson of mankind is that each person in the middle has an opportunity to exceed their own potential – to move up or out, depending on talent, opportunity and good fortune.

Specials events and crisis provide arenas where unlikely individuals can stand out from the crowd. History reveals that we're a lot better… and worse than we tend to think. From great human achievements to ugly acts of depravity, there's a wide spectrum of what we're capable of doing. And just when we think we've seen it all, someone pushes the boundary and goes where no one else ever has.

PERSONALITY

Personality is a defining distinction between different people as it embodies the holistic set of traits making each of us unique. It's a pattern of attributes forming a comprehensive whole. Personality is an inner mix of interacting aspects individuals project outward while responding to external forces in The World. It's a composite of how you think, feel, act and express yourself, including tendencies, attitudes, quirks, style and general disposition. Personality is really a coping filter – both template and screen mediating between inner being and outer world. In a sense it's a transitional membrane buffering a soul from exterior physical reality.

In addition to how we think, feel and act, personality is also character and temperament. This makes it instrumental in determining how we respond to events and circumstances. Human interaction with various situations, especially other people, is what defines us. Attitude and motivation play a big role. Values, beliefs, and emotional balance also come into play, which certainly supports the claim that "Character is destiny".

Personality Pattern Dynamics

ability   mental health   **Language**   culture
   APPEARANCE   Ethnicity   **location**
Religion   gender   **physical health**
   economic class   Hobbies   *age*
education   political affiliation   Race
   Marital Status   OCCUPATION   PARENTS
NATIONALITY   disabilities   *Intelligence*

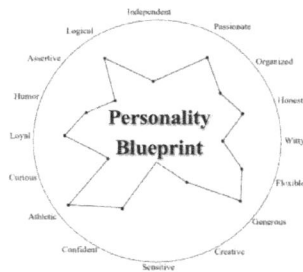

Though people differ greatly with their diverse personalities, there are common threads shared by all. In essence we possess the same general ingredients on the inside, just in different degrees and combinations. It's the unique patterns of traits that create different types of people. Various models capture personality diversity, a popular one being the Myers-Briggs template. Categories of traits are simplified into 4 sets of polar qualities: Introvert/Extrovert, Sense/Intuition, Think/Feel, Judge/Perceive. An individual differs by the degree of each polar trait. Most people lean toward one end of each scale while some fall closer to the middle. The 4 categories produce 16 general variations of the primary traits.

Myers-Briggs Personality Types (I/E – S/N – T/F – P/J combinations)

ISTJ	Logistic – Practical, factual, reliable	**INFJ**	Advocate – Quiet, mystical, idealist
ISTP	Virtuoso – Bold, practical, experimental	**INFP**	Mediator – Poetic, Kind, altruistic
ESTP	Entrepreneur – Energetic, perceptive	**ENFP**	Campaigner – Enthusiastic, creative, social
ESTJ	Executive – Organized, punctual	**ENFJ**	Protagonist – Charismatic, inspiring, leader
ISFJ	Defender – Protective, caring, warm	**INTJ**	Architect – imaginative, strategic planner
ISFP	Adventurer – Artistic, explorer	**INTP**	Logician – Innovative, curious, logical
ESFP	Entertainer – Spontaneous, energy	**ENTP**	Debater – Smart, curious, intellectual
ESFJ	Consul – Caring, social, popular	**ENTJ**	Commander – Bold, imaginative, willed

Another personality perspective considers 5 major Types: Extravert, Agreeable, Openness, Conscientious, Neurotic. They measure if a person is outgoing/social, considerate/cooperative, creative/flexible, self-disciplined, and emotionally stable. Like other models the difference between people is by degree of each trait. Models may simplify the ingredients but like fingerprints, the possible combinations are boundless, so it's rare to find two people with the same exact pattern.

Personality formation is influenced by everything in our immediate environment. Obvious sources are parents, culture, race, gender, religion, etc. While heredity and environment produce inherent biases and predispositions, context is key. The 5 major traits mentioned earlier carry different weight in other cultures. Class structure, social norms, tribal customs and language change the way personality traits are valued and expressed. Yet it still comes down to a combination of what you're born with and what you're exposed to.

Gender roles are influenced by both biology and culture. Being born as a man or women is an internal predisposition which is then influenced externally by cultural conditions. Recent social history has introduced the ideology of gender fluidity and relativity, which further complicates the mix. Still, distinct differences between the sexes have been captured and highlighted in popular culture and expressed in movies, literature, comedy and music all throughout history. This schism is embraced in the humorous "Men are from Mars", Women are from Venus" narrative based on a book by John Gray. The differences between men and women are generalized of course but they are unmistakably present in broad terms, including language, motivation and emotional disposition

Gender Generalizations in Modern Culture

Men	Women
Identify w/machines, systems	Connect w/people's needs
Think of point and make it brief	Talk to arrive at the point
Independently act to problem solve	Share feelings about a problem
Embrace freedom and space	Embrace togetherness
Need to listen and vocalize more	Need to trust and respect more

   Personality differences are also influenced by relationships and relational context. It starts right with the family in how each parent relates to each other and to the child. The same is true with relationships among siblings. Birth sequence plays a role as statistics show tendencies of first-borns oriented more to status and achievement, youngest tend to be more social and creative, while middle-borns tend to be easy going mediators. Extending beyond immediate family, other social relationships play a part, including friends, teammates, romantic partners, coworkers and general group memberships. Groups shape us through defining roles and expectations. From leader-mentor-coach to follower-supporter-participant, each role alters our disposition. Group dynamics stretch, pull and shape us through compromise, dramas, manipulation and power games. We enter relationships carrying preset tendencies which then attract specific types that either reinforce or conflict with them.

   Relationship dramas reveal how personality is subject to situational context. Life is a stage where all kinds of situational conditions emerge in a wide variety of arenas, including entertainment venues, sports platforms, business settings, school, theaters, the wilderness, or any place where there's an audience. True character surfaces during unique conditions, like special events, celebrity status, crisis, or good fortune. Each of these dramatic situations reveal hidden aspects of your personality. Famous celebrity performers often present split personalities, with a false stage persona that's quite different from their "normal" private self. *"He who knows others is wise, but he who knows himself is enlightened"* – Lao Tzu.

   Life's various arenas tend to attract a common spectrum of characters – *universal personality types* popping up in books, plays, movies, folk lore, myths, etc. Television sitcoms routinely recycle similar roles and character types, inserting them into an artificial version of reality. They're the same standard personality types present in literature and storytelling throughout man's history, primarily because human nature doesn't change. Once again, names and faces come and go but the dramas and games are generally the same.

Universal Roles Expressed in 7 Archetypes   *Derivative Roles*

King	– ruler - CEO - president	*The Hero*	*The Villain*
Priest	– holy man - saint	*The Magician*	*The Lover*
Sage	– coach - leader - tutor	*The Mentor*	*The Sidekick*
Scholar	– professor - expert	*The Explorer*	*The Innocent*
Warrior	– soldier - athlete	*The Rebel*	*The Outlaw*
Artisan	– creator - craftsman	*The Shadow*	*The Jester*
Server	– employee - peasant	*The Caregiver*	*The Guardian*

*Differences in uniform/costume, language, bling/accessories, respect, etc.

   Personality coevolves with an individual's process of development, starting with a core structure that gets molded gradually through time. With fixed and variable aspects, the modification part is more significant early on, then diminishes with age as a person becomes set in their ways. The process remains gradual unless a traumatic event occurs. Each stage of life brings with it changes: Infant, toddler, preschool, school, adolescent, young adult, middle aged adult and senior. Individual actions and responses reinforce the personality structure. Recurring habits become resonant patterns that further reinforce it. Habits in general create our destiny, both positive and negative, much like a software program. So, change your program and you change your life. In this manner personality co-evolves with the self's journey of actions and experiences in The World. We initiate change inside-out, then everything outside comes right back in here to reinforce the ongoing process.

Who we are and how we're perceived is completely relative. First there's the way others close to us see us, especially those who we honestly confide with. Then there are acquaintances we associate with who see us differently. Most importantly, there's our own self-perspective. All of these viewpoints create a composite image of who we are, regardless of our true self. Perception can become reality and over time the feedback we accept from others resonates to a point where it shapes our personality structure from the outside-in. And the weaker your own perspective is the more that of others becomes a reality.

The World

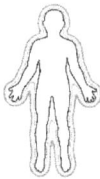

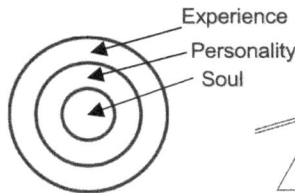

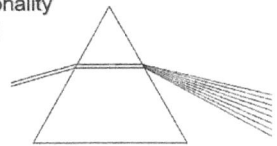

| Soul | Layered Filters | Adaptive Filters | Spirit Personality Self-expression |

Personality serves as a filter between our core essence and the outside world. It's sort of the soul's coping mechanism – an adaptive membrane separating and connecting interior and exterior realms. It's a dynamic relationship subject to the interaction between inner needs and outer demands. As a filter it presents partial elements of our true selves, often skewed by shadow aspects, denial, suppression and self-deception. Multiple parts interact, including mental, emotional, physical, social and moral, each filtered and expressed as a composite. As we mature and develop into more authentic beings, the filtering process diminishes.

PERSONAL DEVELOPMENT

All human beings born into The World enter a realm of nonstop experiences that inform, test and offer choices. Everyone participates in this lifelong journey navigating through ever-changing situations, each bringing opportunities or threats. This continual process of change is normally incremental, but sometimes it can be transformative. Personal development is the byproduct of the journey, where we're all enriched, evolving into something greater than before. It's a dynamic process of continual learning, adapting, growing, coping and developing – comparable to a seed becoming a flower.

The image of a human leaving the womb and entering into The World head first is a fitting metaphor. From first breath we enter the arena of earth, a platform filled with pitfalls, distractions, attachments, surprises…working through games people play, politics, and just simple physical survival. Automatic instinct kicks in as we seek security and comfort, and later transition to more socially focused needs. Soon, traps emerge in the form of material wants; money, power, fame. Basic needs such as security, health and love motivate behavior, all while working through unexpected blockages and complications. Stress, fear, fixation and attachment enter the mix, creating a succession of lessons either learned or missed.

Human development is an experience driven process where the outside world continually molds us. Conflicts and hardships require adapting. Punishments and rewards shape and incentivize proper responses. Instincts kick in, emotions get stirred, and feedback loops let us know which actions feel better or worse. The World continually sculpts us through demanding situations, obligations, rules, restrictions, costs and benefits. Personal growth is the tradeoff we get for dealing with life's inevitable stress and strain head on, not short cuts or easy choices.

Learning is a life-long process. Every day is a new opportunity to expand awareness and build upon what we already know. The tougher the challenge, the more we grow. To stop learning is to stagnate. With age comes a tendency to resist learning new things, however, being in *The World* is an ongoing opportunity for further growth and attainment – anything less is wasted life.

With learning comes increased capacity. Personal development comes through expanded competencies via refined and improved skills or talents, whether they come naturally or newly learned. Each of us possesses a spectrum of various capacities such as artistic, math/logic, music, sensory motor, cognitive, emotional, interpersonal, etc. These same capabilities are applied in an even wider diversity of fields, activities and skill sets, including technical, drawing, speaking, carpentry, gardening, designing, humor, cooking, sports, acting, etc. As noted earlier there's a specific scale of exceptional quality in every human aspect, including these capacities and applied skills. It's the same bell curve pattern with most of us falling in the middle while the exceptional few occupy the outside extremes.

Intelligence is a catch all category for competencies, not just how smart someone is. An intelligence test actually measures ability to learn and solve problems; not how much you know. And it's less about knowledge memorizing but rather the application of that knowledge. As a coping mechanism for success in The World, intelligence functions in 3 general modes: *Analytical*-problem solving,

Creative-innovative aptitude, *Practical*-adapting to environment. Intelligence is a combination of these modes, and the various abilities noted above.

Sample Personality Spectrum of Capacities

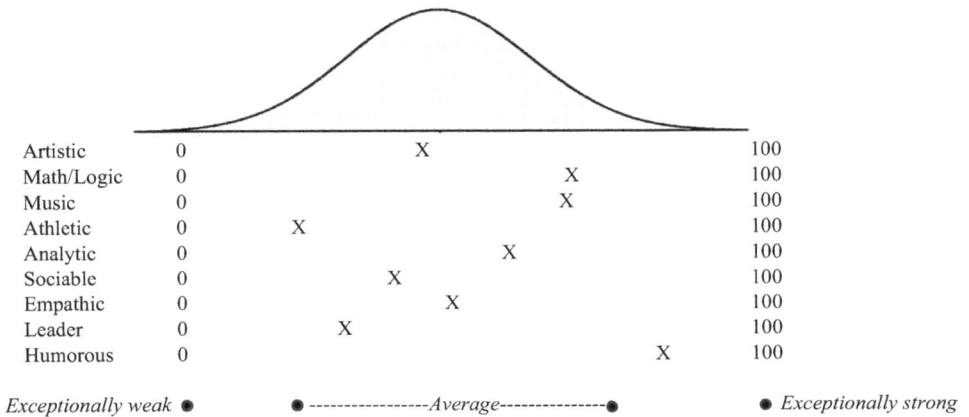

Artistic	0	X	100
Math/Logic	0	X	100
Music	0	X	100
Athletic	0	X	100
Analytic	0	X	100
Sociable	0	X	100
Empathic	0	X	100
Leader	0	X	100
Humorous	0	X	100

Exceptionally weak ●   ● ---------------Average--------------● ● *Exceptionally strong*

Critical thinking is the gold standard for higher intelligence, employing a disciplined approach to learning and understanding in a world where gossip, opinions and false information permeate the social landscape. Accurate conclusions are prioritized over personal preference. Honest fact finding prevails over biased, selective preferred choices. Once a conclusion is formed, alternate interpretations are considered to minimize bias. Critical thinking requires ongoing discipline and vigilance to overcome natural bias.

TYPES OF INTELLIGENCE   *DERIVATIVE FORMS*

Personal	Understanding yourself, wants & needs	Emotional Intelligence
Linguistic	Use oral or verbal language effectively	Poetry, humor, public speech
Kinesthetic	Body control, handle tools	Meditation
Spatial	Imagine and visualize the world in 3D	Abstract design
Musical	Discern tone, timbre and rhythm	Composition
Logical/math	Use numbers effectively, analytical reasoning	Problem Solving, pattern recognition
Interpersonal	Interpret others' feelings and motives	Leadership: others do what you want
Existential	Grasping spiritual concepts of life and death	Philosophy
Naturalistic	Sense nature, plants, animals and environment	Survival instinct

*Intelligence manifests in a variety of aspects including *memory* (capacity to retain and recall), *wittiness* (clever humor, turn a phrase), *dexterity* (multitask competing info), *quick study* (pick up on new things effortlessly)

Emotional Intelligence is equally important for coping in The World. It applies to both inner and outer aspects; self & social. It's a measure of your ability to manage emotions, which translates into appropriate, healthy responses to situations and experiences. Emotionally competent individuals don't blame others, take everything personally or engage in self-sabotage. Nor would they ruminate on the past or worry about the future. They have realistic expectations, let go of bad habits and respond to life's friction with a sense of proportion. Extremes indicate pathology and must be minimized or avoided.

Emotional competence falls within 5 Domains: Self-awareness, managing emotions, motivating self, empathy, managing others' emotions (relationships). Self-awareness is really self-management, fostering discipline and self-control. Social awareness is important in dealing with either a few or many, expressed in empathy and cooperation. It's further refined into social competence, where one attains capacity as a Leader, facilitator, group communicator, or influencer.

Emotional competence is expressed via appropriate responses. It begins with knowing yourself and others. Honest self-awareness is essential. Possessing self-control enables you to be motivated and to manage everyday stress. It can be challenging given our primal survival instincts: fear & aggression, anger & strength, adrenaline fight or flight. Where intelligence is refined in critical thinking, emotional intelligence is codified in language of the heart, expressed in song, art, poetry, dreams, smiles, a good cry, and a multitude of metaphors.

Self-development leading to intelligence and emotional competence is a by-product of life, the cumulative residue of processed experience. To live is to learn, and to learn is to interpret experience. Each event comes with dual aspects: an objective circumstance and a subjective experiential response. Our biases act as filters to objective, neutral events, making them relative to our unique perspectives. Filters and inhibitors limit learning, including biases of culture, age, gender, education, health, wealth, etc., not to mention learning disabilities. Expectations and context also change our interpretation of events and experiences. Attitudes affect perception as well, creating different responses among people experiencing the very same event. Consider how an artist perceives a scenic landscape that most others see as ordinary land features.

Learning and development are greatly dependent on our power of observation. A keen eye and a gift for pattern recognition help immensely. Much of learning is connecting the dots and breaking life's codes. It's also important to go beyond simple hearing; to master the art of listening. And most of all, it's essential to possess a healthy sense of curiosity, the very key to a child's exceptional capacity to soak in everything it can. Ironically, the one natural inhibitor of learning and development is knowledge itself. It seems the more you already know, the less you're interested in learning new things. Childhood wonderment, excitement, adventure and curiosity are a state of being anyone can attain, but it takes greater effort as you go further and deeper into The World. Marcel Proust captured it:
"The real voyage of discovery consists not in seeking new lands but seeing with new eyes"

The human path of development builds upon a sequential variety of universal experiences. We all pass through a series of phases and stages of growth, going through the same things people did in ancient times. Universal experiences include *conflict, joy, happiness, friendship, birth/death, pain/pleasure, hunger, crisis, abundance, travel, love, work, exhaustion, surprise, anger, gain/loss, fear, anxiety, confusion, regret, greed, jealousy, parties, teamwork, family, sacrifice, discovery, sickness, fixation, desire, pride, punishment, excess, and much, much more.* Every experience, event, situation, arena, and emotional response are all universally shared. They've been captured through the centuries in stories, books, songs, and movies. Though we learn greatly from others through literature, videos or stories, some things must be experienced personally to fully appreciate them: anxiety, depression, grief, love, childbirth, assault, embarrassment, insomnia, etc.

Classic Novels Capture the Universal Human Condition

The Grapes of Wrath	The Scarlet Letter	Sense and Sensibility
Catcher in the Rye	Jane Eyre	To Kill a Mockingbird
A Tale of Two Cities	The Count of Monte Cristo	The Great Gatsby
Pride and Prejudice	War and Peace	Crime and Punishment
Of Mice and Men	Wuthering Heights	The Outsiders
Catch-22	Great Expectations	Adventures of Tom Sawyer

The human journey itself has recurring patterns of ebb and flow, progress and regression, twists and turns. Common threads are summarized in universal story plots of literature, plays and movies, including rise and fall, fall and rise, fall-rise-fall, steady fall, etc. All are variations on a few basic patterns. Shakespeare highlighted universal themes of the human journey in his plays about tragedy, comedy, and history, revealing man's power struggles, ambition, jealousy, deception, revenge, fate, betrayal, greed, treachery, etc. These experiences are universal because human nature is a constant. Recurring roles are played by everyone through interpersonal control dramas, power games, and ego conflicts. All of us experience them on either side of the equation: **Aggression:** *Intimidator* (fear) or *Interrogator* (find fault) and **Passive**: *Aloof* (curious) or *Victim* (poor me/guilt). Besides these recurring themes and patterns, all human experience occurs within situations of either Man vs Man, Man vs himself, or man vs Nature.

Personal development progresses through natural rhythms, phases and milestones. Driven by self-interest and the pursuit of happiness, the journey is a process of ebbs and flows, spurts and crawls, and push-pull dynamics where individuals move towards perceived opportunity (attractors) and move away from threats (repellers). Progress gradually moves forward and upward, in steps, levels, incremental changes and transformative shifts. Once a person moves up a level, they settle there, stabilize and become integrated into it before moving up a further level. Development is a ratcheting up building process of successive levels that include all prior levels.

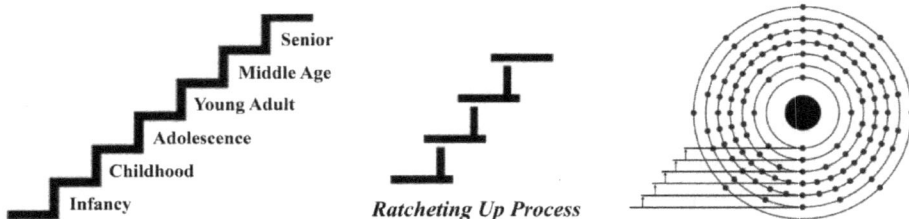

Individual development progresses in stages – similar to electrons moving up into higher levels

Personal evolution is essentially an increase in consciousness/awareness. Experience in The World molds us with tests that temper our inner self. The language of this process is emotion, the source of real motivation and the measure of personal meaning. Pain, suffering and setbacks provide spiritual value, guiding further development. Adversity fosters resilience, competition builds character and stress triggers necessary responsiveness. No pain, no gain. Each transcendent experience changes who we are. An individual must let go of a previous stage to progress – moving forward requires moving beyond. While routine events change us incrementally, emotionally charged experiences transform us. Competence can be taught but higher consciousness and expanded awareness must be experienced.

Life stages follow a universal pattern common in all cultures with minor variations on the same themes. Rites of passage serve as ceremonial gates you must pass through to enter a new phase. Growing up isn't just for kids, it's a life-long ongoing process with multiple thresholds. Stages are determined by the same aspects that define a person's context: age, gender, race, culture, etc. Common milestones include Baptism, Bar Mitzvah, Confirmation, Graduation, Wedding, Parenthood, Promotion, Certification, Retirement, etc.

Development through stages is most critical during early formative years. A healthy foundation comes from loving parents and a safe-secure environment. Other factors include genetics, siblings, friends, and relatives. Major transitions occur between dependent child and independent young adult, then continue with established adult mastery, and finally a diminished senior adult. After the major transformative phase-change from dependency to independence, a further critical transition is the midlife crisis – Ego awareness of mortality, perception of unfulfilled ambitions, goals, dreams and misguided expectations. Men lose hair, women lose beauty. We all begin to tire easily and fall into settled routines, transitioning from growing up to growing old. Until the age of about 45 most are world driven. Afterwards adults transition into an age of mastery (45-65) followed by a shift into an age of integrity, grace & generosity (65-85).

Transition into adulthood required skills: Budgeting, managing relationships, cooking, home repairs, car maintenance, time management, personal hygiene, staying informed, stress management, tech troubleshooting, understanding contracts, basic health care.

Gender differences influence the possible and probable paths chosen by individuals. Men follow a few general path choices: *Transient* – no commitments, exploring and drifting through adult life, *Locked-in* – possessing solid commitment and following their own path, *Type A* – taking risks, go big or go home, *Never married,* and *Integrator* – balance between ambition and family commitment.

General paths for women include: *Caregiver* – focused on a domestic role, *Nurturer* – focused on family, *Integrator* – balancing work and family, *Never married*, and *Transient* – no commitment. In recent decades gender roles have become more fluid with mixing of traits and even role reversals. Age also plays a big part as men transition from competing to connecting and women become more assertive. Aging in general brings transitions of health, competence, alertness, enthusiasm, and social connections.

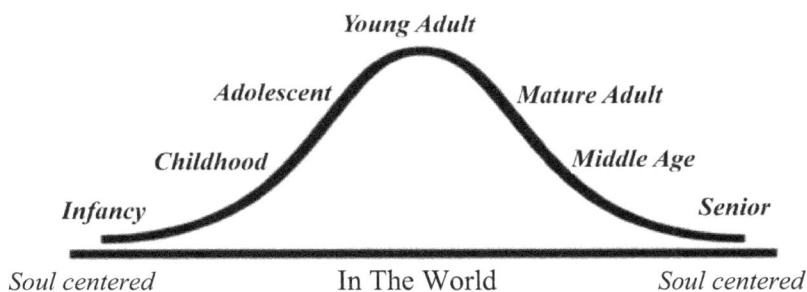

Early and later life are similar: need assistance-dependency, limitations, wear diapers, etc.
**In The World external demands, material values, self-imposed stress, role playing vs authentic self*

Each new stage brings changes in self-identity, perspective, awareness and values; essentially shaping you into a different person. New associations enter the picture, with other friends, coworkers and partners. New personal firsts occur: job, car, crush, BFF, independence, home, costly mistake, long distance trip, family conflict, legal trouble, political interest, serious injury, financial windfall, etc. Different priorities emerge, paralleled by changes in personal interests and concerns. Each new stage brings its own variety of pitfalls, traps, attachments, insecurities, and self-revelations; inner nuances not noticed before.

"Life is what happens to you while you're busy making other plans" – John Lennon

Human beings universally experience common stages as rites of passage. Modern civilization has only changed some of the details and parameters, but the basic essence of each layer hasn't changed much at all. Ancient mythology captures the recurring pattern with the hero's journey template. It's a universal roadmap setting the course we all must abide. Every generation goes through the same process, facing new challenges and tests, masked in different variations on a standard theme. Star Wars provides a classic example where Luke Skywalker is the

Hero following the call to adventure, meeting Obi Wan, facing Darth Vader, finding himself, seizing the reward; a fictional metaphor of our own life journey.

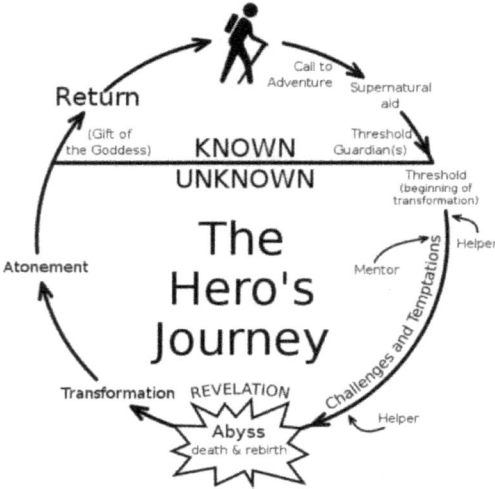

Hero's Journey Stages

Ordinary Life
Call to adventure
Hesitate
Meet mentor-facilitator
Cross threshold
Tests, allies and enemies
Primal ordeal
Pass Big Test
Transformation
Return with reward

*Commitment to the journey requires accepting the unknown, way beyond the edge of your comfort zone…crossing an ambiguous mystical threshold to get the prize.

Stages represent a universal map of forward progress, however in recent decades the life-cycle period length has changed significantly as people live longer, retire earlier, have children later, and those kids mature later. The whole scale is stretching longer – shifting older. The most recent generational cohorts are starting young adulthood later, often into the mid to late twenties, leading to later marriages and family building (from 20 to now 30). Consequentially, middle age starts later (from 45-60 now 55-70) while the golden years shift later (from 55-70 now 70-85). An overall longevity shift makes each new age group seem a decade younger – 50 is the new 40, etc. For the first time in world history roughly one third of all women reach 100. General demographic patterns are becoming more ambiguous with gender role fluidity, longevity and stage periods shifting.

Every twenty years of so a new generational cohort emerges bringing with it changes from the previous one. Differences include common experiences, perspectives, associations, pop culture, inventions, and generational baggage. Each new cohort navigates the same arenas: (Family, Workplace, relationships, competition) but with subtle nuanced unique aspects. Social environments are not the same since shifts in culture affect power structures, public platforms, business environments, language, styles, customs and society as a whole. Laws and informal rules change cohort to cohort, requiring in-kind behavior changes. Each generational cohort literally lives in a different world.

The same human development process with individuals applies at the collective level. There's symmetry between the internal aspects driving a person toward growth and progress, and the dynamics affecting larger groups, including cities, states and nations. Social evolution is a composite result of individuals developing together. A society is as a group with mixed various competencies, combined to form an integrated collective level of development. Group consciousness is essentially a mean average of everyone together. Beck and Cowan's *Spiral Dynamics* identifies groups of people in levels of development (consciousness) with associated general characteristics and color code. It's a natural sequential progression that societies follow, some advancing more than others.

Societal Consciousness Levels

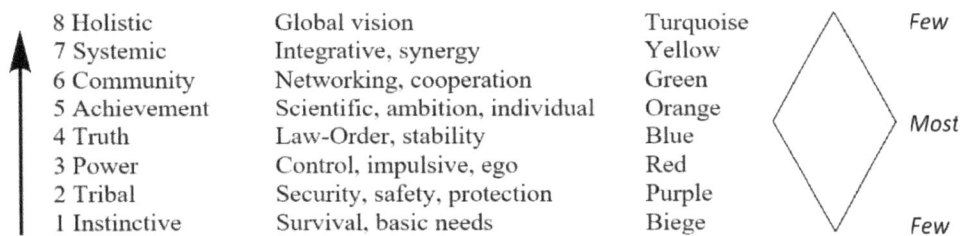

8 Holistic	Global vision	Turquoise	Few
7 Systemic	Integrative, synergy	Yellow	
6 Community	Networking, cooperation	Green	
5 Achievement	Scientific, ambition, individual	Orange	
4 Truth	Law-Order, stability	Blue	Most
3 Power	Control, impulsive, ego	Red	
2 Tribal	Security, safety, protection	Purple	
1 Instinctive	Survival, basic needs	Biege	Few

* Dimond shape mix with most in the middle levels and few at the top-bottom ends
** Which level do you identify with? Which level best describes your country?

Early civilizations developed with concentrations in lower tiers. Over centuries modern societies progressed so levels shifted upward, filling in the middle tiers while few advanced portions rose into higher tiers. As individuals raise their level of consciousness the collective whole rises higher. But no matter how advanced societies become they always include members in lower rungs that don't advance much, are stuck in low-level basic need states and wallow in pools of negativity. We've all encountered those who are several levels below our own and look down upon them while looking up to others who are several levels above us. Each group level has a unique perspective, along with specific values, priorities and higher associations. In any case "You" are always in a middle meme with some way above you and others way below you.

Personal Development is the integrative process of enriching a conscious being through ongoing experiences – navigating through The World in pursuit of self-serving goals and desires. During this process of unfoldment a person is shaped, molded and improved, continually rounding out as a fulfilled being, achieving higher degrees of self-actualization. Everyone you meet is somewhere along their own journey, crisscrossing yours and everyone else's. 8 billion people are currently on paths that cross, zigzag, weave in and out; some joining while most never connect but may run parallel. The human journey is an individual and collective state of perpetual becoming. Each progression integrates prior stages while letting go of prior limitations (one doesn't lose the foot when gaining the yard). Ultimately you discover your authentic self and true purpose. Each day should involve creating, learning, and moving forward and upward. If you're not living, you're dying. You're an ongoing continual work in progress, occasionally moving beyond your comfort zone and into uncharted territory; a tradeoff required for growth. Your upward drive requires steadfast vigilance, maintaining centeredness against a perpetual pull of external demands, distractions, and false destinations.

Along the journey you learn life lessons in how the world works, how to relate to people, how to respond to different situations and how to overcome poor choices. Continual adjustments to circumstances, events, and human arenas temper you and enable further growth. Although lessons come in 3 levels (man vs man, man vs nature, man vs himself), they essentially come down to you vs yourself. The journey itself is the message. It provides the meaning of life – a process creating a difference between who you were and who you are. Successful completion results in Self-discovery and Self-actualization – to find your authentic self and embrace your natural purpose… to realize that once you reach the destination, you're not the same person…the journey has changed you.

For some individuals the meaning of life is really just a purpose; to become a better person. There's no special road map for everyone to follow but there are a few simple things everyone can do: travel more, embrace change, continually learn, make new friends, set personal goals, master the art of listening and intentionally go beyond your comfort zone. And as a bonus, remember that learning from your own mistakes is good… but learning from others is even better, putting you solidly on the path toward greater wisdom.

"The meaning of life is to find your gift. The purpose of life is to give it away" – Pablo Picasso

Remember that all journeys include distracting diversions that can delay or prevent reaching the destination. Some individuals spend their lives pursuing things *out there* while all along the real thing was *in here*. The pursuit of happiness works best when paths are followed. Correct choices become self-evident as doors open up and the wind carries you forward. When you follow your bliss, opportunities unfold. You then transition from ego *driven* to spirit *being,* through a purpose guided life lived daily. Growing up isn't just for kids, it's a life-long endeavor on the path towards higher consciousness. The future is created moment by moment in the choices you make. The trick is to play the game of life without getting sucked into its trappings. Be *in* this world but not *of* it. To put it simply; "life is a journey, enjoy the ride." You'll know you've arrived when you can answer these basic questions: *what do I really want, what is my greatest goal, what makes me happy*, and *how am I different from everyone else*? Once you can answer these honestly, you will reveal your authentic self.

Ikigai – *Japanese for "a reason for living"*

Satisfaction, but feeling of uselessness

WHAT YOU LOVE

Delight and fullness, but not wealth

PASSION

MISSION

WHAT YOU ARE GOOD AT

IKIGAI

WHAT THE WORLD NEEDS

PROFESSION

VOCATION

Comfortable, but feeling of emptiness

WHAT YOU CAN BE PAID FOR

Excitement and complacency, but sense of uncertainty

HUMAN EXPRESSION

We are all living vessels with an inner spirit filtering outward; individual souls interfacing with The World. The constant interaction between inside and outside realms produces various states and dispositions. What filters out has an identifying signature of expression – a unique fingerprint particular to each one of us. Our individual templates are made from the same parameters (health, age, gender, culture, race, knowledge, health, attitude, etc.) but the combinations and compositions are nearly infinite. It's no different than light passing through a prism revealing a spectrum of colors; a clear metaphor of human beings as spirit filters.

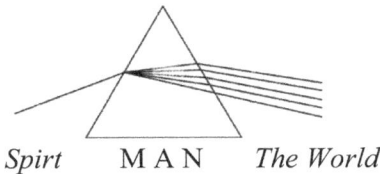

Spirt M A N *The World*

ability mental health **Language** culture
APPEARANCE Ethnicity **location**
Religion gender **physical health**
economic class Hobbies *age*
education political affiliation Race
Marital Status OCCUPATION PARENTS
NATIONALITY disabilities *Intelligence*

Individual Filters

Human expression is always a form of communication. We're naturally social creatures, continually sending-receiving signals in a variety of expressive modes. Physical, emotional and mental inner aspects work together communicating inside to outside. *Physical* includes touch, embrace, fighting, sex or posture, *Emotional* involves non-verbal body language such as facial expressions, tone, inflection, and *Mental* includes verbal language – speaking, writing, singing, etc. Non-verbal communication is often the most important, revealing subtle messages expressed in a particular posture, handshake, stance or tone. Reading faces and eye contact provides essential clues of what's really going on: lying, agreeing, indifference, attractive interest, excitement, discomfort, etc. Micro expressions are hidden, brief mannerisms revealing a contrary response to an outward masked look. Facial expressions convey a broad spectrum of feelings using simple gestures: smile, cry, grimace, stare, laugh, wink, squint, head shake, raised eye brows, pouting mouth, etc. Interestingly, emotional extremes on both ends are similar in expression, such as a grimacing face in both agony & ecstasy.

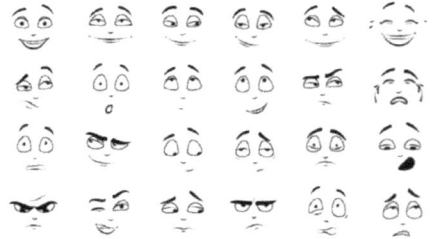

Facial expressions for every emotion   Combinations of just a few features

Thousands of facial configurations to express universal human dispositions and emotions

Our bodies continually provide physical clues to our state of being through a variety of autonomous responses. Something as simple as breathing can reveal a lot about what you're doing, with different rates, rhythms and depth for various activities: exercise, rest, meditation, fear, excitement, sleep, sex, running, strain, giving birth and exhaustion. Each has distinguishing associated breathing patterns.

Human communication is embedded into every activity we engage in, some intentional, the rest incidental. What's conveyed is subject to misinterpretation and bias errors. There's also intentional deception of hiding the truth, which is just simply lying. Of course this comes in degrees, including bending it, omitting parts, partial truths, bold face lies, and artful deception. Barriers to communication involve all the standard aspects of personal context: age, race, gender, culture, education, etc. Gender filters are notoriously common as men and women tend to focus on opposite priorities, speaking two different languages. Men generally emphasize mental responses, focus on work, achievement, and problem solving. They prefer minimal verbiage, seek adventure and desire respect. Women conversely tend to respond from the heart and focus on sharing feelings. They prefer to be close, opening up and listening. They seek romance and desire ongoing reassurance. Not surprisingly, most greeting cards are purchased by women and given to other women. And yes, communication or lack thereof is one of the primary causes of problems in relationships.

You're the only person who's heard everything you've ever said"- Anonymous

Human expression is translated into behavior through the universal modes we engage in, including work, play, producing, creating, giving-taking, competing-cooperating, growing-sustaining and resting, etc. All human behavior is a subset or derivative of these basic modes of expression. During our natural pursuit of self-interest we set goals, fulfill needs and engage in all of the following:

Work – Build, move, change, eliminate, problem solve
Play – Games, leisure, entertain
Create – Arts, crafts, design
Give – Provide, share, (including love)
Take – Receive, confiscate, find, (including love)
Compete/cooperate – Work against or with others
Grow – Learn, accumulate, improve, develop
Sustain – Maintain, repair
Rest – Relax, recharge, sleep

These modes are expressed in jobs, careers, sports, performance, entertainment, travel, romance, training, education, leadership, exercise, governing, exploring, arts and crafts, construction, etc. Each activity is driven by self-interest and personal need. We use a wide variety of approaches, styles, and manners, expressed in combinations of physical, mental, emotional and spiritual aspects. Behavior is the byproduct of dynamic interaction between thoughts, feelings and physical being. Everything we do is guided by how we think or feel about people, things and situations in The World. Healthy behavior comes from a proportional balance between body, mind and soul. Specific circumstances call for temporary emphasis on either rational logic (what makes sense) or emotions (what feels good), but the trick is to avoid extremes on both sides.

Human reality in a practical sense comes down to how we spend our time and energy. Despite a multitude of diverse possible activities to engage in, we really limit it to a narrow few. The average person with a full-time job works 8 hours a day plus commute during the week, and another 8+ hours on sleep and personal care. Doesn't leave too much time for other activities until the weekend. Throw in time spent on your smart phone and watching TV and now we're really narrowing opportunities for other stuff.

Weekday Activities (Average adult)		*Routine Regular Activities*	
Sleep and personal care	9 Hrs	Watching TV	3 Hrs
Work and commute	8 Hrs +	Smart phone	3 Hrs
Leisure and sports	5 Hrs	Exercise	1 Hr
Housework and buying stuff	1 Hr	*Some check smartphone 50+ times/day	
Eating and drinking	1 Hr	**We spend 1/3 of our lives sleeping	

Human behavior is influenced by changing levels of need. Initially we focus on security and survival, followed later by identity, love, and self-actualization.

Higher level needs don't emerge until lower basic needs are first met. Security is normally provided by parents and family. Survival is attained/maintained through material acquisition – work and career. Identity is established through a sense of independence, reinforced through goals, interests, hobbies, or personal attachments. Love is pursued and acquired through caring, sharing, dating, and marriage. Self-mastery is a life-long process developed though perpetual learning, seeking, revealing and expanded competence.

Maslow's Motivational Model

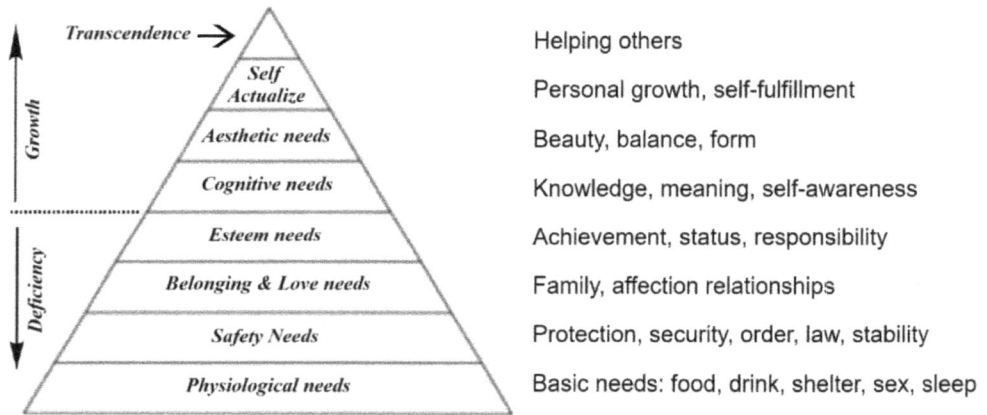

Level	Description
Transcendence →	Helping others
Self Actualize	Personal growth, self-fulfillment
Aesthetic needs	Beauty, balance, form
Cognitive needs	Knowledge, meaning, self-awareness
Esteem needs	Achievement, status, responsibility
Belonging & Love needs	Family, affection relationships
Safety Needs	Protection, security, order, law, stability
Physiological needs	Basic needs: food, drink, shelter, sex, sleep

(Growth — Deficiency axis shown on left)

Each higher level of need brings different aspects expressed in a spectrum of positive and negative experiences as well as nuances and variations of expression leading to reversals. Consider these examples:

*Early child-parent attachment reverses with independent rebellious teen.
*Stable job suddenly jeopardized by demanding boss and envious coworkers
*Romantic relationship ends in heartbreak after one loses interest in the other
*Life-long friendship dissolves as one progresses while the other stagnates.

Human behavior is routinely expressed as a performance. In the normal course of doing things and producing stuff, the what and how covers a broad spectrum with extremes on both ends. This includes quantity and quality – how well we do things and how much we produce. All of it can be captured in a Bell Curve graph where the extreme few emerge at the ends while the majority fall inside the middle area. Every endeavor man engages in follows this universal pattern. Competitive athletics is a prime example where every sport has a legendary star, a few greats and a whole lot of average players: Basketball - *Michael Jordan*, Football - *Tom Brady*, Baseball-*Babe Ruth*, Soccer-*Pele*, etc. Pick any field of endeavor and there will be a similar list of the "Goats" (greatest of all time), the mostly great, and the general notables. There could also be a lesser-known list of "terribles" representing the opposite end of the scale. Comedian George Carlin pointed out that somewhere in the world is the worst doctor or dentist, and right now they're working on some poor patient. The average person has no clue how good or bad the best and worst a person can be on either end. When using the popular 1-10 scale, most of us have never seen a 1 or 10. The Bell Curve is a fundamental aspect of human reality.

A special category of human performance occurs in settings with a live captive audience. There's something magical about this dynamic that brings out the best and most creative potential in a performer. A co-dependent relationship emerges between performer and audience producing a transcendent experience; the whole becomes greater than the sum of the parts. Unexpected abilities surface, enriching the experience and juicing up the creative energy of the performance.

In the prior section on personality we covered universal character types, routinely portrayed in plays, movies, TV sitcoms, novels, etc. These recuring personality patterns become intertwined with universal roles that people act out. They can be found in the workplace, home and family, government, sports, recreation, church, entertainment or any arena where groups of people coagulate. Even in our own family setting we have the ambitious, aloof, comedian/prankster, talker, gossiper, drama queen, attention seeker, victim, etc. The roles just get more defined and dramatic as we go into The World, where all the traditional mythological archetypes are present everywhere. We could classify character types as Enneagrams, which is a combination of personality and world view. The 9 basic types are: *Reformer, Helper, Achiever, Individualist, Investigator, Loyalist, Enthusiast, Challenger, Peacemaker*. Each role has its own subvariants. Any person assuming a leadership role generally tends towards one of four standard modes: Telling, Selling, Persuading, Demonstrating. The same variety of approaches applies to literally every one of the universal roles people assume.

Combinations of diverse character types, unique personalities and specialized roles produce a broad spectrum of human expression with differences in degrees, creating its own spectrum of types. Using the bell curve portrays each category as a scale of averages in the middle with extremes on both ends. We've all experienced characters who are larger than life, whether in a good way or a bad one. Some just express the same emotions with greater flair, often with higher highs and lower lows. These special types are most well-known at the national level as celebrities in pop culture, however, there are plenty of others at the local level too. Every small-town tavern, public square and sports venue has a handful of standout characters that leave a memorable impression on those who have encountered them.

Group dynamics affect our expression the same way it influences our personality. Each member in a group unconsciously projects outward while simultaneously responding to feedback from others. The resulting interaction either reinforces or restrains our natural expression, changing the way we behave in that setting. The effect varies by degree as group size changes – from 1 on 1, to small groups to very large groups. Our relationships are further filters of self-expression.

To be alive is an ongoing act of creativity. Every moment, minute and day involves a creative process of divine spirit flowing in, through, and out each of us. Everyone needs an outlet. Illness and pathology often result from energy blockages or inhibitions of natural expression. Since creative energy is linked to soul-spirit-emotions, free flowing spirit is essential to health. Ironically, our own mind's thinking and thoughts hinder creative flow, leading to writer's block, athlete slumps and problems in the bedroom. The key to creative flow is letting go, minimizing restrictions and throwing out the rules. Creative problem solving improves by thinking outside the box – look at the familiar with a different perspective, mix things up, strip away boundaries and don't hold anything back. It's better when you don't care what others think.

Every human endeavor has a combination of structure and free flow. When restraint is removed and creative juices are allowed to flow uninhibited, magic can happen. Consider musicians free styling/riffing, comedy improv, doodling, passionate love making, joyful dance, flirting, playful water coloring, adventurous exploring, etc. Normally when we think of creativity our first impression is the arts (painting, acting, music, singing, dance), however it's a fundamental aspect connected to all human activity. As beings with dual natures, we channel psychic energy into everything we do. While our daily functioning in The World is driven by conscious action, our inner driver occasionally connects with our outer, leading to temporary moments of special awareness. It manifests as an athlete being in the zone, an out of body experience, a scientist's moment of eureka, tantric sex, a spontaneous inspiration, or suddenly overcoming a debilitating fear. These and other unique situations include a universal *aha* experience that cannot be put into words. Other variations include moments of wonder, laughing to tears, adventure rush, Zen/TAO, warrior spirit, or an exhilarating scream. Similarly, we've all experienced magic moments of awe when solving a complex puzzle, learning a hidden secret, getting chills, breaking a code, or the simple act of child-like discovery. Magic takes place every day in subtle ways during the mundane routine of life, but only if you know where to look. It can be found in daydreams, imaginations, nonconforming behavior, mishaps, oddities, exaggerations, and humor of all types (Jokes, pranks, puns, silly acts). Laughter is healthy because it releases tension and loosens blockages of energy. It can capture an eternal moment, bridge you to a higher realm and transcend The World.

Human expression is powered by emotion, the language of the soul. Everything we do possesses an enriched element of nonverbal communication. Every experience generates some degree of emotional response along with a corresponding facial expression. These subtle gestures are universally recognized across different cultures with few exceptions. Faces can be read like a book and serve as excellent lie detectors if you know what to look for. It's just one aspect of our consciousness filtered externally. We're living templates processing an original pure essence into a kaleidoscope of variations, expressed in body motions, head movements, upper and lower face (brows, eyes, mouth, cheeks), gaze, blinks, etc.

At its highest level, shapeless formless spirit is an essence of pure love, filtered and processed through human beings. Our entrenchment in the physical world encumbers expressions of it, creating unnecessary suffering and emotional dramas. Despite love being a focal point of human experience, it's largely misunderstood, confusing and elusive – an eternal enigma. What we think is love is often merely variations, partial versions and derivatives of the real thing. Its many forms and types are perceived and interpreted differently by everyone. A simple "I love you" can mean broadly different things to different people. When giving love we assume what it means to us is the same thing it means to those receiving it; a perilous assumption. We all filter its source, expressing it in ways that change its pure essence. Human consciousness and ego create a polarized spectrum where spirit-love is expressed in varying degrees and qualities.

Love is a universal theme in human experience with multifaceted forms of expression. We project it as feelings, acts, gestures and states of being. Love is a ubiquitous subject in literature, songs, movies, poetry, the arts, and is the central plot of numerous television programs. It's usually the focus of attention in most human drama, lurking under the surface. All major religions exemplify it. It's a basic human need to seek love and be loved - essential to emotional stability.

Spectrum of Passion – Degree of Emotional Intensity

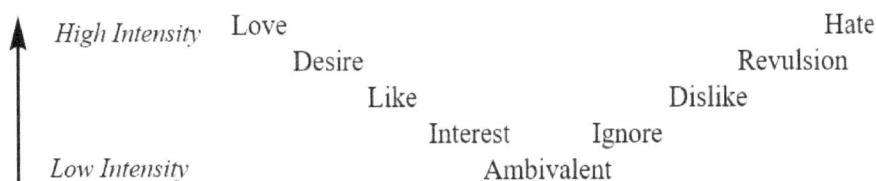

High Intensity	Love					Hate
		Desire			Revulsion	
			Like		Dislike	
			Interest	Ignore		
Low Intensity			Ambivalent			

*Emotional expression is combination of type + intensity, like light (frequency & intensity)
Combination of quantitative/qualitative components (degree of passion and type of feeling)
Extreme opposites: Peace & fear, Joy & pain… the difference between unity and separation

Spirit-Love is experienced and expressed in higher and lower octaves: affection, service, gifts, passion, lust, jealousy, envy. It's conveyed in actions or moments of being: joy of adoration, excitement of discovery, passionate kiss, the thrill of victory and the agony of defeat. It's often expressed as an external projection; onto a person, thing, or activity, or identifying with something *out there*. We can experience love of life, knowledge, material things, an idea, a cause, a special person, or all humanity. Love is a spiritual essence most of us confuse with the filtered emotions we experience. All expression is a derivative form of spirit-love, even hate. Love and hate are closer to each other than to indifference, similar to black and white being closer to each other than any color in between.

Love is associated with passion, *letting it all out*; getting lost in it, which is simply unrestrained expression of spirit. Love is a transcendent essence filtered through constrained beings – expressed positively or negatively, from pure agape to raw lust. Love is spirit manifested in resonant states. Love can be a noun, verb, adjective, adverb. Like water it takes the shape of any container via any filter.

Expressions of love serve as communication modes between you and another person. Distinct languages of love convey its essence: *speaking words of affirmation, spending quality time, acts of service, gifts and giving, and physical touch.* These 5 love languages express the same essential message using different modes of communication. Expressions of love also vary by both degree and type. There are 4 basic levels that capture the nature of love we express as we progress from lower *ego driven* to *non-ego maturity*:

1 *Selfish* – I Win. Desire for pleasure, more stuff, satiation, self-interest (worst: greed-lust-gluttony)
2 *Cooperative* – Win/Win. Attention, respect, recognition (still ego based, feed self-love from others)
3 *Selfless* – You Win. Service to others, sacrifice, embrace others openly
4 *Unconditional* – Universal. Transcendent, exude gratitude, appreciation of everything

Different types of multifaceted love expressed in various modes:

Eros	Erotic, passionate, romantic (body, touch)	
Philia	Friendship, affection, fondness	LOVE
Storge	Parent-Child, familial, unconditional	
Agape	Humanity, selfless, kindness	Like
Ludus	Playful, flirtatious	Lust
Mania	All consuming	
Pragma	Lasting, enduring, committed	Obsession
Philautic	Self love	*Lower-level expressions*

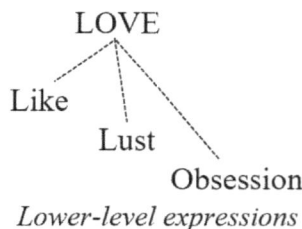

*Other related modes include Companionate, Consummate, Fatuous, Platonic, Infatuated Codependent, Empty, Casual, and Toxic. **Many are unhealthy, low levels expressions.*

Romantic relationships provide further diversity of expressed love, producing their own spectrum of types. They create both interesting and confusing modes of love, including infatuation, sexual, platonic, friendship, puppy love, affectionate, passionate, etc. Each of these forms lead to intense responses: sweaty palms, butterflies in the stomach, nervous, shyness, fondness, fantasizing, free-wheeling, fixating, daydreaming, euphoria, etc.

Navigating the process of romantic love involves dealing with unsettling experiences and awkward situations. The path toward seeking love may include courting rituals, a school crush, teen puberty, young adult dating, bars, night clubs, social media, and the inconvenient friction dealing with distance, transportation, and conflicting work schedules. Traps and pitfalls may involve friends with benefits, rebounds, coveting a friend's partner, hot and cold responses, mixed signals, obsession, stalking, playing hard to get and games of attention. Adding further friction is the language differences between the sexes.

Love relationships possess a blend of the following 3 general aspects at different degrees: ***Intimacy*** – sharing, bonding, baring soul, transparency, ***Passion*** – physical attraction; short lived, in the moment, and ***Commitment*** – Stable, reliable, devotion. Obtaining the right, balanced mix of all 3 is the key to a healthy relationship. Additionally, there should be a right balance between the physical, emotional, mental and spiritual dimensions:

Physical	– Creates desire	Mutual attraction
Mental	– Creates interest	Communicate well and understand each other
Emotional	– Creates affection	Feel good together
Spiritual	– Creates love	Transcendent connection beyond words, feelings

Love is a unifying force pulling us up towards transformative holistic development. Signs of "True love" include spending quality time together without sex, being happy together without friends or entertainment, not being attracted to other potential mates, and being totally honest and open with feelings.

Human Spirit Filters. Human beings are expressive creatures, continually radiating spirit into The World, through unique filters in numerous modes. Every act is an expression, every effort makes a point, and every response sends a message. Living in The World demands continual expression from inner to outer. No two individuals are alike. Everyone emanates a unique expression of spirit – a sort of coded signature with a style and essence revealing who you really are. It's an ongoing process permeating every behavior, activity and performance. It shapes the roles we play and the characters who play them. Human expression is driven by emotion, which is just a derivative form of spirit manifesting into a spectrum of forms. The higher octaves are experienced as love, and it too is expressed in diverse types. Life is never dull with unique beings like ourselves continually filtering inner spirit outward as a collective community of artists creatively illuminating every space, nook and cranny everywhere in The World.

STATES OF BEING

Everything or every *thing* exists in a particular state. All things, living beings, systems and situations attain a relative state of being – a disposition that changes over time. Higher conscious organic beings like ourselves possess the greatest range of possible states and each of them colors our outward expression. States exhibit modes of duality, including being-becoming or passive-active, which determine how we experience reality. Identical activities or events can feel positive or negative, depending on your state. The same task can feel like work or play, again, depending on your internal state. In every situation you can feel either good or bad…depending. The World *out there* isn't really as important as the state *in here*. Both sides change and interact together in a co-created process, but the inner is more influential than the outer in determining our experience.

All human expression is directly tied to states. Change your state and everything about you changes – posture, attitude, the way you laugh or cry, or behavior in general…The World itself changes. And counter intuitively, changing your outward expression can change your inner state. The two work hand in hand as inside and outside realities co-create every experience.

> *"We see the world not as it is, but as we are"* – Stephen Covey

We're beings with multiple integrated aspects and levels working codependently together. The primary levels are *physical, mental, emotional* and *spiritual*, where we sense-think-feel-be. Emotion is the key activating energy shaping your state. Body, mind and soul work together, generally in sync though sometimes skewed out of balance. Each influences the other. *Emotion* is the driving force behind action and behavior. Thoughts are limited until charged with emotion. Even memories are richer and more vivid when attached to emotionally charged experiences. Bodily sensations are magnified by emotional states as well. Any type of performance is greatly enhanced when passion is present. Consider how a warrior spirit can make a smaller army seem invincible even while facing a larger, stronger opponent. "Ordinary" sex can become a transcendent, transformative experience whenever genuine love and passion enrich it.

Emotion originates from the Latin "motere", meaning "to move". Emotion = life energy. Moods are a mode of emotional energy; temporary states that shift in a very non-linear manner. In a healthy state emotions flow naturally – released and expressed externally without hesitation. Bottled up feelings can be destructive. Unfiltered expression can be equally problematic. Negative emotions can build upon each other in a resonating, recursive loop leading to anger, anxiety and a variety of negative states. Balance is essential. Self-awareness even more so. Emotional intelligence is necessary for appropriate expression of competing emotional energies.

States of being act as templates that channel human expression, similar to the prism light filter referenced earlier. Your inside is projected outside through this filter, transforming energy of soul-spirit into corresponding patterns manifested into The World. Expression changes with each state – posture, mood, speech, laugh-cry, facial gestures, etc.

States are really just energy patterns, differentiated by their resonant form of vibration similar to music octaves. Energy types, levels, and patterns create unique qualities. Just as frequency and amplitude produce a wide variety of colors, states have the same effect with living beings. It's the simple reason some people have higher energy than others. Whether it's harmony or discord, resonance or incoherence, the energy pattern makes all the difference. States correlate with the pace and cycle of breathing and heart beats, mirroring the patterns of music tones, timber, and rhythm; there's a song for every mood.

Ancient man intuitively understood energy levels within people and their unique expressions. Nodes of inner energy resonate in lower to higher levels corresponding with particular organs and functions, and also to color frequencies. Your state of being corelates to particular energy levels dominant at the moment. Early human beings operated on a lower, physical-survival level – a lower resonant vibration. As civilization progressed man's collective consciousness elevated with it, raising the vibrational state.

Perennial philosophers identified inside energy levels as *Chakras*, corresponding to 7 primary nodes. Lower frequency energy settles at the base of the spine, associated with basic survival instinctive impulse. Higher frequency energy rises upward along the body's vertical axis, peaking at head's top. The upward path of higher energy forms a vertical spectrum mirroring man's progression in consciousness, from lower to higher – from animal-survival focus up through higher spiritual growth, transitioning from food-fear-selfish-ego to relationships, communication, cooperation and the divine.

CHAKRAS	**Root**	**Sacral**	**Solar Plexus**	**Heart**	**Throat**	**Third Eye**	**Crown**
Color Frequency	Red	Orange	Yellow	Green	Blue	Indigo	Purple
Aspect	Survival	Pleasure	Will Power	Love	Truth	Insight	Cosmic Energy
Blockage	Fear	Guilt	Shame	Grief	Lies	Illusion	Ego attachment
Associated Glands	Adrenal	Gonads	Pancreas	Thymus	Thyroid	Pituitary	Pineal

*Energy resonance at nodal points can be *"gut feeling"*, *"heart felt"*, *"all choked up"*, *"horny"*, *"head case"*, etc.

Higher Vibration

Resonant Energies

Lower Vibration

The upward incremental progress of mankind is mirrored by states representing progressive levels of consciousness. Parallel paths of increasing evolution follow a spectrum or hierarchy of levels, and imbedded nodes (transition points). Primary levels are known as Conscious, Subconscious and Unconscious, along with a Superconscious, plus sublevels within the entire spectrum. It's basically a transition between Waking-sleeping-dreaming-altered states-meditative states, and deep sleep.

Spectrum of Consciousness

Conscious – focused in The World	**EGO**	*Awake* - direct experience, s*ense, act, respond*
Subconscious - partial focus of world	**ID**	*Beneath surface* - desires, fears
Unconscious - unaware of The World	**SuperEgo**	*Sleep (*dreams) – archetypal symbols, *deep well*
Superconscious - aware of the beyond		*Deep Sleep* - psychic, spiritual, cosmic

This spectrum is a sliding scale of degrees with incremental nodal points in between. It's somewhat arbitrary to define where one level ends and the next one begins. There's lots of gray area between states, including *coma, vegetative, non-REM sleep, REM dreams, drowsy wakefulness, daydreaming, awake, high alertness, manic alertness*, etc. It's a little easier to distinguish the differences between the higher and lower ends of the scale. Higher states of consciousness are self-directed – you are the creator, directing your own destiny. Lower states are world directed – you're distracted, pulled, and reactive. Most people resonate to a mid-level, oblivious to others above and below. In general, YOU are in the middle.

Consciousness is more than just a state of being. It *is* being. It's *in* and *of* everything. We are just a living form of it. Every aspect of our being and reality is made of it. Body, emotion, mind and soul are different resonant modes of spirit/consciousness, just as water transitions between solid, liquid, gas or plasma. We tend to focus on the awareness aspect of consciousness and the mental nature of it, but physical and emotional play a significant part. The universe is imbedded with the attribute of mind, from the micro level of atomic particles to the macro perspective of large bodies interacting in space and time. In living beings mind concentrates and divides attention, creating order out of all the internal and external stimuli. It pushes trivial stuff down into a lower subconscious state, like on a long and boring drive or listening to a dull lecture. Since we're bombarded by stimuli constantly, the mind must prioritize. Degrees of conscious awareness follow a daily ebb and flow between routine tasks and interesting, exciting activity. Consider the difference in attention engaging in the following: eating, cleaning, play, reading, climbing, learning a new thing, counseling, surgery or drowning.

Mind is an active mode, *Being* is a passive mode – mute the mind to experience the now… to simply be. To switch from mind to stillness is a change of state. But remember that thinking is just one subset of consciousness out of many: feelings, sense, intuition, experience, etc. So, consciousness really takes on two meanings when we talk about it. In one respect it's a substance of reality taking on multiple forms. In the other it's our level of awareness – to be *fully conscious* is to be awake, aware and present in the moment. This aspect relates to our spectrum of consciousness and the primary modes within it.

Less than fully conscious is the *subconscious* state, where focused attention dissolves, and external awareness is suppressed. Inappropriate thoughts and feelings are submerged here, sometimes bubbling up back to the surface in awkward moments. Desires and fears reside here as well, eventually expressing themselves into conscious behavior patterns. This level is equally active, just not focused on immediate sensory input. It's a middle boundary layer between the constraint of physical reality and the limitless unbounded realm of the unconscious.

The Unconscious state is associated with sleep, completely detached from waking awareness in which the focus is on daily mundane life in The World. This state contains a larger, deeper reality submerged below the conscious level as Mind and thinking are "turned off". It contains a repository of everything we experience, regardless of context. Myths and dreams come from this realm, processed in the language of symbols. Duality is at play as we metaphorically live and die every day and night, falling into a deep unconscious state of being at night only to be resurrected into fully waking consciousness by day. But sleep is an essential requirement, not just physically but mentally and emotionally. During the unconscious, deep sleep state we re-emerge with the primordial source – to re-sync our being after agitation and de-centering from daily stresses in The World. Higher realms exist here, totally beyond our waking awareness.

The superconscious state transcends the others, going beyond physical space and time. It's the deepest level of awareness; a realm of pure reality, ironically obscured by our waking consciousness. What we consider "higher consciousness" in the fully conscious waking state is actually a lower level of awareness, limited to focusing on the physical plane. The superconscious state is deeper and more "aware" than all the others. It's the source of precognition, ESP, telepathy and intuitive insight.

Consciousness is a serpentine essence manifested into every aspect of our being. It's present physically, emotionally, mentally and spiritually. It varies by both degree and kind. It's both the variations of aspects within a state and the overall state itself. It's the ingredients, the meal and the taste of it all at once. Conscious mind is a necessary limiting mechanism of order for coping in a chaotic world. Subconscious mind presides over all the stuff conscious mind submerges, including conflicting feelings, desires and fears. Unconscious mind turns off the outside world and reconnects with the inner source, working through issues via symbols, dreams and coded messages. The superconscious state operates beyond in a spaceless, timeless dimension where infinite boundless spirit originates.

States of being are aways in dynamic flux, ever changing, shifting and flowing, both qualitative and quantitative, both in degree and kind. There's really never a pure state but always a mix of partials within a generalized settled center. It's a dynamic integrative process involving complex interactions between elements at every level. Changes in the mind lead to changes in feelings, leading to further changes in physical sensation and vice versa. Brain waves shift, moods change and glands secrete. Changes in parts translate to changes in the whole. Those who experience temporary highs or lows can't sustain them very long because our "normal" general state is relatively stable and pulls extreme shifts back to the center. Changes in state normally occur in nonlinear fashion, moving by degree without distinct level or stage change, just incremental shifts.

4 Primary Human Aspects	Ideal State of Being	Peak State
Body – Kinesthetics, athletic	Health	Pleasure
Mind – Logic, reasoning	Clarity	Peace
Heart – Emotional, interpersonal	Love	Joy
Spirit – Wisdom, existential	Bliss	Ecstasy

Internal aspects of emotions, thoughts, and intentions interact either in harmony, discord or remain neutral. Inner thoughts and feelings are often in conflict, requiring hard choices. Outer experiences add to the mix with general sensory stimuli (touch, taste, sounds, smells, images). The outside world affects our perception and vice versa. Inner is never separate from outer as we co-create reality. Change of scenery, participation in activities, exposure to new things and traveling are all modes that change our State of being. Consider the difference between traveling with and without an itinerary. Trips with less familiar scenery elevate your vibration. Senses are heightened and the experience becomes more *in the moment*. This same process happens in any situation where there is novelty, surprise, excitement or creativity. Children experience these conditions continually, fostering a sense of wonder and interest. For them, living in the moment is all they know.

Besides inner and outer influences there are differences in things near or far. Attractions and dangers affect us based on proximity – the sense they are approaching or receding. Geography and culture resonate with each of us in different ways, just as visiting certain parts of the country just makes you feel better. The same is true with different groups that just seem to be a better, more comfortable fit. Different people, friends and partners, sync to different states. Like real estate our inner being is significantly influenced by location, location, location.

Self-perception is about how we see ourselves, and it affects how others see us, and vice versa. Perception is not the passive process you assume but one that's continually shaped by internal biases and fluctuating states. Not only do we never really see others for who they actually are, we don't even see ourselves for who we really are. Attaining higher consciousness diminishes false perceptions. I suspect sensory deprivation feels scary because emptiness removes our self-imposed boundaries, opening us up to our true naked selves.

Stress is largely self-induced. Same with fear. External events are actually neutral, only becoming positive or negative based on our subjective perspective. The very same act can make one person laugh and another one cry. Same goes with fear where several people facing the same threat results in some standing tall while others cower in the fetal position. In reality, ALL external events are neutral in nature; meaning is relative and derived from individual responses. Our own inner thoughts, expectations and beliefs create stress, not The World out there. We don't see things as they are, we see them as we are.

States are a mixed conglomeration of aspects and attributes: levels of consciousness, degrees of character traits, moods, attitudes, and a variety of interrelated conditions. An overall state settles into an average middle ground, still linked to multiple sub-levels in sync, the same way vibrating waves form a resonant pattern. And like those many waves forming a collective average, there are always some higher, some lower but most in the middle with a mean, median, and mode value. Though our states of being continually ebb & flow with multiple modes and patterns, they always possess an average composite. The overall form is a settled resonant pattern, reinforced by the way in which we identify with them.

The same general pattern in Individual states of being is also present in collective entities; organizations, groups, towns, cities, or nations. Group consciousness also has average settled levels that define a distinct state. A Variety of diamond shapes can be used to differentiate various groups and their respective collective states. As noted earlier, *Spiral Dynamics* addresses this by identifying group consciousness tiers representing how advanced a particular collective is. We can simplify this concept of societal development as follow:

Societal Collective Consciousness Progression

Post-Modern (*world-centric*)	Inclusive, humanity	*All*	
Modern (*ethno-centric*)	Tribal, clan, nation	*Us*	
Pre-Modern (*ego-centric*)	Self-absorbed	*Me*	

Consciousness

* *Members in any group or society may be higher or lower than the overall group but it's the average composite that matters. Adding members lowers or raises the level of that group.*

States of being are expressed through the unique signature of your body language. Posture and tone are obvious signals of what's going on inside, along with subtle mannerism and verbal cues. Shifts in energy levels also indicate changes of state. Various modes might include hyperactive, active, tense, fidgety, relaxed, drowsy, hypnotized, asleep, comatose, etc. A human body serves as a physical canvas where inner dynamics shape the outer shell.

Physical health depends on a state of balance, achieved through avoidance of extremes. Wellness is a disposition characterized by harmony measured by degrees of fit. Its symptoms include sync, flow, and centeredness. This applies to living organic beings as well as machines, systems and entire societies. Plants, animals and humans experience a range of conditions with a sweet spot in the middle where health is maximized. Mechanical contraptions, complex systems and nations have a similar *Goldilocks* balanced midrange. In healthy states energy flows naturally, stress is manageable, and blockages are minimized. In human beings it means body, mind and soul are in sync. The World pushes and pulls, creates tensions and pressures, requiring continual adaptive management. Proper responses result in health and happiness; weak or ineffective ones lead to sickness and sadness. This applies to each internal level: *physical, mental, emotional.*

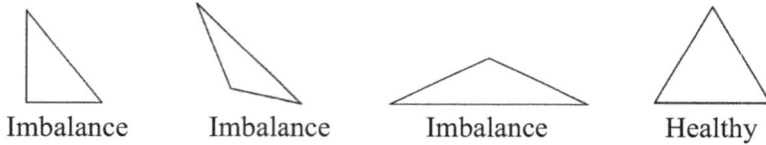

Imbalance   Imbalance   Imbalance   Healthy

Pathology is literally the opposite of health, characterized by the presence of blockage and imbalance. Bodies wear down from stress, minds fixate on trivial minutiae and emotions dwell in puddles of negativity. Fear, anxiety, depression, pain and discomfort become resonant states associated with lower vibrational energy. Blockages are created both externally and internally. Bad habits, addictions, repeated self-defeating patterns all block the flow of positive energy. They start out harmlessly in moderation; social drinking, recreational marijuana, or online porn. Extreme behavior is another culprit, whether it involves pushing physical limits, drug abuse or obsessive-compulsive actions. Addictions become repetitive endless loops, reinforcing a fixed resonant pattern. Emotional dwelling can eventually spiral downward, sometimes sinking through a cathartic sequence of layers, such as the case with the 5 stages of grief.

Stress and pain are ironically essential to health, but only when experienced in manageable portions. Too much of anything ends badly. The Seven Deadly Sins of Antiquity were simply expressions of negativity and imbalance: *Pride, Sloth, Lust, Envy, Anger, Coveting, Gluttony.* Like stress, pain is also a necessary evil that may be unpleasant but provides survival value as it limits harmful behavior. Pain provides an instant clear message shouting "Stop!" Pain like inflammation is either acute or chronic (short term or prolonged). In small dosages it ensures health. In larger dosages or prolonged small amounts, it can cause illness and possibly death.

Misdirected and blocked sexual energy is a universal pathology vexing human beings through every level of development. Like Ego, sex is an evolutionary survival mechanism – a necessary function in moderation but frequently spirals downward if taken to extremes, either too much or too little. Basic human drives can surface at inappropriate times as "civilized" man gets lulled into false pretense that higher consciousness precludes devious behavior. It doesn't. The higher your progress, the more guilt and repression come with it in response to awkward realities. This pathology arises in all classes, types, genders, races and ages. It can hold back development early on and pull one down after great upward progress.

Blocked and misdirected energy in general is the root of most pathologies. Unchecked desires, bad habits, hang ups, lack of self-discipline, etc. all represent misdirection of energy. The mind joins in by injecting denial, rationalization or just plain ignorance. Shakespeare wrote extensively on the human condition, dwelling on human flaws with universal themes that really come down to misguided energy, bedeviling even the most gifted and accomplished individuals throughout history.

Balance is key to health, along with wholistic unity. Health and healing derive from the Greek *Holos*, meaning integrated whole. Blockages, tensions, knotting or leakages anywhere results in pathology, whether it's raw energy or actual physical stuff. Holistic medicine emphasizes this relationship as health is best attained by treating the whole person. Unfortunately, modern medicine emphasizes partiality, specialization and fragmentation – defaulting to drugs and pills as an easy shortcut solution. Treatment must address the whole person; body, mind, heart, and Spirit. Note that *disease* is a measurable physical condition whereas *illness* is a general feeling of being unhealthy. Where disease is centered in the physical domain, illness can be experienced at all levels: physical, emotional, mental and spiritual. Pain and discomfort are languages of illness, providing invaluable feedback.

Health Dimensions

Spiritual	Connected/disconnected	(*free vs material attachments*)
Mental	Positive/Negative	(*free flowing vs blockage*)
Emotional	Joy/Despair	(*free flowing vs dwelling*)
Physical	Functional/Dysfunctional	(*healthy vs sick/injured*)

The Healthy Perspective is centered in the here and now, not dwelling in the past (guilt and regret) or worrying about the future (anxiety and impatience).

Internal mental states are difficult to grasp but they do produce identifiable brain wave patterns that are measurable and distinguishing. Like the spectrum of consciousness brain waves differ by both degree and kind. Changes in feelings, sense or moods manifest in corresponding wave frequencies, serving as a signature or finger print of specific states of being.

Brainwaves – A Spectrum of States respective to Consciousness Hierarchy

Gamma	- High level processing	
Beta	- Normal wake state	
Alpha	- Relaxed, steady mind, reflexive	
Theta	- Deep Meditation & Relaxation	
Delta	- Deep dreamless sleep	

Altering waves changes your state of being - Meditation, chanting, hypnosis, visualization

Memory is connected to mental states, experience, and perceptions in a 2-way integrated process. Prior experiences, images and raw information is stored and retrieved simultaneously with new ongoing experience. Current states of being are influenced by the values we attach to those memories we've stored and retrieved. Information is rarely processed in raw form. It's normally enhanced, enriched, and infused with emotional values, making living beings different from automated AI imitations. The difference is similar to black & white vs color. Our minds process experiences combining mental and emotional language that can be stored and retrieved with meaningful coding. At subconscious and unconscious levels this is symbolic, mythic language, translated to verbal and nonverbal form in the fully conscious state. While memory access is limited in the conscious and semiconscious states, everything ever experienced, observed or recorded is contained deep inside the unconscious realm.

Imagination also plays a significant role in mental processes and states of being. Like perception filters, imagination alters our experience of everyday reality. There's a distinct difference between what happens and what we think happened. Our memories of what happened, both long and short term are distorted by our imaginative free play where we consciously and unconsciously modify it. How we see ourselves and what we do is routinely shaped by wishful thinking, filtered by an imagination that can't help but put the best possible spin on it. Think about how you feel after hearing a recording of yourself, or seeing unflattering selfie pictures or watching embarrassing video of yourself.

Emotional States are responses to the experience of living in The World. Stresses and strains, pressures and tensions, pain and pleasure, produce emotional responses, which serve as a coping mechanism. At a rudimentary primitive level, emotions add survival value. Fear and surprise are built-in reactionary responses prompting fight or flight modes which increase the chances of staying alive. Emotional health is largely about appropriate responses to different situations. It's about maintaining a balanced, tempered disposition regardless of circumstances. Crying or laughing at the wrong times, over and under reacting, or not responding at all are open signs of emotional imbalance. Ideally, we should be joyful but most of us are not, so we seek short term pleasure as a substitute. Those fortunate enough to achieve peace and joy don't need to seek pleasure.

External environments are constantly changing, ebbing and flowing, and likewise produce fluctuating internal emotional responses in kind. This natural process connects us as living beings to everything in the environment of people, places, things and events. Correspondingly, the body and mind follow similar dynamics in sync, where outside changes cause internal interrelated aspects to change together. Lack of sync between body, mind and soul negatively impacts health, which is then expressed as an inappropriate response to the outside world.

Emotional States are resonant energy patterns of consciousness; a type of spiritual communication expressed in real time, in distinct qualities along a definable spectrum. It forms a scale with extremes between positive and negative, with variations of intensity. At a basic level there's polarity between happiness and sadness, "good" and "bad", along with their corresponding physical expressions of pain and pleasure. But that's just the beginning. Within the spectrum are mixtures, variations and nuanced aspects forming an entire kaleidoscope. Emotions emerge out of the dynamic interaction between a soft sensitive soul and a harsh outside world. Several models capture the spectrum of emotion and the relationships between each, including Plutchik's symmetrical wheel of 8 primary emotions: joy, trust, surprise, sadness, anticipation, anger, and disgust. These are delineated as 4 pairs of opposites, listed below next to the comprehensive wheel of emotions. His particular model mirrors the color wheel with primary, secondary, tertiary and compound emotions. Note how both color and emotion express thousands of variations through unique combinations of just a few primary ones.

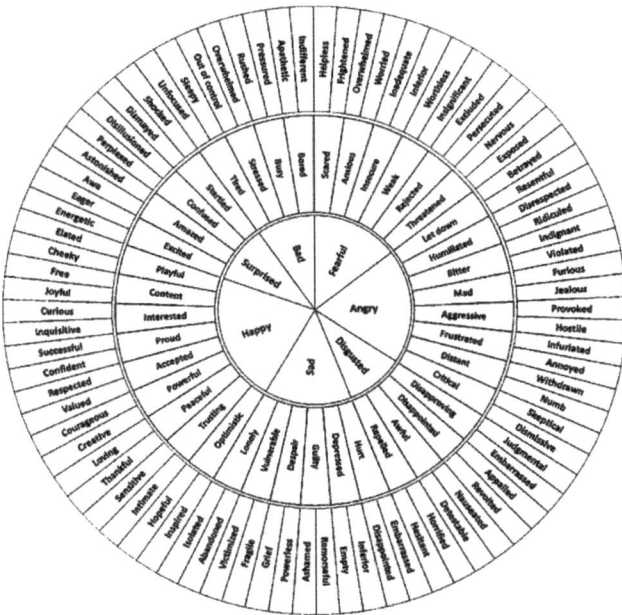

Plutchik Model
Opposites/Complimentary

Joy	Sadness
Trust	Disgust
Fear	Anger
Surprise	Anticipation

Emotion Derivatives

Primary
Secondary
Tertiary
Compounds
Mirrors color spectrum

*7 Root Emotions in this model, all other emotions are derivatives.

Higher states of consciousness translate to positive emotional expression. Joy and love are the natural foundational state of a healthy, centered being. However, "being" in The World means exposure to continual pressures, stresses, blockages and distractions – pulling your level down, pushing your center out, disrupting your natural state. Fear, guilt, and anxiety seeps in, lowering the overall level.

Falling in or out of love is another way of saying change of state. Initially it can be euphoric but never lasts because…the real world persists. The fall is a temporary transcendent experience. Euphoric love is magical in the moment, but not sustainable…freezing out rationality, which leads to friction as you navigate daily reality. It ultimately becomes an unhealthy state – a mania of glazed eyes caught in romanticized illusion doomed to disappoint. The unmistakable look, the glow, the body language, all express a shift in consciousness, dramatic but temporary. In the end, a rebalanced body-mind-soul is necessary to healthy coping.

In a state of love, boundaries between you and everything else dissolve. Love unifies. Attraction is the urge to merge…to get excitedly closer. Sex is the physical pursuit of temporary unity. Oneness and love are synonymous; the connection and connectedness you feel whether with people or things. Symbolically it's a circle with no beginning and no end; a continuous connected line and shared center.

Love, like God, is hard to define. Awkward man-made definitions are simplistic and limiting when defining the transcendent. Despite its highest pure form, love is unavoidably constrained by man. Human beings act as limiting filters transforming and constraining spirit. The love we experience as beings in The World is always merely a derivative form, which we identify with as best we can. Each of us expresses it uniquely and if fortunate enough can resonate with it to momentarily become one. When channeled coherently it becomes a powerful force.

Spiritual states transcend body, mind and emotion, resonating in mystical experience that's only intelligible in mystical states. There's an orientation of higher vibration tuning into an extremely subtle source. It's love centered – shifting away from egocentric fixation, attachment, triviality, moodiness, drama, and fear. You become more connected, stable, balanced, centered and embracing; experiencing unity and atonement (*atONEment*), to transcend space and time. Here you look beyond the physical dimension to seek wisdom and universal truths.

Spirit like emotion emanates from the soul. Its derivative forms are expressed as flowing emotion and patterned thought. It's the direct source of intuition, much less constrained than either thought or emotion which filter through a human psyche. Intuition is a higher sense and a conduit to timeless, spaceless dimensions beyond here and now. Spirit-soul-mystical-divine-ecstasy-passion are all related.

Spiritual states are detached from constraints of the physical world and attachments of living in The World. They're characterized by peacefulness, happiness and general bliss. *Peacefulness* comes down to being at one in the moment, here and now. Acceptance of things as they are. Liberation from time bound constraints, materialism, and externally programmed wants and desires. *Happiness* is more conditional, linked to desires and wishes. It's also more about your perception of progress toward a goal than the actual achievement. It fluctuates during the pursuit of a goal: closing the gap increases happiness while predictable setbacks do the opposite. Finally getting there feels good at first but wears off with time. *Peace* is experienced by simply being, whereas happiness persists in the act of doing. The key to happiness is in actively pursuing one singular thing you wish most for while making everything else secondary. You know you're on the right path when you feel centered, completely living in the moment. Wherever you are you can be there totally.

Spectrum of Spiritual-Emotional States of Being

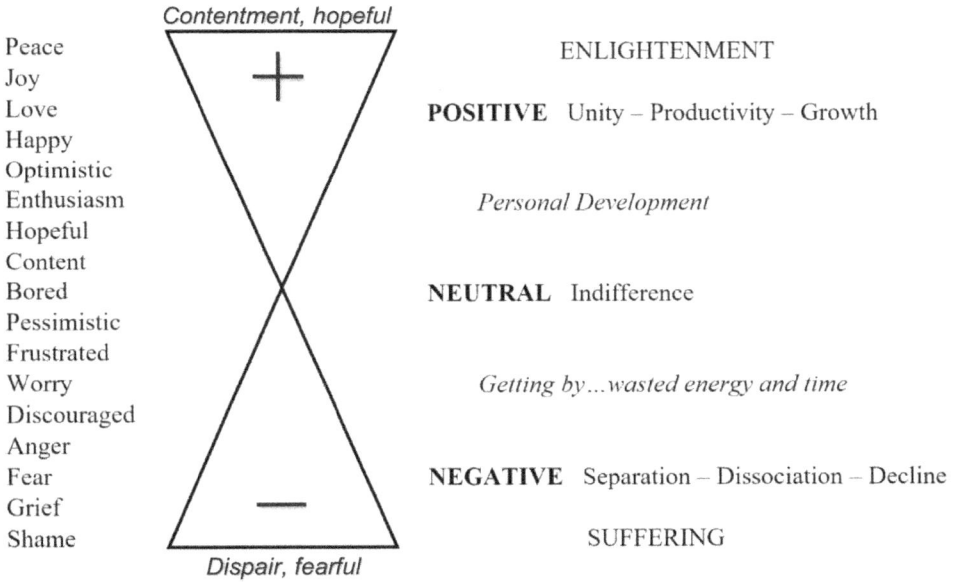

Contentment, hopeful

Peace	ENLIGHTENMENT
Joy	
Love	**POSITIVE** Unity – Productivity – Growth
Happy	
Optimistic	
Enthusiasm	*Personal Development*
Hopeful	
Content	
Bored	**NEUTRAL** Indifference
Pessimistic	
Frustrated	
Worry	*Getting by…wasted energy and time*
Discouraged	
Anger	
Fear	**NEGATIVE** Separation – Dissociation – Decline
Grief	
Shame	SUFFERING

Dispair, fearful

*Peace, Joy are higher than Love: we fixate on Love in the West, missing the higher prize
**Like brainwaves, emotions too have higher and lower vibrations on the spectrum above

Spiritual health or pathology depend on knowing and feeling your sense of purpose as a whole being – in sync, balanced and centered. Spiritual well-being is expressed as a high degree of self-authenticity, no longer egocentric but shifting away from selfish to service, free from external distractions and constraints in The World. Pathology arises from self-delusion and misguided use of energy (spirit). Wrong paths are followed, beset by shiny objects, attachments and short cuts. It's easy to veer off, falling into a separated isolated place, with lack of direction, meaning or purpose. But it's never too late to get back on track, recenter and rediscover your true path and destiny.

"You know what the secret of life is…1 thing". That's great but what's the 1 thing?
"That's what you gotta figure out" – Curly, *City Slickers*

Our state of being spectrum varies by degree and kind. Along the scale are nodal points or resonant structures that form distinct levels. These stable modes can shift in any direction, naturally or via artificial influence. Altered states occur through a variety of stimuli including drugs, music, meditation, physical activity, or major emotional events. They can be self-induced using biofeedback, hypnosis, drugs or trained using repetitive action, visualization, practiced intention (prayer, chanting, dancing, singing) or other intentional rituals that can be supercharged when infused with passionate emotion. The result may lead to a peak experience, religious epiphany, or being *in the zone*. Of course, changing your state can occur by simply trying new things and entering unfamiliar territory. Just push the limits of what you think you can do and enter uncharted realms. Travel to far-away strange places, socialize with new groups and take risks. Real change only takes place near the edge of your comfort zone.

Natural and man-made drugs are a gateway to altered states that even ancient people embraced. Transcendent states were achieved using natural herbs and remedies, many of them rediscovered in our modern era. Unfortunately, drug use today is often practiced by users who are uneducated, unprepared or just plain irresponsible. Higher inner states are insulated from easy access for good reasons – short cuts are hazardous and can lead to costly setbacks, sometimes permanently. The human brain is a supercomputer running all the time, keeping us sane in a world of chaotic stimuli. Drugs interfere with natural mental processes, in a sense, removing built-in safety valves.

Psychoactive drugs including stimulants, sedatives, narcotics and hallucinogens shift states higher or lower (uppers & downers), temporarily mask pain and alter perception. Uppers include Caffeine, Nicotine, Amphetamine, Meth, Cocaine – (all euphoric drugs). Downers are: Alcohol, Valium, Xanax, Barbiturates, Klonopin, Marijuana – (Depressants). Pain Relief drugs include Vicodin, Oxycodone, Fentanyl, Codeine, Morphine, Heroine – (Pain killers). Psychedelics: LSD, Mushrooms, Mescaline, Peyote, DMT, PCP – (Alter Perception).

Every drug can change your state however psychedelics open the doors of perception to realms far beyond ordinary consciousness. Hallucinations change reality in a completely transcendent manner. Straight lines curve, warp and distort; space and time dissolve into a murky Jello-like dimension. Super sensitive experiences emerge with exaggerated images and sounds. Vivid, rich color kaleidoscopes dominate the foreground while once familiar objects appear in weird shapes and sizes. Static things become animated, coming to life with blended attributes; smells, colors, shapes, sounds and forms fuse together. The experience *is* real, *feels* real and makes an unforgettable impression that usually lingers post-hallucination. Repetitive experiences of this type will gradually shift a state of being, opening one to higher levels. While psychedelics may be short cuts to higher consciousness, bypassing natural barriers must be engaged with caution. Note the similarity between dreams and psychedelic states where space and time have no limits, objects are distorted and feelings move the action.

Meditation is a natural process of state change that works by suppressing the mind. Limiting the necessary daily control structure frees one's attention to transcend it, opening up higher levels of awareness. Silencing inner thoughts and external distractions unshackles YOU the observer-experiencer. Meditation is a passive, letting go, whereas religious prayer is an active, ritual based practice. Both approaches require repetition to further progress. Resonant patterns reinforce higher states that persist and become new natural levels. Meditation is just one method of silencing the mind: Mantras, art, sports, rhythmic activity, and music can induce states that transcend mental constraint.

Dreams are a mode of consciousness, bridging the waking state with inner unconscious depths. They're products of duality, linking outer world with the inner – a sort of midpoint between physical and non-physical realms. In one the mind rules, in the other the soul abides. Dreams are confusing because the language seams foreign. Words are replaced by symbols, enriched with psychic emotional value, difficult to translate. Abstract, symbolic and mythic language makes the experience hard to remember. One thing is clear; dreams are always personal to us.

Dreaming is another coping mechanism where we internally process the daily experiences of external stress and conflict, sorting out gaps between wishes/desires and actual reality. Repressed emotions, buried fears and ugly truths hidden from the conscious mind are lurking around in here. Personal issues are processed, worked out and eventually surface, translated from the deep psyche into manifested behavior. Libido dwelling in lower levels is channeled out at some point, positively or negatively. Inner urges drive outer behavior while conversely outer experiences inform inner responses in a 2-way dynamic process – a healthy and vital interplay. External lessons learned have an internal counterpart in the subconscious realm.

Dreams reveal unconscious wishes and desires, routinely disguised in coded language. They involve active dramas within the subconscious, present even during the waking, conscious state. Lucid dreaming is a vivid form where you are aware you're in it. In this state you can consciously manipulate and guide it with intentional participation as opposed to the normal experience of a detached witness. Most dream activity is passive symbolic dialog working out unconscious internal meanings never to be recognized after waking up. It's an essential process for maintaining mental and emotional health.

At the highest states of being transcendental consciousness emerges where all senses are silenced, leaving complete stillness and pure awareness. It's a realm of enlightenment where you detach from yourself and The World and connect with spirit. This higher level may be considered *cosmic consciousness* where you feel deep warmth and one with everything. You begin to experience pure joy-bliss-peace and radiate it outward. You're now fully detached from ego, transitioning from identifying as *you* into a state of *ALL*. Higher states represent potentials available to everyone, though very few successfully navigate all the lower and mid-level stages required to get to this highest state. Lower states cannot comprehend higher ones. It's like a foreign language full of gibberish. Comparing lower states with higher ones produces paradoxical realities.

Higher states correspond to higher levels of awareness, greater depth, clearer understanding, and a more authentic experience. In this realm you're less mind and more soul – a resonant pattern of spirit. You're present in the intersection between a living being and the divine original source of all spirit. Your soul is a localized resonant pattern out of infinite spirit, manifesting into the physical realm here and now bound by space and time. It's beyond culture, race, gender, or any other worldly construct. Your soul is transcendent, informing mind and body without being trapped by either, just as mind transcends the body. Soul enables formless spirit to experience itself, to transition into manifest form, to generate purpose and to establish contextual meaning. It interfaces with the physical realm via organic vehicles we call bodies. Its interaction with experience in The World is expressed as a wide variety of emotions. The closest we can come to describing a soul is through the concept of personality, which is really just a practical interface linking the divine essence of a spiritual being with the constraints of a structured physical world – yet another coping mechanism of a dual reality.

"You don't have a soul…You are a soul. You have a body." – C.S. Lewis

Ever wonder what's it like to experience the highest levels? What changes when you transition from lower ego states to higher detached witness states? For starters you experience greater truth and authentic reality. You experience yourself as you are, not what you wish you were. You see The World as it is, not what you've been programed to believe it is. You feel and project love in its purest form, not lower, conditional variations mired in ego desire, attachment and selfishness. Like dreams and psychedelic realms, these higher states cannot be coherently compared to lower, ordinary everyday experience. You attain a higher vibrational resonance incompatible with lower levels, making the transcendent shift comparable to entering a foreign land. You detach from The World, centered in the present – living in the moment, not past or future. And while detached from The World you can now fully function in it without being of it. And this is the secret cypher that empowers you to change where you are on The World Spectrum – shifting away from the negative, malevolent end and towards the positive benevolent end. YOU become an intentional projective force that overrides and guides the indifferent potentials of the exterior world *out there*.

The few who experience higher states do so on a temporary basis. Whether naturally or induced, higher states must be earned in the same way an athlete persistently trains and reaches an elite level. Momentary peak experiences may occur just as any amateur can suddenly feel in the zone but it rarely lasts. Normal states quickly settle back in and the experience is fleeting. Higher states are transpersonal, including peak experiences, out of body experiences, near death experiences, altered states via drugs or natural causes and fleeting moments of glimpses beyond one's current state of being. Temporary connections with higher consciousness may occur unexpectedly involving awareness beyond the norm. Psychic premonitions may come out of nowhere along with rare cases of déjà vu but they don't last. Sometimes higher states are confused with simply tuning into inner messages from the subconscious mind which is indirectly linked to higher states. Similarly, intuition and aha moments are products of a deeper, timeless realm seeping into the conscious mind during an ordinary state.

So, what's the *highest level* really like? Sorry, but it can't be explained in words; it must be experienced directly. Like mere mortals trying to describe God, it's an exercise in futility. With imagination you might get a vague sense of this incomprehensible state where empty, formless consciousness permeates everything. You observe it as a detached witness, seemingly part of it, connected to everything yet bound by nothing.

Ironically, higher planes aren't somewhere else out there but always within us and around us, ever present. The highest states are similar to out-of-body experiences where you see yourself and unfolding events around you as a detached observer. You live as a participant in your own movie, watching dramas play out without getting sucked into them. You experience just being – living in the moment, transcending subject and object to become one with anything and everything as it happens. The highest level is characterized by enlightenment – a state of unity, wholeness, clarity and fulfilment. Attaining this state changes your outward appearance and behavior – obvious to others as you project love, happiness and goodwill. Your eyes alone are unmistakably mesmerizing as they reveal pure spirit emanating through two round portals – conduits connecting inner and outer realms with pure clarity.

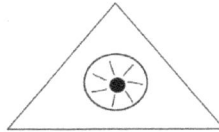

Mental clarity
Emotional balance
Physically healthy
Spiritually connected

MAN THE MACHINE

Human beings are highly complex organic machine processing centers, transforming inputs into directed action and outputs, sustained by food consumption; extracting nutrients and converting matter into energy. The process involves specific ingredients: water, sugar (carbs), protein (amino acids), fat (fatty acids), minerals and oxygen. Energy conversion enables work, growth and maintenance. The human machine integrates various mechanisms as a complex operation: converters – stomach, lungs, liver, etc., plus catalyst/facilitators –

glands, liver, pancreas, microbes, enzymes, bile, saliva. It's all chemically regulated via the intricate endocrine system:

Specialized Glands

Pituitary	– Growth and development
Pineal	– Melatonin (sleep)
Thyroid	– Growth and metabolism
Hypothalamus	– Thirst, fatigue, body temperature
Thymus	– Immunity
Adrenal	– Fight or flight
Pancreas	– Insulin (blood sugar)
Kidney	– Blood filter
Testes/ovaries	– Testosterone, estrogen

Mechanisms are in place to perform very specific functions: *filters* (kidneys, pancreas, intestines, liver), *Valves* (heart, veins, larynx, sphincter), *regulators*, *enablers*, *catalysts* and *inhibitors*. It's all interlinked by blood, which transports oxygen to every cell so energy in the form of sugar can be converted and used. Blood also serves as a conduit for hormones, nutrients, antibodies, coagulants, salts, sugar, fats, and protein. The 5 liters of blood in a human body are just one of three basic fluids supporting the whole, the other two being intercellular (15-20 liters) and cerebrospinal (1/4 liter).

The human body persists in a constant state of dynamic equilibrium, balancing forces to maintain stability. Many of the systems that achieve this occasionally exceed the optimal sweet spot, resulting in temporary imbalance. Symptoms of this include a number of auto reactions such as a sneeze, yawn, cough, cry, goosebumps, shiver, sweat, fever, or allergies. Sometimes regulation breaks down and processes get out of control, as in the case of cancer. Homeostasis is maintained through an elaborate system of feedback mechanisms that tell the body to adjust, either positively or negatively, but always seeking that middle position of healthy stability.

Balance in a human body is based on context, generally set by the external environment. Here our focus is on the physical level, however we must consider everything external, including culture and relationships. Information itself is a key input processed internally and expressed externally, seen in attitudes, beliefs, perception and behavior. Those non-physical aspects must be processed properly every bit as much as food, water, and oxygen. However, man the machine essentially runs on raw fuel.

Nutrition is the primary physical input sustaining a human body, sourced from the earth and sun, processed, then returned right back to the earth. Energy is extracted from food in 5 basic groups: fruits, vegetables, grain, protein and dairy. They're the principal source of nutrition along with other supplemental ingredients:

Carbohydrates – main source and storage of energy
Proteins – tissue growth, hormone production, enzymes, immunity, source of energy
Fat – energy source, protects organs, supports cell growth
Vitamins – help metabolize proteins & carbs, supplement other bodily processes
Minerals – supplement a variety of essential body functions, growth & maintenance
Fiber – enhances gut health and mitigates heart disease
Water – like oil in an engine, conduit facilitating digestion, absorption, circulation, transportation

Consuming food and drink to sustain our bodies is a fundamental daily ritual, routine for most but extravagant for some. Cooking can be an art form beyond simply adding energy in the form of heat to transform a bland meal into a tasty feast. Unlike our animal counterparts that merely eat the raw flesh of their prey, humans chemically manipulate food to enhance an otherwise mundane task and turn it into a pleasurable experience to appreciate. The same goes with drinks where all sorts of alcohol derivatives transform a boring tasteless gulp of water into a variety of effervescent beverages to binge on: beer, wine, whisky, rum, brandy, liqueurs, carbonated soda, juice, milk, coffee, tea, etc. Beyond fine dining we add all sorts of supplements to food to enhance taste, often at the expense of nutritional value: sweeteners, spice, coloring, emulsifiers, preservatives, antioxidants, condiments, oils, grains, powders, salts, seeds, sauces, herbs, etc. Putting taste ahead of health is a short-sighted tradeoff made worse with increasing use of unnatural processed food, synthetics, and preservatives.

Nutrition is interdependent with a variety of ingredients functioning integratively as elements influencing overall physical health. Nutrition must be supplemented and balanced with exercise, including both strength and cardiovascular activity. Exercise must then be balanced with proportional rest and sleep. Each of these physical components must be balanced between general needs and those specific to each individual, and then supplementally to changing circumstances. It's always a moving target so there's never a one-size fits all recipe for nutrition, exercise and rest; it's always individually and situationally dependent. Moreover, these *physical* factors must be balanced with the specific *mental, emotional* and *spiritual* context of the individual.

Health – Fitness Do's	*Unhealthy Don'ts*
Balanced, nutritious diet	Junk food, sugar
Regular exercise (strength & cardio)	Drugs, alcohol
Adequate rest (7-8 hours uninterrupted)	Smoking
Manage Stress, laugh daily	Prolonged viewing video screens
Socialize	Dwelling, obsessing
Move around, go outside, get sunlight	Complacency

Health is a personal responsibility, not something you just leave to a doctor. While advances in medicine are exceptional on the physical level, they overshadow holistic components (mental, emotional, spiritual) and the over reliance on drugs;

handing out pills as a shortcut to actual therapy. Natural remedies can prevent pathology and illness from arising in the first place. Holistic practices such as hatha yoga address breathing, blood circulation, balance and control, plus the nervous system as well. Health is a balance between body, mind, soul and spirit, best attained in combined efforts: diet and exercise (Body), mental stimulation (Mind), dream journaling and acts of kindness (soul), meditation and prayer (spirit).

Achieving Body-Mind-Soul balance navagating through The World requires ongoing vigilance

Engineering. The human body is a marvel of organic, natural engineering with built-in solutions for coping in a physically demanding world of daily survival. Various mechanisms are integrated with electrical and chemical systems in a sophisticated network of mutually supporting parts achieving remarkable synergy. Optimization is achieved through a cantilevered mobile structure serving as a flexible, adjustable shell, with strength and agility in just the right balance. Clever use of shaped joints, levers (feet), hinges (hands), arches (pelvic bone), and flexible tension (cartilage, ligaments & tendons) provides efficient solutions to physical-mechanical demands. Muscles working in pairs convert chemical energy into work. Leveraged bones operate in proportional geometric ratios. Space is maximized through coiling of tubules and spiral wrapping of tissues and organs (lungs, intestines, brain), while optimizing surface area through branching forms (veins, arteries, capillaries, synapses, etc.).

The body is literally sculpted by external forces of nature, achieving parsimony through dynamically balancing physical forces from every direction with an equal counter force, accomplished through geometric form that optimally accommodates those forces. Mathematically, the golden ratio permeates the body structure, reflecting natural imbedded connectedness between ourselves, nature and the cosmos. Symmetry is omnipresent, providing elegant geometry to organic engineering. Bilateral symmetry provides greater value with 2 eyes (binocular vision), 2 ears (directional sound), limbs in pairs that enable balance and function, and even the duality of left and right brain. D'Arcy Thompson, who pioneered the study of organic forms wisely observed "The secret of life isn't heredity: it's engineering, math and physics". He introduced allometry as the analysis of relationships between body size, shape, anatomy, physiology and behavior. This basically connects the dots between life forms and external forces of nature.

Human bodies routinely demonstrate bio-mechanical poetry in motion, from simple, elegant movement in daily activities to the exceptional displays from professional athletes pushing limits in every possible way. Bodies effortlessly flow through time and space in geometric harmony, revealed in balanced movements of various dance forms, art works and synchronized swim. Ballet steps and twirls are mirrored by a tennis player's graceful backhand swing, a diver's twists and turns, a basket baller's flying dunk, or a gymnast's artistic floor routine.

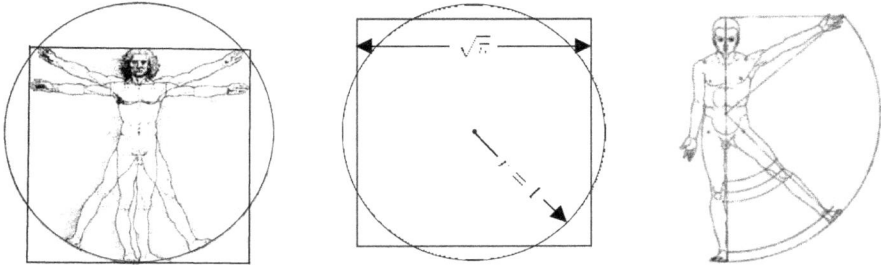

At the other extreme is the body's powerful output demonstrated in field events: sprint, long jump, shot put, high jump, pole vault, hammer throw, etc. Great force is built-up inside and explosively directed outside. This brute force is complimented by a tuned-in athlete's ability to gracefully merge with the apparatus in play. Consider the yin-yang connection between a high jumper literally flowing around the pole, crossing over it while their center of gravity actually passes under it. Notice the similarity between force generating motions of a baseball pitcher, tennis server, batter swinging or golfer's drive. The human body is simultaneously integrated within and without, connecting dynamic internal forces with mirrored external counterforces on all sides. But don't forget this is just the physical aspect – a body is always integrated with mind, soul and spirit. And when all are in sync, like the case of "being in the zone", remarkable physical feats become possible.

Bodies come in a wide variety of sizes, shapes and weights, demonstrating the versatility and adaptability of its organic design. Extremes come in every dimension: tall/short, fat/thin, squared/curvy, heavy/petite, flabby/athletic, or particular exaggerations of slender/stocky, muscular/chiseled, etc. Body types are influenced by a combination of age, gender, race, heredity, diet and activity. Health and attractiveness are based on the goldilocks principle of proportionality; not too much in either extreme.

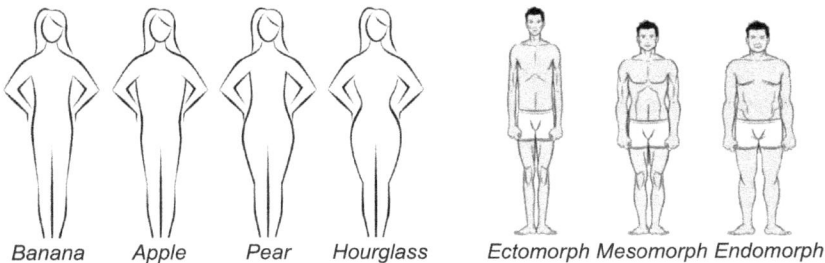

Banana   Apple   Pear   Hourglass   Ectomorph Mesomorph Endomorph

Process Nature. Human bodies are more fluid than stable form; an ongoing process giving the appearance of a persistent structure. It's just an illusion. You're less of a noun and more of a verb – a work in progress that just keeps going. But because forces and counterforces remain in a delicate, near-balance state, you appear to be a static form. Ironically, living beings achieve stability through balancing dynamic, chaotic forces.

Homeostasis is the biologically stable state resulting from counterbalanced internal pressures and forces – a sweet spot where living bodies facing tumultuous demands maintain a healthy disposition. It's a perpetual dynamic balance enabled by continual cell replacement, fuel conversion, temperature regulation, injury repair, air intake and expulsion, blood fluid circulation and waste material disposal. Metabolism is just one subset of this balancing state where food is processed for energy, work, growth and sustaining processes. Another continual balancing aspect of the body is the skeletal-muscle tension dynamic that fluidly shape-shifts a human form between moments of stillness and periods of great athletic activity.

The body is a complex process system maintained by integrated regulatory mechanisms. Much of the order is directed and controlled chemically through glands in the endocrine network. Built-in sensors send chemical signals that provide feedback loops to regulate all the ongoing processes. Maintaining balance is the prime goal. Mechanisms to achieve that include filters, valves, pumps and sieves enabling fluid filtration, absorption and excretion.

As activity increases the body adjusts in kind. Breathing pace goes up, heart beats faster, blood sugar level changes, hormones are released, pupils dilate…all adjusting in synchronized response. Balance is never just about a single state – it includes multiple conditions, dispositions and changing states of being. It's always a moving target, accommodating a broad range of physical limits where the body dynamically adjusts to an ever-changing external environment.

Processes and rates of change occur in a variety of cycles, the most common being the 24-hour circadian rhythm. Daily cycles in the body include a wake-sleep rhythm, body temperature changes, hormonal cycles, blood count and much more. Other natural cycles are shorter or longer than a day yet play equally important roles. Cycles reflect the rhythmic nature of life's ongoing changes in activity, intensity, energy levels and states of being. Biological clocks keep us connected to nature to maximize our ability to harmonize with the environment, essential to our health. Natural cycles are expressed in weather, tides, moods, seasons, and social movements. Cycles connect all life and are embedded in the universe as a whole. The simple act of a heart beating or breathing in and out mirrors the divine process of spirit creating life, moment by moment, from nothingness into somethingness. But most importantly, cycles are a key aspect of maintaining balance in any physically dynamic equilibrium system, and our bodies are no exception.

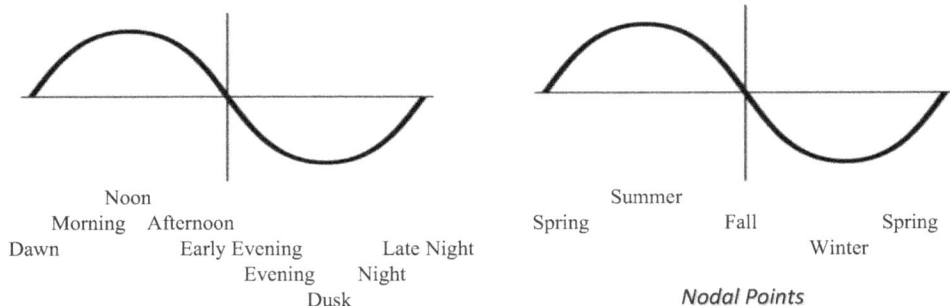

Noon
Morning   Afternoon
Dawn   Early Evening   Late Night
                        Evening   Night
                            Dusk

Summer
Spring   Fall   Spring
                                        Winter

Nodal Points

Biological processes work to establish homeostasis, optimal metabolism, and bodily health. Homeostasis is just a fancy way of saying "balance", which the ancient Greeks identified as the key to a healthy life thousands of years ago. They espoused the "golden mean" in everything and recommended avoiding extremes. Our bodies naturally abide this wisdom, coping with stresses and strains of daily life while avoiding too little or too much of it. Whenever imbalance emerges the body readily adjusts. We're built to adapt to ongoing change. Stress and illness are normal and generally self-correcting. Sickness is really a feedback language, providing a necessary signal for the body to adapt. Most medicine simply treats symptoms while the body itself does the heavy lifting.

Anatomy. A human body's essential parts are wholes with subsets of much smaller components. Inside each of us are trillions of cells, about 600 muscles, 200 bones, and 80 rgans. Despite our solid appearance much of the human body is composed of various liquids in the form of juices, lymphs, plasmas, blood, bile, acids, enzymes, tears and miscellaneous lubricants. And the parts that seem to be solid are actually quite malleable, even bones. Your stable "solid" body is more of a flowing process – a fluid work in progress. Ongoing cell generation literally pushes the body outward, including hair growth, nails, teeth, and skin cells pushing upward to replace dead ones. Similarly, the body is continually releasing fluids and gases to regulate temperature or release waste toxins. Your body's fluid nature is accentuated by the continual process of self-destruction and reconstruction where every day, hundreds of billions of cells routinely die and get replaced.

Cells are the fundamental substance supporting the entire system. Every part, mechanism and organ are composed of cells that function both independently and collectively. Specialization of cells includes red blood cells that have no nucleus, nerves which carry electric signals, adipose to store fat, inner ear hair to sense vibration, muscle cells which contract to move bones, epithelial that form barriers, eye cells operating as photoreceptors, sperm and egg made of unique reproductive cells, and many others serving specific functions unlike the rest. Somehow, it's all orchestrated as one interconnected, harmonious whole, a wonderous enigma.

Specialized cells collectively form to create specialized organs. They in turn form larger interconnected networks functioning as specialized systems. A human body is a supersystem integrating individual parts into a cooperatively functioning collective supersystem of systems.

HUMAN SYSTEM OF SYSTEMS

Skeletal. Supportive structure – Maximizes tradeoff between strength and mobility. Interdependent combination of bones and muscles – structure and movement. 206 bones in a human body. Engineering function provides strength, flexibility, protection and blood cell production. 600+ muscles that can only retract work in unison with each other (force-counter force) enabling all ranges of motion. System of levers and joints efficiently maximize lines of stress, tension and compression. Variety of joints creatively meet the needs of a system requiring strength and agility, including ball and socket, gliding, saddle, pivot, hinge and ellipsoidal. The skeletal array of a human body is one of nature's shining achievements in biological engineering, using hinges for fingers and toes, a rib cage and skull for protective enclosure, a foot designed cantilever bridge, and a ball-socket hip joint to enable a great range of motion while supporting heavy weight stress. Most impressively, bones are living structure that continually break down and self-renew – adapting shape and proportion to a growing body subject to the ongoing stresses and tension of daily load-bearing forces.

Circulatory. Simple fluid transport system moving supplies and waste back and forth. Blood's a key element; an evolutionary offshoot of primordial ooze, consisting mostly of water to transport oxygen, nutrients, waste, hormones, antibodies and anything else needed by organs, tissues and billions of cells. Blood moves via pump (heart muscle) and collects oxygen via breathing lungs (diaphragm muscle). The human circulatory system is a liquid 2-way transport network of arteries and veins, most side by side like a modern traffic highway with vessels and capillaries similar to shipping ports

Endocrine. Complex network of glands and organs regulating subsystems within the body. These are the hubs where the circulatory system transports chemical messengers to control metabolism, energy, growth, development, reproduction and response to injury, stress and mood. Glands in the head, neck and torso are all specialized, operating through a feedback mechanism of positive and negative responses, which either promote or inhibit. Regulatory hormones and chemical messengers of the endocrine system could be compared to money in an economy where profits and taxes incentivize or limit business activity.

Digestion. Energy must be processed continually to support human growth and activity. Food is transformed into chemical energy where nutrients are ingested, moved, digested, absorbed and leftover waste expelled. Once food is eaten the process of breaking down solids into useable energy is continuous and layered in stages, including the mouth, pharynx, esophagus, stomach, small and large intestine, rectum and anus. Each step of the way food is broken down with assistance from glands in the mouth and liver, plus pancreas and gall bladder. The stomach caries most of the load and to accomplish its task it requires a lining of hydrochloric acid, which is replaced every 3 days.

Nervous. Human nervous system is a complex communication network for sending and receiving messages. In addition to chemical messengers, electrical impulses are used to signal information. The network comprehensively extends throughout the entire body in a meridian matrix; a conduit for ongoing communication. It's basically a 3-part system consisting of a central control hub brain, lines of communication, and sensors/receptors that send/receive impulses.

The human **brain** is the most complex organic "machine" ever created. It evolved in stages, beginning with a primitive reptilian form focused on basic survival needs and automatic functions – breathing, heartbeat, temperature regulation, etc. Next the limbic brain developed, expanding the ability to respond to environmental stress and threats. Part of this progress involved the emergence of emotional sense and response, adding further value as a means to cope in The World. Finally, the neocortex evolved, greatly expanding man's ability to rationally think and learn. It's no accident that this cerebrum is by far the largest part of the brain and sets human beings apart from all other organic life forms. All three parts function in an integrative manner as well as independently.

Human brains are further divided into two halves, Left and Right brain. Duality is embedded right into the structure, accommodating thoughts and feelings, science and art, objective and subjective reality. Moreover, the brain is hardwired at birth yet flexible to change with life experience – physically malleable to adjust via learning. Brain development confirms that size is not as important as the interconnections forming new nodes and passageways that emerge from engagement in an ever-changing environment.

*The whole human body is an intricate communication system: electric nervous system, chemical gland system, sensory receivers, mouth transmitter, DNA coding language, body language, feedback mechanisms, eyes-ears-nose-tongue-skin, tears, eyebrows, etc.

Senses. The brain and nervous system rely on sensors to provide internal and external information updates. 5 primary senses include sight, hearing, taste, smell and touch, are supplemented by a variety of secondary senses such as balance, time, movement, pain, temperature, acceleration, etc. A multitude of sensors accommodate a wide range of stimuli: Eyes perceive light between 400–700 nanometer wavelengths, seeing with more megapixels than the most advanced man-made camera. Ears hear sound between 20-20000 vibrations per second. Your nose can distinguish up to 1 trillion different odors within 10 basic categories: Fragrant, woody, fruity, citrus, sweet, minty, toasted/nutty, chemical, pungent, decayed. Your tongue distinguishes a range of tastes between 6 particular types: sweet, bitter, salty, sour, umami, ammonium. And touch includes a subset of different perceptions including pressure, temperature, tickle, tension, pain/pleasure, etc.

All senses interact and work together to provide the brain with a consistent, reinforcing interpretation of external reality. Comprehensive information pours in constantly with great variation yet it all breaks down to simple binary impulses, like a computer where every data point is either a 1 or 0, yes or no. Everything we experience and perceive is ultimately an interpreted reality by the computer we call a brain.

OTHER SUBSYSTEMS

Outer membrane. Skin is the largest organ of the human body, accounting for about 15% of your weight. Functioning as a protective armor, our exterior layer shields us from all sorts of micro and macro threats. The outer cover includes hair and nails, and maintains a resilient, tough yet flexible surface that replenishes itself through constant cell replacement. It also "breathes" air on the surface and lets liquid pass through pores, releasing toxins and enabling cooling off through evaporation. Skin is further adaptive in how it changes color to mitigate UV rays and heals surface wounds in response to injury.

Defense.   The body has numerous built-in defense mechanisms. Skin is just the first level, secreting anti-biotics to inhibit fungi and bacteria. Exterior level defense includes mucous membranes in the mouth, nose and genitals. Further inside, the immune and lymphatic system fend off a daily attack from bacteria, viruses, protozoans and a variety of foreign substances. Interstitial fluid filters and removes diseased agents. Antibodies and antigens recognize invaders, kill with extreme prejudice and dispose them. White blood cells attack invaders, clean up damaged cells and dispatch waste material. Healing is a supplemental defensive process where blood coagulation enables closure and repair of damaged tissue.

Reproduction.   Human survival includes both maintaining individual life and preserving continued collective life. Unlike other organic forms that duplicate without variation via asexual means, humans mix genes via sexual reproduction, enabling greater adaption to changing environments. Male and female bring equal assortments of traits carrying 23 chromosomes each to combine and create a unique composite whole. Once gamete formation and fertilization occur it's followed up by a mysterious process of cellular division, differentiation, specialization and reorganization. After 4 weeks the embryo is 500 times larger than it began. The mother carrying the new human being is both connected and separated from it using a placenta filter allowing nourishment without intermingling blood. Further connected nourishment continues after birth via the mother's mammary glands.

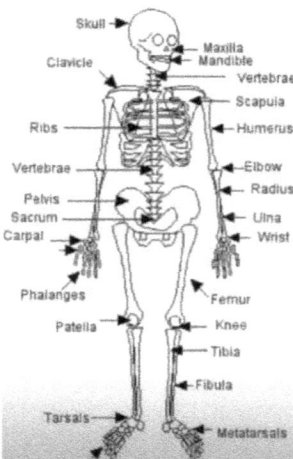

Skeletal system   Nervous system   Endocrine system

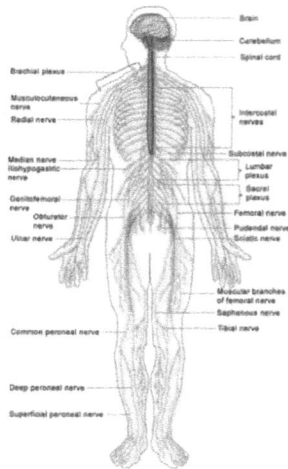

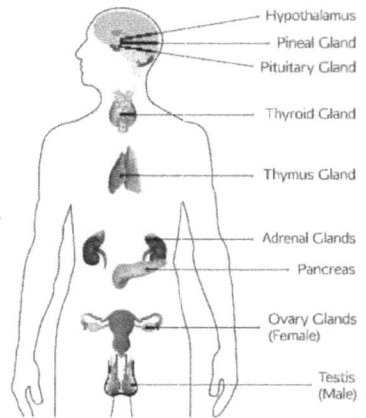

Summary. Human beings are a comprehensive supersystem of systems operating as a holistic integrated machine with both independent and collective functioning parts. They're fragile souls encased in a rugged physical vehicle produced from basic materials proportionally derived from the earth. They're temporary patterns in time and space, mirroring the external environment from which they emerged. Parts are specialized, functional and independent yet work cooperatively in unison with the collective whole. Human beings are dynamic processing center which transform energy, maintain stability and generate power to act, grow and adapt to the outside world; a unique living microcosm serving as a small cog in the greater systems of systems we call the cosmos.

Assimilation – Movement – Growth
Metabolism – Senses – Fuel conversion

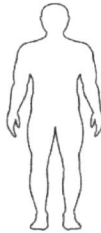

Reproductive system   Muscular system
Circulatory system   Integumentary system
Endocrine glandular system   Lymphatic system
Digestive system   Urinary system
Nervous system   Immune system
Skeletal system   Cardiovascular system

Transportation – Excretion – Homeostasis
Reproduction – Communication – Defense

Comprehensive integrated supersystem of systems with dynamic equilibrium processes

MAN *THE* MICROCOSM

Man occupies the center of the Universe, in the middle of Everything; in other words, YOU are in the middle of Everything. Human reality is the experience of everything inside and outside us, above and below, before and after, visible and hidden. We're like the center cog of a giant machine where each gear turns in sync with the others. Think of yourself as a single cell in a living universe where all cells are connected parts of a larger single organism. In each case we find a series of parts and wholes, micro and macro versions of the same essence interconnected with everything else.

Our universe is a self-reflecting mirror where form and contents are repeated at every level. Any being present in such a construct must contain the same essences from which it came from. Man is a mimicking template – a comprehensive model representing a compact microcosm of the greater whole. The universe is literally imbedded in man. Every omnipresent essence and attribute of physical reality is naturally found and expressed within and through human beings.

In review, the five primary essences of reality are *Substance, Order, Purpose, Process and Relation*. Each of these are ubiquitous in human reality, at every level, state, expression and experience:

Substance. The stuff of man includes the body, component parts, cells, tissue, organs, fluids, fuel (food), proteins, carbs, blood, bones, muscles, etc. Raw materials in man are present in symmetrical proportion to mother earth, where percentages of trace elements and the ratio of water to land are mirrored within our physical form. Everything inside you was first made within stars, making us a derivative product of the central feature of the universe.

Order. A human brain is no doubt the most complex computer ever created, serving as the fundamental source of order within us. The mind filters and manages nearly infinite stimuli, providing sense out of a fragmented external world. It manages internal systems, makes intelligent decisions, learns, recalls memories and imagines possible futures. Order is expressed in networks within the body, organic structures, bilateral symmetry, internal mechanisms and regulatory functions.

Purpose. Human beings are driven by needs, desires, motivations and values, with higher consciousness seeking growth and development with goals and a sense of direction. External behavior driven by internal motivation becomes purposeful action. Needs and wants charged by emotional energy pursue desired ends.

Process. Man is literally a process center, churning raw materials internally into productive energy for growth and sustainment. Non-stop internal dynamics include operating bodily systems, thinking and intuition, emotional responses, physical reactions and changing states of being. External pressures from The World require constant adaption. Your journey through it all shapes and transforms you with each demanding new experience; an ongoing evolutionary process of learning, coping, growing, developing and ultimately transforming.

Relation. To comprehend man is to see the whole being. Personality is your composite whole linking a multitude of aspects like age, gender, race, culture, education, language and many other factors. Identity is context based, dependent on relation to everything else. Who you are depends on external relationships to others; friends, family, partners, situations and changing conditions. Every one of us is a unique being whose identity is based on relational reality.

Universal Attributes of Physical Reality Imbedded within Man.

Beyond the primary essences imbedded in everything there are plenty of other universal attributes present as well. These subsets of the higher essences are also built into the fabric of physical reality, and therefore, built right into YOU.

Hierarchies. More accurately defined as Holons, there are sets and subsets of whole-parts found everywhere we look. This universal pattern/structure emerges in man as the organic hierarchy of cells-tissues-organs-systems; each aspect being both part and whole at succeeding levels. And then Whole man becomes just another part in further hierarchies of family-clan-town-city-state-nation-world.

Systems. Like Hierarchies/Holons, systems are an integrative principle connecting parts in a synchronized manner creating synergy. Man is a supersystem of systems with sets and subsets of parts, mechanisms, processes, networks, and synchronized operations – a literal mini version of the universe.

States of Being. Everything in physical reality changes state, either short or long term. Man perpetually changes state in a continual fluid manner with emotions, thoughts, sensations and general disposition. It's a continual flux as moods shift, attitudes change, and energy oscillates – both physical and spiritual. Health is never fixed, always varying by degrees. Your internal state is a moving target along with the outer environment where conditions change with equal fluidity.

Intelligence. The mental quality we identify with the human brain/mind is a universal attribute imbedded within everything. Even particles possess a rudimentary intelligence directing specific actions, leading to preferential responses. It's the subtle force directing particles to form into atoms, which in turn form into molecules, mixtures, organic cells and complex structures – the very foundation of physical reality. It's linked to the subtle force working inside us, especially the nervous system and its most concentrated complex part, the brain.

Symmetry. All of nature's mathematic ratios and relational patterns are present in man including π, spirals, the golden ratio, branching, meanders, packing, explosions, radial and bilateral symmetry, etc. The Fibonacci series is ubiquitous in our bodies in a variety of forms. Balance and order are visually obvious in facial symmetry, shape, proportions and mirroring aspects of most body parts.

Duality. Everything possesses dual attributes of the physical and spiritual. Our universe expresses this with matter and energy at the lowest level, living beings at higher levels. Man is a 2-sided creature with feet on the ground and head in the sky, balancing two competing aspects at all times. Daily experience in The World is mirrored by a shadow dream realm, manifested in a waking/sleeping framework of conscious/unconscious states. As a soul incarnate man must balance inner spiritual needs with external physical demands. Coping and adjusting are continual, sorting through dichotomies of mind/body, logic/feelings and objective/subjective perception; a process supported by a built-in Left/Right brain structure. Ongoing tension and rebalancing persist as earthly attachments and distractions conflict with spiritual fulfillment, through alternating cycles of being and becoming.

Human life is a process of working through Duality Dynamics, where inner forces clash, adjust, respond and filter through outside pressures. When inner-outer forces align in harmony, moments of synchronicity emerge, confirming you're on the right path. Just as planets and stars follow a natural rhythm within the cosmos, so too does man, periodically achieving centered balanced internal synchronicity.

Spectrum of Human Reality
A scale of universal experiences

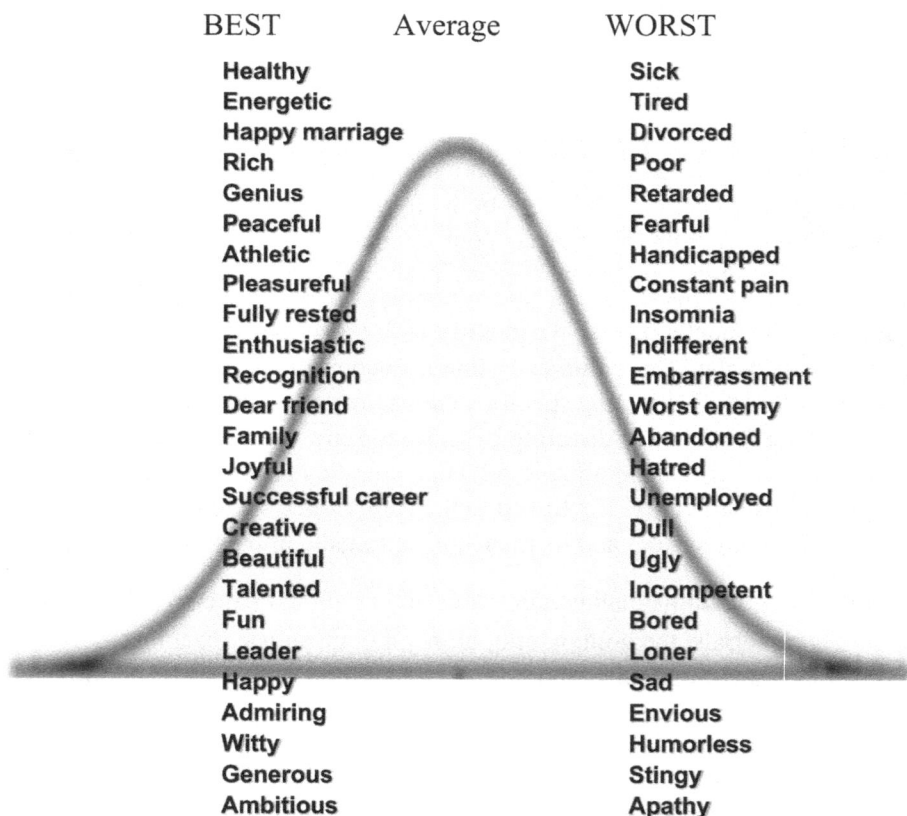

BEST	Average	WORST
Healthy		Sick
Energetic		Tired
Happy marriage		Divorced
Rich		Poor
Genius		Retarded
Peaceful		Fearful
Athletic		Handicapped
Pleasureful		Constant pain
Fully rested		Insomnia
Enthusiastic		Indifferent
Recognition		Embarrassment
Dear friend		Worst enemy
Family		Abandoned
Joyful		Hatred
Successful career		Unemployed
Creative		Dull
Beautiful		Ugly
Talented		Incompetent
Fun		Bored
Leader		Loner
Happy		Sad
Admiring		Envious
Witty		Humorless
Generous		Stingy
Ambitious		Apathy

Everyone experiences these dual aspects at different degrees. Most all of us fall somewhere in the middle average area most of the time. A very small percentage of people represent the extremes on either end of this scale most of the time. Human experience always produces a bell curve pattern.

Man's Universal Connectedness. Man is a perfect microcosm of the universe, anchored in the middle; a part of the whole and a whole with its own parts, made from the same ingredients. An inseparable living vessel within a larger biosphere interlinked by life forms of all types, including trillions of microorganisms inside every human body and millions of organic species permeating our world, sharing the same raw materials that built, feed and sustain the whole network.

Man may be thought of as a single cell in the living earth-being called Gaia, or a single synapse in the collective mind of humanity, or further yet, a subtle singular essence within the formless spirit realm of the collective unconscious. Consider a human brain with 100 billion cells mirroring the galaxy with its 100 billion stars, or the universe with 100 billion galaxies – each serving as microcosms of greater wholes. Man is formed and shaped by the same forces governing the larger whole, weaving the same symmetric patterns at every turn. Parsimony is unmistakable at every level where laws of nature create optimal forms. The human body is sculpted by a myriad combination of environmental forces that make our shape, size, and structure what it is; no other outcome is possible. The universe and our local habitat on earth created organic products symmetrical with itself. Man is a perfect template mirroring the original whole.

"Man is a synthesis of the infinite and the finite, of the temporal and the eternal" – Soren Kierkegaard

Our dynamic cosmos is fluid and ever changing, likewise incorporating the very same dynamic attributes present within each human being. Man is a comprehensive process center always operating in flowing, fluctuating states: growing, adjusting, transforming, producing, moving, creating and sustaining through dynamic equilibrium. Breathing in and out every second mirrors the universal pattern of life, death and rebirth. A beating heart follows the same principle, mimicking the quantum act of creation where physical reality is born moment by moment. Similarly, the life cycle is universal where birth is followed by growth, ascendency, decline and death. A human life is layered with multiple sets and subsets of cycles, including daily wake/sleep rhythms, monthly fertility periodicity, life stages, developmental milestones, the spiral life path pattern, and cycles of health, illness and renewed wellness. The evolution of the physical universe as a whole is mirrored by an individual human life which follows an identical evolutionary forward progress. Man's journey toward higher consciousness is essentially the cosmos finding its way back home.

YOU are the center of the universe, and closer to home, the center of The World, which is after all just a collective projection of individual YOU's. And changing that world is simple – just change yourself and everything else changes. All those qualities you attribute to things "out there" are also in You. All reality begins with what's inside You; consciousness creates the rest. Man operates in a participatory universe as thoughts, desires, attachments and perceptions determine experience, all originating inside the center. What you see is what you get. Heaven and hell aren't "out there" but are in here. We're tethered to the world we created ourselves, both individually and collectively. Our necessary sense of identity is built around a false perception of separation from everything else. You are both an individual and a connected part in the whole, never completely separated or apart. Man is symbiotically connected to The World, the universe and everything conceivable. Both aspects of the part and whole change in unison and are co-created in a reality that emerges moment by quantum moment, inside and out.

"Do not try and bend the spoon, that's impossible. Instead, only try to realize the truth... there is no spoon. Then you'll see that it is not the spoon that bends, it is only yourself." – The Matrix

Human reality consists of everything highlighted in this chapter working together in a complex, dynamic interaction of attributes and processes. As a vehicle for spirit incarnate, man navigates a physical environment that perpetually challenges the soul. The perceived world "out there" operates in a filtering process with tests, gates, traps, rewards and punishments. Desires and attachments bring life lessons through recurring themes. Duality in our world brings challenges providing opportunities for learning, growth and development as fully conscious beings. Living in The World is like attending a soul school where the fundamental competency is balancing your eternal spiritual needs with temporary material attachments and distractions. Man experiences this ongoing process through fluid states of being, unique self-expression and developmental growth – various modes of being, doing and becoming.

The Greek Protagoras understood the nature of Man-centered reality thousands of years ago when he proclaimed "Man is the measure of all things". We can update this concept and call it You-centered reality, for each and every one of us is at the center of Everything. And all reality essentially depends on its relationship to You. Yes, YOU are in the middle of it all, between the macro and micro, inside and outside, conscious and unconscious. You are a perfect inverse of the universe; the key to its lock, the hand to its glove, the plug to its outlet. YOU are a perfect template of the universe: A machine, a process center, a supersystem of systems. YOU possess the very same ingredients that made the world, stars and all the content in the cosmos, with similar proportions. YOU are the microcosm of the greater whole we might more accurately call the *Youniverse*.

      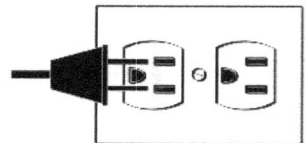

Key-Lock   You-Universe   Plug-Outlet

THE YOUNIVERSE

6   TEMPORAL REALITY

Human experience is dependent on Time. Time is life, life is time, living takes time. It's precious, priceless and limited. We often don't realize just how precious it is until it's gone. It's all there is and nothing exists without it. You are in the center, the middle point of space and time. Everything else is within and beyond – above and below, inside or outside. Everything that ever has or will happen is before and after you. You're the center of here and now, tethered to the flowing sensation of time as your future emerges and your past recedes, moment by moment. Time is the very essence of your experience, turning the noun of life into the verb of living. It powers your personal transformation from who you were to who you are.

MAN IN TIME

Our perception of time is always biased to the present, despite the greater reality of everything that happened before or will happen later. It's only natural to focus on what's in front of you now. But with age, experience and wisdom comes expanded awareness of longer timeframes, both past and future. Managers and strategic leaders succeed by seeing further ahead; anticipating requirements rather than reacting to crises. Leadership vision is all about seeing the big picture and managing time. The larger the organization the longer planning timelines need to be made. CEOs are future biased planners while plant foremen are present biased, dealing with day-to-day challenges. Friends and family are past biased when relating to us, perceiving us largely by who we've been, while coaches and mentors are future biased, seeing the potential we're capable of.

Time gives life a musical quality through its rhythmic, cyclic nature. It contains a temporal code, whose keys are like notes in a melody. Life has its ups and downs, shifts in tempo and an oscillating nature resembling a music score. Each person lives a unique song with different tones, beats and melodies, playing out over the course of a lifetime. In an abstract way you could say time and music shape life in the same manner space shapes architecture.

Life Path.   Our individual journeys through The World are full of twists and turns, distractions, sudden changes of direction and unexpected destinations. It's like any trip you might take but here the vehicle is you. There's no motion through space, just experience in time. Every life path consists of a spiritual being housed in a physical body navigating the material plane, where the journey is an unfolding process of discovery and learning. Each experience adds a new layer, molds us, and enriches our awareness of how things work. Like burning fuel or expending a battery, we exhaust physical time in exchange for an excellent adventure through that soul school we call The World.

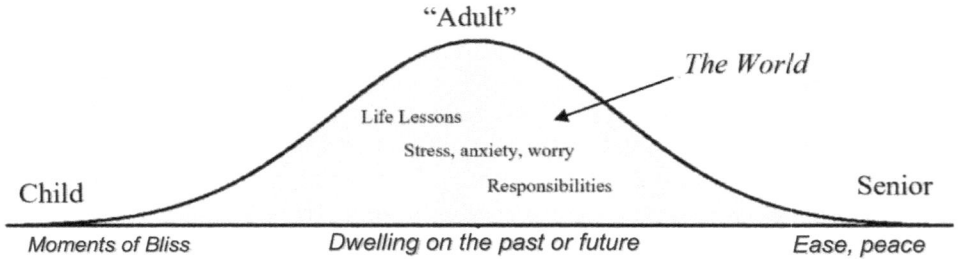

Human Timelines. Life paths include a variety of phases, passages and milestones. We all follow a general sequence of universal stages progressing along our personal life journey, each associated with particular ages and stages. These phases are connected mini timelines which together form a complete biography. Of course, the major timeline is simply Birth-Life-Death, segmented by numerous sub timelines within. As we cross paths with other people our journeys form a tapestry of individual and shared timelines that run parallel at times and diverge at others.

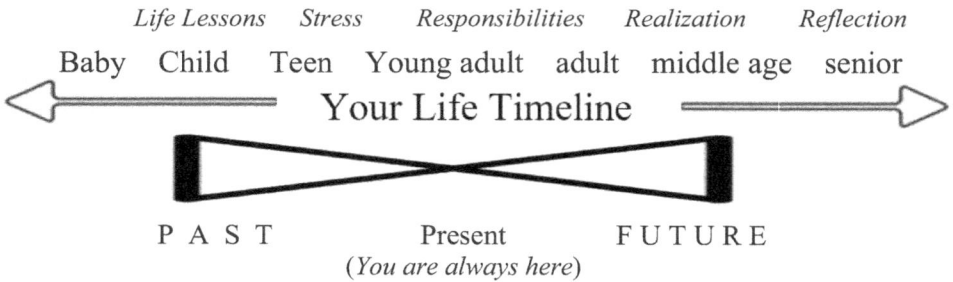

As we move forward in time along our path through life we adjust to the outside while maintaining an invariant inner core. Looking back at each stage and milestone crossed we see a different person than compared to who you are today. So then which is the real You? The answer is all of them. You're both the person you see on the outside and the core essence inside. They're two sides of the same coin. In every phase it was always just you. It's wearing a mask and appearing different but remaining the same. Your personal identity is a temporal node that shifts along each timeline. Like any adventure it's not the pit stops or destination that counts but rather the journey itself. It's the process of enrichening experience that leads to greater wisdom and fulfillment, driven by the passage of time.

Biography. Your life is one long timeline with general periods occasionally interrupted by single events - some special but most are ordinary. Your biography includes all of it as a journey through time in The World. The path is not just a linear timeline but also a rhythmic wave of cyclic experience. Your path of development includes ups and downs, progress and retreat, trials, tribulations, exhaustion and renewal. Biography is the timeline of any single human lifetime.

How We Spend Our Time

Sleeping	230,000 Hours
Entertaining	140,000 Hrs
Working	120,000 Hrs
Television	80,000 Hrs
Eating	40,000 Hrs
Driving	38,000 Hrs
Education	20,000 Hrs
Exercise	10,000 Hrs
"Free Time"	40,000 Hrs

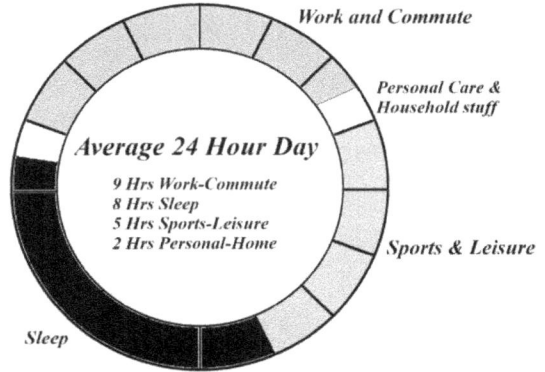

Average 24 Hour Day

Work and Commute

Personal Care & Household stuff

Sports & Leisure

Sleep

9 Hrs Work-Commute
8 Hrs Sleep
5 Hrs Sports-Leisure
2 Hrs Personal-Home

Today, life spans range up to 90-100 years on the high side. Measuring recent history on this scale you could think of our country as being roughly 3 lifetimes old, not very long from a broad historical perspective. The Civil War was just 2 lifetimes ago and World War 2 was only 1 lifetime ago.

Perceiving History in Various Time Frames	*Upper Range of Life Spans*	
Nation's lifespan about 250 years	Turtle	150 Years
Individual lifespan up to 90-100 years	Man	100 Years
Generational time period 20-25 years	Dog	20 Years
Elected leader's term 2-4 years	Fly	25 Days

Generational Cohorts. Each of us shares a timeline with others born in the same period. We collectively share similar experiences with common cultural values and societal memes. Generations flow through temporal waves of social history associated with shared qualities of the times – a *zeitgeist* unique to their period. Each wave has its own fads, hot issues, noteworthy events, TV shows and iconic movies, music styles, politics, tech products, celebrities, trends, and key public figures. Sometimes innovative products are introduced greatly influencing collective behavior. The combined effect of all these various attributes makes each cohort wave a unique experience to those living in that period of history.

Recent Generations defined   Iconic Tech

Silent	1925-45	Traditional, private, hard work, face to face,	Radio
Boomers	1946-64	Me gen, anti-rules, embrace change, idealist	TV, cassette tapes
X	1965-80	Tech gen, info access, skeptical, independent	Cable TV, VCR
Y (Millennials)	1981-95	Tolerance, social connection, work-life balance	PC's, smart phone
Z	1996-09	Social media, network, isolated, SJW's	Facebook, TikTok,
Alpha	2010-24	*Health maladies, high cost of living, ???????*	*Streaming Video, AI*

*Post-war spike of baby boomers cohort greatly influenced social trends passing through decades

Cohorts are generally portrayed as distinct bands but realistically they blend together beyond the arbitrary band limits. YOU have your own specific band with the same roughly 20-year range but it's 10 above and 10 below your birth year. Unless you're born in the middle of the official band period, you're likely to have a blend of aspects from the one above and the one below. And each of us is born in the middle of cultural history where all the popular movies, songs, fads and events that came before us or come after us seem foreign and less familiar.

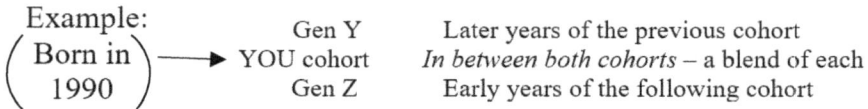

Example:
$$\left(\begin{matrix} \text{Born in} \\ 1990 \end{matrix}\right) \longrightarrow$$
Gen Y — Later years of the previous cohort
YOU cohort — *In between both cohorts* – a blend of each
Gen Z — Early years of the following cohort

Time Sense. We take for granted our built-in sense of time perceived as separate intervals; distinct dimensions of experience. Though time passes uniformly we see it as a multifaceted serpentine essence, changing in length and tempo. Our grasp of temporal reality involves the unconscious practice of seeing events and processes in terms of durations and rates of progression. It all takes place in our imagination since time itself is just an abstract concept conjured inside your mind.

Human Time Sense

Short – Minutes, hours (Minimum: Seconds, split second)
Med – Day, Weeks
Long – Month, Year (Maximum: 1 lifetime, approx. 90-100 Years)
*Imagination enables conceiving time intervals way beyond our limited life span

Human Attention Span

Short – 5-10 minute presentation, conversation
Med – 20-30 minute formal meeting. (over 30 minutes loses attention)
Long – 90 min – 2 Hr: TV program, movie, sports, concert (over 3 hours loses attention)
*4-8 Hour seminar requires frequent breaks

We judge almost everything we do by its relation in time. When there's no hurry we take our time, bide our time, take it one day at a time or just do it in our spare time. If we're stuck in one place we mark time, and if isolated we may do hard time. Time flies when we're in a hurry, so we rush because time is money and time waits for no one. Urgency leads to crunch time, so we get it done just in the nick of time and then slow down and kill time. We repeat this insanity time after time because we understand time heals all wounds. ☺

Timing Sense. Some wise person once proclaimed *timing is everything*. This applies to every possible act, event or occasion. The trick is having a sense for it. Nowhere is this more important than the arena of human live performance, where a simple act or response can have a completely different outcome depending on when it's done. Comedy, politics, risk taking, business dealing, sports, romantic advances; all obvious examples where timing is critical – where a few seconds can make or break an outcome. The spoken word is a clear example where the same sentence can have completely different responses depending on when it's said; just ask a comedian or politician. Sometimes things just seem to click. Other times it may feel as though everything you try falls flat. It's simply a question of timing.

Ancient man understood the importance of timing, seeking direction from mythic gods before doing anything important. Some early civilizations based their entire social system on the harmonics of heavenly bodies, tracking and plotting recurring patterns observed in the night sky. Greeks viewed time having 2 aspects: Quantitative and Qualitative, the latter was tied to a sense of good timing, knowing when it was best to do something. Ecclesiastes 3 says it all: "There is a time for everything and a season for every activity under the heavens", including all of the following: *A time to be born, die, plant, uproot, kill, heal, tear down, build, weep, laugh, mourn, dance, scatter stones, gather them, embrace, refrain from embrace, to search, to give up, to keep, to throw away, tear, mend, be silent, speak, to love, to hate, a time for war, and a time for peace.* Yes, timing *is* everything, but the trick is to have a natural feel for it. It seems some people are born with it however it can be mastered with practice and repeated experience.

Time Management. Time is the most precious limited resource we have; thus, success requires making the best use of it. Leaders who can't manage time eventually fail. Planning and organizing are universal requirements, whether individual or collective. Your quality of life literally depends on how well you make use of your limited time so it's essential to take it seriously. Who can honestly look back into the past and say they never squandered significant amounts of their time? Each of us regularly engages in time management whether we're aware of it or not, it's just that some are much better at it than others. One thing's certain; we all are guilty of wasting time. Leading offenses include too much TV, social media, video games, browsing your "smart phone", waiting, worrying, over sleeping, gossiping, interruptions, distractions, and repetitive bad habits. Much can be avoided by limiting screen time and engaging in effective time management:

Make the Most of Your Limited Time (individual or group)

*Get organized and plan. Set a realistic schedule. Set goals. Plan for tomorrow tonight.
*Set reminders. Use apps and notes. Manage scheduled breaks.
*Prioritize. Do the most important things first. Delegate when possible.
*Remove distractions. Avoid procrastination. Learn how to say NO.
*Know yourself: when to multitask or focus, what time you operate most effectively.
Goals should be specific, measurable, achievable, relevant and time considered

Aging. Time wears out everything. From the moment anything is created it begins a long, unavoidable process of decay. Physical objects break apart, machines and systems run down, living beings decline, wither and finally perish. There's a subtle duality to time as it offers the gift of life on one hand but then trades off that living process by gradually taking it back. To live is to age. To work is to run down. A battery is only useful when it's running down.

No one can escape the universal experience of growing older. It's a necessary evil of living – trading off your finite number of days a little at a time. From the moment after birth each of us begins the life-long process of dying, day by day. Most don't notice the slow, gradual process as we're busy living, but there comes a time in the home stretch where suddenly there are fewer days left than those already used up. The end becomes a real, tangible thing that is approaching uncomfortably sooner than later. Along the way more and more old acquaintances or loved ones suddenly disappear – some naturally, others unexpectedly.

As we age, we experience a gradual, continual diminishment of physical abilities. Over time we learn to accept the reality that we can no longer do the same things that a short while ago seemed easy. Some can reduce or slow down the aging process by simply remaining young at heart. Others can manage their new reduced capacities while many fail to adapt and simply recede into a more isolated, acquiescent version of their former selves. Aging is part of the universal journey all human beings share and requires a realistic acceptance of it as a healthy response to living in The World.

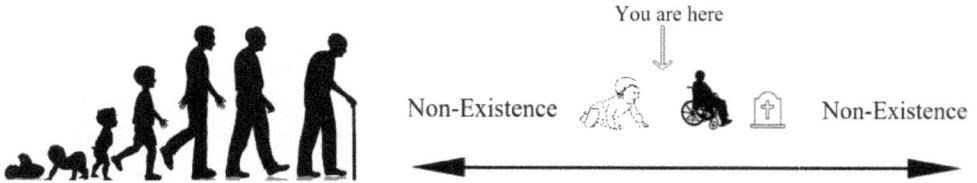

You are here

Non-Existence   Non-Existence

Death.   New products wear out or become obsolete, inevitably becoming garbage. Living beings survive by replacing millions of dying cells with new ones. Over time, fewer new cells are produced as organisms gradually diminish, run down and die. Products end up in landfills, people end up in cemeteries. Aging is really just slow dying. Time eventually runs out. Your timeline ends. Every other timeline continues. From birth and first breath you've been running down and running out of time. Each of us gets a limited amount and a timeline with the same simple pattern: Birth-Life-Death. And while it's everything to us it's rather miniscule to everything else. Greater reality is the nonlife, non-existence before our birth & after our death. The world before and after our brief, relatively insignificant appearance and disappearance barely even noticed us. Most people fear death and what comes after they're gone but never contemplate life before their birth.

DYNAMICS OF TIME

Space is static, time flows. It's an emergent phenomenon that keeps on going, creating all reality moment by moment. Time proceeds in a sequence of irreversible events and experiences in a never ceasing incremental progression. Every moment is unique, never to be experienced the same way again. Our sense of it is purely conceptual. The way we measure and record time is simply a human construct. Clock time is completely arbitrary. Our relationship to time is similar to a fish in water – we live in it, move through it and drift about. Comprehending the attributes of time requires appreciation of its macro and micro level attributes. As always, man is in the middle, relating to larger and smaller scales, abiding both big picture and smaller scope. Each level beyond our middle position has dynamic aspects affecting our experience and perception of it.

"Time is what prevents everything from happening all at once" – John Wheeler

MACRO & MICRO ASPECTS OF TIME

Macro analysis applies to the broad picture, with large scale timelines that are identified on our universal matrix of everything. They represent the 3 primary levels – Social-culture, Biological-human, and Physical-nature, each correlating to history, biography and evolution respectively. The other macro level aspect of time is the structure of past-present-future and how they relate to each other.

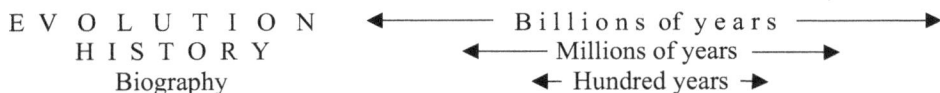

```
E V O L U T I O N    ◄──────── Billions of years ────────►
    H I S T O R Y        ◄─────── Millions of years ───────►
      Biography            ◄─ Hundred years ─►
```

Micro analysis captures mechanics of time in a closer look at its essence, attributes, process nature, and how it basically works, including increments, tempo, cyclic aspects, physics and numerous relative qualities.

MACRO LEVEL – *Past, Present, Future.* Reality unfolds with man in the middle and fittingly, time unfolds with the same exact pattern. Here, present is always in the middle with a relative past – future related to it. Both man and the present are centered, occupying central points in the universe. Everything else relates to that center position, in time and space. Since there are multiple centers, everything becomes relative to it. Space and time, here and now, past and future, are all relative constructs.

PAST. Everything that's ever happened before represents the greater portion of reality, containing all the vast amounts of history, much more than our very limited present time. The past just continues to grow with each passing day, adding to the whole. But all that's a pittance compared to what could have happened. The past as we know it is merely the limited, finite collection of manifested moments chosen from an infinite number of possible moments. The difference between what could have happened and what did happen is beyond comprehension.

Ironically, and counterintuitively, the Past is Present dependent. It's the residue of *now* and is only recognizable by things left behind, some of it tangible, most of it not. It only exists in present memory, and our contextual view of it. What we call the past is continually modified as our records are updated and our perceptions of it change over time. Negative views of our past create regret and guilt. Positive views foster nostalgia and reminiscing. The past isn't as set and permanent as you think, it's actually quite malleable. As it recedes further away, day by day, it becomes more and more an imaginary reality, a dream that seems real but elusive; a ghost lingering in the background no longer of this realm.

"The past is a foreign country; they do things differently there" – L.P. Hartley (1953)

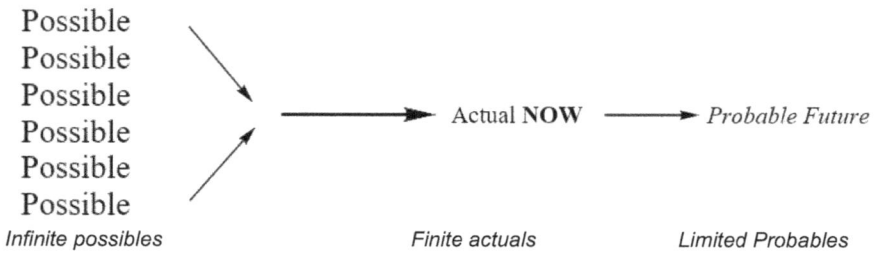

Possible
Possible
Possible
Possible
Possible
Possible

                    Actual **NOW**   *Probable Future*

Infinite possibles   Finite actuals   Limited Probables

FUTURE. Similar to the past, the future consists of virtually infinite possibilities but a finite, limited line of actualities. It already exists as a probable, potential reality. Everything that can happen already exists in a realm beyond time. The present is simply the moment-by-moment choice from possible but probable futures, no different than the quantum wave collapse in physics that fixes a moving particle in position from infinite possible positions. Visions of the future are images of probable realities. In a dream state you can experience any possible future and sometimes make it a greater likely one.

Prediction is hard because of so many influencing variables and few constants. The biggest variable is human behavior, never certain and often chaotic. Prediction in the far future is practically impossible due to emergence of new properties, conditions, and unforeseeable consequences with each new development. Consider what's happened in just the last 100 years and try to guess what might happen in the next 100. Good luck. Think about society 100 years before the printing press, mass communication, global internet, mobile smart phones, streaming movies, etc. Predictions have fractal like branches generating greater complexity with each additional time increment. It's complicated by the *butterfly effect* where sensitivity to initial conditions produces enormous differences later on.

Our impression of the future determines our sense of security in this moment. *Better* = hopeful and optimistic, *Worse* = fear and anxiety. Fear is a future construct created entirely in your own present mind. So is worry, anxiety, and nervousness. Dwelling on the future creates a waiting experience. Obsessive focus on the future increases pain in the moment. Impatience is lack of embracing the Now. It's a symptom of focusing on the destination at the expense of the journey.

Future Focus Creates State of Impatience

Short = waiting lines (Post office, DMV, traffic, grocery checkout)
Medium = anticipating weekly TV show, tomorrow's online date, weekend off, coming holiday.
Long = turning 18, graduation, wedding engagement, retirement, funeral arrangements

Future always exists as a potential in a timeless realm free of physical constraint yet influenced by the present. Thoughts influence future probabilities. Visualization resonates potential futures and when combined with emotional attachment becomes a powerful influence. Thoughts, feelings, desires and affirmations here and now incentivize probable futures; our conscious attention attracts futures. What already happened in the past also influences probable futures, depending on how recent. Manifested events here and now limit what can happen next. Anything that might happen in the future is always a statistical aspect that changes based on variables in the far past, recent past, and present. History is important as it influences the future. As actual future unfolds previous possibilities narrow. Rare events may become more likely while likely events may become unlikely. All futures begin today, and yours is no different. The World always moves forward into uncertain, uncharted futures, with increasing complexity. Looking at current trends offers insights for making best guesses. Today, we can see waves of parallel and intersecting paths extending through the internet of things, transforming society at every level. Digital technology continues its omnipresent saturation into every aspect of human life. Artificial intelligence is only in its infancy stage, ready for exponential expansion in every possible application. And like all lines moving into an uncertain future, there will be a mix of positive and negative residual effects. Will life get easier or more complicated? Will wonderous advances in video processing be offset by exploitation of deepfakes and pornography? Will increased access to powerful digital technology for all enrich our lives or be exploited by nefarious characters? With so many possibilities, there are still definite probabilities, each becoming clearer as we pass further into the future, day by day, moment by moment.

PRESENT. Now is what really matters. Past and future depend on it and connect to it. Time is both unique to each observer and a shared simultaneous experience with others. To live is to be…in the present. Being is presence in the Now. Thrill seekers are 100% in the now, same with athletes in the zone or passionate lovers. Zen is all about now. Being in the moment is achieved through laughter, surprise, witnessing, magic, dance, making love, creating art, or simply relaxing in stillness. Kids and pets operate in the Now, as do all animals in nature. Adults are more concerned with where they came from and where they're going. Being fully in the Now removes any past or future. Ironically, time only flows in the here and now. Experience of it is timeless. The present moment is a freewill open window, creating reality moment by moment. The Now is both eternal and ever changing.

Our perception of reality is always delayed in a present that already happened. It takes time for sounds and light to reach us. The sun we look at is already 8 minutes in the past, other stars much longer, thousands of years in the past. We never see the universe as it is. Our delayed Now is where life occurs. Dwelling in either future or past is detrimental: Future = anxiety, worry, unnecessary stress, Past = regret, guilt, and depression. We dwell on past or future at the expense of being in the moment.

Past-Present-Future Interrelationship. Past and future are simple derivatives of the present. They are relative, timeless constructs, while the only temporal reality is here and now. Time is an experience created by our memory and imagination, where the past is a memory and future is imaginary, both conjured up in this Now. Past and future are man-made constructs, influenced by thoughts, emotions and experience in the present. Nothing happens in the past or the future, everything happens now. The real You only exists now and is timeless. Every You from past or future exists in the Now of that moment.

"One life; a little gleam of time between two eternities" – Thomas Carlyle

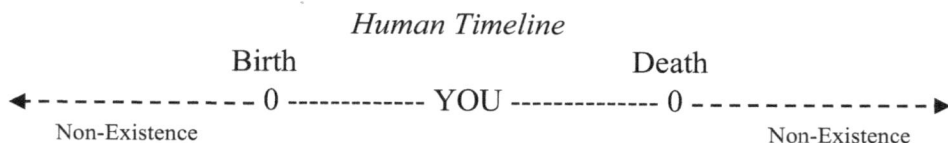

Human Timeline

          Birth   Death

◄ - - - - - - - - - - - 0 ------------- YOU --------------- 0 - - - - - - - - - - - - ►

     Non-Existence   Non-Existence

* *There is no time in a timeline: It's simply YOU, being in the now; always was and is.*
  Space is a difference between you and everything else. Time is a difference between You and...You.

Past and future are like octaves of the present, connected symmetrically and symbiotically. Past - present - future all exist at once. It's like a video hard drive containing all the information but reads data bit by bit, playing a movie frame by frame in the moment. It's all there but can only be experienced bit by bit, moment by moment in the now. Past, present, future all change as we change – Everything changes as You change. Past and future have attributes but no properties, just a specter or shadow without substance.

MICRO LEVEL – *The Mechanics of Time*. Plato saw time as an illusion; "The moving image of this changeless eternity". The essence of time is difficult to define since it's emergent, possesses a process nature and operates in a relative manner. It's entirely enigmatic; on the one hand seeming fundamental to everything, yet on the other appears to be a quizzical illusion. One thing is certain – time is more verb than noun. It exists as a constant creative becoming – a series of moment-by-moment births, without before and after, just now. Like a ball having no front or back, rolling along without directional identity, just eternal motion absent any past or memory. Time is the abstract difference between then and now – between where you were and where you are. Time flows like a river where the water you're standing in only passes by once and never again. Time is like a shadow; it isn't a real thing but rather a mental construct.

Being fully present in the Now frees us from the time bound state where we transcend it. As a Zen master noted, "It is believed by most that time passes; in actual fact it stays where it is". The eternal now is a state of *Brahma*. Every point in space touches every point in time. Everything is everywhere and always. Every passing moment is connected to the next, each almost identical but not quite. Every moment is an eternal moment, in the here & now, not out there, not anywhere. Photos or snapshots capture eternal moments, frozen in timelessness, not in time. A song on a disc is frozen in timelessness until a laser reading its tracks brings it to life, creating an eternal live moment. A movie on a reel is frozen in timelessness until a projector brings it to life, frame by frame in the eternal moment of now. Words in a book are frozen in timelessness, until reading them one by one, sentence by sentence enlivens them.

Man in the Middle may be physically trapped in the time bound present, but can transcend into a timeless past via *memory,* or the timeless future via *imagination.* Everything we think, feel and experience is in a moment linked to a series of connected moments, through various levels and parallel timelines. Time is like motion, speed, or distance, just another man-made reference point. It's a measurement of duration and traces sequence, causality, and change. Though constrained by physical space-time dimensions, man can transcend time in spiritual realms, dreams, imagination, psychic states, or out of body experiences. Transcendence is possible through higher conscious and unconscious realms – where the timeless past and future actually reside.

Increments.   Time is a flowing, unbroken chain of sequential moments. Early man yearned to measure and track it in an orderly, systematic scheme. This led to establishing a standard set of intervals capturing small, medium and long periods of time. Tracking time's passage was accomplished through a variety of time keeping systems and devices, essentially calendars and clocks. The prime increment is a *day*, running from sun up to sun down. Longer periods were measured by lunar and solar cycles. Ancients followed moon cycles – New, quarter, half, full. At just under 30 days in length, the lunar period almost fit into a year 12 times, but there's always a partial gap left over. The difference required constant adjustment, either by adding lots of extra days, or after several years the gap would require an extra month to fill it. Egyptian and later Roman calendars achieved a closer fit. Finally, Julius Caesar updated the system, adding a leap year day adjustment every 4 years. Romans also introduced the 7-day week. Egyptians divided days and nights into 12 segments each, creating the hour length. Babylonians who used a counting system based on 60 instead of our familiar 10, divided the hour into 60 minutes, and each of those into 60 seconds. Even their calendar was based on 360 days.

The earliest time keeping devices were sundials, invented over 5000 years ago in Mesopotamia. They were limited to daytime only, using the sun's shadows to track time. Eventually water clocks were introduced to measure time after sunset. Mechanical clocks first appeared about 800 years ago, but didn't have the dials or hands we're used to seeing. Several hundred years later the pendulum clock was invented, creating a fairly accurate time piece with true regularity. And it just so happens that a swinging pendulum takes 1 second to do its thing…how convenient. These modern working clocks made time keeping an integral part of western culture where hours, days, weeks and months became the structured ordering principle embedded into society.

Increment	Association	Impact
Split Second	Athletic feat	gold vs silver medal
Second	Heartbeat, breath,	avoid accident
Minute	Phone call	late for meeting
Hour	Lunch break	missed flight
Day	Primary living period	missed work quota
Week	Paycheck	missed deadline
Month	Ovulation	premature birth
Year	School, seasons	Foreclosed house

Time increment names have astronomical and mythical origins. *Sun*day, *Moon*day, *Satur*n day are obvious examples. Tuesday, thru Friday come from mythic gods associated with planets, including *Thor*sday and *Woden*sday. Similar to Monday, our 30-day period is a *moon*th.

As civilization spread and the world became a smaller, more interconnected place it would become necessary to establish time zones. Like hours, the world was divided into 24 segments. This standardized relative time for travel and reduced the chaos of transcontinental differences. Still, there remained discrepancies between nations and cultures where calendars and time systems vary from the standard western version. Muslim countries base their calendar on the moon cycle, creating a disconnect with annual seasons. Many cultures downplay the structured nature of western time. The Piraha tribe of the Amazon only embraces the present and their language has no past tense. The Hopi tribe and some Native American tribes don't use time-verb tenses and don't reference linear time. Man's system of managing time is arbitrary, relative and culturally varied.

Tempo.   Motion, Time and speed are all connected. Time = Distance/Speed. Tempo is pacing; the rate of movement, the speed at which things happen. Life is all about ongoing change, made all the more interesting through variety and mixtures of different tempos. Human experience is really a collection of events happening at different rates; most are low tempo, routine non-eventful experiences, separated by brief intervals of higher tempo, exciting activity. Variations in tempo change our perception of time. Things seem to take forever when we're bored but appear to speed up once we're excited, yet time itself doesn't change. Consider this; the difference between a slow burning candle and an explosion is tempo.

Speed is an attribute of tempo, or the rate at which something changes position. Shifting the previous equation, we get Speed = Distance/Time. It's a measure of how fast or slow something happens. Speed is normally a constant rate of motion. Acceleration is a change in speed – an increase/decrease in the rate of change. Both measures are completely time dependent

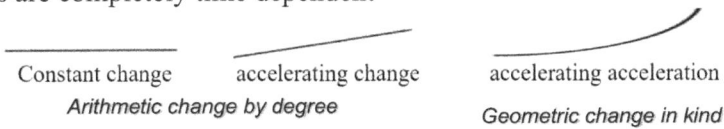

Constant change	accelerating change	accelerating acceleration
Arithmetic change by degree		*Geometric change in kind*

Tempo is essential to power, which is a combination of strength and speed. Applying the same force in a shorter time produces greater power. Motion, Speed, Force, Time and Power are interdependent. Power = Work performed/Time. Simple acts of strength are much more impressive when done quickly, in a higher tempo. Time is the special element that transforms strength into power.

We all have a very limited perception range of tempos, or rates of change. We can easily grasp the passing of time in the moment, moving vehicles or a relay race. We observe the slow passing of days, weeks, months and years. However, our comprehension of tempos is limited when the pace is too fast or too slow. Our range falls somewhere between observing ice melting on the slow end and formula race driving on the higher end. Beyond this range are extremes such as the pace of glaciers, canyons, mountain formations, continental drift, organic evolution, etc. and on the faster end of the scale, we can't grasp the action of an explosion, bee's wings, a moving bullet or the speed of light. Our built-in senses limit us to the experience of time tempos in a very narrow range (man in the middle).

Tracking Moments *vs Movements*

STATIC	*MOVING*
Photo	*Video*
Event	*Trend*
Balance Sheet	*Income Statement*
Stock price	*Stock chart*

Cycles.   Time has both a linear and circular nature. Increments can be either line lengths or cyclical periods. Cyclic change and regularity are omnipresent qualities built into everything. Besides obvious cosmic and natural cycles all around us, there's an equal presence within us and all living organisms. In a universe of comprehensive interconnection this makes perfect sense. Living systems naturally synchronize internal clocks to their external environment. Animals, plants, and living organisms in general possess internal biological clocks in sync with nature, daily rhythms, sunlight and darkness, summer and winter, etc. Bio clocks exhibit remarkable precision. Humans follow a remarkably regular pattern of internal cycles, including waking-sleeping, body temperature, activity-rest, heart beats, breathing, cell division, a 90-minute sleep sub cycle, brain wave cycles, etc. Travelers suffer from jet lag due to this regularity of internal body cycles.

All living beings follow a life cycle of growth and decay, progress & setbacks. This is true with individual organisms and group collectives of organisms (Ant colony, bee hive, etc.). Cycles of organic growth and decline mirror the rhythmic pattern of environmental ebb and flow, whether it's daily, weekly, monthly, or the life span as a whole. The more common ones are daily rhythms (solar), monthly (lunar) and seasonal. Collective cycles include farming, hunting, harvesting, animal migrations, predator-prey populations, and even a 17-year cycle in cicada activity. Natures cycles range from tiny (microseconds) to grand (geological & cosmic scales). All living creatures are constrained prisoners of time, linked to the physical environment they came from and then fated to an inescapable limited life cycle to be born, rundown, and die.

Biological Cycle Lengths

Ultradian (Shorter than 1 day)
Circadian (Daily)
Infradian (Longer than 1 day)
Diurnal   (Night & Day)
Annual   (1 Year)
Seasonal (1/4 Year)

The cyclic nature of passing time is both a physical phenomenon and a psychological one. Repetitions of events resonate with us, becoming an actual attribute of our sense of time. Experiencing seasons coming and going with consistent regularity creates a template in our minds that affects how we perceive the flow of time, and our judgement of past and future events. Where ancient man sensed rhythmic time through natures seasons, now modern man perceives it through circular clocks and repeating calendars.

Our daily lives are repetitious cycles of activities engaged in recurring incremental patterns. Each day traces a familiar sequence of repeated intervals that form a standard schedule. Wake-dress-eat-travel-work-rest-eat-leisure-sleep. It's a common pattern followed by many with some variations and differences in interval lengths. This repeating cycle is mirrored by the round clock on the wall that we use to time those very intervals

Physics and Time. Aristotle's early theory of time connected it with motion and distance, seeing each defining the other. Galileo applied scientific methods to analyze velocity, inertia, gravity, motion and time. Newton introduced laws to define motion and time, with special emphasis on acceleration. These early pioneers all shared a perspective of linear, absolute time. Einstein completely changed that paradigm by first joining time with space as a single interconnected unity, then completely removed all absolute values of time, relegating it to only a local, observer-based reality. He combined space and time dimensionally, adding time as a 4th dimension, supplementing the original 3 of space. He linked them in a complimentary union, the same way he linked matter and energy, mathematically expressed as $E=mc^2$. What this linkage means is each person experiences the same physical event differently. Past, present, future, motion, distance, speed…ALL become dependent on a local observer, none are ever absolute.

Changes in space translate to changes in time. The only place in the entire universe where space-time can be measured in absolutes is where You're standing. Again, man is in the middle, the center of reality where everything relates to. Both time and space are consistent here and now, but only for You and no one else. Fortunately, in everyday life relative differences are miniscule, practically identical. But on a cosmic scale they become very different. Since space has 3 perpendicular dimensions how would time fit in? Since spatial dimensions are each perpendicular to the other, time is perpendicular to all 3 of them. To visualize this transcendent relationship, we can use an abstract 4-dimensional hypercube.

 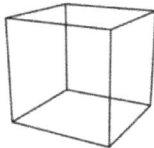 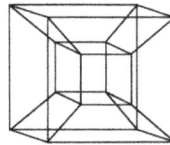

Each side is equal length Can't be visualized in our limited 3D perspective

| 2D Square | 3D Cube | 4D Hypercube |

Time's flow is experienced both linearly and cyclically. We sense the temporal process by observing motion, events and experiences. Our own calendar is a tool to trace the linear passing of days and the cyclic rhythm of seasons. Much of life is expressed in harmonic periods, from our heartbeat to the daily sleep cycle. Rhythm is a built-in quality of living processes where life goes in and out of balance, grows and decays, oscillates between periods of activity and rest.
Time itself isn't rhythmic; it's just associated with the abstract cyclic pattern of life. Western culture is founded upon a structured application of time where rules and regulations keep the trains moving on time. It works just fine because Einstein's relativity of time doesn't matter much in everyday situations.

Past = Classical time, Present = Quantum moment, Future = Quantum potential

t = Time
v = Velocity
c = Light speed

$$t = \frac{t_o}{\sqrt{1-\dfrac{v^2}{c^2}}}$$

Speed (V) approaching light makes V/C reach "1" and 1-1= 0 so Sq Rt of 0 implies timeless value
Relativity represents continuous time *(analog).*
Quantum represents discrete increments *(digital)*

Quantum Mechanics emerged after Einstein's relativity bombshell and introduced yet another revolutionary paradigm in physics. Where relativity destroyed any notion of absolutes in motion, speed and time, quantum mechanics equally blew science's assumption of certainty. Previous models of distinct particles making up the structure of physical matter suddenly became obsolete, replaced by abstract clouds with statistical probabilities and uncertainties. The very act of measuring such things alters their physical state and behavior, limiting our ability to know the precise state of anything.

Quantum mechanics provides a clear model of how time works at the micro level. Consider an atom, which is no longer viewed as a simple nucleus with a ring of electron orbits but rather an ambiguous cloud of movements. Electrons are actually swirling around in nearly infinite possible positions and directions, only to be determined the very moment they're measured or observed. Time works exactly the same way. The cloud in this example represents possible futures with the present now becoming actual upon human interaction, or more specifically, the conscious act of observation or measurement. It's the dividing line between free will and determinism. This is how the timeless realm transitions into the time-bound realm at the micro level. On a more macro scale, it's the cosmic pulse of physical reality created moment by moment, from potential to manifest – a creative process at every level producing novelty everywhere, everywhen.

Entropy is the continual increase of disorder taking place over time, which scientists claim indicates the universe is running down. Some also claim entropy is the actual source of temporal reality, producing an asymmetrical 1-way direction in time. However, entropy and its increasing disorder is only half of the equation. Synergy is an opposite process leading to increasing order over time. You are a perfect example of synergy – how evolution produced greater order from unorganized raw materials. Where entropy results in uniformity, randomness and loss of structure, synergy produces differentiation, complexity and increasing order. Entropy is not the driver of time but rather is driven by time.

RELATIVE NATURE OF TIME

Temporal reality is an enigmatic phenomenon having no absolute value, being both explicit and ambiguous simultaneously. On multiple levels time exhibits different qualities and is relative in its essence. This is true at the physical level, the experiential level, culturally and consciously.

Physical Relativity. Time is never absolute. Einstein unmistakenly confirmed it to be a subjective observer-based reality that's different for everyone. If you're still and someone else moves it can easily appear to them that they're still and you're moving. That's only a difference of perspective; time hasn't actually changed. However, Einstein demonstrated that *time itself* does change based on one's speed of motion. Physical relativity means that objects moving at speeds approaching light will compress time, eventually achieving timelessness. And being observer based, a person moving fast would not notice physical time changing because the very space they occupy changes with it. Increasing gravity caused by a large mass (planetoid) also shrinks time because space itself is compressed. Space-time warped by large masses has the same effect as a fast-moving observer. In our local reality everything is moving and time is flowing relative to every observer with NO absolutes. Physical time, experienced time and perceived time are all subject to the same relativity pattern. Remember that clock time is a man-made construct, a practical tool for use as a local time reference point only (local to the observer).

Non-Physical, Experiential Relativity. Time is relative both physically and *consciously*. Man perceives time differently based on changing internal-external factors and shifting states of consciousness. As an infant there is no time, there is only moment to moment experience. As a person ages time becomes a real thing, moving slowly at first and then appearing to accelerate with old age. When excited or threatened, time shrinks, even creating a momentary experience of timelessness. During periods of bliss time disappears, felt by artists creating, athletes playing, children exploring, lovers embracing, or a singer performing. When bored or uncomfortable time expands, crawling, slowing down into sluggish *timefullness*.

Time experience is relative to age. A young person's time rate is slower as each day seems to add lots of promise. As you get older each day becomes a smaller and smaller part of your life and seems to pass faster and faster. Man's time perception is influenced by the limitations of a finite life span. The decade between 20 and 30 may seem shorter than the 4 years between 14 and 18. Lifespans also influence perception, previously ranging around 50-60 years are now 70-80 and even higher. Each stage of life comes with a unique perception of time associated with it (Child, Teen, Young Adult, Mid age, Senior). Attentions spans, long term goals and patience levels all change with each passing stage. Relativity of time perception also applies to different species. Animal life spans range from mere hours to several hundred years, certainly producing relative values of time.

*Release of the Star Wars Trilogy is closer to World War 2 than today
**President Biden's birth is closer to Lincoln's inauguration than his own inauguration
***Tyrannosaurus Rex is closer to us than to Stegosaurus dinosaurs

Cultural Time Relativity. Countries around the world differ significantly in how time is perceived and experienced. Western time aligns with written language read from left to right. Israel's Hebrew is read right to left. China's Mandarin language is read vertically top-down. People living in other cultures have a different sense of tempo in their daily lives. It's noticeable in body language, response to events and general level of patience. Cultural differences in time are also reflected in rigidity of schedules, attitudes about work or recreation, and reaction to missed appointments. Most sports events are regulated with fixed time periods limiting the action – like life time runs down except in soccer where it runs up.

Eastern culture perceives time as an eternal circular cycle without structure or linearity. Concepts of karma and reincarnation support time as a conduit where what goes around comes around. This perspective relates to a person floating in a river of time, where one's back is facing the future and the present is constantly moving past them. Events and experiences seem to follow a recurring pattern of yin-yang, beginnings and endings where things return to an original state. Eastern religions perceive time as a process of perpetual recycling. Hindu, Chinese, Hopi Indians and others also perceive time as less linear, more circular. Compare this view with the western linearity of forward moving increments. The symbolic circle of time in the east has been converted into a mechanical round watch in the west, where time is sliced, diced and precise – reduced to increments that control the order of societal progression. In the east, time is neither change or changeless – it's the emergent seam where emptiness becomes form. Heraclitus captured it best:

"No man ever steps in the same river twice, for it's not the same river and he's not the same man"

Consciousness. Our experience of time is completely dependent on our state of consciousness. We create the experience of time *out there* based on consciousness *in here*. Our sense of time provides continuity to everyday life as it informs consciousness and creates order out of chaos. In the beginning a child soon begins to sense a sequence nature to their experiences, along with spatial separation between themself and everything else. In this manner, consciousness creates a space-time reality. Mind's necessary function is to provide order by dividing, limiting, separating and projecting form. This organizing process separates into structure and relationships what was previously a unified whole. Thus, past and future emerge out of the singular present. An infant child operates in a timeless mode until perceptions of separateness and linearity emerge and persist.

States of being influence our experience of time. Moods, excitement, boredom, drug induced and altered states all change our temporal perception. As a mental construct it's perceived differently when we're isolated, distracted, tired or in an emotionally charged state. The mind's projection of the future determines our emotional state; with positive expectation we feel good and hopeful, when negative we feel anxiety and fear. Without mind we would just *Be* in the moment. Fixation on either past memories or future possibilities is an escape from an undesirable or uncomfortable present.

Subjectivity equals perception bias. Everything is filtered through our internal consciousness lens. Consider the odd experience of seeing a favorite actor age over the years in movies while we feel as though we haven't changed. Or the confusing feeling watching a 20-year-old movie that seems only a few years old and then seeing that lead actor today, aged in a sudden time warp.

People are present biased because we overemphasize our current time in history. Top 10 lists invariably skew to recent periods, often overlooking past examples. Consider how the presidential candidates in 2020 (Biden & Trump) were born closer to the Spanish American War than to the present. Our present bias makes this difficult to comprehend. Perception of time is always YOU biased. We naturally perceive the beginning of recent history as the date we were born. Everything we temporally relate to is distorted by our own limited place on a very large time scale. We inescapably view everything through temporal blinders. When we think about the past it's a projection from our present state of consciousness. Both past and future are formed in the here and now. Both are simple extensions of Before and After, relative to our conscious Now.

Memory. Our memories are conscious creations, colored by our current state of being. As conscious states change, impressions of memories can change as well. Error and misinterpretation creep in along the way, beginning with the first impression and continuing through our ongoing development. As we change our self-image there are corresponding changes in how we relate to our own memories. This gap also applies when multiple people experience the same event yet later recall memories of it differently.

Memories of dull periods are recalled in a way that makes those periods seem shorter than they were. Alternatively, exciting events may be short in duration but how we remember them makes them seem longer than they were. These recall biases distort the original experiences, making long periods seem shorter and vice versa. It's probably because slow, boring periods have fewer specific moments to recall while exciting times have more memorable moments to recall. Memory is essential to our perception of time. Without it there's only the eternal Now. Memory gives structure to a formless, abstract experience of time.

The earth itself is one giant timepiece, providing a trove of recorded information in a variety of physical and biological forms. Geological layers of sediment tell a clear story about earth's past. Rocks, ridges, valleys and tunnels all persist as open clues to events separated merely by intervals of time. Radioactive decay and carbon dating provides a precise template for tracing back in time, revealing how, what and why things happened in the past. The fossil record is literally an entire biological compendium of recorded history – an information resource just as useful as any modern digital format.

Time and memory are codependent, each necessary for the other to exist. Memory is built into the infrastructure of the universe – an omnipresent attribute found in everything at different degrees. Human beings have developed a very sophisticated version of it and mistakenly assume it's the only form. However, it lurks all around us hiding in plain sight. Consider a large stone push up a hill, storing potential energy waiting to be released some future day when it rolls back down to a level resting place. Or consider a foot path made visible, revealing past travel from preceding pedestrian traffic. Memory is present wherever information is stored in some form. It's the substance crime investigators thrive on. Examples can be found everywhere at every level; physical, biological and social:

Memory Built into the Universe at Every Level

Physical	Stretched rubber band, dented bumper, crime scene, Grand Canyon
Biological	DNA, tooth cavity, body scar, skeleton, antibodies, belly button, bruise
Social	Artifacts, ruins, statues, plaques, history book, stories, trails, legacies

Our view above the earth is equally rich with stored information; the night sky. Patterns in constellations tell a subtle story about the formation of the universe billions of years ago. So too does the color shift of stars, their apparent motions and measured distances from each other. It's no different from a crime scene where subtle clues capture past events with surprising precision, contrary to what the untrained eye cannot see. When observing the night sky we're simply looking back into the far, far past – a setting that existed many thousands of years ago we're just now seeing. Cosmic history is perfectly preserved in this manner through an aspect of physical memory, where everything that ever happened is embedded into everything else. Without memory we would have no concept of time. Time is memory dependent and both are consciousness dependent.

Metaphysical Relativity. Human beings are subject to the effects of duality, always connected to both external physical reality and inner spiritual dimensions. Accordingly, time is experienced in two ways: Physical time bound and non-physical timelessness. We are conscious beings simultaneously navigating through time in both physical and non-physical realms. As original newborns, there was no awareness of time, just being, moment to moment. Soon our experience in the world expanded, gradually becoming bound by its limitations. Mind creates a reality of objects separate from us; "I" and everything else *out there*, similar to how western religion separated God from within us. Our experience of time is always two-fold; outside and inside. On the outside we follow clock time with great regularity. It constrains our entire life activity. Inside, our perception is shaped by states of being and inner aspects of sense (mental, emotional, spiritual). In a variety of altered states time is both distorted and transcended.

Duality results in time affecting us in two general dimensions – physical and non-physical. In the material domain, bodies experience cell development, growth, death and replacement. Your heart beats, lungs breath in and out, muscles contract and relax. You wake, work and rest. In a non-physical dimension your mind thinks, the soul feels and imagination creates. Intuitions emerge from a timeless domain transcending physical constraint. Clairvoyance, precognition and psychic visions are possible since the future already exists in the timeless realm. All future possibilities already exist prior to actual manifestation in the physical world.

Time Relativity in Human Dimensions

Physical	Sense in the *present* Now
Mental	Abstract linear construct, *Past and Future*
Emotional	*Non-linear*, Feelings weave around, in and out
Spiritual	Intuition, insights *transcend time-constrained physical realm*

As beings present in dual realms our experience in time is both fluid and fixed. Living in the world molds us as we navigate through time in the physical environment of external reality. Along the way we develop and grow as individuals yet deep within we possess the same eternal soul, invariant and beyond worldly drama. Respectively, our experience with the passage of time involves dynamics linking the outer world, inner being and deeper spirituality. This includes both linear and non-linear aspects of temporal reality.

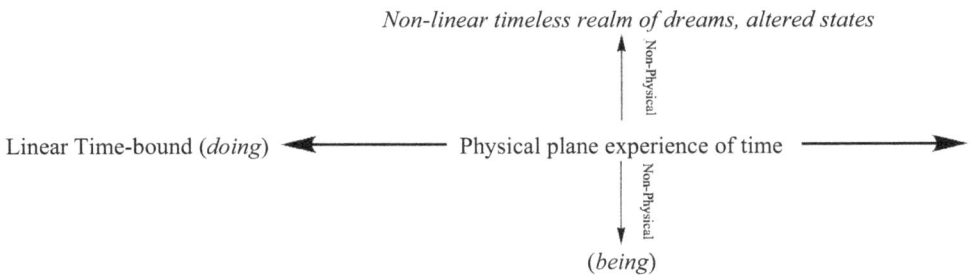

Non-linear timeless realm of dreams, altered states

Linear Time-bound (*doing*) ⟵——————— Physical plane experience of time ———————⟶

Non-Physical

Non-Physical

(*being*)

Timelessness. The standard concept of eternity is backwards and upside down. It's not endless time, but rather no time. Timeless vs all time. A snapshot photo is eternal yet absent of time. The speed of light is timeless. If a person could actually travel at that speed, time would literally stop. At the speed of light, the time bound reality we're use to is transcended. Similarly, the quantum level is timeless, where atomic particles exist as probability clouds without direction or position. Micro level dynamics create reality moment by moment. Emergent time manifests micro moment by micro moment – the creative seeding of everything in an ongoing continual infinite string of sequential eternal moments. The physical world emerges from a timeless quantum level where particles form patterns and structured relationships with other particles. It's a transformational process where manifested matter acquires timebound constraint.

The timeless energy of light transitions into ordered patterns of timebound matter – a transcendent process of change in kind. Eternity implies never ending time. If so, anything that could happen will and must happen, regardless of how improbable. When someone says "That wouldn't happen in a million years", it really means it WILL happen, probably after a million years. This is the secret sauce of evolution, which takes place over such a long time span that it makes even the most improbable possibility an absolute certainty.

Our actual experience of time is always here and now yet when we look back at each passed moment, captured in photos or video, it's perceived as an event frozen in time. A snapshot picture of you is not stuck in time but rather beyond time, residing in the realm of timelessness. Same with videos or movies. A film reel is an inanimate media strip that seems to come to life when played but the whole sequence of frames are individual, static images not constrained by physical time. The whole experience of watching movies is an illusion that fools your mind into seeing motion when all there is are static, lifeless frames. You as a physical being may be trapped in space-time but pictures and mental perceptions transcend those limitations of your time-bound reality.

Human perception of eternity always seems to be something out there and beyond here, a separate existence. However, eternity can be experienced now. As Joseph Campbell put it; *"Eternity is not the hereafter...this is it. If you don't get it here, you won't get it anywhere"*. Religions that espouse immortality are referencing the timeless aspect of spiritual transcendence. Unfortunately, most confuse it with non-ending time instead of timelessness.

Eternity is an infinite number of finite Nows – where the stillness of eternal presence abides. It's commonly portrayed as an unending straight line that goes on forever however, it actually resembles the circumference of a circle, with motion along a line that goes everywhere and nowhere.   * 0 and ∞

GEOMETRY OF TIME
Infinity is derived by adding dimensions

-Point becomes an infinite line
-Line becomes an infinite plane
-Plane becomes an infinite 3D cube
-Cube becomes an infinite 4D hypercube

Circle represents the eternal
Never-ending timelessness:
A line that never ends
Infinity or ∞ (circle to circle)

Time Travel. The fantasy of moving through time in either direction has been widely covered in pop culture where movies routinely portray individuals hopping in and out of past or future time periods like tourists on vacation. Let's just set the record straight once and for all. There will NEVER be a case where a human being travels forward or backward in time. It's simply not physically possible. While trapped in a universe constrained by physical laws we are inescapably limited by temporal reality. The best anyone can do is change the rate of time experienced but nothing more. Time, like space, is a difference, and there's no such thing as a negative difference – it's always a magnitude or absolute value. The idea of negative space or time is just a concept but not a real, physical thing. Any discussion of going back in time is nonsense in the physical realm. Same with going into the future. The only way to transcend temporal reality is to enter a non-physical, transcendent realm, such as a dream state or unconscious mode. Since time doesn't exist in those realms it opens up relationships with the past and future.

Moving exceptionally fast can slow the rate of time. Moving at light speed can stop the rate of time. But no object can ever reach that speed due to inertial limits. Higher gravity can slow the rate of time, and I suppose being in a black hole would stop time. But in either case the only difference is changing the rate of time. Traveling back to the past or forward to the future is just not *physically* possible.

The closest we can come to time travel is time transcendence, which is possible in a non-physical realm. Dreaming is a state of being not subject to limitations of the physical world, so naturally we can experience the *feeling* of time travel. The same goes with imagination. But these are not cases of time travel but rather an experience of time transcendence – consciously operating beyond the limitations of a time bound world. Time travel creates paradoxes and confusion because of duality, mixing aspects of one realm with another.

Conclusion. Time is a process essence revealed in our TOE model. It manifests at the 3 primary levels as Evolution, History and Biography (Nature, Society and Man). You are centered in the here and now, a conscious being of the present aware of a relative past and future. You are in the middle; *Spatially* between above and below, inside and outside, and *Temporally* between before and after, past and future. You are conscious in the present moment, the Now of temporal reality. That's the center of everything. Our consciousness creates the experience of time. As Christopher Hill explained "Time is a self-created abstraction perceived, not in sensory experience, but as the relationship between experiences."

Time Reflects the Process Essence of our Universal Matrix (TOE)

Social	History	Timelines of collective development
Bio (Man)	Biography	Timelines of individual development
Physical	Evolution	Timelines of nature & organic development

*Each level subject to dynamics of change: slow-steady, moderate, quick, accelerated or sudden

Life is a continual process flow; not the temporary event patterns we perceive here and now. Each individual being experiences a unique timeline journey – a string of nows with both incremental and fluid aspects (digital and analog). Physical reality is more about dynamic processes of becoming, moment by moment, not the static forms left behind. Time is the abstract trace of those processes, which continually change everything, including you. We're more fluid than fixed; beings present in the eternal now flowing along a journey of experience moment by moment. We're timeless souls inhabiting physical bodies constrained by time, born into the world, exit upon death. A duality with 2 aspects; one eternal, the other temporal – a composite being, embedded in both time and timelessness.

DYNAMICS OF TIME SPECTRUM

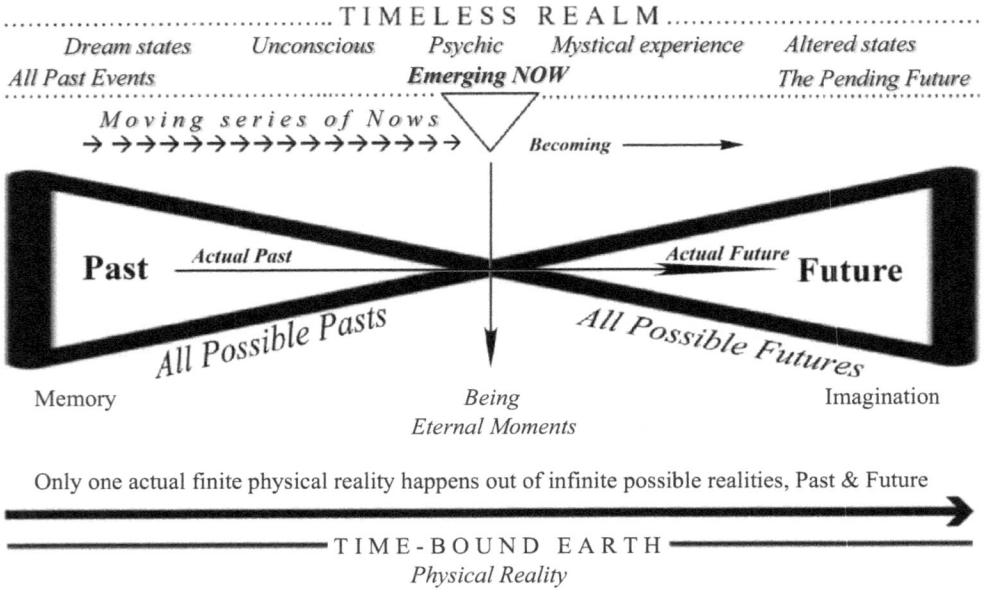

....T I M E L E S S   R E A L M..

| Dream states | Unconscious | Psychic | Mystical experience | Altered states |

All Past Events   **Emerging NOW**   *The Pending Future*

M o v i n g   s e r i e s   o f   N o w s
→ →→→→→→→→→→→→→→→→→   *Becoming* ⟶

Past   *Actual Past* ⟶   *Actual Future* ⟶ **Future**

All Possible Pasts   *All Possible Futures*

Memory   *Being*   Imagination
          Eternal Moments

Only one actual finite physical reality happens out of infinite possible realities, Past & Future

⟶

T I M E - B O U N D   E A R T H
Physical Reality

HISTORY is the record of time at the social level in a tapestry of timelines;

Local, regional, and around the world, each with parallel tracks and scale depths.
History connects humans' collective activities, with repeating patterns and cycles
where the glue tying it all together is human nature. The meaning isn't so much
about names and dates but what actually happened. After all, someone was bound
to do it. Out of infinite possible realities history is the record of the actual
manifested one. Streaming backward from the here & now, we trace everything
that happened with parallel and cross current threads. It's a sequential record of
every now that ever happened – the comprehensive memory of past moments.

A closer analysis of history reveals common recurring themes woven through
various periods and conditions. The consistency of human nature produces
predictable patterns in how things happen, regardless of the time, place or setting.
As Shakespeare extensively chronicled, use and abuse of power is a common
theme for those who rule, and their subjugated masses of haves and have nots.
Cities, states and nations continually face challenges of economics, religion,
resource allocation, security threats, health limitations, food scarcity and internal
disputes. Population demographic change pressures cultures to adjust, frequently
leading to civil strife. Modern societies struggle with government regulation,
taxation, racism, exploitation, corruption, epidemics, drought, foreign threats…the
same challenges that beset every society back in time…even ancient periods.

Political history reveals ongoing shifts in concentrations of wealth and power,
driven by differences in motivation, abilities and fortune. Good intentions by

organized political agents are usually short sighted, creating worse conditions through unintended consequences. Attempts at redistribution are rarely equitable, fostering further discontent. Democracy is a very new development in world history, compared with long established past periods of monarchy or minority rule.

   History records the continual transient nature of the world at large. In earlier times native settlement was the norm where people generally lived in the same place. With the modern era came greater mobility, locally and abroad. It's also ushered in technology generated transience where products and processes quickly become obsolete, replaced in a manner that changed the landscape. With a longer timeframe you can see how world maps are merely temporary snapshots of ever-changing boundaries and country names. Nations like the Congo have changed their name multiple times in just the last century. Our natural bias towards our own limited place in history creates the false perspective of a static world map.

   History is always interpreted, subject to degrees of inaccuracy or incompleteness which get worse over time. Inaccuracy creeps in from limits and bias. The vast majority of human history (99%) occurred before the advent of writing. Embedded clues reveal the flavor of past times, veiled in subtle forms: old movies, literature, folklore, ruins and artifacts. Impressions are obscured by both *Present bias* and *history bias*. In the latter, we overlook all the possible events and developments that could have happened but didn't, many very close to happening. Nearly infinite possibilities that were equally likely simply didn't occur. Present bias occurs when we look at the past from a current perspective, judging prior periods in time with a mindset tied to conditions today. Values, standards and living conditions are very different between those living today and the great majority who've lived through human history. Few really appreciate our privileged position on the forward edge of time's arrow, living in a unique period unlike any other in the past. You occupy the leading edge of history, here and now, a teeny sliver compared to the vast majority of people who ever lived on the planet, left behind in the ongoing wake.

PERSPECTIVES OF HISTORY

Surveying the past is a far-reaching task given the scope and depth of human history. It's not just a straight line going back from here to there but an interconnected matrix of parallel lines moving in different directions and tempos with crisscrossing pathways. It's a comprehensive tapestry involving multiple dimensions with levels, scales, scope and depth.

Scales of History: How far do you want to view the past? The range is huge so it's necessary to limit perspective. For simplicity we'll use 3 general scale ranges:

Near   Recent past (few generations)
Mid   Intermediate (multiple generations)
Far   Ancient past (*looooong* line of generations)

Scales of Human History Compared

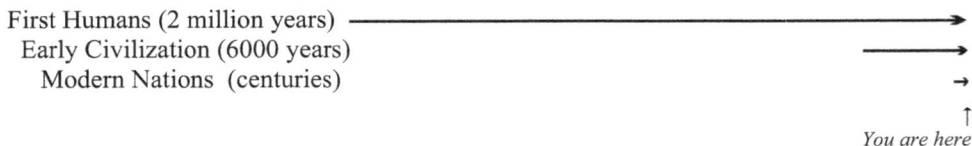

First Humans (2 million years) ⟶
Early Civilization (6000 years) ⟶
Modern Nations (centuries) →

↑
You are here

Levels of History. Every region, nation or society contains multiple levels or hierarchies, each with its own autonomous history. Global history actually contains a tapestry of nested subsets, each with their own unique timelines and events. For simplicity we'll use 3 general levels: Local, Intermediate and broad. Town or City is local, state or province is intermediate, and national or regional are broad levels.

Nested Levels of History

Cosmic (Evolutionary level, not *history*)

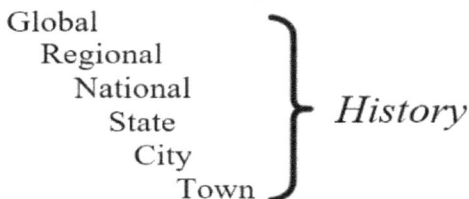

Global
    Regional
        National
            State   } *History*
            City
        Town
Family – Family tree, generational
You (Biographical, not *history*)

Global   — Geological events. Rise and progress of hominids. Post-evolution thru Pre-societal. Ice ages, Events that affect the world as a whole (World War, Climate change, Pandemic, worldwide web, Olympics)

Regional — Events specific to regions such as Asian, Middle East, Sub-Saharan Africa, European, East-West, Artic, Mediterranean, etc. (1st civilizations, advanced civilizations, Renaissance, Chinese dynasties, oil production, alliances, natural disasters, war)

National — 195 Countries in the world. Each nation's own history, internal and external events. Major players in recent history: America, China, Germany, Japan, England, Russia, India, France, Canada, Italy

State   — Events and key figures specific to states. (government and politics, economic, cultural)

City   — Events and key figures specific to cities. (Similar elements as the state level: Sports, schools, traditional events, infrastructure developments). Major historic World cities: New York, London, Paris, Tokyo, Beijing, Istanbul, Berlin, Rome, Moscow

Town   — Events and key figures. (new buildings, fires, church events, parades, weekend festival)

Family — Tribe, clan. History as lineage of generations. Great, great, great, grandparents. Common names (Smith, Jones, Miller, Rodriguez, Garcia)

Scope of History: Human activity and events occur in a variety of different areas or categories. History proceeds at every level across all of the following arenas:

political, business, art, music, sports, entertainment, science, technology etc. Each category has its own timeline of change over time.

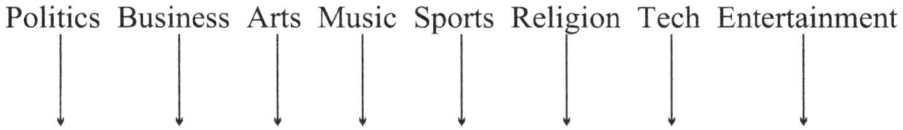

Politics Business Arts Music Sports Religion Tech Entertainment

Depth of History: Every region, nation, culture, etc. has its own independent historical timeline. This includes 7 Continents, Major regions (East, Middle East, West, Northern hemisphere, Southern hemisphere) and 195 countries in the world.

History at the National Level

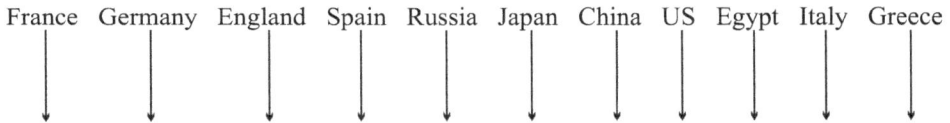

France Germany England Spain Russia Japan China US Egypt Italy Greece

Combining Levels, Scales, Scope and Depth produces a rich, comprehensive perspective of social history – a complex tapestry of various time lengths, categories and areas of interest. A more practical way to examine it is to simplify history by dividing its parts into general periods or phases:

3 General Periods: *Far-Mid-Near* (Ancient, Middle, Recent)
    Far   Earliest civilization thru Roman Empire & Birth of Christ (4000 BC – 500 AD)
    Mid   Medieval Period (Dark Ages) thru Modern Era (500 AD – 1900 AD)
    Near   Modern & Post-Modern Era (1900's – Present)

3 General Periods of Perspective:
    Pre-Modern   Mythic, Magic, Gods
    Modern Rational   *Objective reality*, Scientific revolution
    Post-Modern   *Subjective reality* (Recent history, roughly last 50 years)

Another simple approach is to delineate history into transitional or transformational periods. Here's a breakdown of history based on major economic periods of transformation: Foraging, Horticulture, Agriculture, Industry, Information, AI.

Economic Transitions in History

Tech-Robotic	Trans Modern	AI, Cybernetics, Bots	*Androids*
Information	Post Modern	Electronics, Computers	*Networking*
Industrial	Modern	Factories, Assembly lines	*Machines*
Agrarian	Pre Modern	Male Gods, Food Surplus, Empires	*Animal Power*
Horticulture	Ancient	Female Gods, Mother Earth, Nature	Man Power
Foraging	Pre History	Early Hominids-Caveman, Primitive	Man Power

*Transition periods often include newly introduced materials (stone, wood, metal, steel, ceramics, plastics, composites) or new foods & beverages (beer, wine, spirits, coffee, tea, coca cola)
**Emerging Transmodern era is rapidly going above and beyond the Post-Modern paradigm*

PRIMARY TIMELINES.

History consists of timelines with various lengths, locations and categories. It appears to proceed in linear fashion but to make sense of it we must divide portions into meaningful increments. Timelines can be separated between event lines (vertical) and periods intervals (horizontal).

```
------XXX-----|-----|-----|-|--------XXXX-------XXXXX---|--------XXX---------XX-----|--->
      Event              P E R I O D
```

It's really a choice between event-based snapshots or period-based zones. In either case events and periods serve as nodal points where identifiable change or transition takes place. There's an overemphasis on using decades to delineate periods or eras with unique characteristics. The 10-year period approach is fine but the beginning and ending set to 01 and 10 of a decade rarely match the times. It's better to use actual milestones near the beginning or ending of a decade since they accentuate unique attributes of a period. For instance, the "fabulous 50's" really describe the period between 1952-1963 (innocence of Ike & JFK until the assassination). The "swinging sixties" are represented more accurately between 1964-1969 (Beatles, hippie love & moon landing). And the 90's were more realistically the period between 1989 – 2001, set between the fall of the berlin wall November 1989 and the terrorist attacks in September 2001 (11-9 & 9-11).

Longest Known Timeline. Science estimates the big bang occurred just under 14 billion years ago, marking the start of a Cosmic Scale of Time that predates biological evolution and human history. Natural and cosmic events are functions of evolution, not human history. Evolution connects us to the first life forms, organic elements, DNA, and all ancestors. History is a transcendent derivative of evolution – a social essence transcending the physical one. YOU are a transcendent derivative of history, with a unique biographical story. Your life stages, milestones and overall journey are comparable to history's periods, eras and phases.

A Cosmic Scale is difficult to comprehend. Our limited life span relates to days, weeks, months and years. Anything beyond a few centuries becomes an abstract, imaginary construct. Sometimes the only way to make sense of broad timeframes is through condensed comparative models. The PBS series *Cosmos* graphically represented a cosmic timeframe using the familiar 12-month calendar format:

Cosmic Calendar - *A Perspective of Time*

Jan 1st		Big Bang
May		Milky Way forms
Sept		Sun & Moon form
Dec 31st	10:30 pm	1st Humans appear
Dec 31st	11:59 pm	– (last 12 seconds of last minute) Early civilizations form
Dec 31st	11:59 pm	– (very last second) Modern history begins

first humans appear ↓

●——→
Big Bang   Sun forms   *You are here* ↑

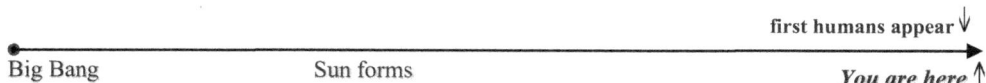

Evolutionary Scale.   After the big bang and the subsequent billions of years for stars and galaxies to form, planets eventually followed. Once formed it was just a matter of time, a very long time, before organic life emerged. It took a few billion more years for multicellular life to develop, eventually into mammals, then early hominids. So even before we get to human history, we must appreciate the vast swaths of preceding time passed at the cosmic and evolutionary scale. Earth's *looooong* developmental periods are measured in Eons, Eras, Periods, Epochs and ages; timeframes between a few million and a billion years. These Geological time frames are beyond comprehension, as foreign and unintelligible as a nanosecond.

Geological Time

Supereon – several billion years
Eon   – 1 billion years
Era   – 100's of millions
Period   – 10's of millions
Epoch   – several million years

Prehistory.   The human experience begins with early man diverging from an ape species roughly 7 million years ago, splitting into 3 separate branches: Gorillas, Chimps and Humans. 4 million years ago humans adopted upright posture. A few million years later man's brain size began to increase significantly. Humans went from cavemen to hunter-gatherers to agriculture and herding. 50 thousand years ago man progressed past stone tools, incorporated fish hooks and crude weapons.

Man's Appearance begins History
Ages: Ancient and Pre-History
Stone Age (2.5 million BC – 3000 BC)
Bronze Age (3000 BC – 1200 BC)
Iron Age   (1200 BC – 500 BC)
*Ages vary by regions of the world

Major Stages-Periods-Eras
Prehistory   History
Paleolithic
Neolithic
Ancient
Medieval
Modern

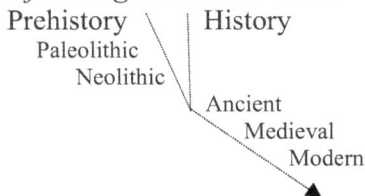

Ancient History.   Human history proper begins with early developments of civilization. After thousands of years in prehistory humans cluster beyond just clans and tribes, forming organized assemblies with rudimentary elements of managed organization. One defining attribute of this transitional phase was the innovation of writing. Despite crude materials, written language fostered extensive documentation of human activity and enabled preservation and expansion of accumulated knowledge.

Timeline of Earlier Civilizations		*Timeline of Ancient Innovations*	
3500 BC	Mesopotamian	2,600,000 BC	Cutting tools, stone chips, edge tools
3200	Egyptian	100,000 BC	Art (45,000 BC cave paintings)
2500	Indian (Indus Valley)	65 000 BC	Bow & Arrow
1750	Chinese	50,000 BC	Rope
750	Greek (500 Classical, 335 Hellenistic)	20,000 BC	Agriculture, livestock
500	Roman (Republic)	11,000 BC	Domesticated animals
330 AD	Byzantine (Eastern extended Rome)	8,000 BC	Boats
800	Feudalism	7,500 BC	Bricks
1300	Mongols	6,000 BC	Textiles
1200	Ottoman	4,000 BC	Wheel (Mesopotamia potter's wheel)
1300	Aztec	4,000 BC	Glass
1436	Inca	3.500 BC	Wheeled vehicles

Early Development of First Civilizations occurred in 3 Areas of the World:

MIDDLE EAST
Nile Valley: Egyptian
Mesopotamia: Sumerian, Akkadian, Babylonian
   EAST
   Indus River region (India): Harappan
   Yellow River Region (China): Huang Ho
      WEST
      Central America region: Mayan, Aztec
      Andean Region (East Brazil): Inca

Later Development of Advanced Civilizations and Empires:

MIDDLE EAST
Greek, Roman, Persian, Ottoman
   EAST
   Chinese Dynasties, Japanese Kingdoms
      WEST
      European powers, England, America

Advancement of Early Civilization. Increasing size and density of populations required innovative ways to produce and distribute food. Solutions included domestication of animals for pulling plows, water channeling networks, markets, food storage, and division of labor. It's no coincidence the largest centers of early civilization emerged by major rivers and large bodies of water. Domestication of animals included horse, goat, pig, sheep, cow, etc. with human directed breeding and feeding. Cultural progress led to advancements in social order with a system of laws, political control and organized government. Economic progress followed with improvements in production and distribution of goods, market places and systems of currency. Increased security was needed, leading to steel swords, lances, daggers, small arms, and artillery. Things got real once swords were replaced by guns. *Great civilizations emerged next to major rivers:
Tigress & Euphrates (Mid-East), *Yellow* (China), *Nile* (Egypt), *Indus* (India)

Most of human history consisted of savages and barbarians until the very recent emergence of civilized culture. People just fell into 1 of 3 classes: Ruling elites (exclusive few), craftsmen, traders, bureaucrats, and the vast majority of farmers and commoners. Despite these class divisions there were still opportunities with expanded specialization of functions and roles to support larger communities.

Civilization developed in a layering process as populations increased, incentivizing individuals to assume specialized roles, leading to a division of social classes. Over time, villages become towns, then cities, then states. Logistical support systems emerged with streets first, then rudimentary plumbing, fuel, transportation, and communication modes. Increasing size and sophistication then required formation of organized government and establishment of standard rules, laws and customs. This process is repeated at each successive level upward with added nuances driven by greater demands and scope. Eventually a nation is formed unifying all the layered parts. Note that most governments throughout history have been run by a minority of privileged elites or single men (monarchs and dictators). Democracy is a very recent development.

Catalysts of Developing Civilizations

Animal powered machines:	Horse, cattle, donkey – plows, grain grinding wheels, water wheels
Sea Transport:	Sailing ships, navigational tools – sextants, magnetic compass
Political organization:	Bureaucrats, hierarchies of control, religious authority
Tech progress waves:	Copper, Bronze, Iron, mechanical innovations

The germination of Pre-Modern civilizations surfaced in the Mid-East Mediterranean region, setting the table for European/Western modern and post-modern advancement – progressions through Sumerians, Egyptians, Greeks, Romans, Byzantines, and Ottomans. The early ancient and later modern eras were separated mid-way by the Medieval period, roughly 1000 years of relatively slow progress due to disjointed fragmentation of prior organized hubs; a once integrated network of cities and towns degenerated into disconnected, unstable, isolated pockets of activity…until an awakening of cultural consciousness emerged known as the renaissance, meaning *rebirth*. A collective coalescing and reconnecting fostered a burst of creative energy and innovation, expressed in sciences, arts and intellectual pursuits – reintroducing classical culture. This period too was followed about a century later with the enlightenment wave, further advancing science but with an added flavor of human centered philosophy, overriding rigid dogma.

History of the Human Condition.   A key takeaway from history is appreciating how living conditions have changed over time. Progress has been painstakingly slow despite advancements in recent years. Simply put, anyone alive today is far better off than the over 100 billion people that lived before. During 99% of human history, living conditions have been primitive, uncomfortable and inconvenient. Those who lived through those times didn't dwell on it much since it was simply normal. During the last several thousand years conditions improved mildly yet nowhere near the immense improvements made in just the previous few centuries, especially the last one. Cities used to be dirty, smelly filthy cauldrons of horse manure and sewage. The 20ᵗʰ century transformed everyday life so dramatically that very few people have any clue how harsh life was like just a short time ago.

PHASES OF CIVILIZED MAN

Ancient Period (6000 BC – 750 BC) – Early civilizations with magical and mythic orientation. Gods lord over the earth and events. Kings rule territories.
Classical Period (750 BC – 500 AD) – (Greek & Roman) Intellectual advance. Rational approach. Logic, math, engineering, medicine, philosophy, music
Medieval Period   (500 – 1300) – (Feudalism) Isolated, religion directed, oral tradition, magic-Myth, local vernacular vs prior Latin language standard
Pre-Modern   (1300 – 1800) – Rebirth of classical Greek and Roman roots.
Renaissance Period   (1300 – 1600) – Intellectual, science, writing, philosophy, creative arts
Reformation Period   (1500 – 1555) – Catholic church split/Protestant sect. Rejects hierarchy rule.
Scientific Revolution (1555 – 1700) – Significant progress - math, physics, chemistry, biology, etc.
Enlightenment Period (1685 – 1815) – Age of reason, skepticism and elevation of individual man.
Modern   (1800 – 1970) – Industrial Revolution: Transition to manufacturing process
Enigmatic period   (1905 – 1945) – Paradigm shifts in science, physics, psychology and arts
Post Modern   (1967 – 2000) – Subjective, relative perspectives. Multiple valid viewpoints
Trans Modern   (2000 – present) – Tech driven, internet integrated
All periods overlap during their transitional phases

Tracing the advancement of the human condition is challenging and misleading due to the uneven paths and regions affected. Progress is often limited to specific regions while others remain stuck in the past. Western culture advanced greatly during the last few centuries while the rest of the world changed much less or not at all. A surprising number of nations and populations still experience living conditions that were common 1000 years ago. Also obscuring the real picture are differences within any population due to separations of classes and wealth. Types of slavery still exist throughout the world today, affecting millions of poor subjects coerced by malevolent governments. Living conditions are broadly different within any country between the haves and have nots, regardless of technological progress. When we consider the inventions and innovations that make life easier, keep in mind that it doesn't apply evenly to everyone, everywhere.

Recent history is unique given the variety of amenities and comforts making everyday life better. These include things we consider essentials today: adequate housing, quality and quantity of food, running water, heating/cooling, electricity, healthcare, education, etc. For most of human history these "essentials" were questionable at best. With the rise of early civilizations living conditions gradually improved somewhat by introducing basic sustaining systems. Eventually homes transitioned from huts and animal skin tents to mud structures, stones, brick and mortar, to wood and glass windows. Running water piped right to one's house was a huge convenience; then eventually adding a sewer system. Candlelight and gas lamps made nighttime a little more manageable for thousands of years. Outhouses were a civilized solution during that same timeframe until quite recently as indoor flushing toilets were introduced. Wood and coal burning stoves were used since ancient times yet continue today, though gas and electric versions took over just a few hundred years ago. Anyone born in the mid 1800's lived under the same somewhat primitive conditions that others did for thousands of years. Compare that with someone born in this past century and it seems like two different planets.

Recent Transformation of the Human Condition, in just the Last Century

Widespread Use of Modern Amenities:

Electricity	Cell Phones, Pagers	Running water	Radio
Supermarkets	Airconditioned homes	Immunization	Paved roads
Gas/electric stove	Electric Washer/Dryer	Movies/theaters	Smart Phones
Automobiles	Compact Film Cameras	Personal Computers	Television
Photocopier	Refrigerated home food	Interstate highways	Internet

Dramatic Improvement in the Human Condition Globally in just 100 years

Poverty rate	– People living in extreme poverty: From 84% to 9%
Basic Education	– Uneducated people: From 83% to 14%
Literacy	– Unable to read: From 88% to 14%
Vaccination	– (diphtheria, tetanus, whooping cough): From 0 to 86%
Child Mortality	– Surviving first 5 years: From 57% to 96%
Democracy	– 1% of the world living in democracy: Today 56%

```
O----------------------------------------------------------------------------------------------|-O
< ---------------99% of human history people lived in primitive conditions-------------------->|
< No indoor toilet, no refrigerated food, no air conditioning, no laundry, no soft bed, no lighting >|
```
Last 100 years

DYNAMICS OF HISTORY

The passage of time brings change to everything in a progressive, iterative process. History as a subset of time follows the very same dynamics but with added complexities of human nature interacting at the collective societal level. Man in time is now mankind in time, including everyone who ever lived and every city, state and nation of the world. The process now involves both the mechanics of temporal change and the dynamics of human beings interacting socially.

Mechanisms of Collective Temporal Change. History emerges in the same quantum manner as time itself. From many possible events only one happens. Some are more probable than others, but nothing is ever certain. And as soon as something does happen it changes the likelihood of the next possible thing that might happen. Day by day, year by year, events unfold in a probabilistic manner. Events that do occur soon become the past, influencing the new present which in turn influences the future. History is the long wake of many, many moments that actually happened, with no trace of the unlimited possibilities that could have happened but didn't. Generally, what does happen are events that were most probable. But over extended periods, less probable things can and will occur. Given enough time, anything even remotely possible will and must happen. Unlikely possibilities become mathematical certainties when the scale of time is enormous, which is why evolution shouldn't surprise anyone.

The course of history is shaped by mechanisms of change and progress. The process of social advancement is both linear and nonlinear, cyclical and spiral. Things can proceed smoothly until a tipping point is reached, bringing sudden disruption. A combination of predictable and unpredictable events steers the course of history at every turn. Mechanisms contributing to change include weather events, economic panic, political revolts, cultural fads, epidemics, new technology, charismatic leadership and outbreaks of war. History, like time itself, is a process subject to ongoing emergent developments influenced by a variety of interrelated and interconnected elements that only appear obvious after the fact.

Interacting Elements Mutually Influencing the Course of History:

Environmental	Geography, Weather, Climate
Political	Power, authority, governance
Economic	Production, distribution, consumption
Cultural	Music, Art, Religion
Intellectual	Science, technology, innovation

History generally proceeds in slow, gradual pathways thought the pace varies between slower and faster periods. Occasionally we see a perfect storm where everything lines up, leading to a quickening – a time when change shifts from incremental to transformational. Major transitions in societal economic systems occurred in successive overlapping waves: Hunting, Agriculture, Industry, Information, and high-tech – each phase advancing quicker and lasting shorter.

Key Transitional Advancement Periods in History

First Civilizations emerge
Greco-Roman Intellectual advancement
Renaissance-Enlightenment further advancement
Industrial revolution
Electronic, Tech transformation
Internet information explosion

Watershed events divide history's timelines in half. Discoveries, innovations, tragedy, cataclysms, black swan events, etc. all create split timelines – before and after an impactful event – a watershed point where things are never the same again. On a certain level the world has changed in the way things are done or the way we perceive them. This includes inventions like the light bulb, automobile, and television, or innovations like home plumbing, electricity and air conditioning, and events like the assassination of Archduke Ferdinand, Pearl Harbor and 911. The same dynamic applies to your life as an individual. Your personal timeline has numerous watershed points that changed the trajectory of your life path. These might include meeting an impactful person, landing an unexpected job opportunity, a traffic accident, winning a lottery, birth of a child or a debilitating illness.

Mechanisms of Societal Change. History proceeds via social interaction where internal and external forces continually clash. Patterns of build-up and release in collective tension leads to waves of change. When conditions are just right a tipping point is reached, transitioning from gradual to large scale change. Shifts often follow a pendulum like swing between opposite extremes, bringing change in a back and forth, cyclic manner. Collective groups become agitated for various reasons: perception of injustice, lack of resources, or regulation and financial burden. Basic human flaws such a greed, envy, disdain and hatred can resonate at the collective level, resulting in disruptive events.

Internally a population has ups and downs of satisfaction, with peaks and valleys of agitation. Attitudes changes are expressed in numerous forms; general dissent, peaceful protest, and armed revolt. In extreme conditions transformational change leads to dramatic revolution – government take over and replacement. Externally, foreign interaction creates similar dynamics resulting in shifts of satisfaction or agitation. These are expressed in trade disputes, border conflicts, diplomacy, treaties, alliances, or in extreme conditions, war.

Catalysts and Cataclysms. History changes in different tempos, ranging between slow, steady, moderate, quick or sudden, like a stream of water that can flow at different rates or crash and splash. Slower prolonged changes are called periods; sudden ones we call events. Tempos are directed by elements mentioned earlier, each serving as a catalyst (positive or negative) or a cataclysm (negative) that influences the nature of change.

Natural – Cataclysmic events of nature may seem random and rare but occur quite regularly over time. On any given day there's a major storm somewhere on earth. In any given year there's a life changing storm somewhere and given enough time there's a history changing storm somewhere on the planet. Natural disasters come in a variety of forms: hurricane, earthquake, volcano eruption, monsoon rain, tsunami, flash floods, meteor impacts, etc. Consider the meteorite strike at

Chicxulub Mexico 66 million years ago that wiped out dinosaurs and nearly three quarters of plant and animal species worldwide. On a smaller scale volcanos and hurricanes have been responsible for wiping out whole towns and cities. We can include viral outbreaks as natural disasters with their sudden fast spreading tendencies and deadly effects. The Bubonic plague killed nearly half of the Europe's population in the 14th century, lingering on for almost seven years, taking the lives of 100-200 million people.

In addition to sudden natural events there are longer term transitions of equal or greater impact. Climate change can cause extreme cold or heat, moisture or dryness, impacting migrations on a large scale. It's believed by some that a 300-year drought contributed to decline of the Greek empire. The dust bowl of the 1930's led to mass migrations in several US States. It's quite possible the harsh Russian winter of 1941 stopped the Nazis from winning the eastern campaign and the war as a whole. No one knows how much history could be different given the ongoing impacts of weather, climate and sudden unpredictable acts of nature.

Man-Made – Nature's destructive impact is often matched by man-made cataclysms, purposefully, accidentally or through plain stupidity. Going to extremes is the usual culprit and given human nature it's only a matter of time before some reckless behavior, individual or collective, intentional or unintentional leads to a history changing disaster. Poor central planning can result in mass famine. Exuberance in financial markets can lead to panic and economic collapse. Careless sanitation behavior can open the door to disease and widespread pandemics. And a single individual with warped delusions can upset a stable social fabric with one consequential assassination. Man-made disasters are easier to identify when they result in sudden catastrophe but they're equally disruptive when the damage culminates slowly over longer periods of time.

Catalysts of historic change come in a variety of forms and mechanisms, including transitions in resource availability (food, water, fuel), innovations, political discourse, cultural creativity, economic collapse, religious zealotry or scientific paradigm shifts. Technology improvements have been very influential, both in general and in specific sectors, such as communication, transportation, medical or military. Environmental change is a constant. Nations and civilizations thrive when they adjust to the demands of that ongoing change. Eventually, complacency sets in and those in charge fail to meet new challenges, resulting in decline and predictable crisis or downfall. War is the ultimate challenge and catalyst, resulting in winners and losers, either of which changes the course of a nation's history. Victors create the new world order, for better or worse.

Theories on What Drives History

Marx	-Materialistic dialectic, reduces it all to a human contest for resources & power
Spengler	-Cycles of rise and fall, growth and decay
Weber	-Focus on religious movements
Carlyle	-Great man theory, certain individuals are born to lead and make great change
Toynbee	-Dynamics of challenges and response

Other theories focus on evolutionary aspects, use of force, divine right and social contract. All of these are valid as influences steering the course of history. Ultimately a number of factors work together to create resonant tipping points forcing transitions as events unfold. Similar to organic species (99% are extinct), so too are world cultures, most gone forever.

Economic. Human nature expresses itself collectively as a competition for goods, resources and power. As Marx emphasized, there's always a "struggle" among individuals and groups, between classes of haves and have nots. However, where he saw only negative, capitalists see positive. It's natural for individuals to pursue happiness through personal gain and acquisition of things, just as nations do the same on a higher level. As Adam Smith observed, engagement in self-serving actions with others doing the same creates a balanced market place of supply and demand, occasionally skewing more to one side of the equation. Whenever human nature is involved, imbalances will occur, including excessive supply or demand, over regulation and bad government policy leading to disruptions, including recession or worse, depression. Nations often fall when their economies fail.

Political. Competition for power in society or between societies is an ever present dynamic, both internally and externally. It's expressed in a variety of ways: diplomacy, mediation, legal action, deception, corruption, alliances, coercion, and armed aggression. War is simply politics by violent means. In any society of significant size there are competing forces that clash. Normally opposing interests resolve differences in a civilized, healthy manner. But even democracies face periods of crisis where a faction feels so disenfranchised that it creates chaos beyond peaceful resolution. Political and socioeconomic turmoil can transform a nation, through revolution or civil war. Well-known examples include the US in 1775, France 1789, Russia 1917, China 1949, Europe 1848, Cuba 1959, Iran 1979.

External relations with neighboring states or nations routinely clash with opposing interests, either resolved peacefully or through war. Diplomacy is politics at the international level, with similar tactics at play, engaging in either mediation or coercion. Though war is usually expensive for opposing sides, it's generally embraced with misguided assumptions of easy victory. Short term gain is the incentive, driven by any number of rationalities including economic, territorial, religious, defensive or nationalistic conquest. Outcomes are usually transformational, sometimes existential.

Battles that Changed the Course of History

Battle of Britain	– Insured Germany would face a 2-front war
Battle of Hastings	– Established Normans as rulers of England
Siege of Baghdad	– Ended the 500-year Abbasid Caliphate
Fall of Constantinople	– Expansion of Ottoman Empire into East Europe
Battle of Cajamarca	– Spanish conquest of Inca Empire
Battle of Yorktown	– American colonies independence
Battle of Waterloo	– Ended French domination of Europe
Battle of Marne	– Stopped German advance into France
Battle of Gettysburg	– Turning point of US Civil War
Battle of Stalingrad	– Turning point of German-Russian war front
Battle of Marathon	– Elevated Greeks as equals to Persian Empire

Nationalism is a prime driver of war typically based on *we're better than you*. It's also rationalized as a necessary clash of cultures. Nationalism often transitions into imperialism, conquest and colonialism. Numerous despots and conquerors became individual self-induced cataclysms: Attila the Hun, Ivan the Terrible, Genghis Khan, Timur, Caligula, Vlad the Impaler, Mehmet Talat Pasa, Vladimir Lenin, Benito Mussolini, Maximilien Robespierre, Adolph Hitler, Joseph Stalin, Nicolae Ceausescu, Mao Zedong, Pol Pot, Idi Amin. Kim Il Sung, Mengistu Haile Mariam, etc. History is littered with individuals who changed history at great cost.

Religion. Every area of the globe has been significantly impacted by religion. Three of the world's great religions emerged in the Middle East region: Christianity, Judaism, Islam, including their associated languages: Aramaic, Hebrew and Arabic. History is greatly influenced by these and many other lesser religions, through both internal order within societies and external conflicts between societies. While religion has played at least a partial role in many wars, some have been waged solely because of it. They can be violent and prolonged, then lie dormant, then surface again later. Its deep-rooted nature makes peaceful resolution difficult, so wars of this type linger on. Sometimes conflict arises within the same religion due to differences in interpretation of sacred texts. The worldwide war on terror contains both political and religious elements.

Demographic Dynamics. History is certainly driven by population patterns involving size, strata, age, race, gender, density, etc. Early civilizations emerged as groups of people collectively settled areas. They soon developed, expanded and became more concentrated, with positive and negative effects. One positive was increased specialization enabling economic progress. On the negative side, population density created havens for the spread of disease and pandemics. Similarly, pioneers and explorative nations unwittingly carried their parasitic germ baggage with them to foreign indigenous peoples and wiped-out tens of thousands of unprepared victims, ultimately erasing those civilizations from history.

Populations generally grow over time, leading to thresholds that impact and change a country's development. Catastrophes and wars have briefly interrupted the continual growth of populations only to see those numbers expand to record levels afterwards. In a surprising twist no one saw coming, world population numbers are expected to decline, largely due to urbanization and lower birth rates. Demographics play a major role in how individual nations progress or regress.

Great Leaders. Single individuals can change the course of history for better or worse. Some special people rise to the occasion when they're most needed. There's disagreement on whether the times choose the leader or a leader makes the times, but periods of crisis generally attract the strongest leaders. A partial list of impactful leaders during wartime, economic crisis, or political and cultural chaos includes Gandhi, Mandela, Churchill, Lincoln, Napoleon, Washington, Caesar, Alexander, Roosevelt, Stalin, Mao, Hammurabi, Charlemagne, Genghis Khan, Catherine. Notice how almost all are recognizable with only one name. Also note how many have a negative association. But let's not overlook other impactful leaders in various categories such as religion, which would certainly include figures like Christ, Buddha, Mohammed, Confucius and the Dalai Lama. There appears to be a reciprocal relationship between crisis and exceptional leaders.

Technology and Innovative Ideas. Ongoing progress of mankind is paralleled by continuing improvements in things and processes. Many innovations have led to gradual advancement while others have made major impacts on social progress. History is often changed by the single introduction of an invention that achieves widespread use. In ancient times there was the wheel, fire and improvised tools. In modern times the list of innovations is long, ratcheting up human advancement at a breathtaking pace. Things really started to take off during the 1700's and have only accelerated since, with the rapid advance of science and technology.

Innovative Ideas and Concepts – Furthering Human Progress

Scientific method	Laws of thermodynamics	Sanitation
Writing & language	Religion	Quantum mechanics
Domestication of animals	Music	Natural Selection
Heliocentrism	Immunization	Psychology
Agriculture	Currency	Capitalism
Mathematics	Concept of Zero	Relativity
Laws of motion	Democratic government	Systems theory
Human rights	Atomic theory	Artificial intelligence

Technology Timeline

3.3 Mil BC Stone Tools	1765 Steam engine	1888 Kodak camera	1948 Holograph
1.1 Mil BC Fire	1776 Submarine	1889 Dish washer	1953 Polio vaccine
6000 BC Irrigation	1783 Hot air balloon	1891 Escalator	1953 Videotape recorder
4000 BC Sailing	1791 Gas engine	1892 Electric oven	1957 Orbital satellite
3800 BC Wheel	1794 Cotton gin	1892 AC motor	1959 Photocopier
3500 BC Writing	1797 small pox VAX	1893 Motion pictures	1960 Laser
2100 BC Calendar	1804 Locomotive train	1895 X ray radiography	1960 Heart pacemaker
1500 BC Glass	1825 Electromagnet	1901 Radio	1963 Cassette tape
1400 BC Rubber	1826 Photography	1903 Airplane	1969 Moon landing
1300 BC Concrete	1826 Gas stove	1907 Ford assembly line	1970 Pocket calculator
1200 BC Iron	1829 Sewing machine	1914 Military tank	1974 Personal computer
300 BC Geometry	1830 Electric motor	1916 Helicopter	1976 Mars landing
850 Gunpowder	1836 Revolver	1916 Stainless steel	1978 Test tube baby
950 Windmill	1842 Anesthesia	1918 Sonar	1978 GPS
1040 Compass	1844 Telegraph	1922 Air conditioner	1981 Space shuttle
1250 Mechanical clock	1856 Plastic	1922 Radar	1982 Laptop computer
1436 Printing press	1862 Machine gun	1926 Talking movies	1982 Music CD
1568 Mercator map	1866 Dynamite	1926 Electric refrigerator	1984 Cell phone
1592 Flush Toilet	1873 Typewriter	1926 Liquid fuel rocket	1985 Genetic engineering
1608 Telescope	1876 Telephone	1927 Television	1989 Internet
1617 Adding machine	1877 Microphone	1928 Penicillin	1995 DVD
1656 Pendulum clock	1879 Electric light	1931 Electron microscope	1997 MP3
1705 Newcomen engine	1884 Steam turbine	1937 Electric computer	1999 DVR
1709 Thermometer	1884 Fountain pen	1944 Jet airplane	2000 Blue-ray
1755 Refrigeration	1855 Automobile	1945 Atomic bomb	2007 Iphone
1764 Spinning jenny	1856 Electric fan	1947 Transistor	2016 Neuralink

Inventions and the introduction of new products aren't as important as their widespread use. The lightbulb invented in 1880 wasn't really a part of the social infrastructure until the early 1900's. In 1925 only half of US households had electric power. It's more meaningful to associate the lightbulb with the 1920's rather than 1880. This distinction applies to all technology and innovation.

Rise of Empires. Most nations in the world play a small part influencing the course of history, compared with the impact of a select few reaching the level of empire. Over the centuries various regions contained the right combination of elements enabling growth of powerful societies. These dominant regional players spread their culture over vast territories, controlling other nations in their sphere of influence. Empires all possessed great economic, political and military power, with significant increases in military sophistication and organized professional armies.

Many lesser empires impacted world history around the globe, including Macedonian, Qing Dynasty, French Colonial, Yuan Dynasty, Nazis, Portuguese, Japanese, and numerous others. An empire's total duration is often less important than the periods of peak strength and influence. Most continue on after losing position and stature. Some attained incredible dominance for a short period while others achieved moderate stature but held it for incredibly long periods. Empires come and go but their culture generally lives on in one form or another.

Timeline of Major Empires

Persian Empire	550 BC	(230 years)
Han Dynasty	206	(400+ years)
Roman Empire	27	(1000+ years) *longest in history
Umayyad Caliphate	632 AD	(90- years)
Mongol Empire	1206	(150+ years) *second largest in history
Ottoman Empire	1299	(600+ years)
Spanish Empire	1492	(450- years)
British Empire	1585	(400+ years) *largest in history
Russian Empire	1721	(200- years)
American Empire	1776	(250 years) *America IS an empire

As nation states spread across the globe their borders and spheres of influence increasingly overlapped, leading to either conflict or cooperation. Since war is expensive, wise elders avoid the shortsighted attraction of armed conflict, choosing instead to form fruitful alliances. Cooperative agreements pay dividends, offering cheaper security for all. Yet even those arrangements frequently fall apart over time and then backfire, such as the alliances that incentivized World War I.

World Dominance linked to National Global Reserve Currency

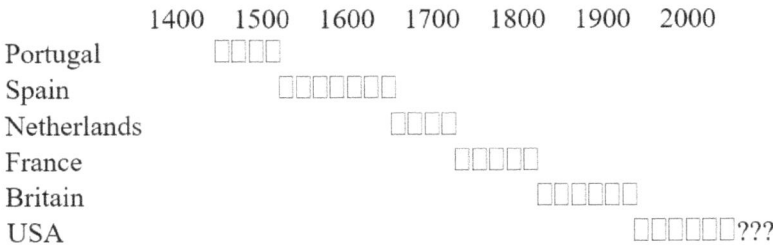

	1400	1500	1600	1700	1800	1900	2000
Portugal	□□□□						
Spain		□□□□□□□					
Netherlands			□□□□				
France				□□□□□			
Britain					□□□□□□		
USA						□□□□□□???	

PATTERNS IN HISTORY

In the short run it may seem history is somewhat random or even chaotic but over longer timeframes we can see recurring themes, repeated outcomes, and synchronized developments with connected cause-effect relationships to man's collective activities. Time proceeds forward in a broad and deep tapestry of parallel developments, creating an interconnected web of change. The process has a rhythmic quality about it, producing waves with peaks and valleys, shifting tempos and pulsating intensities. Like a music score, history produces flowing periods of even measure, punctuated periodically by distinct note-like events. The back-and-forth nature of societal change is a recurring theme in history; one that could be compared to a metronome or musical pendulum. Repetitions of history also resemble a sine wave with nodal points. The rise and fall of empires could be plotted graphically in such a manner, along with the path every society follows exhibiting inevitable armed conflicts, revolutions, financial crises, etc., all of which are simply nodes on history's timeline.

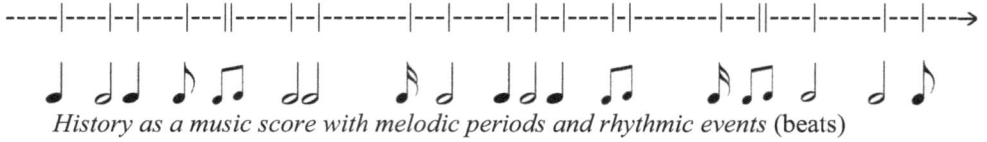

History as a music score with melodic periods and rhythmic events (beats)

Two very distinct patterns in the tapestry of history are the connections that occur within single periods and those that cross multiple periods (scope and depth). First, **patterns in Scope** appear as common themes, where a single, resonant zeitgeist permeates the entire social climate. This creates a general synchronized influence on each of the major categories of human expression. Collective values, tastes, attitudes and mindsets trend across the entire spectrum: Arts, Music, Tech, Literature, Philosophy, Religion, Politics, Economics, and Science. All these separate venues become influenced by resonating with general conditions and attitudes of the times. Consider the *Renaissance period* where rational thought replaced mythology, superstition and magical perspectives. After centuries of story telling the oral tradition gave way to the written word. Religious dogma lost ground to sense and reason. In the arts, abstract forms were replaced with representational realism. Flat paintings transformed into 3-Dimensional linear perspective. Creative arts in general were swept by the same spirit of inquiry and exploration sweeping across the sciences. The zeitgeist of the period added a flavor of humanism into every facet of interest and activity.

Another example of this pattern in scope can be seen in the early twentieth century, which could be called the *enigmatic period*. The early 1900's began with a series of perplexing theories and puzzling developments that swept across various unrelated fields concurrently. Einstein's theory of relativity in 1905 upended concepts of physical reality, turning a once simple, fixed view of objects in motion into a variable, non-uniform reality with no absolute value. Time and space were merged into a single essence. Suddenly everyday objects could change in mass and shape, warping in length near the speed of light where dimensions of time and space no longer have rational meaning.

In the arts, similar bizarre concepts were introduced. Cubism was a fad showing multiple viewpoints of a single object where front, side and back could be seen at once. Fauvism emphasized attributes of light, favoring the bold and vibrant with opposite colors juxtaposed. Suddenly the subject of a painting wasn't the object but the flavor of light within and around it. Futurism further influenced this period, manipulating the essence of time in works of art showing simultaneous actions fused together. Successive movements would be superimposed to transcend the element of time. So, 3 different innovative art fads introduced concepts warping elements of space, time and light – parallel with radical developments in physics.

Similar innovative, disruptive nuances surfaced in music, with the grounding concept of a keynote being manipulated and then transitioned to multiple keys and changeable tones. Normal harmony was set aside for purposeful embrace of dissonance. In Jazz it was suddenly common to separate and shift melodies apart from the whole. Riffing would just emerge, taking a music piece into a whole new direction without notice. In this same enigmatic period Sigmund Freud radically changed the field of psychology; deep dives into the dream world, psychoanalysis, and a theory of the human unconscious. Then right on cue, Salvador Dali was painting dreamscapes filled with warped objects, symbols of the collective unconscious, and melting clocks. Just like physics, the art world was tuning into realms of timeless, spaceless realities in sync with the zeitgeist of the era. Every period in history shares this commonality aspect where attitudes, approaches and creative energies synchronize across various unrelated fields of endeavor.

Patterns in Depth occur with recurring themes and outcomes take place over multiple time periods, revealing themselves as repeating cycles and lessons of history. The most common pattern is a back-and-forth shift between opposite extremes, cycling like a pendulum. "History repeats itself" is an ancient truism proven over and over again. "Nothing new under the sun" is a claim with similar spirit. Nothing can be written, spoken or done that hasn't in essence already been. Most "new" stuff is really just redone or repackaged from something prior. Most of life is a variation on a common theme. To borrow a popular slogan, we might playfully say "Same stuff, different millennium". There's repetitious sports interview cliches', pre-event hype, sales presentations, coach's speeches, politician promises. It's all been done before. Today's NFL Football stadium is just a modern version of the Roman gladiators in their colosseum. America's Wild west of the 1800's is now seen in big city gangs across the country. The Wells Fargo Wagon of heartland Americana is now the UPS and FEDX Truck. Ancient wizards and dragons have resurfaced as Jedi knights and death stars.

"We didn't start the fire; it was always burning since the world's been turning" – Billy Joel

Omnipresent Pattern of Cycles. Every field of human interest exhibits cyclic behavior, both in short term and longer periods of time. The most common cause is the tendency to pursue a course to its extreme and then retreat back in the opposite direction. It's expressed in alternating periods of tension and release, constraint and freedom, conservative and progressive. It's just a pendulum pattern of back and forth – pushing too far and coming back too far.

History may repeat in cycles but always with an added nuance, creating a spiral pattern. This is why prediction is difficult; cycles aren't static but rather dynamic, modified with updated conditions over time. You can never perfectly relive a past moment, but you can certainly come full circle back to the place you started, always with a slight difference. Spiral patterns come from circular repetition combined with forward progress through time.

Universal Pattern of Growth and Decline. One of the more obvious cycles in history is the rise and fall of everything. This includes societies as a whole and all the sub-components within. The pattern is simple: Slow growth, accelerating growth, peak, slow decline, rapid decline, reset. The latter "reset" stage can be a moderate revolution or an extreme, destructive dissolution, but in either case it's always transformative. This cycle is a lot like life and death.

Patterns of growth and decline can be long and gradual or fairly rapid. Modern advanced societies tend to progress through cycles more rapidly than those in earlier history when the tempo of life was slower. The phase periods may change but the sequential pattern is identical.

Cycle of Societal Progress and Decline

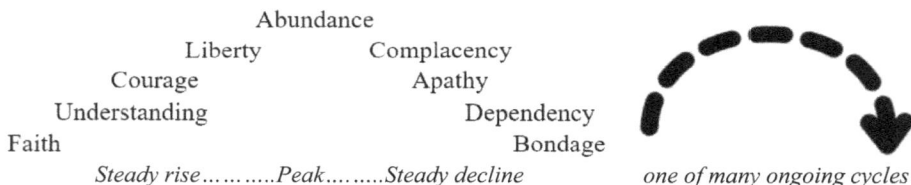

	Abundance	
Liberty		Complacency
Courage		Apathy
Understanding		Dependency
Faith		Bondage

Steady rise..........Peak..........Steady decline

one of many ongoing cycles

Cycles of Rise and Fall in Civilizations

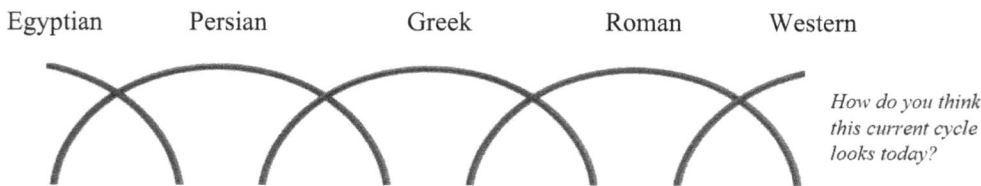

Egyptian   Persian   Greek   Roman   Western

How do you think this current cycle looks today?

History itself has a zeitgeist following patterns of shifting hot spots where human activity and energy are greatest in particular places in time. These temporary hubs are the places where social progress is peaking, for the moment, until a new hub becomes *the* hot spot. This ebb and flow pattern occurs at every level, (local-state-nation). It even applies to your hometown hangouts that become the *in place* or the happening hot spot here and now. Enjoy it while it lasts because it never does.

HISTORY CYCLES: Political, Economic, Cultural and War

Cycles may take the form of waves where social change advances in phases. This pattern is similar to generational cohorts where each successive band is categorized by a new nuanced flavor, a contrasting quality. Waves of social activity have been connected to various cycles in nature, including sun spot where an 11-year pattern seems to corelate to peaks and lows of various human expressions: creativity, rebelliousness, excitement, productivity, unrest, etc.

Political cycles reflect societal states of being, often changing in waves that parallel the 20-year generational cohort pattern. Elected presidents, governors and state leaders possess qualities temporarily in favor now, while those very same qualities would have been rejected a few years earlier. The back-and-forth cycle in political preference is revealed by the oscillation between conservative and progressive ideology. Pendulum swings between both poles push to either side, eventually creating dissatisfaction that leads to a swing back cycle. It's manifested in the back-and-forth sequence of congressional control and US presidents, Democrat and Republican. Rarely does one party occupy the Whitehouse for more than 2 consecutive terms.

| **DEMOCRAT** | Truman | | Kennedy, Johnson | | Carter | | Clinton | Obama | | Biden |
| **REPUBLICAN** | | Eisenhower | | Nixon, Ford | | Reagan, Bush | | Bush | Trump | |

Other political cycles are longer or shorter than the 20-year generational change increment. Swings between conservativism and progressivism shift frequently, matching closer to the periods of several election cycles. On the longer side there are particular swings in pubic unrest that emerge over multiple decades. Consider the 50-year cycle in protest-rioting seen in America peaking in the 1870's, 1920's, 1970's and 2020's – first 3 periods saw a parallel surge in the KKK while the last one produced Antifa and BLM groups. Other cyclical periods in American history included Federalists vs decentralist, slavery vs abolitionists, the 1930's New Deal, 1960's Great Society, 1980's Reagan Revolution, and countless others.

"Hard times create strong men, strong men create good times, good times create weak men, and weak men create hard times." – G. Michael Hopf

Cultural tastes shift back-and-forth just like political cycles, swinging between opposite styles of traditional (conservative) and Avant Gard (progressive). This manifests in every mode of artistic expression including: apparel styles, home décor, music styles, television programing, skirt heights, architecture, performance art, etc. The same social wave extends beyond culture to include political and economic spheres, most notably in the swings between individual rights and government restrictions. The pendulum pattern is symmetrical with the cycle of new generations rebelling against the status quo, who later become the new conformists that subsequent generations rebel against.

"Classical" refers to anything old-fashioned, once in vogue, likely to reemerge later

The universal cycle of growth and decline is most obvious in economic waves of boom and bust, with basically 4 phases: expansion, peak, decline, trough. All societies experience this pattern, ranging between short term growth and recession, and long-term cycles of expansion and depression. It's the same natural cycle that living beings go through, whether it's one's health, career or romantic life. A close comparison would be an athlete's cycle of training development, improvement, peak performance and eventual regression. The pattern of progress and setback is quite natural; the more one advances the harder it is to continue or even just maintain the progress. It's true of economic boom and bust, war and peace, progress and setbacks, all inevitable and natural. It's also quite mathematic; higher progress improves by smaller percentages whereas lagging growth can increase by much bigger percentages. That's why it's better to start coaching a losing sports team since just a few more wins translate to a greater improvement percentage.

Cycles of Western Economic Crisis	*Modern US Recessions*
1772 Credit Crisis – Bank of England	1945 Post World War 2
1789 Copper Panic – US	1948 Fall in fixed investments
1802 Recession – US	1953 Post Korean War
1812 Recession – US	1957 declining consumer confidence
1825 Panic – Bank of England	1960 Fiscal monetary tightening
1837 Panic – US Banking System	1969 Fed rate hikes
1873 Panic/Depression – Europe and US	1973 Oil embargo
1893 Panic/Depression – US	1980 Fiscal monetary tightening
1907 Financial Crisis – World wide	1981/1982 Anti-inflation rate hikes
1920 Monetary Collapse, Deep Recession – US	1990 Gulf War oil crisis
1929 Sock Crash, Great Depression – US	2001 Dot com bubble
1937 Depression recovery setback – US	2007 Great Recession, mortgage crisis
1973 Oil Crisis, Recession – US	2020 Covid Recession

*Various business cycles have particular timeframes: Inventory 3-5 years, Fixed Investment 7-11 years, Infrastructural Investment 15-25 years, Paradigm shifting Tech Innovation 45-60 years

Cycles between war and peace are universally present world-wide. Some nations have engaged in war longer than being at peace. And what we think of as "peace" is often just a compromising façade kept up to avoid the unpleasantness and expense of addressing actual grievances. This is true even in the most stable democracies. America itself has rarely had a decade without some armed conflict.

The key takeaway is that war is an equally "normal" state as peace. This pattern of war and peace is more pronounced in nations with elevated status in the world. Dominant countries or empires are naturally more active with greater presence and more at stake, leading to increased participation in armed conflict. During periods of "peace" there are often non-violent wars taking place, usually in economic and diplomatic arenas, along with culture wars, where the battles take place in the hearts and minds of populations. The basic difference between peace time and war time is productivity in the former and physical destruction in the latter. Cycles of war always catch participants off guard as they approach the next one based on the last one, overlooking nuanced changes along the way (circle vs spiral again).

United States at War

Indian Wars	(1622-1890)	*268 years*
American Revolution	(1775-1783)	*8 years*
War of 1812	(1812-1815)	*3 years*
Mexican War	(1846-1848)	*2 years*
Civil War	(1861-1865)	*4 years*
American War	(1898-1902)	*5 years*
WWI	(1917-1918)	*2 years*
WWII	(1941-1945)	*4 years*
Korean War	(1950-1953)	*3 years*
Vietnam War	(1964-1975)	*10 years*
Desert Shield/Storm	(1990-1991)	*6 months*
Afghanistan	(2001-2021)	*20 years*
Iraq	(2003-2011)	*8 years*
Global War on Terror	(2001-2021)	*20 years*
Cold War with USSR	(1945-1991)	*45 years*

The patterns of history are similar to those in individual lives, just projected to a collective level. Both individual and collective paths have the same predictable ups and downs, peaks and plateaus, and shifts back and forth. Sometimes lessons are learned along the way, often they're dismissed. As long as human nature remains the same, this pattern will persist. Repetitious cycles and waves will continue as each generation relearns what the previous one already experienced. Tastes and preferences will swing back and forth; new leaders will implement policies that initially work until they're pushed too far and backfire – then an opposing approach will be tried until it fails and the cycle goes full circle.

"What's past is prologue" – Shakespear (*The Tempest*)

Human history traces the throughput of different people playing the same roles over and over through time. It's a record of forward progress of societies carrying the torch from predecessors, in much the same way genes are passed on in individuals from one generation to the next. Established structures remain in place as different people come and go, cycle in and out, over and over again. Waves of human beings pass through the same schools, classrooms, businesses, government positions, etc. Sports teams and theater troupes process waves of new talent coming and going each generation. It's no different than a factory assembly line – insert generic worker here, replace when necessary, rinse and repeat. Imagine viewing this recurring process over hundreds of years in high-speed motion; seeing something resembling a living creature meandering, growing, changing, evolving.

History as Continual Human Throughput

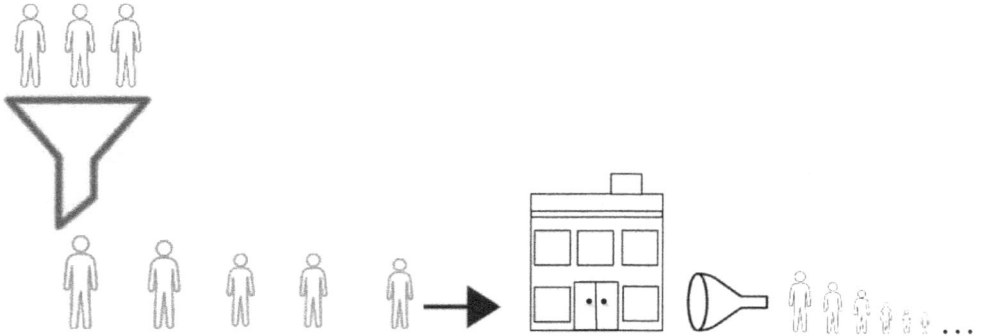

Centuries of people recycled through schools, businesses, factories, governments, sports teams...

HISTORY'S WAKE AND RESIDUE

The passage of time at any level consumes the present, leaving behind remains, without which we could never confirm anything ever happened. In our individual lives we leave traces of our actions everywhere. CSI teams routinely solve cases based using hidden clues left behind. History provides similar evidence, some visible, most not. Fortunately, we can extrapolate quite a lot of information from physical traces remaining after thousands of years.

Legacies include all the leftovers from prior time periods in both physical and non-physical form. Relics, ruins, artifacts, and vestiges of all kinds litter the landscape where previous peoples lived and died. Non-physical legacies are equally omnipresent, including words-languages, customs, rituals, superstitions, myths, tales and holidays. Legacies, relics and ruins warp our perception of history as a whole. The durability of things left behind gives those things prominence, focusing our attention that creates built-in bias. Things we can no longer see are forgotten and given less attention, regardless of how significant they once were. It's similar to the way modern news focuses on negativity and extreme events, creating a false impression that things are worse than they actually are.

Archeologists examine the ruins and remains of past civilizations, investigating ancient sites like modern day forensic detectives. Historians are comparable, perusing old records, written materials, physical markings and derivative items that provide clues as to how and why they were made. Who knows what future people in a thousand years will find looking back on those of us living today? Shouldn't be too difficult with the volumes of digital media that overload our information systems, assuming that media survives long enough. It's uncertain given how brick and mortar structures have demonstrated much greater durability than video tape or hard drives. And what will future generations think of us when they sift through extensive piles of garbage we've left behind, buried in landfills across the globe?

20th Century Relics: Film camera, phone booth, floppy disc, drive-in theater, fax machine, polaroid, typewriter, record player, encyclopedia, cassette tape, alarm clock, phonebook, VCR, maps, etc.

Legacies persist as layered increments, built up on successive prior layers. The process of geological sediment layering is mirrored by societies where products and structures of preceding civilizations are left behind, buried beneath current ones. History is preserved in this manner, with deeper and deeper layers over time. Remnants reveal out of place add-ons that no longer serve a purpose. Like vestigial limbs, parts of structures often hang around despite outliving their usefulness. These odd forms persist because it's often easier to build around and on top of things instead of starting from scratch. That why it's not uncommon to find homes with odd extra rooms, wires that go nowhere or wall plates that cover the past sins of a lazy handyman. It's also why there are dead end streets, empty or repurposed buildings, and ghost towns were life once thrived.

Human legacy is billions of skeletons buried beneath the earth. 117 Billion people have lived on earth

Legacies are the posthumous rewards of dominant cultures whose success and staying power lives on, often the victors of wars. Nations of greater strength, influence and longevity leave greater legacies. Recency of history also plays a part in what's left over today. The last empire to claim regions all over the globe was Great Britain, and thus English has been the dominant language world-wide. The American empire is really just a derivative of its English roots, further sustaining the language as a global standard. A similar dynamic took place in history with Romans carrying Greek traditions forward. Language legacies lead to lingering words that no longer mean anything to current generations. Words with ancient origins persist for centuries as meanings getting lost in translation. Lots of western words derived from premodern Latin still dominate the fields of medicine and law.

Languages considered extinct still live on in their multiple branches and derivative forms. Latin and Sanskrit are prime examples. Similarly, religion and general customs carry on, branching into multiple variations along the way. Past life continues forward, living on in persistent societal expressions. Eventually there comes a time when past cultural and ideological legacies clash with modern, newly established societies, occasionally resulting in regional war. Outcomes merely establish a new baseline for future cultural clashes which never fully subside.

Customs and traditions are legacies as equally impactful as physical relics. They're found in many forms, such as a handshake, a bow, or a tip of the hat, performed without much thought on their origins. Wearing various ceremonial caps or oppositely, prohibiting the wearing of them are further examples – traditions practiced today because someone said so long ago. Superstitions are passed on generation after generation, often based on conditions that no longer apply. Wearing a rabbit's foot, finding a 4-leaf clover, breaking a wishbone or fear of the number 13 seem odd but are still practiced by a significant number of people. Myths, fairytales, parables, nursery rhymes and storytelling reach far back into ancient times. The calendar is filled with holidays and designated special days that recognize events from long ago whose details are long forgotten. And most of the words and sayings that make up our language today are leftover relics from a variety of other languages and cultures born in a time and place far, far away.

Language Evolution and Diverse Origins (European slice)

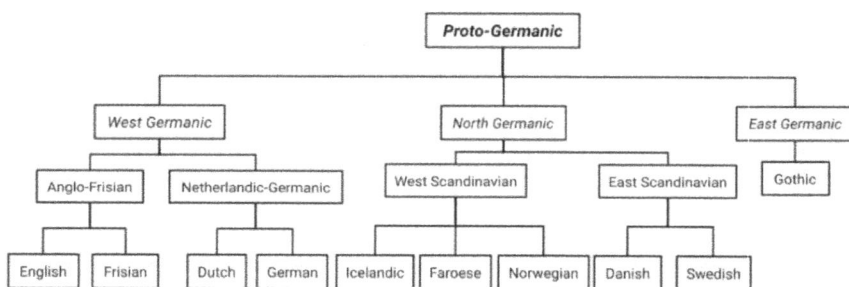

For any civilization to advance it must pass its essence onto a new generation, similar to individuals passing on genes. All the customs, rules, programs, details and institutional processes must be learned anew. Today this transition is handled through our comprehensive education system, no different than a young person learning how to become fully functional adults. It's a steady-state process of transferring information producing a cyclic upward spiral of newly generated progress. Just as science moves forward building upon predecessors, societies advance by passing on and expanding customs, values and standards. Where society is the fixed hardware, culture is the software, requiring continual downloading through each successive generation.

Material relics, ruins and artifacts aren't the most important societal legacies. Much more value resides with inherited lessons learned. History repeats itself, offering opportunities to avoid mistakes or bad choices for those wise enough to take heed. Consequences of human nature can be overcome if the simple patterns produced from generational repetition are recognized, appreciated and incorporated into current action plans. Unfortunately, this is the exception, not the rule.

Universal Lessons. As Henry St. John noted, "*History is philosophy teaching by examples*". Future success merely requires appreciation of all those recurring themes of the past. It's not complicated. You can gloss over the details and the vast record of history as long as general concepts are grasped and acknowledged. Simply recognize the few essential patterns in each of the major areas of society:

Economic

Cycles of boom and bust influenced by excesses, regulatory interference & over reaction
Resources are limited and routinely squandered
Prosperity is always temporary

Political

Despots appear in every generation. Appeasement never ends well
Government policies with good intentions often bring harmful unintended consequences.
Government policies are biased to the short term at the expense of the long term
Political leaders over promise and under-perform...and are rarely held accountable
Populations always contain dissatisfied factions who challenge and disrupt the majority
Societies thrive when freedom and regulation are balanced.
Revolutions typically replace the existing power structure with a worse set of scoundrels
War is much more likely when peaceful nations fail to respond to acts of aggression

Leadership

Leaders and visionaries make the difference between successful nations and all the rest
Nations behave a lot like the individuals who run them
Leaders best represent people's interests when they share the same roots and experiences

Culture

Human *behavior* changes over time and circumstances, human *nature* remains constant
We're individually and collectively better than we think and worse than we think.
Preceding generations had it way way way worse than recent or current ones.

History connects us to prior civilizations through artifacts, passed stories, recorded documents, language, accumulated knowledge, technology, customs, traditions and lessons learned. We're all connected to the original pioneers of mankind who preceded us long ago and left behind omnipresent traces of what they did and how they lived. Legacies, relics, remnants, ruins, and vestigial left overs are just the visible links that overlap and tie every life that ever lived together in an unbroken chain. History's residue and wake pattern contain a treasure trove of information revealing who we were, who we are and where we're heading.

Accelerating Tempo. After eons of excruciating slow, slow, slow geological change and then continued slow change in the evolution of organic life, the first humans finally appeared. Progress crawled ever so gradually, changing over long periods of time in arithmetic increments. Eventually a tipping point was reached where early civilizations emerged, gradually speeding up the pace of collective progress. Then another tipping point occurred, just a few centuries ago, further accelerating the tempo upward. Rates of change suddenly transitioned from arithmetic to geometric. On a graph this would look like a slightly angled straight line suddenly curving dramatically up. Technology coalesced, built upon itself like a recursive loop, then exploded like a chain reaction. Innovation from unrelated fields merged together stacking previous knowledge higher and higher. This self-reinforcing process passed through successive generations where new innovators stood on shoulders of preceding ones, seeing higher and farther than ever before.

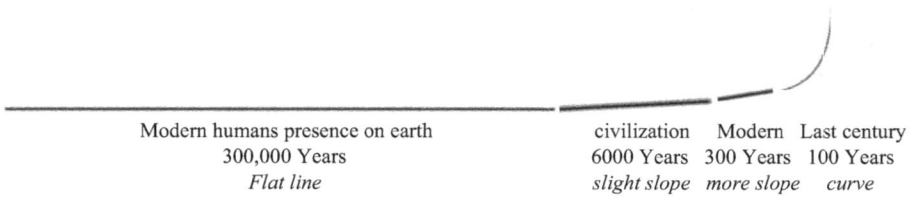

Modern humans presence on earth	civilization	Modern	Last century
300,000 Years	6000 Years	300 Years	100 Years
Flat line	*slight slope*	*more slope*	*curve*

Accelerating Tempo Catalysts

Pre Modern (1300-1800)   Printing press > mass communication
Modern (1800-1970)   Industrial manufacturing > mass production
Post Modern (1970-2000)   Electronics > Computational explosion
Trans Modern (2000-present)   Internet > Information explosion

Transportation: Railroad, Steamboat, Automobile, Airplane, Jets, Spacecraft
Communication: Horses, Telegraph, Telephone, Radio, Television, Satcom, Fiberoptic
Computation: Abacus, Slide-rule, Mainframe Computer, PC, Smartphone, AI
 * *Miniaturization of everything: microchips, hard drives, mobile devices*
** *Ever increasing computational speed, bandwidth and storage*

   The accelerating pace of change in recent history makes today's living environment uniquely different from all other past periods. For most of human history people lived under conditions that were virtually identical to preceding and following generations (there was no concept of cohorts with unique values). For thousands of years your parents before you and your children after you basically lived with the same set of household things, rules, education and entertainment. Then suddenly, a few hundred years ago, the pace of societal progress increased and carried with it innovations in tools, information, implements and creature comforts. New products and better ways of doing things were introduced on a regular basis. Hard, manual labor gave way to comfortable tasks and expanded leisure time. Your life routinely becomes easier and more fulfilling than what your parents experienced. Concepts like retirement and leisure travel are radically new. General optimism and opportunity is also a relatively new condition compared with the historical default of threats, fear and survival focus. This recent transition is dramatic and unprecedented yet most people living today aren't even aware of it.
   Our "recent bias" is equally matched by our "progress bias" – the false perception that modern civilization is a world-wide standard. It isn't. Today is no different from early times where cultures progressed unevenly; some move ahead while others stagnate or get passed by. Lots of places in our modern world remain very backward, primitive and undeveloped. Many countries lack readily accessible drinking water, electricity, modern medical equipment or adequate food resources.

Some secluded tribes are living in the same conditions that existed centuries ago. Human advancement may move forward continually but it progresses in an uneven narrow pathway. As a general rule, whatever your conditions are at this moment, much of the world has it far worse off than you.

History as the World's Greatest Story Ever Told. Nothing in literature is as comprehensive as the tale spanning human history, which includes every possible theme, story plot, suspenseful climax and teachable moment. Everything you need to know about the nature of man, human achievement and acts of folly are here, in one ongoing story. It's an interesting, fateful, often comedic yarn that includes lots of drama, tragedy, violence, romance, grandeur and the utterly ridiculous. It's really a documentary but contains the flavor of both fiction and non-fiction, the mythological, surrealism and borderline fantasy. The saga of mankind covers it all. There are numerous timelines of every scale, with both lasting periods and punctuated sudden events. Characters of all types come and go on a regular basis, along with their collective societies. Some represent the worst in man, such as despots and malevolent empires while others reveal our highest potential – visionaries and transformative movements. It's like a non-stop Shakespearian play with a *Game of Thrones* dynamic of power plays and shenanigans exposing human greatness and flaws, cunning and intrigue, triumph and tragedy.

The syntax of our great story consists of mechanisms of change and the dynamics of collective human interaction. Subplots are driven by economics, politics, religion, demographics, leaders and technology. The story holds up with reinforcing themes running through every segment in a tapestry of interconnected elements – some in parallel, others on tangents. Patterns and recurring themes are present throughout, revealed in repetitious acts, constant ups and downs, back and forth extremes, and unexpected yet inevitable reversals of fortune.

Our great story may never end but it does have a definable conclusion. Lessons learned are plentiful for those who pay attention. The long, endless odyssey of mankind's ongoing journey reveals how previous generations lived and how you today are blessed with an extremely privileged condition. Man's history is a mere sliver; a blip on the grand scale of time, arriving on the scene very, very late (the last minute on Dec 31st of our cosmic calendar). Your life as an individual on the scale of human history is also a sliver; a blip, a fraction of the last second (11:59 and 59 seconds on Dec 31st). History is the record of collective human experience, tracing back extremely long spans of time which none of us can truly comprehend. Our world is in constant change; difficult to notice in the near term, unmistakably obvious in a longer perspective. The very place you call home that seems secure and semipermanent didn't even exist a short while ago and is changing all the time in subtle ways. The town you grew up in likely looks very different after just a few decades passed. How did it look 100 years ago? How about 1000 years ago? These are miniscule time frames in world history, yet they seem like eternities to us.

Where time itself is *the* universal process, history is a partial subset tracing our collective human progress. Time and history connect everything in a vast tapestry weaving within periods and between successive events. You and everything that happened before you are connected in time, regardless of how recent or distant. Time is nothing but a consecutive series of nows, including the one you're experiencing at this moment, the same as it did for everyone before you...the same as it will for everyone after you. History is simply the collective record of those nows for every successive generation of human beings that ever lived.

SPECTRUM OF TIME

Cosmic ($\approx$14 billion years) Longest measured time. **Evolution** *begins*. What happened before the big bang?
    Earth (4.5 billion)
       Organic Life (1 billion) first multicell life
        Humans (2 million) first appeared in Africa
         - - - - - - - - - - - - *dividing line where* **history** *begins* - - - - - - - - - - - - - - - -
         Civilization (6000) Mesopotamia first
          Recorded History (5000) Egypt
           Empires
            Nations
             People - **YOU** (90-100) – *biography*
               Decade (*zeitgeist cycle*)
                 Year (*Societal cycle*)
                  Month (*menstrual cycle*)
                   Week (*work cycle*)
                    Day *(life cycle)*
                     Hour (*pay schedule*)
                      Minute (hold breath)
                       Second (*heart beat*)
                        Split second (*blink*)
                         Millisecond – 10^{-3} seconds(*explosion*)
                        Microsecond – 10^{-6} seconds
                        Nanosecond – 10^{-9} seconds
                        Picosecond – 10^{-12} seconds
*shortest ever recorded (trillionth of billionth seconds) → Zeptosecond -10^{-21} seconds
                        Plank Time – 10^{-43} seconds

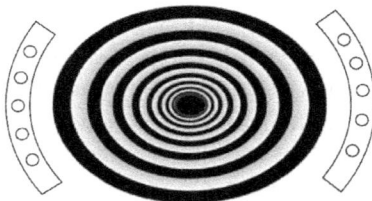

The Time Tunnel – 1966 television series
Just one of many time travel concepts, all which are *PURE* Fiction & Fantasy

7   HIDDEN REALITY

A comprehensive Theory of Everything must cover the full boundaries of what we know *AND* acknowledge limitations of what we don't know, beginning with an appreciation of the uncertainty built into physical reality along with natural limitations of human perception. Donald Rumsfeld famously pointed out there are known knowns, known unknowns and unknown unknowns – things we know, things we don't know and things we don't even know that we don't know. Hidden reality consists of the latter two cases. They include things beyond our perception, comprehension, awareness and actual blind spots. Hidden Reality is a combination of Known Unknowns and Unknown Unknowns. This natural limitation of the human experience was infamously embraced by Socrates who realized acknowledging that inconvenient fact was the first step towards true wisdom. Proclaiming "I know that I know nothing" set him free and sets the table for our category of known unknowns. Individual and collective knowledge is finite, incomplete and subject to a variety of biases and faulty logic. This recognition plus insatiable curiosity opens the door to higher understanding and awareness.

The World as man experiences it puts us all in a middle position, confined by natura limits of sense and perception. It's a world of practical scope in-between everything – a range with particular knowns we can somewhat easily grasp. But there are realms beyond; above and below, before and after, that are outside our awareness. There's an immense unfathomable cosmos beyond and a wonderous hidden microscopic landscape below, both ends deep in levels of hidden realities.

MAN IN THE MIDDLE – *Limitations of Perception*

Man's middle position narrows perception to the here and now, front and center, at the expense of higher and lower levels. Focusing is a practical necessity – a tradeoff where tunnel vision brings greater clarity but excludes the big picture. There's built in value to this limitation since the constant stimuli bombarding us from all directions must be perceptively managed, or we'd quickly lose our minds. Filtering and discriminating are necessary constraints for coping and comprehending. Ironically, the opposite, complete absence of stimuli experienced in an isolation chamber is equally hard to cope with.

While focused attention provides clarity it's still subject to natural biases. Perceptions can be warped by physical, mental and emotional filters that unintentionally distort reality. Errors creep in from poor focus, distraction, expectations and misinterpretations. The mind can play games with how we see and hear things simply based on our changing states of being.

Much of what we perceive on a routine basis is formed by mental constructs that change reality to fit our senses. Consider how distant objects look smaller than they actually are. Fast moving, oscillating cycles appear as steady forms.

Individual frames in a movie reel create the illusion of steady motion projected on a blank screen. Colors appear differently based on neighboring colors or surrounding light. The color of objects is actually just the reflected light, not the light that's absorbed into the object itself. Something that looks red has no red in it. Optical illusions emerge by changing angles, lighting, shadows, and context – often making simple forms appear very different than they actually are.

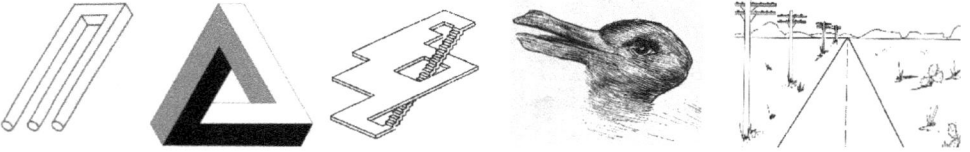

Three-dimensional impossible objects   *Duck or Rabbit?*   *Warped 3D illusion*

To maintain our sanity, we cling to the notion that space, time and motion are absolute. They aren't. All motion is relative and there are no absolute speeds or directions for anything. It all depends on You, the individual observer, perceiving objects and their motions in relation to yourself. Adding to the confusion, motion looks the same regardless of what's moving, the object or you. Moreover, size, shape, color and quantity all appear differently based on context of surroundings. All values and perceptions are context determined. There's no such thing as big or small without your individual perspective connected to it. The same is true with subjective values such as good and evil, positive and negative.

Our physical environment is literally saturated with light we can't see. Even the light we can see may appear as pure "white" but there are lots of variations based on context, filters and color temperature. Speaking of colors, the light we perceive as pure white contains all the primary colors and more within it – but they're invisible, beyond our awareness until filtered.

Solid objects we perceive are essentially an illusion. The material world is mostly empty space, shaped by structured vibrating energy fields in resonant form. Since individual atoms are mostly empty space, every *thing* must also be. The solid stable forms we perceive all around us are actually in a constant state of flux. Dynamic equilibrium fools us into thinking there is permanence when the opposite is true. Timelapse helps expose this flux reality, revealing the hidden life of slow moving "inanimate objects". "Buildings" you're used to seeing everyday are actually *crumblings* – structures requiring continual maintenance to preserve their integrity. This was captured nicely in the series "*Life After People*", where ghost towns and abandoned cities provide visual evidence of how our perceived "permanent" structures steadily break down and decompose over long unattended periods. Timelapse also reveals the hidden life of microorganisms, visualized in a cut apple quickly browning. As we speed up the movie of life many seemingly static, inanimate things come to life. At the right speed you could witness the

Grand Canyon forming, mountain ranges emerging, glaciers receding and towns growing into megacities.

Everything we experience and perceive is from a vantage point at the center of the universe. The vast majority of things are bigger and smaller than each of us. Scales of space and time stretch far beyond our small slice in the middle. Naturally our senses are honed to a very small, focused range allowing us to adapt and survive in the world, but it comes with a tradeoff. Enriching our sensitivity to the here and now excludes awareness of everything else at higher and lower levels. A microscope or telescope does the same thing in reverse – increasing clarity on the ends of the spectrum while completely missing the rest. Focusing somewhere, anywhere, results in fuzziness everywhere else.

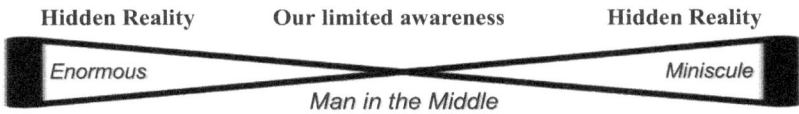

Hidden Reality	**Our limited awareness**	**Hidden Reality**
Enormous		Miniscule
	Man in the Middle	

SIZE	Cosmos	6 Ft height, 180 Lbs. weight	Atomic particles
TIME	Universe age	1 second heartbeat, 85 year life span	Nanoseconds
SPEED	Light speed	3mph walk, 27mph sprint (world record)	Glacier

Spatial Perception declines as we move further away from our cozy middle position. On the low end is a microscopic world we're completely oblivious to. If you could shrink in size many thousands of times smaller, you'd be in awe of the wonderous activity going on at the atomic and molecular level. Particles dancing and interacting everywhere in unbelievable synchronicity, switching places with others and swirling around in harmonic cycles. Energy transformations exhibiting patterns that would make any artist salivate. The parts and wholes working together in an orchestrated manner yet you would swear there was a high degree of randomness also. Immediately you would realize the classic concept of an orderly atom model was not present. All you'd see is an endless pattern of spherical particle clouds dualistically interacting in a nebulous but synchronized manner.

Modern science has revealed that atomic activity is not the simple structure we once assumed it to be but rather an unpredictable indeterminate wave form. Particles at this level do not have absolute measurable locations or pathways but potential values determined by probabilities. And since the very act of measuring them changes their disposition, we must conclude as Heisenberg did that uncertainty is built into physical reality. Thus, the very foundation of our physical realm consists less of actual substance and more of a murky cloud of uncertainty, making hidden reality the norm.

Beginning our ascent back to original size and perspective we might first pause at a level in the middle between atoms and us. Here we would find the world of microscopic organisms, again far too tiny to see but teaming with activity all around, everywhere we look. Life at this level includes a wide variety of tiny creatures and a myriad of microorganisms saturating the entire environment. The larger ones are soil and skin creatures lurking around every crack and crevice while on the smaller end are legions of bacterium and tiny viruses. These invisible germs are an essential part of the entire ecosystem. Omnipresent microbes are so ubiquitous they possess 20 times more mass than all the animals, including humans, on earth. Insects which are still tiny compared to us but much larger than their microbe counterparts are clearly visible so we can observe them on the periphery of our perception range. And like microscopic germs, they also saturate the ecosystem and are equally visually unpleasant.

Microscopic Scale of Life

Amoeba	.5 millimeter (*Large object on microscopic level*)
Neuron	.1 millimeter
Red Blood Cell	.01 millimeter
Staphylococcus Bacteria	.001 millimeter
Influenza Virus	.0001 millimeter
DNA Molecule	.000001 millimeter (1/millionth)
Atom	.0000001 millimeter
Nucleus	1/100,000th of an Atom
Quark	1/100,000th of a Nucleus
Planck length (*Smallest size we know*)	1.6×10^{-35} meters

*Microbes Types: Viruses, Bacteria, Archaea, Fungi, Protozoa, Algae, Lichens, Prions, Slime molds
**Mycoplasma gallicepticum: smallest single cell "living organism" on earth. (2/1000th of a millimeter)
***Planck length is so small YOU are closer to the size of the Universe than the teeny tiny Planck length

Just above the invisible level of microbial life forms is another layer of hidden pseudo-life: the realm of microscopic residue. This alien microworld of odd remnants permeates our exterior bodies, clothing, living spaces and every object we interact with. Layers of particles on the ground or floor are kicked up as we move around, readily becoming airborne. Carpets, curtains, rugs, mattresses, clothing and the skin of our bodies are lively jungles briming with activity, including a variety of grotesque-looking microscopic creatures. It may sound a little unnerving to realize there are over a million microscopic mites covering your body right now, most of which look like horror movie creatures. It's just part of this hidden world of multifaced micro-objects – an eclectic mix of unrelated stuff scattered everywhere, clinging to everything. This realm includes skin flakes, rock flecks, clothes threads, insect remains, sand, dirt, lint, pollen, dander, soot, spores, fibers, dust and miscellaneous debris. You certainly do your part by shedding a million skin cells every day. In addition to these particles there's residue from airborne vapor, smoke and water moisture. Now throw in dust mites and other parasitic micro creatures and it forms a bountiful cornucopia of unpleasant junk. No wonder we see a steady increase in allergies and asthma with all this hidden residue permeating our breathing habitat.

Your "clean" living space is actually just a particle playground where every day micro-debris is stirred up, moved around and added to the mix. There's a secret life of dust involving both human residue and environmental particulates, both natural and man-made. The structures we live in are constantly breaking down, needing maintenance and repair, leaving a constant build-up of microscopic leftovers that get into everything. Daily shedding of skin flakes and hairs (tens of thousands of flakes) tend to add up. The view at this level reveals human skin looking like a bumpy, porous landscape, checkered with ridges and a forest of long rods sticking up (hair), strewn with microbial parasites glomming on. Just walking in from the outside provides a steady stream of dirt your shoes track in. Smoking anything will steadily saturate every wall and ceiling. A simple cough or sneeze is really just a forceful outward projectile of unwanted debris. A hardcore germophobe might remind you of how much unwanted residue is exchanged in a friendly handshake.

Beyond your personal space there's plenty of hidden stuff in the general environment that's similarly unwanted, especially air pollution. Some streets are still laced with poisonous lead from the period of time when cars were fueled by leaded gasoline. Water is often contaminated with unwanted germs or particulates, usually within acceptable nonlife-threatening levels (enormous amounts of bacteria in a "clear" glass of water). The money you use on a daily basis is likely covered in germs and frequently has traces of cocaine from illicit drug traffic. The hidden world of residue at the microscopic level is a lot like sausage making: we're probably better off not knowing the actual details.

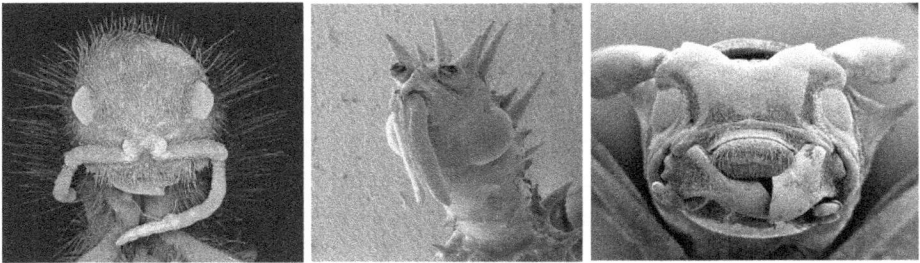

Ugly microscopic squatters living rent free in your skin. *(from the Oklahoma Microscopic Society)*

Macroscopic Scale of Life. Back to our original human vantage point we now tilt our heads up, gazing towards the realm of big things. Above us a multitude of ordinary large objects, including homes, trees, buildings and public structures. Most manmade things are well within the range of comprehension, though some push perception to the limit. Ponder the volume of concrete used to construct the Hoover Dam, or think about turning off every light switch inside the Pentagon; the largest office building in the world. Skyscrapers tower over our puny selves when standing at the base looking straight up. Modern sports arenas with tens of thousands of seats are wonderous to take in from any vantage point. Large, long suspension bridges, deep tunnels, modern cruise ships and aircraft carriers all inspire awe. And let's not forget ancient man's impressive constructions, like the great pyramids of Egypt. But man-made structures are miniscule compared to nature's own macroscopic wonders. Consider the giant hole in the ground at Grand

Canyon or standing by Mount Everest looking up at the peak. Picture the enormous volume of water flowing over Niagara Falls every minute, hour and day or the unfathomable size and depth of the Pacific Ocean. Earth's surface offers countless examples of large-scale features that make any one individual seem insignificant by comparison. But that's just the beginning of the macroscopic realm.

Extending our view above ground level we experience the first taste of awe-inspiring bigness as the atmosphere itself. It's over 6000 miles thick with multiple layers extending into outer space. Though generally composed of thin gases the sheer volume of it all creates a downward pressure of about 15 pounds per square inch at our level. Air may appear invisible and empty, yet it contains a variety of different gases; the largest percentage being Nitrogen at 78%. Also hidden is life sustaining water vapor dwelling in the atmosphere, totaling over 332,500 cubic miles of water within it world-wide (3.75 million billion gallons of water). You might sense this reality on days when humidity is extremely high and the surrounding air touching you is literally saturated with water.

Moving beyond earth and its atmosphere membrane we enter the true realm of macroscopic enormity in outer space, where man's tiny existence on the planet suddenly shrinks to insignificance. At this level objects are massive in size and distances are so vast they must be measured in light years. Our own sun, massive to us but quite ordinary compared to others, is so far away it takes over 8 minutes for its light to reach us. The nearest star beyond is much much farther, taking over 4 years for its light to reach us. Other stars take hundreds and even thousands of years to be seen here despite traveling the maximum speed possible. But that's miniscule compared to the distance from the nearest galaxy – Andromeda. Light from there takes two and a half million years to make the journey, and that's the *nearest* one.

MICRO LEVEL	MACRO LEVEL	PHYSICAL SIZE SCALE
(BELOW YOU)	(ABOVE YOU)	
Organ	City	10^{27} M Observable universe
Cell	Country	
Chromosome	Earth – 25 K mile circumference	
DNA Molecule	Moon – 239 Thousand miles away	
Atom	Sun – 93 Million miles away	10^1 M *Man in the Middle*
Nucleus	Alpha Centauri – 4.3 Light years away	
Proton	Andromeda – 110,000 Light years wide	
Quark	Furthest star – 28 Billion Light years away	10^{-35} M Planck length

Relative size: If atoms were blown up to marble size, softballs would be the size of earth.

Equally incomprehensible are the vast sizes and masses of celestial objects. Jupiter is over 1000 times the size of earth but so what: the sun is over a million times larger than dinky earth. Imagine how much bigger the largest stars are and then multiply that exponentially to gauge the size of massive black holes. The largest one detected has a mass 66 billion times greater than our sun. Beyond massive size and distances, we can add seemingly infinite quantities of stuff in this cosmic domain. Considering how enormous stars are, it's bewildering to think there are 10 sextillion stars in the known universe. That's 10 followed by 21 zeros. Don't even try to imagine how many atoms and subatomic particles are present in that jaw dropping total. No one can really know how much more there could be given our observational limitations dealing with enormous distances beyond. Nor can we be sure of what we do see since everything viewed here and now in the night sky happened hundreds and thousands of years in the past.

Scale of Relative Mass - *Diminishing Human Awareness*

MICRO LEVEL		Diminishing Awareness	MACRO LEVEL	
Human	80,000 Grams (175Lb)		Human	80 Kg
Bread loaf	450 Grams (1Lb)		Elephant	5000 Kg
Paperclip	1 Gram		Dinosaur	73000 Kg
Hair	.003 Gram		Titanic	52 Million Kg
Human Cell	10^{-9} Gm		Great Pyramid	5 Billion Kg
E Coli Bacteria	10^{-12} Gm		Earth	6 Septillion Kg (10^{24})
HIV-1 Virus	10^{-15} Gm		Sun	2 Nonillion Kg (10^{33})
Ave protein	10^{-21} Gm		Ton 618 Black Hole	130 Duodecillion Kg (10^{39})
Water Molecule	10^{-24} Gm		Milky Way Galaxy	1.2 Tredecillion Kg (10^{42})
Electron	10^{-28} Gm		Virgo Supercluster	2 Quattuordecillion Kg (10^{45})
			Observable Universe	44 Sexdecillion Kg (10^{51})
				(10 followed by 51 zeros)

Observable universe is 93 billion lightyears in diameter. Actual universe is 250 times bigger with 2 Trillion galaxies we can't even see, and 200 Billion Trillion stars that we know of.

Temporal Perception is equally limited by man's narrow mid-range of awareness. Besides size and distance extremes there are also microscopic and macroscopic intervals of time. Each end of the spectrum reveals a hidden, obscure realm of life beyond our awareness. Our practical range of time falls between seconds, minutes, hours, days and years. On the lower end a second is a convenient base unit as it mirrors a heartbeat, providing a natural reference point for sensing time. On the higher end a year works well as it matches the cycle of seasons. So, seconds and years establish a basis for perceiving time at our level. Beyond that middle region is a reality that eludes our conscious awareness and comprehension.

The lower end includes things very short in duration that happen very quickly – micro increments that are generally undetectable by our senses. While the blink of an eye is a mere fraction of a second, it's still detectable. Just a little further beyond that increment things become invisible. Insect wings or even bird wings flap so fast they appear stationary, like the strobe effect of a fast-turning wheel. A humming bird's wings move up and down about 80 times per second, an imperceptible action. Video monitors refresh the screen at least 60 times per second, sometimes 120 or 240 times (60hz, 120hz. 240hz). Neither is detectable though the faster the cycle the smoother the effect, producing a more pleasing picture. Light bulbs operate in a similar manner where electric current is cycling back and forth at a rate of either 50, 60 or 120 times per second. All we see is a steady light source, not the hidden reality of constant flicker.

Going further down the realm of microscopic time we get to tiny fractions of a second. Slow motion filming allows us to see what our senses cannot: the dynamics of a water droplet crashing and expanding with great symmetry, the fast motion of an engine piston cycling furiously over 100 times per second, computer processors cycling through *billions* of instructions per second, the sudden bursting disruption outward from a high-powered explosion. All of these exist in an alternate reality. Much of what we see and experience are simply interpreted illusions. Watching a movie seems perfectly normal despite the fact it's really just multiple snapshot frames consecutively moving rapidly, experienced as continuous smooth motion.

There's almost no limit to how small a time interval can be…almost. The same is true with fast speeds, which similarly make comprehension impossible. As we get further into the realm of super tiny or super quick it becomes necessary to use fractions of a second including billionth, trillionth, quadrillionth, quintillionth, sextillionth, or septillionth (10- 24th). Interestingly, we can comprehend the speed of sound (barely) but we have absolutely zero comprehension of the speed of light – it operates on a completely foreign level. Human perception is firmly anchored to local time ranges, oblivious to microscopic ones. The shortest time ever recorded was 247 *zeptoseconds* (10^{-21} seconds), but the shortest interval physically possible is Planck time, fixed at 10^{-43} seconds (1 followed by 43 zeros).

*Shortest local time: Traffic light turning green and the driver behind you honking their horn
**Longest local time: Waiting lines at the Department of Motor Vehicles

On the higher end of the time spectrum, we're use to periods from a year up to several decades. Though humans can live up to 100 plus, it's difficult for the average person to really comprehend that time span. Movies may familiarize us with long-past eras but they don't make it any easier to grasp the passing of hundreds or thousands of years or more. We can certainly connect with our parents and grandparents but beyond that things get a little cloudy. Relics left behind during periods your great grandparents lived in seem like alien objects today.

The previous chapter referenced macroscopic time periods ranging far beyond what we're used to. Time spans extending from thousands of years to millions and billions of years seem equally imaginary or unreal. It's really a foreign realm; where planets form (earth is 4.5 billion years old), where organic life emerges (1 billion years to form), where human beings develop (2 million years ago), and civilizations take shape (8000 years ago). We didn't even start recording history until about 5000 years ago in ancient Egypt. Think about how little you know about people who lived just a few hundred years ago. A macroscopic timeframe may describe geology, continental drift, glaciers, canyons and evolution itself but it remains an elusive dimension beyond ordinary comprehension.

Near-range history is understandable but mid-range and far-range not so much. Events from 500 years ago or further is vague at best and subject to recording errors and misinterpretations. Even recent history is routinely rewritten due to errors and false narratives. Imagine how inaccurate things get going back a just a few thousand years. Our notion of what happened in the far past gets exponentially more doubtful with time spans reaching millions and billions of years. And there's only so far we can go back; *13.8 billion years* – the span between the big bang and now. That initial spark of creation sets a hard limit on the knowable universe.

Hidden Reality within the Scale of Time

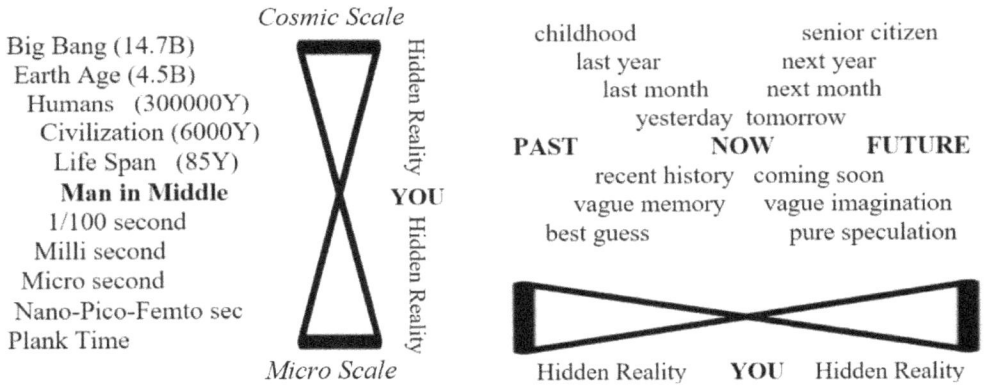

Cosmic Scale

Big Bang (14.7B)
Earth Age (4.5B)
 Humans   (300000Y)
 Civilization (6000Y)
 Life Span (85Y)
 Man in Middle   YOU
 1/100 second
 Milli second
 Micro second
Nano-Pico-Femto sec
Plank Time

Hidden Reality — Hidden Reality

Micro Scale

childhood   senior citizen
   last year   next year
    last month   next month
         yesterday tomorrow
PAST   **NOW**   **FUTURE**
   recent history   coming soon
    vague memory   vague imagination
   best guess   pure speculation

Hidden Reality   **YOU**   Hidden Reality

Macro or Microscopic time limits include speed and tempo, both interdependent. Our normal range of relatable tempos falls within a narrow region beyond which things become imperceptible. On the fast side we have explosions, bullets, chemical reactions, sound and light. On the opposite, slow end there's human aging, glaciers, coastal erosion, plant growth, decomposition and government response to problems. We see the same face in the mirror every day but don't really notice the ongoing continual changes. Photos of earlier periods tell a different story. A human life span is a work in progress experienced moment by moment, day by day, connected along a lengthy sequence beyond our present awareness. We're occasionally reminded of the subtle process when we run into people we haven't seen in years and suddenly think, wow, they have changed a lot, or, where has the time gone?

Scale of Relative Speed

Walking	3 Mph
Running	15 Mph
Biking	25 Mph
Automobile	70 Mph
Racing Car	200 Mph
Commercial Jet	610 Mph
SR-71 Blackbird	2,200 Mph
Space Shuttle	17,500 Mph
Earth Orbit	66,500 Mph
Lightening	270,000 Mph
Fastest Star	5,112,000 Mph
Light	670,760,099 Mph

Scale of Relative Tempo

Gamma Radiation	3 quintillion cycles per second
Visible light	500 trillion cycles per second
Radio wave	1 billion cycles per second
Supersonic sound	20,000 cycles per second
Electricity	60 cycles per second
Engine pistons (highway)	40 cycles per second
Heart beat at rest	1 cycle per second
Snail moving	3 inches per minute
Grass growth	1 inch per week
Hair growth	½ inch per month
Tree growth	2 feet per year
Canyon erosion	1 foot per 200 years

Extremely long periods of time produce hidden history, with forgotten peoples, events and civilizations. This happens at every level; national, world and local. Countries come and go, replaced by new ones in the same location over and over again. Entire civilizations are buried underground, often discovered by accident. Others remain in tales of folklore, stories of ancient people passed on in vague, sketchy depictions, with characters and legends who lived as pseudo-celebrities, the talk of the town right where you live but long since forgotten. These past larger than life characters faded away with the memories of your great, grandparents who died long ago, lost in hidden history. Of the 117 billion people who have lived on earth, almost all are forgotten with no record or trace. They share the same fate as most organic species; 99.9% of all that ever lived are dead, gone and forgotten.

Some processes are masked by subtle tempos operating outside our perception, including the pseudo life of collective groups taking on a life of their own. Consider the modern city captured in timelapse video resembling a living system bustling with activity: stuff flowing in and out, down and around, continual building, moving, changing, rearranging, cycles of action and rest, traffic back and forth, etc. It's eerily similar to an ant colony showing signs of life as a collective, integrated whole. Traced over years and decades a city evolves like a living being. It's unmistakable when the video is sped up but otherwise remains a hidden reality.

HIDDEN REALITY DIMENSIONAL LIMITS

	Extreme Scale End	Man in the Middle	Extreme Scale End
SPEED	Fast (Light speed)		Slow (Glacier motion)
TEMPO	Fast (Atomic dynamics)		Slow (Plant Growth)
SIZE	Large (Star, Galaxy)		Tiny (Quark, Electron)
DISTANCE	Far (Light year)		Near (Planck length)

YOU

ENERGY PERCEPTION LIMITS

Life is essentially a dynamic process driven by an ongoing continual transfer of energy. It's an invisible process embedded within everything, expressed in various forms. Energy exists as mechanical, chemical, electrical, thermal, and nuclear, along with derivative forms such as magnetic, sound, and gravity. Energy exchange is a fundamental aspect of reality lurking in the background beyond our awareness. It's transferred between everything in various modes: radiation, motion, thermal, static discharge, electric induction, and waves. In addition to these physical forms and modes there are non-physical states. The human body processes bio-chemical energy, but it's also connected with spiritual energy channeled into thoughts, feelings, will power, a surrounding aura, and perhaps morphic fields. While energy has multiple forms and modes it's universally expressed as vibration. This is the fundamental essence at the core of all creation and dynamic action.

Vibrational Reality. Our environment is completely immersed in vibrational energy, characterized by parts in constant motion oscillating at various frequencies. Vibration is simply rhythmic energy – a repeating pattern in time. What differentiates it is a space-time ratio. Speed, intensity and length are all variable, producing infinite varieties and combinations of pulsating motion. Anything that repeats with regularity is essentially a vibration. Cyclic activity is vibrational, including both your heartbeat and your daily commute to work and back.

Everything around us and within us is vibrating. All sound and light, all physical objects and anything with energy passing through it vibrates. The earth itself hums at an inaudible level. Our bodies are filled with a multitude of vibrating parts, from brainwaves and heartbeats to cycles of respiration, digestion and temperature change. Communication is vibration based; internally with vocal cords and ear drums, and mechanically with microphones and speakers. Living is a process of repeated daily patterns of breathing, eating, sleeping, waking, working, and every other repetitious activity. Sex is essentially two individuals vibrating together. Chewing, blinking, walking, talking and clapping are just more of the same. These continuous intraday cycles are supplemented by other repeated activities done on a weekly, monthly or annual basis. Cycles and vibration are two sides of the same coin expressing the same essence in subtly different manners.

Biorhythm Cycle Types

Ultradian	Shorter than 1 day	Circulation, breathing, digestion, pulse, blinking, REM
Circadian	Cycle equals 1 day	Sleep, body temp, immune system, hormone regulation
Infradian	Longer than 1 day	Pregnancy, menstruation, seasonal depression

Vibration is the fundamental hidden dynamic of physical reality. Every "thing" is composed of hyper fast-moving parts within. Molecules in "solid" ice vibrate a million times per second. Atoms making up everything vibrate a quadrillion times per second. Though electrons possess a wave-particle duality, they're essentially a vibrating cloud of energy. Matter or materiality is just an illusion, presenting the appearance of solid objects and substance but actually composed of vibrating bbn energy patterns. String theory in physics is founded upon this same principle where the absolute smallest stuff in the universe is just vibrating loops of energy.

The stuff that makes up our structured world is simply temporary forms held together as knotted-up energy fields – resonant lattice structures. Energy is perpetually flowing power in motion, usually tethered by electric charge or pulsating fields; its natural disposition is vibration in space. Atoms are basically tethered particles. Molecules are tethered atoms. You, me and every object are composed of a whole lot of vibrating tethered energy parts. We may appear as solid physical beings, but the actual hidden reality is far from it.

Omnipresent Vibration Expressed at All Levels

Physical	Atomic particles, electric current, gyroscope
Biological	Brain waves, moods, metabolism
Social	Generations, fads, history waves

Light is a special form of vibration with two component waves; electric and magnetic. All radiation is a form of light, omnipresent in our environment, elusive from our conscious awareness. Light is life. Light is spirit manifested into physical reality. It's energy in motion whose patterns create almost everything we experience. All that we see is filtered, reflected and expressed in infinite combinations that create our visual world. Light is pure vibration emerging in countless forms, lengths, intensities and energies.

Surprisingly matter and light are directly connected. Einstein equated matter and energy with his famous $E=MC^2$ equation. Matter is essentially structured light. Atoms are a cloud-like system of particles and energy fields buzzing about, absorbing and radiating light energy. Though we perceive solids, they're actually lattice patterns of knotted webbed energy, forming a field structure of largely empty space. The essence of matter is not the physical stuff we're tricked into sensing but rather an immaterial abstract pattern.

The vast majority of light in the universe cannot be seen by us. What we can see and know is just a small band on a broad electromagnetic spectrum of waves. These range from short Gamma rays to long radio waves. Light that is visible to us is rich in information, composed of numerous wave lengths and frequencies that tell a lot about where it came from and how it got here. It contains all the colors of the rainbow and thousands of variations in between.

Electromagnetic Scale – All Just Various Forms of Light

| AM Radio | FM Radio | Televison | Cell Phones | Radar | TV Remote | Light Bulb | Sun | X-rays | Radioactive |

Micro-Low Frequency R a d i o   W a v e s   M i c r o w a v e s   I n f r a r e d   "Visible light"   Ultraviolet X-rays | Gamma Rays

(Building size)   (Atomic size)

L O W   E N E R G Y   *HIGH ENERGY*

*Comprehensive light spectrum has 70 known octaves, 69 of which are invisible

Cycles are motion in place while waves are motion through space. All radiation is expressed in waves of energy tethered to space, just like waves traversing water or a wheat field. Energy is transferred and passes through each of these mediums while the medium itself remains in place. All radiation is light, mostly invisible. As it transfers energy through waves it's subject to a number of dynamic interactions including reflection, refraction, interference, resonance, filtration, impedance, diffraction, absorption, etc. The amount of energy involved depends on the frequency and amplitude of the wave.

Light has a divine, pseudo-spiritual quality originating *above* while its lower vibrational counterpart, **Sound**, resides down *below,* within the physical world. The former is immaterial and heavenly, the latter is tangible, earthly and physically felt. Both forms of pure vibration reside on infinite scales where only a limited band in the middle is accessible to our senses. Sound like light is also a simple essence of wavelength-frequency and amplitude. Where light is enriched by color, sound is enriched by timbre or tonal quality. Most importantly the two share the same functional value by serving as linking conduits in a connected environment.

Sound is a great feedback mechanism revealing what's going on around us. All activity, motion and change are openly advertised to any and all equipped with listening sensors. Vibrations are disseminated in all directions, penetrating various mediums that connect everything within range. It's a physical communication process where codified reality is interpreted from constant messaging of filtered sound. This process takes place in human communication where vocal cords vibrate to send a coded message to others. The environment or medium is both connector and filter, such as an earthquake's vibrational wave pattern measured by a seismograph after it filters through rock and magma.

Sound like light is not "emitted". It's a vibrational disturbance oscillating the air, creating a wave effect. Consider the disturbance of dropping a rock in water, producing circles of traversing wave motion. A floating bobber going up and down reveals that water is not moving laterally, just up and down. The only part actually traveling is the energy itself. Like the wind wave on a wheat field, stuff is not moving away but rather oscillating while tethered. Dominos fall in place one by one, none moving anywhere while creating the illusion of lateral movement. The key here is that all vibrations are a local disturbance in place, not emission or traversing motion which is what the energy actually does invisibly.

From the micro level of buzzing atoms to the macrocosm of earthly sounds beating, pulsating and reverberating all across the globe, our world is a non-stop sound machine. We're one small part of a greater symphony of life including birds, insects, crickets, woodpeckers, howling wolves, bees, singing whales, all harmonizing with nature's creeks, jungles, rainforests and all manner of wildlife. Other living creatures with greater audio sensitivity can hear a lot more than we can with our very limited human ears.

Everything vibrates to a particular frequency. Like the spectroscopic pattern in atoms there are unique acoustic signatures in every action, process and thing, both physical objects and organic beings. Radionics is a science analyzing vibratory signatures of living things. Man's inner organs and subsystems resonate to particular frequencies while the entire body synchronizes as a composite whole. Brain waves vary with different states of being, between waking (Beta), day dreaming (Alpha), sleeping (Theta) deep, profound meditation (Delta). Human beings are like tuning forks resonating to particular frequencies, responding in kind (attractions, attachments, desires). When healthy and stable, a state of resonant synchronization emerges. When out of sorts, dissonance increases. Same applies to physical processes and social circumstances. Vibration is *the* fundamental process imbedded within physical reality yet 99% of it resonates at frequencies above and below human perception – the very definition of hidden reality.

Atoms	20 Octaves *above* (treble)	MAN	20 Octaves *below* (bass)	*infrasound*
◄Inaudible		Audible		Inaudible►

HIDDEN   REALITY   *HIDDEN   REALITY*

Resonance is simply synchronized vibration – a dynamic that produces temporary stable structures and form in an otherwise chaotic flux of disorder. It creates the illusion of stability in the here and now. Think of a marque sign whose pattern of blinking lights appears to spell out words. Repetitious motion or action tends to induce a steady state of presence – a momentary stable form. It's the pattern we see when a spinning top won't tip over due to inertial resonance.

Harmonics are resonant qualities that emerge at intervals on a spectrum. Harmonized frequencies are complimentary and self-reinforce. They're like nodal points similar to repeating keys on a piano that synchronize in octaves. Harmonic sound is pleasant to hear and has a calming, healing effect. It's present in chanting, prayer and mantras. The human voice has a resonant quality that when channeled with passion by great orators has the power to move the masses.

Human Sound Range – *7 bands*

Bass	Sub-bass	16-60Hz
	Bass	60-250Hz
Mid-range	Lower Mid-range	250-500Hz
	Mid-range	500-2000Hz
	Higher Mid-range	2000Hz-4000Hz
High	Presence	4000Hz-6000Hz
	Brilliance	6000Hz-20000Hz

Loudness-Amplitude (*Decibels*)

0 Db Silence
20 Radio, TV Studio Recording
60 Normal conversation, business office
70 Average street noise
80 Inside car driving
90 Inside truck driving
100 Electric saw
110 Orchestra music
120 Rock music
130 Artillery fire (*threshold of pain*)
150 Ear Damage (deafening)

Hidden reality limits in Frequency: 16Hz – 20,000Hz (all others above or below are hidden)
in Amplitude: 0db – 130db (silence and deafening)
**Hidden reality in both light and sound = invisible and inaudible*

MENTAL & EXPERIENTIAL LIMITATIONS – *Limits on Processing*

Hidden reality includes anything we don't know, regardless of why. Our natural limits of senses and perception could be considered a hardware issue, but there's also the software side. Much of what we don't know is simply due to ignorance or general lack of awareness. Both ignorance and non-truths in any form equal hidden reality, including urban myths, inadvertent falsehoods, misinformation and bold-faced lies. Here we must embrace the wisdom of Socrates and acknowledge we don't know a lot, largely because we haven't been exposed to certain essential information. Public education helps remedy our information shortfalls, but ultimately, it's an individual's responsibility. Exposure to a variety of disciplines and experiences provide a fast track to improving one's lack of knowledge.

Most of what we know consists of partial truths, information by degrees. Unless you do a deep dive on a given subject it's a safe bet you don't see the whole complete picture. Even so-called experts rarely know every possible detail about a subject. That's why science is always evolving, and points of view continually shift as new information surfaces. The vast majority of public opinion is based on partial truths, both individually and collectively. But understand some "truths" are more partial than others since it all differs by degree, so what's important is knowing which are more truthful or less partial.

Other limitations include processing errors due to natural biases, cognitive constraints or misinterpretation. While we're drowning in information from *out there*, how much of it is really meaningful? At least a portion of our ignorance is due to false knowledge, mistaken guidance and confusing facts with opinions. Ongoing vigilance is necessary to separate relevant information from noise. We make matters worse by naturally overestimating what we know and underestimating what we don't know. Accuracy is diminished by each of the following: the Dunning-Krueger effect, confirmation bias, preference for familiarity and the tendency to be fooled by random disconnected events. We enthusiastically embrace content we agree with while avoiding contrary data, creating a self-fulfilling prophecy of cognitive dissonance, disconnected from objective fact. Simply put, *ALL* of us know less than we think we do.

Then there's the Gell-Mann amnesia effect where individuals criticize media reports in areas they're personally knowledgeable about yet continue to trust that very same media in other areas despite their questionable track record. The same applies to people within our own intimate circles.

Human Cognitive Bias Limitations

Confirmation bias	Dunning-Krueger effect	Bandwagon effect	Selective perception
Functional Fixedness	Anchoring bias	Reactance	Placebo effect
Clustering illusion	Not Invented Here bias	Ostrich effect	Stereotyping
Status Quo bias	Exposure effect	Overconfidence	Recency bias

We need to add a supplemental category here of ***Mis-knowns***: things we think
we know but get wrong. This includes misunderstandings, mis-interpretations,
biased conclusions, uninformed opinions, illusions and just plain ignorance.
Learning isn't just acquiring new knowledge but clarifying and correcting our false
knowledge. Mis-knowns plague individuals and larger groups – even society as a
whole. They're often more dangerous than ignorance, especially in the case of
"experts" who give others authoritative direction. Will Rogers understood it well;
"It isn't what we don't know that gives us trouble...It's what we know that ain't so."

Mis-knowns are reinforced by pop culture where ignorance can be spread to the
masses. Movies are a major source: people get shot and shrug it off, spaceships
make noise, car crashes cause big explosions, knock a person's head and they fall
unconscious, fist fights leave no facial wounds, hackers go on a keyboard and
break into security in 20-30 seconds, etc. Out of natural laziness we just assume
stuff without looking at facts. Most of us don't have accurate impressions of our
own home country. Very few understand the lay of the land or where populations
actually reside, but detailed maps will quickly dispel such false perspectives.

Mis-Known Derivatives

-Misinformation: False information we just get wrong
-Disinformation: False information intentionally put out to deceive
-Malinformation: True information intentionally exaggerated/misused to deceive

The Mandela Effect is a type of mis-known where we're convinced we know an
obscure cultural fact that just isn't true. They're surprising given how certain we
are despite being completely wrong. It's like getting details confused about a
movie you've seen multiple times or false recollections of how a familiar object
looks. See if you know the following facts that your memory might disagree with:

-In Star Wars C-3P0 has a silver leg
-Tony the Tiger has a blue nose, not a black one
-It's Jif peanut butter, not Jiffy
-The monopoly guy doesn't have a monocle
-Below the equator the globe is 80% water.
-It's not Oscar Meyer...it's Oscar Mayer
-Africa is bigger than China, US, Europe, India, Japan, Italy, England combined!

Known unknowns are in some cases intentional mis-knowns; things that we can
know but would rather not know. Sometimes you're just better off not knowing.
Otto von Bismarck warned "Laws are like sausages, it's better not to see them
being made". This applies to many things. You probably don't want to know
what's in someone else's basement, what's behind the performance stage, what bar
people look like when the lights turn up, and what's going on in fast food kitchens.

Equal Populations – *concentrated spots & coasts vs the heartland*

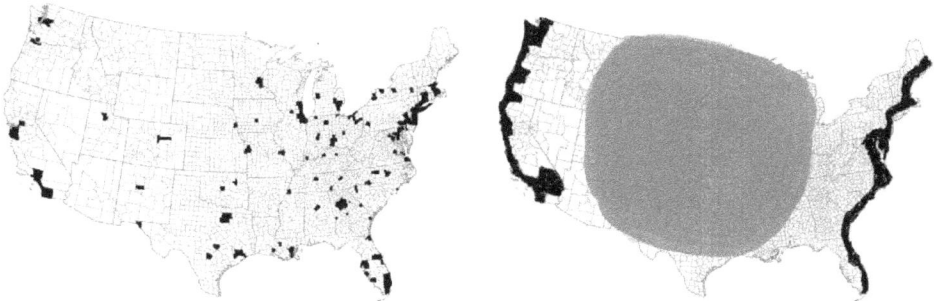

Polarized population: 1/2 live in condensed pockets. Coastal population = The Heartland

*8 states have over half the US population
**6 out of 200 countries have over half the world land mass
***90% of the global population lives in the northern hemisphere

Lack of awareness is natural in a world full of diverse possibilities, multiple options and private interests. The journey we make navigating the world resembles a mouse scurrying through a maze looking for cheese. Metaphorical walls are everywhere along with closed off inaccessible pathways. We sense something is out there, perhaps right around the corner that cannot be seen. Hacks are an example of hidden methods that could save us all a lot of time if we weren't so ignorant of them. A lot of what we don't know is a case of inside information where others have access that we don't. Insiders include the wealthy, powerful, celebrities, scholars, scientists or specialists who have access to areas unavailable to the rest of us. Layers and gates prevent exposure to the majority, leaving access to a privileged few. This is true in the corporate world, government, institutions of higher learning, the entertainment world and a wide variety of special interest groups. The uninitiated are always kept in the dark and in a sense, we are all uninitiated from various exclusive groups. We're each limited by our own specific class, race, sex, religion, etc. minimizing our exposure to every other category. Fortunately, internet sites like YouTube, X, Facebook, Instagram, and others provide a little more access to what otherwise would be totally inside information.

The World is full of big, intimidating things that far exceed the grasp of any one individual. Besides big natural features like immense mountain ranges, vast oceans and boundless deserts, big man-made constructions can be similarly intimidating: skyscrapers, coliseums, 100-mile bridges, 35-mile tunnels and other man-made megastructures. The US Government comes to mind when you think about mammoth scales, and that's just the federal level, let alone all the state, city and local branches. There are approximately 450 federal agencies, not to mention all the numerous subagencies. Just the Department of Defense itself is a colossal complex, starting with the Pentagon, extending across the entire continental states and then spanning the entire globe. There are roughly 800 military bases world-wide plus a variety of supplemental facilities and support sites. Who can keep track of it all? How can anyone account for everything going on at one time? It's like a separate nation with a culture, language and territory all of its own.

Picture the warehouse end scene in *Raiders of the Lost Ark* with thousands of crates and hidden containers housed in a government facility, probably common in locations across the entire globe. Besides just government depots and giant logistic hubs there are even more such facilities in the business sector containing a colossal inventory of manufactured parts. It's like Home Depot multiplied by a thousand, considering all the machine parts, automotive end items, electronic components, construction materials, plastics, textiles, metals, petroleum products, etc. Your car is made up of over 30,000 parts. Think of the millions of different parts and accessories needing to be stored, processed and tracked. How difficult would it be without modern computers? If your head doesn't hurt thinking about it, you're not really thinking about it.

Pondering even more mind-boggling aspects of obscure reality, think about how we're all living on a planet spinning on axis 1000 miles per hour while moving through space at 60,000 miles per hour. That same planet is part of a solar system spiraling through the Milky Way Galaxy at over 500,000 miles per hour. Just being here, looking down on this flat piece of land, it's hard to accept that I'm standing on a giant curved surface with millions of people standing upside down on the other side, all racing through space. We can extend that same lack of awareness to how bad living conditions are in the rest of world – where billions of people live in poverty, have poor health care, experience daily fear or anxiety, are dirty, hungry and uncomfortable. It's perfectly natural to base our reality on here and now in our own small corner of the world.

I can only ever know what I see or experience and no one can ever really know what you see or experience. What we do share are similar limitations of perception, seeing forward with a field of vision around 180 degrees, somewhat blurry on the peripheral sides and a small blind spot near the middle. But the bigger "blind spot" is the 180 degrees behind us that we don't see. Or all the other possible perspectives at different angles that are revealed in a funhouse mirror room. And think of how your life looks from either the ground up or an aerial viewpoint.

Ignorance and lack of awareness apply to both individuals and collectives. The little bit we each know can be mitigated and leveraged in groups where your shortfall is offset by someone filling the gap. We assume a group will always be more informed than an individual, however subtle dynamics emerge that cancel the advantage. Sometimes false information surfaces and is quickly shared by the group, making everyone worse off. This occurs with myths, urban legends, group think and misinformed leaders. It a case of more is less, the opposite of synergy.

Each of us occupies a middle position within the limits of a bell curve, between extremes on both ends. This is true with all of our attributes, such as raw talent, intelligence, size, age, attractiveness, strength, etc. Naturally we embrace the familiar, becoming much more aware of qualities and attributes in our middle region. But it means we're generally unaware of cases existing at the extreme ends. On the common 1-10 scale of attractiveness how many of us can say we really know what a 1 or 10 looks like? How about intelligence or strength? How many can say they know what a 1 or 10 is? This is true of everything we experience in life where there is a best and worst, highest and lowest, largest and smallest.

HIDDEN, OBSCURE AND INVISIBLE REALITY

Hidden reality covers a lot of ground, including things we cannot see, hear or feel – invisible, inaudible and intangible. Things we do see are far exceeded by stuff that's either covered, buried, cloaked, veiled, masked or secluded. Our environment has a built-in degree of obscurity hiding more than it reveals. That same light of our world brings with it shadows and darkness. Sun is complimented by moon, which spends most of its time partially or fully masked by shadow (half of it is never seen). These dark veils that cloak "real" things emerge as parasites of the object covered. Shadows bring with them a negative connotation, an anti-form or a diminishing evil. The same is true with blackness – the simple absence of color or light. Yet they're complimentary and necessary to wholeness, fundamental to visual context, depth, and identity. Shadows go hand in glove with an object, the flip side of a single whole. But they have a transitory nature, changing in shape and size, distorting with different angles, perspectives, and lighting. They can suddenly disappear and reemerge, possessing an elusive pseudo life. Shadows are a close relative of silhouettes – an empty outline of a whole object. Light and shadow are both derivative forms in our perception. Like shadows, most of the light we see is either reflected, filtered or distorted by the objects it connects with. Just be aware that shadows *out there* are mirrored by our own shadows buried *inside*. Our conscious perspective is merely 10% of our total awareness, greatly overshadowed by the subconscious (60%) and the unconscious (30%).

Much of life's key processes go on behind a veiled curtain, never to be seen. A human body's internal activity is a rich complex of integrated processes churning continually. Same with machines, engines, computers, and the infrastructure of the house or office building you're in, with all its plumbing, electric and HVAC conduits. The entire animal kingdom engages in a masked daily death dynamic where members of every species kill or get killed just to survive. While most of it goes on in distant places, some of it happens right in your own neighborhood living spaces, mostly unseen. And where do all the dead animals go? It's another case of sausage making; we're better off not seeing or knowing what's really going on.

Our daily environment is a playground of hidden clues reminiscent of a CSI crime scene. With the right tools an entire realm of invisible artifacts can be uncovered, linking past events and activities, causes and effects. All the clues of what happened in the past are right in front of us – coded information temporarily disguised and concealed. In many ways the invisible world is more "real" than the visible one we're familiar with. Every day we're obscured from the majestic night sky with its billions of stars, masked by the oversaturation of daytime light. Which perspective do you think is closer to reality?

The cosmos is filled with a variety of known unknowns that intrigue scientists eagerly seeking answers to what's really going in the background. Dark matter and dark energy seem to play a significant role in the universe but they're essentially beyond direct measurement. Black holes are plentiful, immense in size, power, and degree of mystery. Their gravity is so strong nothing can escape, including light. So what happens inside one? Do laws of physics cease to exist? Does the trapped mass inside a singularity go somewhere else beyond our local universe? These massive enigmas produce more questions than answers and since nothing can ever escape their force, we'll never know with any precision just what the heck they are.

Religion, spiritualism and the occult are fundamentally based on hidden reality. Religious practices focus on intangible spirit inside and immaterial deities outside. The term "occult" comes from the Latin *hidden*. While we associate It with mystical, cryptic, or shadowy nature it's essentially a version of CSI directed within. The occult deals with unseen energies and influences beyond our conscious awareness. Astrology, Tarot, Numerology, etc. use pseudo-science approaches to analyzing immaterial mystical aspects, which unlike actual science are unquantifiable phenomena. Clairvoyance, ESP, necromancy, as para-science involve the same aspects and limitations.

The passage of time produces layers of insulation separating us from everything that happened long ago, as well as everything that will happen far in the future. In our middle position residing along a vast time spectrum, we're obscured from everything else that ever happened. Even within a single lifetime much is forgotten or buried in mainstream culture; in the 1970's it was normal for kids to be outside all day unsupervised, frolic on concrete playgrounds with 30-foot-high monkey bars, ride in the front seat of cars without restrains, ride bicycles without helmets and jump those bikes high over friends lying on the ground, and then swim in pools without life guards (almost every last one of them survived).

Most of history is obscured from our awareness as civilized man only recently began documenting events. Unless it's preserved in some physical form on the planet's surface, almost anything that happened over 10,000 years ago is either long forgotten or completely unknown. Excluding the millions of years in man's prehistory, most of what happened in just the last 50,000 years is largely lost. Hidden history includes an immense comprehensive record of mass meteor impacts wiping out great swaths of life. Lost civilizations, undocumented events and stories of what went on long ago are essentially gone forever. But who knows, maybe we'll eventually find some definitive evidence of Atlantis, excavations of buried civilizations in the Amazon, or consensus proof of how the pyramids were built.

Hidden history includes lots of harsh events, bad years, crappy eras. 536 is considered the worst year with excessive volcanic eruptions causing widespread dark skies lingering over 18 months, leading to crop failures and plague that killed millions over a decade long period.

Built-in Uncertainty. At the very foundation of physical reality, we can never really know the true nature of the parts that make up everything. It comes from the Heisenberg principle where the mere act of measuring things changes its disposition. At best a physicist can capture either speed or position of a particle but not both. This constraint is just the beginning of built-in uncertainty created by quantum mechanics, where reality at the micro level is a function of probabilities – no absolutes. At the macro level a similar aspect of uncertainty emerges with Einstein's relativity, where time, space and motion are never absolute but relative to each observer. Uncertainty is further confirmed by mathematics, where prime numbers cannot be predicted and have no antecedents. It's also reinforced by Gödel's incompleteness theorem where any finite system cannot prove its own consistency. Any theory of the world cannot account for truths outside of it. Axioms & propositions within a system cannot be proven by that same system but must rely on things above and beyond.

Good science assumes ignorance and uncertainty up front, restricting theories to mere assertions – premises that can't be proven but only temporarily accepted until they can't be disproven. Scientific method embraces a sort of guilty until proven innocent posture. Mathematic axioms and postulates are treated the same way. It's simply paying homage to Socrates again, following healthy skepticism and critical thinking. Every science has its own unsolved mysteries. While most are just temporary unknowns, others will likely never be solved, as is often the case with physics, solving current mysteries can just open up new ones, continuing down a never-ending rabbit hole. And there will always be a number of unexplainable phenomena, such as ball lightening, near death experiences, quantum entanglement, Nazca lines, Taos hum, the placebo effect, etc.

Upside down earth: Maybe our true reality?   There is no absolute reference point

Uncertainty just begins at the micro-physical level of reality and continues higher up to the biological and social levels of expression. Human beings are the posterchildren of unpredictability, both individually and collectively. Even when we think we know someone well we're often surprised by what they're capable of. Those you're sure will succeed may fail while others you underestimate may pleasantly surprise you. Group behavior is equally difficult to predict. Flash mobs are a perfect example of collective uncertainty at the social level. The same effect is present at the highest levels of society where culture, politics and economic activity move in ways that are not anticipated by those supposedly directing them.

Peripheral Ambiguity. As we shift our focus away from our familiar, comfortable middle position perceptions become fuzzier and clarity disappears. We saw this in the physics of small and large spaces, both short and long time frames, and the lack of absolutes with quantum mechanics and cosmic relativity. There are also complex motions of fluids and gases that follow turbulent unpredictable paths where properties of emergence produce sudden tipping points leading to unforeseeable large-scale change. The process of emergence adds a whole new layer of uncertainty to the mix: how does 1 human cell become a whole complex being? More questions arise as we enter the realm of complex processes, abstract realities, and theoretical mental constructs. This built-in ambiguity is omnipresent in mathematics with logarithmic scales, complex numbers, irrational numbers and fractal repeating patterns. The concept of infinity rears its ugly head, expressed in various forms: points on a line, never ending number sequences, eternal time, paradoxical values like π, and in geometry we get the unbounded circle without beginning or end. Powers of 10 are a helpful math shortcut to describing large numbers but even they have limitations. Goodluck trying to comprehend a Google (10 followed by 100 zeros) or a Googleplex (10 followed by a Google of zeros).

Math is just one form of coded information. Lots of other material is captured in coded form, packaged with similar ambiguity. Maps are a prime example including all of the following varieties: road, flight, military, demographic, weather, naval, astronomy, terrain, etc. Coded information is everywhere in both nature and man-made forms: blueprints, foreign language, binary code, horoscope, crime scene, braille, etc. All the meters and gauges in your car represent coded data that serves as feedback on the vehicle's status. These are information tools that help us sort meaningful data from a sea of noise. The opposite of this is esoteric language, the kind you'd hear from a lawyer, scientist, doctor, mechanic or politician, each of whom make simple concepts ambiguous, whether intentional or not.

Increased sophistication and miniaturization of manufactured products adds a significant element of ambiguity to the human condition. Household appliances and vehicles of all kinds are now brimming with circuit boards and microchips, diminishing our ability to troubleshoot and perform simple maintenance. Classic cars that were once the pride of youthful independent mechanics have been replaced with compact engines interlinked to computers and micro circuits. That same trend swept across entire industries, rendering routine service on machines and systems impractical, further requiring dependence on outside specialists.

It's no coincidence that the key essences of life are all ambiguous by nature. What is more fundamental than time itself yet so difficult to define with its serpentine quality that can't be easily explained. The same is true with love in all of its many forms, versions and expressions – how could anyone claim to fully understand or explain it? How about life in general? Who can say for sure when it begins or when it ends? And is organic life the only version that exists? Consider the different degrees and states of things that are deemed to be alive, some obvious, others not so much. Energy is equally nebulous with numerous forms and varying qualities that are difficult to pin down. Have you noticed that energy is never still but always moving, flowing, eternally changing? How about energy in the form of light that radiates everywhere, reflects, refracts, interferes and resonates? Sometimes it looks pure white while other times it's transformed into a beautiful array of complimentary colors. And the ultimate essence within you, consciousness – the very thing that elevates us above all other life yet remains extremely difficult to explain, define or understand.

Infinity and Eternity. Concepts of infinity and eternity are ambiguous in and of themselves, but they are also mistakenly used interchangeably. Eternity is a temporal state of perpetual timelessness where there is no beginning or end. Infinity relates to a magnitude or quantity of something that is limitless, going on or extending without boundary. Eternity is timeless, infinity is boundless.

Mystery. With all that we don't know, some is only temporary while the rest may never be known. Life's paradoxes and puzzles do make things interesting. Magic and mysticism are sort of exciting, with a promise of something special beyond. Mysterious things come in so many forms such as an enigma, paradox, or puzzle, and include unsolved crimes, hidden treasure, tunnels, holes, caverns, buried artifacts, lost items, strange phenomena, and anything unexplainable. Ancient man developed myths to account for things not understood as well as religion, serving as a source for answering life's mysteries. Both mythology and religion find explanations in a higher, divine realm, as do practices in the occult. Down here in the mundane physical world answers to mysteries are pursued by investigators, scientists, explorers, auditors and psychiatrists.

There are really 3 types of mysteries: Those temporarily unsolved, those never to be solved, and hoaxes. Even though some will be solved and others exposed as scams, the effect is still the same: unknown circumstances creating hidden reality. Secrets on the other hand are not quite mysteries because they are known by someone. Coded information might fall in between those two, depending on who may already know it. Confusion is sort of similar, however it relates to something already known. What matters is whether we don't know it or can't solve it.

Ironically, physics as the king of hard science seems to produce the most mysteries, even after solving previous ones. For every layer revealed there seems to be 2 more deeper within. The physics quest for a Theory of Everything that ties all the known forces together in a nice, neat equation remains as fleeting as ever. As instruments reach further into space, we seem to get no closer to determining where the universe is going. Will it expand forever, continue as is, or will it contract back again to a recurring big bang event? This puzzle requires knowledge of how much matter fills the universe, but that's even more problematic with the notion of dark matter, which is difficult to measure and not entirely understood. Dark matter and black holes are really just theorized - especially black holes which can't fully be known since light doesn't escape it. Physics also runs into a puzzling conundrum dealing with the smallest particles defined as quarks, which like dark matter is impossible to see and perplexing to measure.

Back to earth, the mysteries most common folk ponder are about life, death, after life and even ghosts. What is life and how did it begin? What's the meaning of life and why are we here? On the flip side, what is death like and is there anything beyond it? Is there heaven or a cycle of reincarnation? Perhaps we're asking the wrong questions and should take note of Soren Kierkegaard who said "Life is not a problem to be solved, but a reality to be experienced". Beyond our own lives, what other kinds of existence are there? Do aliens and UFOs exist and is there life on other planets or are we all alone? What about lost civilizations and stories of Atlantis? Who built Stone henge or made the giant stone statues on Easter Island, and how did they do it? How were the great stone pyramids built over 4500 years ago without modern equipment or technology? These are mysteries everyone can relate to but remain uncertain or unknown.

What about the immense uncharted regions of the world's oceans where 70% of all life on earth resides? Its depths contain great unseen mountain ranges, canyons and over 3 million shipwrecks and several million human bodies. Only 5% of the ocean has been explored and 94% of oceans are of pitch-black darkness. A dormant volcano taller than Mount Everest, Mauna Kea, is anchored to the bottom of the ocean. Denmark Strait Cataract is the world's largest waterfall (11,500 ft) hidden below the ocean surface. The vast majority of deep-sea life remains undiscovered. What wonders are yet to be found within the deep abyss?

Perhaps the most profound mystery of all, deeper than the oceans and greater than the expanse of the cosmos is the nature of consciousness. It's imbedded in everything at different degrees, exhibiting a variety of subtle aspects, levels, states and types. It's expressed in everything through subject-object relationships, action and interactions, being and doing. It's spirit manifested through infinite forms, patterns and temporary structures. We normally relate to it as states of being within ourselves (conscious, subconscious, unconscious) and levels of awareness. Some think it's merely a function in the brain, a residual effect of biochemistry in neural networks. Others see the mind as a transcendent, emergent phenomenon beyond the physical body. Others sense something grander – a universal phenomenon omnipresent in the universe, an intrinsic attribute of everything, living or not. No one can deny its mysterious, ambiguous nature is extremely difficult to define.

Spiritual Realms. Consciousness and spirit are equivalent forms of the same essence, perhaps representing the difference between being and doing. They permeate the universe in content and dynamic creative action. As immaterial, transcendent essences they're likely more accessible through religious and metaphysical practices. Despite man's quest to measure and quantify everything, this domain will have none of it. Spirit and mystical realms are beyond reductionist science, possessing qualities and attributes expressed through mythology, philosophy, religion, art, poetry, music and the occult. This hidden realm is linked to the universal concept of GOD, present in almost all cultures worldwide throughout human history. But whatever one may think of GOD there's always an ambiguous, mysterious aspect attached to it. The silly notion that we could define or describe GOD is ridiculous given how incredibly inferior we are by comparison. Attempts to describe GOD in our own image and limited language is lame at best and often downright embarrassing.

Invisible, unquantifiable spirit pervades and defines individuals, their states of being and their collective group associations. Humanity as a linked whole forms a collective consciousness with shared values, beliefs and experiences, which then define each period in history. There's also a transcendent aspect of it identified by Carl Jung called the collective unconscious, which is a subtle pattern of knowledge and imagery each generation inherits – born with psychic memories from past ancestral experiences shared through human history. These patterns are submerged deep in our unconscious but surface occasionally and express themselves as universal instincts or impressions such as phobias (fear of snakes, falling) or archetypes (hero, trickster, wise guru). Consciousness and spirit dynamically weave through time and space, filtering within man and without mankind, always subtle, ambiguous, persistent and profoundly mysterious.

Alternate Realities. Both science theorists and fiction authors have proposed the possibility of higher dimensions beyond those we currently understand. Our 4-dimensional space-time reality may be just the first level of many multiple dimensions of both space and time existing outside our limited perception. String theory incorporates the idea of multiple, additional dimensions, hidden and tucked within those we're familiar with. An intriguing proposition is a parallel universe where everything here is duplicated there with slight differences. You might have an identical twin there making different choices in that reality. Every moment would have a quantum potential aspect where 1 path is followed out of infinite possibilities. This would lead to enormous complexity as moment-by-moment new decision branches are created with infinite possibilities chosen differently in each case. Going deeper than this concept is the scenario where multiple dimensions create different realities that we're tied to but completely unaware of. Here you would be living many separate timelines but only aware of this one. Another version of you would be doing the same yet unaware of you. Both of these hypothetical realities have a quantum nature where finite actualities come from multiple possibilities, creating the reality you experience in the here and now.

Other theories involve the concept of a multiverse, containing many individual universes distinct and different from the one you're in. This could include realities where the laws of physics are not the same. Some have suggested universes within universes like a series of nested eggs, each one subsumed by the other. Proposals of an Omniverse, Metaverse or Megaverse are really just variations on the Multiverse theme. While these concepts provide interesting possibilities, the labels are a little misguided since Universe still implies one large whole. So, the multi or omni versions would just be variations or subparts within a single all-encompassing whole, which is still a *universe*. What these hypothetical theories have in common is their hidden, unknowable nature. Like the quantum choice selecting one out of many, we'll never know what the many could have been. Or like hidden history where only 1 event occurs out of many potential ones, we'll never know what possible events and continuing branches might have easily happened but didn't.

Hidden reality can be temporarily accessed through altered states where our normal awareness is changed in a manner that opens doors of perception, enabling higher sensitivity. Our mind's primary function is to provide order in a chaotic world and in doing so insulates us from deeper reality. Drugs can change this by limiting normal mental processes and opening the flood gates. It starts with mild drugs like alcohol, caffeine, nicotine, and marijuana, then gets more hardcore with heroin, cocaine, and amphetamines. Drug induced states are common today, facilitated by over-the-counter pharmacological medicines and psychotropic drugs. It's fairly routine to prescribe Xanax, Zoloft, Celexa, Prozac, Ativan, etc. Psychedelics and hallucinogenic substances like LCD, DMT acid and mushrooms take it to a higher level yet. Users describe a profound experience that can't be put into words, with an expanded awareness far beyond the mundane world we're accustomed to. Psychedelics transform and transport one's experience of being into a foreign reality. Maybe the closest analogy for the uninitiated would be a dream state of wild distortions, like a dramatic nightmare or the bizarre hallucinations experienced after periods of extended sleep deprivation.

The ultimate hidden reality scenario is one where we're really just living in a virtual reality construct, similar to The Matrix, where everything we do and experience is not real. In this framework our consciousness *in here* is participating in a virtual reality *out there*, one that comes to life as we direct our consciousness to it. It's like a graphic video game that we're playing in real time with an avatar in it representing our physical body in virtual space. This framework is similar to the reality we experience between the dream world and being *awake* – how can you ever know which is real? In both these cases of virtual reality and drug induced altered states of being, you're faced with a choice: do you want to simply accept the illusion of your normal state of ignorant bliss (blue pill) or do you yearn to know the truth and go down the rabbit hole that most likely never ends (red pill)?

"Reality is merely an illusion, but a very persistent one" – Albert Einstein

Nothing. You would think the opposite of reality is non-reality, but wouldn't that in itself still be a kind of reality? It's like the concept of nothingness or *the void*. It's referred to as emptiness in eastern perennial philosophy where *nothing* is the fundamental original state of everything. In western culture we associate nothing with holes, darkness, space and the empty set in math. Physical science reveals the universe is 99% empty, which also applies to most of the so called "solid" stuff in between, including you. Atoms are mostly empty so everything of substance anywhere and everywhere is also largely just empty space. Religion and mythology all embrace a creation concept that begins with nothing transforming through an act of creation. Eastern counterparts follow a similar narrative where an original void is transformed from nothing into something and then everything. To understand this framework, you must first accept nothing is actually a something – a potential to be or become. Nothing may be *no thing*, but it *is* an essence – an unrealized budding possibility. Even empty space is not nothing for it has the capacity to limit light and be warped by matter. Just as electricity is slowed down by impedance in a wire, light traveling through space as electromagnetic radiation is slowed down by permittivity of space, a sort of impedance of the universe, which ironically supports the old notion of classical physics; space as an ether.

There's a subtle symmetry between "0" and infinity, nothing and everything. Any number divided by 0 equals infinity, any number divided by infinity equals 0. There's a similar relationship between losing one's identity in a self-less state and being *at one* with the universe. And consider the relationship between darkness and light – two similar opposites, one empty and one full that complement each other. Hegel noted that being and nothingness were opposite and equal.

Nothing is an uncreated something. Everything is in nothing. Nothing contains all things. Nothing is outside the world. Nothing is Everywhere. Everything came from nothing. Anyone claiming to be a devout religious practitioner cannot refute this. Whether it's Christianity, Hinduism, Judaism or Islam, all major world religions offer a creation narrative similar to genesis where in the beginning there was nothing. Even after creation there is still nothing within everything. This is the essence one connects with during states of deep meditation. This inner experience of nothing is expressed as stillness, emptiness, and *no thing*. Space itself is part of the hidden plenum where nothing becomes something, the two being opposite sides of the same coin and a fundamental facet of eastern philosophy.

Nothing is the backdrop providing opportunity for creativity. Consider a blank piece of paper representing emptiness, rich with infinite potential. Suddenly a pencil mark is made, and a shape takes form. More marks follow and the rest is just variations on a theme. Or ponder a prolonged silence where nothing can be heard in the moment. Suddenly a word is spoken and out of nothing there is something, with infinite variations of spoken words that may follow. This is the simple process that transforms nothingness into somethingness.

The perfect symbol for nothing is 0, a number shaped like a circle with no beginning or end. It resembles the original seed that became a giant sequoia tree. Or the first cell that transforms into a complex organic being. When tiny in size "O" becomes just an abstract point, perhaps the beginning of an infinite sequence. When large in size that same O resembles a whole; the capacity to contain all. Mysteriously, nothing becomes something, then everything. The process where 0 becomes 00 (infinite) is transcendent, like the perspective vanishing point on a painting with depth. You're stuck in the middle of these two opposite and equal extremes. You and everyone else living in the here and now are connected to nothing within and everything without. Every act you commit is like a new mark on a blank piece of paper. Every word you speak is a new sound disturbing silence. Everything you do and experience is an act of creation, producing a new something where there once was nothing.

CONCLUSION

Known Unknowns reside at the extreme ranges of our perception of space and time: big-small, near-far, fast-slow. They include the micro world of tiny creatures and residue surrounding us and the hidden realm of tiny atomic particles buzzing, humming and interacting inside everything. They include a macro cosmic expanse so large and far we can never know most of it. There's hidden energy dynamics of omnipresent vibrational interaction underneath the surface imbedded in everything around us. Coded information is also omnipresent, connecting everything in subtle patterns. Physics reveals a built-in uncertainty and randomness at the quantum level translating into every act, event and branches thereof. Supplementing this lack of certainty are the hidden, obscure, and the mysterious, along with a complimentary shadow essence lurking and clinging to anything: light expressed in darkness, black holes, emptiness and nothingness. Further unknowns dwell within unexplored domains including oceans, space, subterranean earth and death. There's so much of reality that's invisible, inaudible, intangible and incomprehensible; things we can't see, hear, feel or think about. Some hidden realms are temporarily revealed through altered states of consciousness – opening up perceptions to higher and deeper domains where experiences are not of this world. Virtual reality frameworks go further down the rabbit hole, similar to a waking dream state where it's nearly impossible to distinguish what's real and what's illusion.

The bell curve captures our ignorance of things on either extreme of us as we're confined to the familiar middle ground in between everything – timeframes, sizeframes and mindframes. Most of us have no idea what's possible on either extreme of any perception spectrum (1-10 scale of attractiveness, athletic excellence, intelligence, etc.). Our grasp of reality gets diminished when complexity, ambiguity and uncertainty are introduced. Built in perceptive and cognitive limitations further cloud our sense of what's real and what isn't. As the Hindu's explain, "The World is *Maya*" – it's not as it appears or as we experience it but consists of cloaks, distortions and illusions.

Hidden reality may be a fundamental limitation of the physical world; however, we should appreciate the way mysterious, hidden aspects of life make things much more interesting. How dull would it be if everything was obvious, clearly defined, predictable and known by all? There'd be no excitement, adventure, discovery or romance. Obscurity and distraction provide a welcomed buffer from all the negative, ugly things going on continually. Pain, suffering, tragedy and death are daily realities experienced by people across the globe, even right in your own local neighborhood. Who wants to be aware of that 24-7? Who wants to know what their exact future holds down to the last detail? Just embrace the wisdom of Socrates and acknowledge that there's so much you don't know, can't know and will never know. Ignorance really can be bliss.

SPECTRUM OF HIDDEN REALITY

KNOWNS	MIS-KNOWNS	KNOWN UNKNOWNS	UNKNOWN UNKNOWNS
Things we know	Things we get wrong	things we know we don't know	things we can never know
Human Knowledge	Illusions	Extreme small/large, fast/slow	?
Science	Mis-interpretations	Energy dynamics – Vibration	?
Facts	Biased conclusions	Hidden history	?
Experience	Ignorance	Coded information	?
Observations	Complexity	Obscurity, shadows, mystery	?
Critical thinking	Ambiguity	Unexplored frontiers	?
Focused scope	Poor memory	Alternate, virtual realities	?

HIDDEN REALITY

PART TWO: *F I T S*

8   THE UNITY OF KNOWLEDGE

Since the dawn of mankind humans have engaged in a never-ending quest to understand *The World*, how it works and why things are the way they are. Early humans operated in a chaotic, dangerous environment with mysteries lurking around every corner. Rational man eventually transcended natural fear with a burning desire to make sense out of chaos, establish order, and create a structured, predictable way of life. This new mode elevated human beings above the animal kingdom and set the stage for further progression into advanced civilized societies.

Progress simply came down to acquiring knowledge – a process of gathering facts, information and knowing how to do something. It's not just the data but also the software. It's an informed way of taking action. Greater knowledge begets better methods. It's acquired primarily through experience and education. Knowledge expansion is a process of converting unknowns into knowns; revealing mysteries and uncovering previously unrecognized realities. With greater knowledge comes a need to organize it and distinguish between its multiple aspects, dimensions and branches. This includes mental aspects, skills, techniques, individual vs. collective, science, research, social fields, and subjective disciplines. Despite the great breadth and diversity of knowledge there are common themes, universal principles and similarities between different fields. With a strategic perspective we can piece together the fragmented, isolated, individual areas of knowledge and unify them into a single, collective whole.

HISTORY OF KNOWLEDGE

Accumulated knowledge accelerated the ascent of man, assisted by documentation; systemic recordings and experiences passed onto succeeding generations. Ancients relied upon learned wisdom expressed through storytelling and rudimentary recording materials. Later civilizations grew in power, paralleled by their degree of accumulated knowledge. Babylonians, Egyptians, Greeks and Romans further extended man's progress, setting the stage for western civilization's emergence as a culture of science. Meanwhile, eastern cultures were grounded in non-scientific spiritualism, steeped in subjective indefinites or transcendent reality.

The scientific approach began with Egyptians thousands of years ago was later perfected by Greeks and then Romans. This ancient time in human history referred to as the classical period, spans between 3000 BC to about 500 BC. After steadily raising the bar of knowledge for centuries, the fall of Rome stifled momentum, putting progress on hold for nearly 1000 years as a medieval period often called the dark ages. By the mid 1300's, progress resumed in the renaissance period – a new era with a revival of classical knowledge that had been lost or forgotten for so long, punctuated by the invention of the Gutenberg printing press. Lasting about two centuries, this stage of renewed progress transitioned higher into the enlightenment period, further expanding collective knowledge across many fields. In this *Age of Reason*, advances were rapidly made in every science along with growth of philosophical-intellectual movements. Mythical and mystical explanations were discarded and replaced with scientific reason. Religious dogma and state control were replaced with emphasis on human freedom and individual rights. Organized, systematized information had come of age, setting the stage for the coming industrial revolution, information revolution and the tech revolution.

Progressing into the modern period (1800's-1900's) science enriched the world with great leaps in physics, biology, psychology and a series of life-changing inventions. Then around 1970 a post-modern shift emerged where structured objective science faced a cultural backlash characterized by subjectivity, skepticism, relativism and resistance to absolute values.

Contemporary civilization is founded on structured categorized knowledge layered in a tapestry of individual sciences. In classical times there were just 7 Liberal Arts composed of grammar, rhetoric, and logic (the trivium) plus geometry, mathematics, music, and astronomy (the quadrivium). Today we have an endless spectrum of specific disciplines spanning the entire range of physical, biological and social domains. This comprehensive assortment makes even the most organized library seem confusing. The Dewey decimal system divided it all into 10 groups with 100 subdivisions each. It may look like a nice, neat format but it masks the unified symmetry present within.

In past centuries public knowledge was lorded over by privileged scholars. Emergence of the printing press democratized information, enabling transformative expansion of available knowledge on a mass scale. Widespread dissemination of information continued with the establishment of libraries, state schools, colleges and specialized institutions. Modern civilized nations prioritize education, making it an important function of government. Today it's a systematic comprehensive structure, accommodating children (K-12), young adults and adults. Its facilities are maintained at every level of government – local (grade school & high school), State (university system), and National (institutes). This comprehensive framework of schools is supported by enormous sums of money (US Department of Education budgets are well over 200 *Billion* dollars annually).

Evolution of knowledge began with a shift from ancient mysticism to systematic methodical science. Centuries of pseudo-knowledge based on false gods and forces in nature gave the masses a sense of false but necessary order. With advances in real science, the best parts of false science were extracted and turned into the real thing. Ancient alchemy became chemistry, astrology morphed into astronomy, and mythology transitioned into organized religion. Mystical, imaginative explanations were replaced with empirical evidence. Gods of nature became just nature, and a pantheon of deities evolved into a single, unitary God. Interestingly, science and religion followed similar paths of evolution; both began with multiple laws and Gods, then transitioned towards single, unitary laws and a single God. Science is limited by paradigms while religion is constrained by dogmas.

Recent history brought a transition from reductionist science to a more holistic, integrative approach. Instead of just breaking down everything into parts there's more emphasis on systems thinking, identifying synergistic principles where the sum is greater than the parts. Analysis now elevates context, relationships and connectedness, thus replacing micro-reductionism. Where *classical science* is rational, analytical, linear and quantity/object focused, the *holistic systems* perspective is intuitive, synthesizing, nonlinear, quality/relationship oriented.

Transitions in development of knowledge are greatly influenced by mediums used to transmit information. Early man used oral traditions with rough images and artifacts. Next came refined languages so ideas could be captured in written form and widely disseminated. Later came movable type and printing presses; the dawn of mass media. Shifting from industrial age to electronic age brought digital bits and bytes; material media was replaced by immaterial electronic information.

Advancements in Computing Capabilities

Primitive rocks, sticks, dirt figures
Tablets, scroll markings, abacus
Mechanical gears, levers, pulleys
Electronic vacuum tubes, punch cards
Digital electronics, transistors
Mainframe computers, magnetic tape
Microprocessors, calculators, PC's
AI, Quantum computing, atomic

The evolution of knowledge can be traced by simply looking at media forms in every stage of transition. Whether it's scrolls, books, photos or movies, the medium is often the message. From early stone tablets and cave wall drawings to videotape and hard drives, the means of capturing information were an integral factor in the process of expanded capacity. Each media form requires its own method of distribution, both evolving in parallel fashion. Books begat libraries, movies begat theaters, and digital media begat computers and the internet. Modern formats emerged including periodicals, papers, cartridges, discs, film, videotape, etc. The latest transformative revolution is the world wide web, with its game changing emergence bringing comprehensive knowledge and information to the masses in real time. It serves as an equalizer providing democratic free access to all the accumulated knowledge of human history. There's no longer an excuse for ignorance if you're truly interested in knowing anything.

Renaissance	*Enlightenment*	*Modern*	*Post Modern*
Revisit Classical era	Science and rationalism	Objective science	Indeterminate, relative
Retain mystical sense	Stamp out mysticism	Coherence, order	Fragmentation, uncertainty
Naturism	Humanism	Can know everything	Can't know anything

Premodern = Dogma   Modern = rational progress   Post-modern = subjective relativism. Sources of wisdom went from storytelling to literature, then philosophers, scientists...now TV personalities

Between 1600 – 1900, science evolved dramatically, producing a new stage in human civilization; *The Enlightenment*. Objective science produced universal truths about the world while humanism proliferated, following a shift from mythical gods to the preeminence of man. But with great scientific progress came a loss of connection with the sacred. Newton virtually removed the divine essence from the world by reducing it to mathematical laws, transforming nature into a mechanical machine. The enlightenment period eventually culminated with a great industrial revolution as western reductionist science acted as a catalyst for an explosion in manufacturing through division of labor, specialization and the assembly line.

Significant Developments in the Evolution of Knowledge

Babylonian – first written language, codes of law, astronomy, maps, irrigation, trigonometry
Egyptian – calendar, plow, medicine, ink, agriculture, fermentation, architecture
Greek – philosophy, science, geometry, water mill, cartography, algebra, bridges, democracy
Roman – engineering, roads, concrete buildings, aqueducts, Julian calendar, art
Chinese – papermaking, abacus, compass, gunpowder, printing, silk, porcelain, rockets
Renaissance – mechanical printing press, microscope, telescope, slide rule, glasses, pendulum
Enlightenment – methodical science, vaccines, calculus, empiricism, democratic government
Industrial revolution – academic libraries, leveraging machine power, widespread innovation
Electronic revolution – radio, television, telecommunication, mainframe computers, videotape
Digital revolution – Internet, PCs computers, hard drive storage, smart phones, artificial intelligence

DYNAMICS OF KNOWLEDGE AND INFORMATION

Knowledge continually progresses forward as succeeding generations build upon the growing heap of stuff produced by everyone prior. With expansion comes added branches and subdivisions where distinctly different areas take shape and fill out, very much like a tree growing fuller with larger limbs separating branches and clusters of leaves. Each new discovery and innovation leads to new possibilities and further branching; a ratcheting up process where knowledge expands in a stair case progression. Continual growth proceeds in both slow, gradual periods followed by punctuated, rapid shifts – drips and drabs plus revolutionary breakthroughs followed by long periods of consolidation and acceptance. Thomas Kuhn describes these dramatic changes as *paradigm shifts* where accepted dogma is turned on its head and replaced with a radical new concept. They periodically occur in science, art, religion, philosophy, social discourse, etc. Knowledge revolutions run parallel with great sages and intellectuals at the leading edge of the wave. They include great philosophers (Socrates, Plato, Aristotle), great sages (Buddha, Lau Tzu, Confucious), Great inventors (Davinci, Franklin, Edison) and great scientists (Galileo, Newton, Einstein).

Knowledge is not a finite quantity but rather an ever-changing, ongoing evolving domain. Discoveries (intentional and unintentional) generate new laws, which may provide answers to old questions but invariably open up new ones. Occasional cross fertilization of separate fields contributes to the incremental build-up. Day to day studies supplement current knowledge incrementally until a tipping point is reached where enough new data leads to new level of perspective. Collective consensus is achieved once enough information forms a consistent pattern leading to self-evident conclusions. However, new ways of seeing things sometimes reestablish old themes that fell out of favor, reemerging in repeated cycles of perspective. This too occurs in science, art, religion, philosophy, etc.

What we know and understand is really a multi-dimensional process of acquiring information in a variety of modes with multiple aspects. We know stuff though mental, emotional, and sensory-motor experiences in combinations of thoughts, intuition, imagination, memory, and feelings. Our knowledge can be practical or theoretical, general or specific, deductive or inductive, logical or abstract. We know most things through individual perceptions and experiences, providing personal lessons learned the hard way. But we have access to public knowledge as well via books, libraries, videos; lessons learned the easy way. Some of what we know is considered common sense – basic precepts universally understood which almost everyone grasps. Other things require extensive thought and contemplation; abstractions less known by most people. And what we know can be either limited to a narrow area or span a broad, diverse range of fields.

Types of Knowledge

Explicit	Codified, documented, data
Implicit	Applying explicit knowledge gained
Tacit	Acquired over time
Procedural	How to do something
Coded	Languages based on symbols
Expert	Subject matter specialist
Empirical	Sensory based
Posteriori	Personal experiences
Reason	Doesn't require actual experience

*Ignorance is simply lack of knowledge, not lack of intelligence. All of us are knowledgeable in certain areas and ignorant in others, without relation to how "smart" we are.

Culture influences both personal and collective knowledge. An East-West duality affects the way people see the world and know things. Westerners generally default to mechanistic concrete secular science while their eastern counterparts embrace a mystical connected wholeness. Two opposite but complimentary approaches to seeking truth, both valid yet partial. The west perspective is devoid of the divine, a legacy from the biblical story of Eve eating forbidden fruit from the tree of knowledge that cut access to the Garden of Eden in mankind's first red pill moment. Western science took a great leap by rising above ancient mythology, magic and superstition but in doing so, threw the baby out with the bathwater by removing the sacred elements of experience. Man became just another part in a cosmic machine, stripped of divine essence, separated from nature.

WEST	EAST
Science	Perennial Philosophy
Logic	Intuition
Reductionist	Holism
Dismiss spiritual	Embrace higher realms
Separate from nature	Connected with nature
Experimentation	Meditation, contemplation

Sources of knowledge are plentiful, accessible in a variety of types and modes. Information abounds literally everywhere around us; the work place, school, home, books, radio, TV, friends, strangers, computers and smart phones. We acquire knowledge through a non-stop process involving formal education, personal experiences, demonstration, testimony, social discourse, mimicking, instinct, reading, studying, and investigating. Learning stuff can come through first-hand knowledge (direct personal experience/self-taught) or second-hand derivative sources, passed on by others, such as direct instruction from an informed individual operating as either a teacher, coach, mentor or subject matter expert. The process is qualitatively dependent on the source of instruction, the environment and the receptivity of the student.

Observing and interpreting facts begins with senses that acquire it. Information is simultaneously perceived and received through multiple filters – physical senses, mental impressions, emotional feelings and spiritual divination. Man discerns knowledge through these multiple modes, personalizing the entire process of interpretation. Cultural filters play a role (language, social structure, value system) as well as a variety of personal biases.

Acquiring knowledge requires openness to new things and experiences. Clinging to current beliefs comes at the expense of new insights. Curiosity is the cure. Compulsory instruction is much more effective when students are interested in it. With curiosity comes a questioning mindset, persistently seeking answers until things make sense. Public schools tend to force feed subjects, emphasizing memorization, expecting compliance with minimal interaction, and teaching *what* to think instead of *how* to think. Socrates would not be impressed. There's over emphasis on reductionist, fragmented instruction, with little attention given to holistic synthesis and analysis.

Knowledge and learning are symbiotic, both fueled by curiosity while impeded by dogma and rigidity. Both are enhanced through good listening skills and attention to detail. Hearing and listening are two very different things. Good listening skills must be supplemented by attention span and the ability to focus. Effective leaning is all about receptivity and degree of interest.

Teaching effectively is a two-way street requiring effort from both teacher and student. A variety of techniques work, depending on the subject and audience: facilitation, collaboration, demonstration, inquiry, or simple lecture. The key is achieving and maintaining interest – not easy with smartphones and soundbites.

Sometimes the best way to learn is to become the teacher yourself. Even when you think you know something it can be quite humbling to try explaining it to someone who doesn't. Renown physicist Richard Feynman's approach was rudimentary; be able to explain any subject to a child, using simple language and be brief. Strengthen your weakest areas, identify gaps and close them. Ironically, another famous physicist captured the very same sentiment; "if you can't explain it simply, you don't understand it well enough" – *Albert Einstein*.

Elements of Superior Learning	*Effective Teaching Methods*
Receptivity	Grab attention early and hold it
Listening skills	Ask open ended questions
Attention to detail	Relate to the student/audience
Attention span	Keep it simple, avoid jargon
Curiosity	Strengthen your weakest areas
Asking essential questions	Promote independent thinking

Inductive learning takes particular experiences and forms a general conclusion from them. The opposite approach, *deductive learning*, is grasping a general principle and applying it to a specific case. For example, "All women are beautiful", Amy is a woman, therefore Amy must be beautiful. Inductive reasoning is used in science where particular facts are gathered to establish a pattern and produce a hypothesis. *Inferential thinking* is used by detectives; deriving a cause from effect or looking at an artifact and determining how it was made.

Learning is a life-long process necessary to survival and progress. It's essential for you to take an active interest in your own self-development, which begins and ends with simply learning new stuff. Regardless of the source, the quality of someone's life is directly related to what they know and how much they want to know more. Every time someone engages in small talk with a neighbor or stranger, they're bound to pick up some tidbit of information they didn't know before. Every time you browse news headlines there's a pretty good chance you'll see something new. But the best way to learn anything is through vicarious experience – to learn from others' experiences, mistakes and successes. By simply listening with intent and paying attention to key details, you can acquire not only knowledge but true wisdom without even having to pay the price of learning it the hard way through personal trial and error.

Knowledge is not Intelligence. It's the stuff we know versus our ability to know it. Intelligence is about mental capacity and the aptitude to reason, plan, and solve problems. Previously we addressed the spectrum of Intelligence with its multiple types (Linguistic, mathematic, music, sensory motor, crafts, etc.). Intelligence involves how quickly you can grasp a new concept or acquire new skills. It also includes mental dexterity and wittiness. Knowledge is just the accumulated stuff acquired through intelligence, and its opposite is ignorance, which is simply the lack of knowledge. Transcending both is wisdom, which is an internal insight attained from synthesizing accumulated knowledge.

What we know is only as useful as how we apply it. The trick is to follow good thought processes while avoiding mental traps and bad habits. **Critical Thinking** is essential to effectively analyze information and form intelligent judgments. To think critically, you must be aware of your own biases and assumptions and apply consistent standards when evaluating information. Be objective, consider multiple perspectives, and analyze judicially. Critical thought clarifies understanding of complex topics while avoiding errors or misconceptions, leading to better informed opinions. It's a process of carefully considering a problem, question, or situation to determine the best solution.

"You can tell whether a man is clever by his answers. You can tell whether a man is wise by his questions." — Naguib Mahfouz

The most important step of problem-solving is clearly identifying the actual problem. Many "solutions" fail because the real problem isn't recognized. For instance, in a country collecting trillions of dollars in taxes but still faces deficits, the problem probably isn't too few taxes. Same with you if your income is high but you can't make ends meet your gambling habit. Clearly identifying the root problem up front saves lots of time and effort that won't be wasted pursuing false ends. Once the actual problem is clearly identified and a solution is chosen, the process isn't over yet; monitor and adjust as new or contrary information surfaces.

Critical Thinking	In Science	To Be Effective	Intellectual Standards
1 Identify the problem	Is it logical?	Identify the problem	Clarity – understandable?
2 Gather information	Is it falsifiable?	Define the context	Accuracy – is it really true?
3 Review evidence	Is it comprehensive?	Determine options	Precision – adequate detail
4 Drop least valuable info	Are we honest?	Supporting points	Relevance – real problem
5 Summarize the evidence	Is it replicable?	Choose best solution	Depth/Breadth – complete
6 Select key points	Is it sufficient?	Self-correct as needed	Logic – does it make sense?
7 Evaluate decisions			
8 Decide			
9 Implement decision			
10 Test results			

The opposite of critical thinking is biased, subjective or emotional reaction. It's counterproductive to stick to your position after new information refutes it. Don't wait for someone else to prove you wrong, scrutinize your own position and test your own claims. Remove emotion from your analysis and don't take the results personally. Critical thinking demands you be honest, openmined, self aware, skeptical, welcome criticism, encourage different viewpoints, strive for clarity and precision, and remain humble. The worst kind of knowledge lies in those things you think you know but don't. By knowing yourself first you can avoid mistaking facts for opinions, and present points of view that are not solely based on beliefs, faith or imagination. By knowing others, you can avoid following the advice from those without qualification, including celebrities or TV talking heads who may be likeable and communicate well but don't really know much about the topic at hand.

Creative Thinking is the flip side of critical thinking, supplementally combining mental flexibility with imagination. Value is added by loosening the natural structure of mental processes. It's thinking outside the box and stepping back from your own self-imposed constraints and assumptions. Great discoveries and revolutionary ideas came from intellectual pioneers who balanced rational thought with intuitive insight, mixing reason with unconstrained contemplation.

CrEaTiVe ThInKiNg leverages a fresh perspective, fostering unique, original solutions. It's an inventive thought process producing unexpected conclusions and improved ways of doing things. Creative thinking can be aided by brainstorming or lateral thinking to generate innovative ideas. It's surprisingly effective in response to complex, wicked problems. It opens you to see possibilities that others don't (visionary). It's asking questions other may think are silly or ridiculous. Consensus opinion or status quo is a barrier to fresh, original thinking.

Creative Thinker	*Barriers*	*Critical* vs	*Creative*
Challenge Status Quo	Ego	Analyze	Imagine
Avoid Assumptions	Group bias	Reason	Playful
Naturally curious	Conformity	Objective	Subjective
Explore all possibilities	Coercive	Linear	Non-linear
Vivid imagination	Prejudice	Convergent	Divergent
Don't believe in an ultimate idea	False assumption	Left Brain	Right Brain
Never think anything impossible	Wishful thinking	Probability	Possibility
Take risks	Subjectivity	Focus	Diffuse
Adapt to changing circumstances	Confirmation bias	Classify	Analogize
Look beyond the first 'right idea'	Jump conclusions	Breakdown	Invent
Visual thinkers	All or nothing view	Compare	Combine

Problem solving is a universal endeavor everyone engages in daily, whether it's a simple task or a complex dilemma, individual or collective. Most of us do it subconsciously, working through mundane requirements but occasionally needing to deliberately weigh pros and cons when approaching bigger problems. Organizations, businesses and governments face the same daily challenge, often involving how to best manage people, resources and time to accomplish goals. Layers of staff spend the majority of their time working through options, usually under pressing time constraints. Leaders who demonstrate competence as problem solvers move higher up the chain. It's a key discriminator between executives and bureaucrats. Problem solving is *most effective* when both critical and creative thinking are employed in tandem, creating synergy when combined. They're the Yin/Yang equivalent of Art and Science. Problem solving is *least effective* when too much over-thinking occurs resulting in analysis paralysis.

Quality of Knowledge.   Information is not just quantitative but also qualitative, containing value at different degrees which depend on qualifications: Is it valid, credible, reliable and free from bias? Is there evidence to support it? Is it meaningful, adding value to a need? What degree of correlation does it hold to other pieces of information? What degree of certainty do you have accepting it as factual? When it comes to knowledge and information, equality is never a given.

Bits and bytes of information are not equal. There's a significant difference between good data and noise. It's a qualitative difference affecting all information. In the modern world of mass information overload, it's essential to filter out the meaningful from the meaningless. Most "new" information is duplicate rehashing of previous stuff. Thousands of years ago a bible passage proclaimed "there's nothing new under the sun", which applies more than ever today. We see it in pop culture where recurring movie and song remakes dominate (Titanic alone has over 20 versions and remakes). Compounding the noise and clutter are omnipresent low-level information sources whose continual output adds very little value to our collective knowledge. Consider repetitious cycles of sports, soap operas, special interest magazines, cheap entertainment, and other pop culture rehashes, not to mention never ending political theater and economic data saturation.

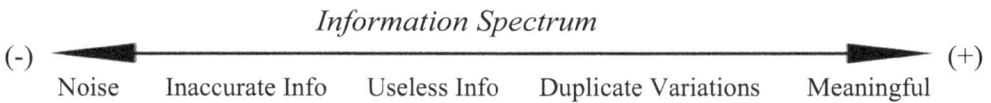

Information Spectrum

(-) ◄───────────────────────────────────────► (+)

| Noise | Inaccurate Info | Useless Info | Duplicate Variations | Meaningful |

Different areas of knowledge delineated in the Dewey decimal categories are also unequal in qualitative value. Comparing the subjects in 300's, 500's and 700's shows how different the quality of information is. Certainly, natural and social sciences provide more meaningful knowledge than arts and recreation. This reality doesn't diminish the need for those categories; it is what it is. The enormous daily information overload requires a vigilant effort to sift through the noise and extract the meaningful. As Knowledge expands, a good deal of data noise accompanies it. We're literally drowning in raw generic information, most of which is extraneous.

Qualitative Differences in Knowledge

WISDOM	Applied knowledge	Integrative, principles, patterns, reflective	Quality
KNOWLEDGE	Context based	Analyzed, synthesized, interpreted	
INFORMATION	Meaningful data	Processed, organized, compared	
DATA	Raw	Individual facts, observations, measurements	Quantity

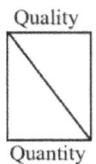

Problem solving is data management: Collect, organize, analyze, conclude, decide on possible options

WISDOM	Book	System	Joining wholes	Why, purpose
KNOWLEDGE	Chapter	Axiom	Whole	How, context
INFORMATION	Sentence	Equation	Connected parts	Relationship
DATA	Words	Numbers	Parts	Who, what, where, when

Societies abide by balancing collective partial truths. Each of us clings to our own individual partial truths. All political parties get some things right and others wrong. Very few people actually grasp universal absolute truth. In a world of collective partial truths individuals must pursue ones that are less partial than others. Recognizing your own areas of temporary ignorance opens a path toward greater wisdom. This includes acknowledging mis-knowns (things you think you know but are wrong about).

"If you don't read the newspapers, you are uninformed. If you do, you are misinformed."– Mark Twain

Quality of knowledge generally increases when information is shared and combined within a group. Two heads are better than one, and so on. Individual knowledge is limited by built-in personal biases. Whether it's physical senses (touching, seeing, hearing) or social exchange, the result is subjective knowledge. Collaborating with others tends to filter out some of these individual biases, producing a cleaner, more truthful perspective. It also serves to fill in gaps where some individuals are lacking in one area while other don't.

Knowledge and Truth become Relative and Partial via the Human Lens

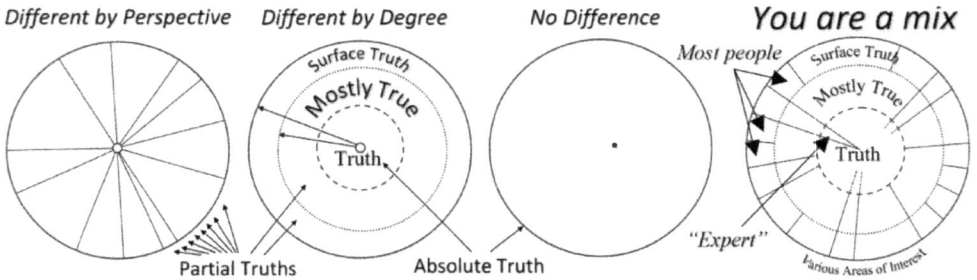

Individuals occupy a slice on the left pie circle, seeing reality from a limited, biased point of view. The same applies to all sciences, religions, political ideologies, etc. We all know some things truer than others; specialists and "experts" know greater truth in 1 area only

Knowledge expansion is never a straight line but more of a zig zag with incremental bits and periodic setbacks. Hidden knowledge results when sources become obscure; books are no longer in print, governments classify documents, video and film deteriorate, program reels go missing, etc. Lost knowledge can occur in catastrophic events such as a library burning down. Hundreds of languages have gone extinct, along with the cultures who spoke them. Skilled crafts vanished with the transition to machine processing. Countless inventions that couldn't reach public markets disappeared. Undoubtedly the vast amount of human history was never documented or recorded, just lost in the ether. Much of recent history is only partially documented and incorrectly portrayed. We still don't know precisely how the pyramids were built. With great reliance on high-tech solutions, many low-tech, old-school techniques are abandoned and eventually forgotten.

Lost Knowledge Though History (most common cause of tragic loss: fire)

Polish Libraries	16 million volumes	18 Terabytes
German Libraries	10 million volumes	12 Terabytes
Library of Bosnia/Herzegovina	1.5 million books	1.7 Terabytes
Library of Serbia	500,000 volumes	581 Gigabytes
Library of Alexandria	500,000 scrolls	571 Gigabytes

*Millions of original tablets, codices, manuscripts, books and sacred texts have been lost all over the world. These losses transform known knowns into known unknowns, further adding to hidden reality

DIVERSITY OF KNOWLEDGE AND INFORMATION

Everything we know fits into a broad range of different categories and organizing it all poses quite the challenge, but you have to start somewhere. We mentioned the Dewey decimal system earlier where everything is broken down into 10 major classifications with hundreds of further subcategories. A similar approach is followed in the Library of Congress Classification system which uses 21 major categories; same stuff, more headings. Other systems exist with their own somewhat arbitrary scheme. Virtually every field of interest and academic discipline falls somewhere within these organizing systems. However, none of them is based on how different subjects relate to each other.

We can simplify our mountain of knowledge into 2 broad categories: science and non-science. Each of those can be further subdivided with branches and subcategories. Science includes hard and soft branches (objective and subjective), ranging from fields based on precise quantifiable measurement and other fields with less predictable outcomes. Non-science covers a separate range of disciplines, mostly dealing with the human condition. The extensive compilation of mankind's knowledge breaks down into the following major categories: *science*: natural and unnatural, *non-science*: parascience and pseudo-science, *humanities*: philosophy, history, literature and language, mythology and religion, and the *Arts*. Each field of study reflects a different attribute of knowledge, adding value to the collective whole. Note that the humanities category is a major subset of non-science.

Philosophy is like the physics of thought and reason, serving as the foundational discipline supporting all others. It examines all aspects of knowledge in pursuit of universal truths. Philosophy searches for the *why* after science explains the *how*. Both explore the nature of reality but from different perspectives and dimensions. One deals with the mind, the other with the physical. Philosophy is the methodical study of existence, knowledge, values, reason, mind and language. There are 6 major branches of philosophy: Metaphysics, Epistemology, Ethics, Logic, Politics, and Aesthetics – specific disciplines requiring serious attention and thought.

Philosophy addresses a multitude of concerns most people take for granted, questioning the very essence of what everyone experiences in life. Nothing is left off the table, everything is subject to close scrutiny and comprehensive examination. This culminates with a series of *ultimate questions*: What is the meaning of life? Why is there evil in the world? Do we have free will or is everything predetermined? What is consciousness? How does one explain the mind-body duality? Philosophy isn't for the timid as it pushes you to the limits of your intellectual comfort zone, probing relentlessly into your world perspectives, making you question everything you've been taught. Sometimes all it takes is a simple question; is punishing a child for playing with fire, good or bad? And after you answer with smug confidence, a nuanced conditional change in the scenario can suddenly make your answer look ridiculous.

The Meaning of Life – *Philosophical Definitions of the Ultimate Question*

To have pleasure (Hedonism), To be logical and minimize suffering (Stoicism)
To defend individual liberty (Libertarianism), Do unto others what you'd have them do (Kantism)
It doesn't matter (Nihilism), Do God's will (Theism), Make decisions-be positive (Existentialism)
Stop making sense of life, just live (Absurdism), Act in self-interest, common good (Humanism)
Free yourself from pain (Epicureanism), Bring the most good to humans (Pragmatism)
Love impartially (Mohism), Live an ordinary life (Confucianism), Learn the practical (Legalism)

*This ultimate question is too simplistic to answer in a short sound bite. The meaning of life is less about **doing** and more about **being**…experiencing life to the fullest by being fully in the present moment.*

Good philosophical debate leads to better understanding of how the world works and our relationship to it. The process begins with healthy skepticism about a particular subject with the same spirit of Socrates who questioned everything. This leads to thought arguments where you state your case, then provide supporting details and facts. A dialectic process follows where point-counterpoint ensues, culminating with a synthesized better perspective than either of the original opposing positions. It all begins with an inquisitive mental attitude and a desire to seek higher truth. Philosophia is a Greek word meaning "love of wisdom".

Metaphysics is examination of ultimate reality; the science of first principles and the nature of being (Ontology), founded by the Greeks seeking answers about their very existence. Their analysis began with going beyond simple sense perception and delving into mental processes. Looking out into the world, they sought to define the essences of reality, including substance, being, knowing, identity, causality, purpose, universals, and physical relationships. They were first to grasp and define differences between quantity and quality, potential and actual, universal and particular. Their metaphysical inquiry investigated what is what, structures of objects, and their classification, along with questions about the nature of the universe, God, and life after death.

Metaphysics is a specialized subset of philosophy focused on the micro nature of reality. Greeks identified essences, first causes, material attributes and qualities, seeking to discover irreducible parts contained in everything – culminating with a fairly accurate concept by the atomists. They considered numerous primary elements, including *earth, air, water, fire, atoms, forms, mind, etc.* Pythagoras introduced the metaphysical connection of order and harmony as the key elements of life, rooted in symmetry patterns of music and geometry (music of the spheres).

The fundamental quest to understand reality comes down to examining *life, being, matter and spirit*. Religion enters the picture with science of God (theodicy) as first cause or creator, and the search for truth via religious investigation. A major debate emerges with the question of whether God simply created (deism) or actively continues to interact in the world (theism). Some ambitious philosophers take it one step further with futile attempts to prove the existence of God, using abstract conceptual models, tortured word play or mathematical equations.

Epistemology is the specialized discipline of knowing, which addresses the reliability of human knowledge and truth, including methodology, validity, and scope. *Metaphysics* deals with *Whats*: what things are, what a person is, what our world is, while *epistemology* deals with *Hows*: How do we know what we know, what kind of person we are, or what kind of world we live in. One deals with *being*, the other with *knowing*. How do we know whether the waking state is real and not the dream state? Human knowledge will always be limited by sense perception and inherent biases such as personal values and cultural language. The trap is seeing apparent truth *out there* without consideration of our own conscious and unconscious filters *in here*.

Sources of What We Know	*Philosophical Perspectives of Knowing*
Sense Perception	*Empiricism* – external experience
Reason	*Idealism* – mental experience
Imagination	*Rationalism* – mentally interpret external experience
Intuition	
Memory	
Language	
Emotion	
Faith	

ETHICS deals with moral codes, honesty, integrity, justice and equality, as well as human acts and what is deemed to be *right* behavior. Ultimately it asks what is good and bad, right and wrong and considers what ought to be. A major debate within ethics is whether morality is absolute or relative. In other words, is it a universal, unchanging standard or a variable, context dependent essence, situationally based on time, place, or a particular group. Important distinctions include *wisdom* – knowing what is right and true, and *virtue* – acting in accordance with wisdom (One *being*, the other *doing*). Teaching virtue or how to live an ethical life is less about instruction and more about guidance, modeling and storytelling. Wise gurus avoid active preaching and instead, teach or mentor passively using parables, analogies and situational metaphors. They tend to be respected role models who walk the walk.

"Goodness is easier to recognize than to define" -W.H Auden
"The meaning of good and bad, of better and worse, is simply helping or hurting" – Emerson
"Character is destiny" – Heraclitus

Morality simplified is about things you shouldn't do; 10 commandments being one example. But morality is more than just a constraining code of behavior – it's about living the best kind of life, with virtue, honor and dignity. It's also about just becoming a better person. Issue oriented topics of morality include euthanasia, abortion, capital punishment, pornography, etc. Living a *good* life can include treating others well, raising a moral family, providing for those in need, following the rules, and being of good character. The latter means serving as a role model, being consistently honest, loyal, dependable, kind, and maintaining a positive attitude. In short, daily acts of service that make the world a better place.

General morality and specific morals are difficult to define as they fall somewhere within a spectrum. On opposite ends they are absolute yet in the middle grey area they become relative due to human subjectivity. Some things are universally accepted as good, while others are not.

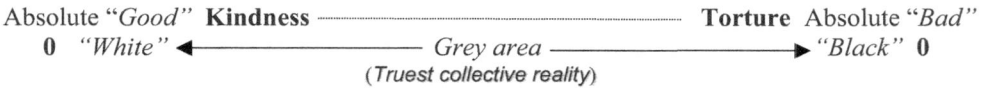

Absolute *"Good"* **Kindness** ⸻⸻⸻⸻⸻⸻⸻⸻ **Torture** Absolute *"Bad"*
   0   *"White"* ◄⸻⸻⸻⸻ *Grey area* ⸻⸻⸻► *"Black"* **0**
                        (*Truest collective reality*)

LOGIC is the discipline involving laws of correct reasoning, with emphasis on intellectual arguments based on inferences. It's the micro level of how to think, with structured procedures and rules. Arguments are statements based on a set of premises leading to a conclusion. Logic determines whether arguments are correct. Simply put, do premises support the conclusion? Logic is derived from the Greek word "logos", relating to reason, discourse, and language. It can be formal or informal and uses data to make inferences. Logic plays a central role in many fields, such as philosophy, mathematics, computer science, and linguistics.

Logic began in ancient times with Aristotle, who classified subjects as either *universal, particular, indefinite*, or *singular.* He based his reasoning on the use of syllogisms, where 2 statements sharing a common element are synthesized to produce a 3rd concluding statement. For example; All dogs are animals, all animals have 4 legs, therefore all dogs have 4 legs. This system served as the foundation of western logic for over 2000 years until roughly the mid 1800's when mathematics drove the development of formal symbolic logic; replacing concrete expressions with abstract symbols.

Logic often results in a true of false conclusion, with no room for something in between (law of excluded middle). It's rules of procedure include propositions, inferences, axioms, postulates, and hypothesis.

If-Then syllogisms are grounded in connected, sequential causal relationships. Universals and particulars describe general qualities and their specific applications. *Deductive* reasoning involves subject and predicate; the person or object of interest and what they're doing (noun and verb).

The most common, everyday use of logic is the informal type, where critical thinking is used to make arguments based on circumstantial evidence. Arguments can be either correct or incorrect, depending on whether premises support the conclusion. Deductive arguments are the strongest; if the premises are true then their conclusion must also be true. *Inductive* arguments are statistical generalizations, inferring that all in a group share the same trait based on individual observations within that group. *Abductive* conclusions are practical deductions drawn from limited, incomplete evidence. If proven incorrect they reduce to becoming mere fallacies.

POLITICS is the branch of philosophy concerned with how to regulate society and the communal life of human beings. It investigates what rights are absolute and whether government must protect such rights. It deals with questions of individual freedom vs collective conformity, liberty and governmental coercion, and how to balance competing interests. Political philosophy integrates the material and spiritual elements of social life, expressed in other branches of philosophy, including ethics, aesthetics, law, economics and sociology.

Numerous governmental systems have been tried since antiquity, ranging from republics to democratic city states to centrally controlled frameworks. Questions of power, equity, justice and individual freedom are central to determining how a society will be run and managed, and what "right" behavior will be tolerated by the ruling class. State control and the organizational structure of power affect all aspects of society. Intertwined within a political system are economic, religious and cultural elements, varying by degree.

AESTHETICS is the philosophy of beauty developed by the ancient Greeks dealing with the nature of art, taste, psychological aspects of appreciating beauty, and the relative nature of what is and isn't beautiful. Aristotle classified "beauty" into types while later philosophers analyzed the experience people have in response to beauty. Aesthetics goes beyond just art, exploring the dynamics that determine what makes something beautiful. There's the question of whether it's intrinsic (universal) versus subjective ("beauty is in the eye of the beholder"). Art, beauty and taste are generally context dependent, injecting relativity into the evaluation process which is affected by factors such as culture, psychology and individual taste.

Aesthetics investigates the psychological nature of how people see, hear, imagine, and react to works of art plus the breathtaking experience of witnessing a beautiful work of nature (landscape, mountain range, sun rise & set, majestic snow caps and colorful flower fields). A subject is perceived as beautiful if experiencing it evokes aesthetic pleasure. The feeling is similar when you witness a great performance and feel deeply moved: an elegant ballet dance, dramatic acting scene, stunning athletic feat, heroic acts or a sublime poetry reading. Something takes place within the human psyche that's difficult to quantify or qualify. Our biased judgment of what is and isn't beautiful is affected by subconscious elements (psychic baggage) connect to sensory, emotional and intellectual aspects simultaneously. Beauty has the power to inspire, lift us up, change our moods, and touch our soul.

WEST VS EAST

Philosophical perspectives vary between cultures and regions of the world. Our western perspective is deeply rooted in Greek heritage where love of wisdom and culture laid the foundation of philosophy. First, Socrates acknowledged ignorance as an honest, sober starting point. His skeptical methodology employed a series of disarming questions that shook the uninitiated, dismantled hollow dogmatic structures and forced weak positions to be reconsidered. Plato took it further with his own dialectic process where dialogue between two opposing views leads to a synthesized higher truth. He introduced a theory of forms where ultimate reality exists outside of space and time, with original archetypes serving as templates for everything manifested into the material world. Aristotle raised the bar further with his systematic approach to philosophy and metaphysics; classifying, ordering, analyzing and synthesizing *everything*. His theory of change established *first causes* and principles which trace all events and objects back to an original, unmoved mover. Greeks also sought *invariants* – universal unchanging qualities, which led to developing geometry axioms, atom theory, laws of causality and other revolutionary concepts.

Early Greek philosophy retained a spirit of mysticism mixed in with their rudimentary science. Nature, the world and everything within it were considered alive and connected, with no separation between spirit and matter. Their science emphasized reductionism, seeking to dissect everything into the smallest parts. Over time this led to an inevitable separation between spirit and matter. Western science carried this perspective through the centuries that followed, leaving the legacy of a machine world devoid of any divine spirit.

On the other side of the world metaphysics was grounded in perennial philosophy. Eastern culture embraced holistic connectedness in a unified reality of cosmic harmony. Where the west separates and divides nature into dualities (good and evil, mind and body, being and not being), the east sees only singular divine oneness, perpetually changing in form only. Chinese philosophy produced Taoism where the world operates in a continuous process of spontaneous change, driven by opposing forces of Yin and Yang. The natural state of the cosmos is eternal imbalance always seeking balance but never quite getting or staying there. Emphasis is on flow, process, change and time, with little regard given to material substance. The universe is an everchanging connected whole that's literally alive, not the dull western landscape of separate objects. Over time Confucianism emerged, representing a counter perspective to the TAO, focused on humanism and right behavior in a worldly environment. Though dissimilar, the two approaches are actually complimentary to each other.

Buddhist tradition of *dharma* contains aspects of both the west and east. Human experience is a combination of physical and mental aspects, similar to the West's mind-body duality, while also defining spirit as non-substance emptiness, making reality a groundless plenum with no ultimate essence. Like Confucianism, it too is based on right living where wisdom can be achieved thought liberation from the physical world, ultimately attaining enlightenment.

Western systems promote the individual (religion, philosophy, Science, Economics, Government) while Eastern systems elevate collective society. This manifests in the economics of capitalism (self-interest), government (individual rights and freedom), religion (personal salvation), and science (individual relationship with physical laws). In Confucianism, moral acts are those that are in harmony with society.

PHILOSOPHY SCHOOLS OF THOUGHT

The history of philosophy traces an ongoing pendulum cycle of shifting perspectives. One school of thought rises to prominence until a new one makes a strong counter claim, builds consensus, and replaces the previous one…until the next emergent "new" opposing school of thought. This process occurs in nearly every branch and area of interest. A recent example of such a shift can be seen in the post-modern transition from emphasizing reductionist particulars to holistic synthesis; from focusing on isolated parts to embracing integrative systems.

Schools of thought wax and wane in cycles of acceptance and rejection, getting replaced with variations on a theme. Many are revised "new and improved" versions of the original: classical-neoclassical, structuralism-deconstructionism, modern-postmodern. Some are nearly black and white opposites, such as Idealism versus materialism. Some are anchored in specific disciplines, such as economics: *Monetarists, Supply side, Demand side* (Keynesian*), Capitalism, Market socialism, Marxist, Malthusian*, etc. These schools of economic theory also wax and wane in acceptance, despite the clear track record of success and failure of each because politics trumps economics. And they also have their variants over time: *Supply side/Neoclassical supply side, Keynesian/neo-Keynesian*, etc. Likewise, most branches of science, philosophy and metaphysics produce original schools of thought that evolve in dialectic cycles of thesis, antithesis and synthesis, swinging back and forth like a pendulum of opposing perspectives.

Historic Perspectives in Philosophy

Empiricism – knowledge comes from experience (Hume, Locke)
Rationalism – mind acquires knowledge independent of experience (Descartes, Leibniz)
Idealism – outside world is a product of internal mind (Plato, Kant, Hegel, Fichte)
Positivism – objective science preeminent over metaphysical influence (August Comte)
Stoicism – wisdom comes from cosmic knowledge. A self-reliant life (Cicero, Seneca, Zeno, Marcus Aurelius)
Structuralism – relationships and patterns are key, not things (Levi Strauss)
Phenomenology – meaning derived from perceived lived experience (Husserl, Merleau-Ponty, Sartre)
Materialism – matter over spirit, things over ideas, possessions (Epicurus, Marx)
Existentialism – individuals create own purpose and meaning (Pascal, Kierkegaard, Sartre, Camus, Heidegger)
Skepticism – knowledge of the outside world is always uncertain (Hume, Berkeley, Diogenes, Laertius)
Cynicism – distrust of social conventions and authoritative decrees (Diogenes)
Pragmatism – practical, verifiable knowledge is best vs abstract ideas, concepts (James, Porty, Putnam, Dewey)
Romanticism – subjective individual over society, freedom from rigid control (Hegel, Schelling, Fichte)
Nihilism – no knowable truth, especially from societal authorities. Morals & values just made up (Nietzsche, Carr)
Post Modern – no dogma or absolutes. All is subjective and relative, multiple perspectives (Baudrillard, Hicks)

Perhaps the most iconic debate in philosophy over time is the nature of reality being either experienced based or reason based. It's expressed in different ways including left-right brain, materialism-idealism, physical sense-mental interpretation, mechanistic vs consciousness, and objective vs subjective. The preeminent philosophical question today centers on the nature of consciousness, which too has led to an equal dichotomy of opposing perspectives.

Multiple Dimensions of Reality Expressed in the Human Experience:

Intellectual level	*Truth*
Aesthetic level	*Beauty*
Ethical level	*Goodness*
Spiritual level	*Unity, wholeness*

Philosophy in general is considered the highest intellectual discipline, providing vision and purpose that science cannot. It reminds us that wisdom is only achieved through perpetual inquiry and life-long vigilance in pursuit of higher truths. In a world of noise, clutter and misinformation it offers clarity, precision and resolution. In our quest to discover more about reality *out there*, we inevitably learn more about who we are *in here*. Learning is a perpetual process that never ends, fueled by insatiable curiosity. And to those who don't understand why we persistently question everything, I refer you to Socrates who said it best; "The unexamined life is not worth living".

SCIENCE essentially began with Greek civilization where logic and reason both displaced religion and mythology. Having a practical alphabet of 24 symbols helped greatly, replacing cumbersome dependency on hundreds of different images. They developed a structured analysis of space with the introduction of geometry, creating systematized abstract concepts. The illusive notion of Beauty was codified as order, proportion and relational limits. Greek progress still contained elements of mythic Gods but significantly turned the corner with Aristotle's rational approach and his concrete system of laws.

Romans built upon Greek scientific progress, introducing their own advanced processes and methodologies to move civilization forward. After Rome's collapse, science stagnated but then regained its momentum during the renaissance and enlightenment periods, which finally replaced mysticism with sense and inductive reasoning, making science preeminent, setting the stage for a scientific revolution.

Western civilization began to thrive, paralleled by the rise of academic disciplines in Europe where hard science spread in scope and depth. Scientists and intellectuals normalized a new methodology, employing objective quantitative analysis and willingly excluding subjective preference or personal bias. Physics lead the way followed later by astronomy, chemistry, biology, geology and then many others. Scientific method was so successful that it extended into areas of soft science and non-science disciplines; fields where human subjectivity plays a part.

Success in the west was directly tied to its cultural embrace of science. Order and organization became essential. The world and universe were captured through a taxonomy; a categorizing of everything into sets of objects and types. Western science produced newfound celebrities with every impactful discovery and impressive invention. Engineering progress enabled spectacular construction projects, transportation improvements dramatically increased mobility and commerce, and assembly lines greatly expanded economic production. Science was now symbolized by instruments of measurement, including calipers, gauges, tweezers, levels, quadrants, magnifier and slide rule; the most iconic being the telescope-microscope tandem. But while western science celebrated the specialists who made all this possible, it did so at the expense of holistic thinkers, neglecting integrator-synthesizers.

Good Science	*Bad Science*
Assumptions challenged	Sample not representative
Suspended judgement	Correlation confused with causation
Revised with new evidence	No blind testing
Looking for what others missed	Results are not repeatable
Controlled experiments	Not Peer Reviewed
Precision, quantifiable data	Misinterpreted results
Testable predictions	Unsupported conclusions
Objective analysis	Selective reporting of results

Hard Science (*Includes natural science, formal science and applied science*)
What elevates hard science as a superior system is its objective approach gathering pure data, performing organized methodical experiments and testing through repeatable measurable results. It's essentially reductionist; taking physical reality apart, analyzing the pieces and coming to a rational conclusion, often in the form of a new theory. It's really quite simple: form a hypothesis and then test it to validate or reject it. Accurate results lead to expanded knowledge. Precise language is required, which is both a blessing and a curse. Accuracy is prioritized over clarity and understanding, so meaning is diminished by confusing esoteric jargon. Science is always easier to disprove than validate. Axioms produced are not laws and cannot be demonstrated; they can only be tested for validity, and if disproved must be modified or replaced. Our theories, descriptions and models of reality are always approximate at best with the goal of getting closer and closer to actual reality. You'll know good science theory when you see it: it's compact and meaningful; explains the most with the least verbiage, covers the broadest scope, is consistent with other disciplines and conforms to predictability.

Natural Science includes physics, chemistry, biology, astronomy (space science) and earth science. It's grounded in the tangible, with definitive disciplines that apply objective investigation, methodical experimentation and unbiased analysis. Each hard science contains multiple subbranches. Earth sciences include geography, geology, meteorology, minerology, hydrology, oceanology, etc. Life sciences are a further subset of natural science to include biochemistry, microbiology, botany, zoology, ecology, oceanography, etc.

Formal Science consists of mathematics, logic, computer science, artificial intelligence, systems theory, and several others. It deals with abstract structures using language formats in context with formal systems, unlike natural science that involves physical systems.

Applied Science includes agriculture, engineering, health and medicine, nutrition, architecture, computer science, management, military, and many more. Applied social sciences are a subset, including social work, public health, urban planning, and public administration. This category of disciplines uses the methodology of science to produce practical solutions to real world problems.

Soft Science differs from hard science by incorporating the human-social element. Included are economics, sociology, political science, psychology, law, linguistics, anthropology, and communication studies. This category of specialized disciplines still incorporates methods of hard science but focuses on human behavior rather than physical, predictable systems. Fields in this area rely on subjective interpretation, so results don't match the precision standards of hard science. Living beings are subject to thoughts, feelings and biases. Collective behavior can be tracked as a whole and analyzed for macro level patterns, but the human element precludes consistent reliable results.

HARD SCIENCE	SOFT SCIENCE
Clock time	Psychological time
Objective	Subjective
Pure data	Interpretive
Physical world	Social sphere
Left Brain	Right Brain

NON-SCEINCES don't depend on precise measurable results as sciences do but focus on patterns of human behavior and general themes within the human experience. These disciplines are not grounded in scientific procedure or quantitative results; however, they do follow methodical approaches using systems of analysis that may give the appearance of science without actually being so. They're not bad science per se, they're just mistakenly presented as science, especially true of parascience and pseudoscience.

Humanities are a special subset of non-science dealing with human culture, values and expression of the human spirit. They include History, Literature, Philosophy, Religion, and the Creative Arts. Philosophy serves as sort of a bridge between science and religion, addressing questions of higher reality in a rational manner. The humanities occupy a middle ground between hard and soft science. On one end pure science is based in objective, structured results, often incorporating mathematical formulas and relationships. On the other end, soft science maintains the methodical structure of hard science but injects the human element, which prohibits precision and predictability. The humanities transcend science; welcoming subjective interpretation, loosely defined parameters and fully embraces the human spirit. Where science is focused on the *how*, humanities consider the *why*.

Para Science involves investigation into areas that can't be explained by conventional science. It's focused on *Paranormal* phenomena-events, supported primarily by beliefs that are beyond the scope of normal scientific understanding. *Normal* would be those things explained by science while *para-normal* extends to things above, beyond or contrary to the normal. This includes psychic ability, extrasensory perception (ESP), telepathy, telekinesis, psychometry, precognition, clairvoyance, ghosts, poltergeists, necromancy, human auras, out of body experiences and near death experiences. Supernatural and paranormal operate outside the bounds of established science, operating primarily through anecdotes, testimony, and suspicion. Attempts at methodical research and experimentation are usually thin on concrete evidence with non-repeatable outcomes. In rare occasions where scientific-like results are produced they inevitably suffer from cognitive biases, clustering illusion, non-falsifiable claims, and confirmation bias.

What makes paranormal phenomena intriguing are recurring common stories from different eyewitnesses describing similar experiences. Take the case of near-death experiences where so many people describe the same things happening. There's the sensation of being outside the physical body, traveling through a tunnel leading to a bright light, a feeling of weightlessness, peace and joy, and then a realization that "it's not your time to go yet". It's uncanny how many different people from all walks of life seem to share identical experiences.

Paranormal events don't conform to conventional expectations of nature. Popularized accounts of weird, unnatural activity are suspiciously difficult to confirm, including crop circles, spontaneous combustion, levitation, unaccountable noises and explosions, mysterious appearances and disappearances, or stories of UAP's, UFO's and alien abduction. Without empirical evidence, verification is dependent on the credibility of witnesses. Skeptics explain these phenomena as misinterpretation, misunderstanding, and psychological responses within one's own brain. Regardless, polls indicate over half the population believes in the paranormal, some of which likely do because they want it to be real.

Alternative medicine could be considered a cousin of parascience with its non-scientific practices. It's not quite mainstream but it's increasingly used to supplement scientific based medicine. Examples are Qigong, Acupuncture, Ayurveda, Homeopathy, Naturopathy and Energy Therapies which focus on fields existing in and around one's body. Energy therapies included Magnetic Field Therapy, Reiki, and Therapeutic "Healing" Touch. Alternative medicine is an expanding field of practices that often do help but can't be proven scientifically.

The Occult traces back to ancient times of magical and mystical explanations. Anything involving hidden phenomenon can technically be considered occult, whose actual meaning is literally "hidden", or knowledge of the hidden. Most cultures have placed a negative stigma on anything remotely considered occult, stoking fear and revulsion at the mere mention of it. Over time, it's become an ambiguous term, leading to broader associations into areas it really doesn't apply to. Magnetism, gravity and religion were once lumped into that category along with other scientific disciplines that certainly don't belong. It really comes down to phenomenon that are supernatural, involving mysticism, various forms of divination and other worldly stuff. With the emergence of modern science many borderline practices have since been diminished to occult status. Today this category could include the para sciences along with magic, witchcraft, superstition, curiosities and wonders, and some aspects of the New Age movement.

Pseudo Sciences include astrology, tarot, numerology, palmistry, I-Ching, Feng Shui, acupuncture, and several others. These practices are surprisingly analytically structured, involving established procedures and standardized methods. In this respect they resemble credible science, but where they differ is in the results. While other soft-science fields deal with human subjects, pseudoscience includes transcendent aspects where objective physical measurements and outcomes are not consistently predictable, so instead, they focus on subjective interpretation. The focal point is man and his relation to the cosmic environment, navigating through a world above, below and within. These disciplines identify what we're brought into the world with: skills, tendencies, strengths, desires, etc. They address your life path and destiny. Most importantly they describe the dynamic process nature of how your inner being interacts with the outside, expressed in various dramas involving relationships, struggles, achievements, and lessons learned.

Pseudoscience operates on subconscious and unconscious levels. Revelations surface by tuning into the subtle non-physical ether permeating all life. Similar to the creative arts, a connection is made between spirit and matter, expressed in symbolic patterns, mythic archetypes and primal images.

Practitioners in each field learn their subject matter through a combination of intuition, observation and years of direct personal experience. There are no short cuts in this domain where the primary school of knowledge is life itself, and the more difficult and challenging, the better.

Astrology as a practice is the closest thing resembling science in this genre, despite its lack of quantifiable results. Originating in Mesopotamia (Middle East) then later spread to India and Greece, it combines astronomy, geometry, and mythological archetypes. It methodically examines 12 correlated signs, houses, and planets, supplemented by personality types (4 elements & 3 qualities), geometric aspects, and daily influences of change (transits). Like other pseudo sciences, there are plenty of duality patterns; Sun/moon, outer/inner planets, chart halves, energy genders, etc. By integrating relationships, geometric patterns and archetypes it produces both a snapshot horoscope, and a projection of ongoing cycles of change.

What's key is geometric relationships between planets, house placements and timing cycles. Subtle energy interactions create dynamics influencing your life path, personality, appearance and temperament. The *12 Signs* are like filters of energy producing a spectrum of types and traits. *12 houses* are the environments or situations it's channeled through. Geometric relationships between planets and their signs determine how energy is expressed; either free flowing, tempered, mixed or blocked. For every individual the process begins the day, minute and second, you're born, establishing a sort of cosmic blueprint with your innate abilities, disposition and potential. Cycles of life play out over time as planets cycle around, leading to angular patterns that either assist or challenge you. Like meteorology, it's a matter of probabilities, not certainties. A horoscope chart provides insight on your potential and likelihood of personal events. But like playing cards, it's still up to you as to how you play your own hand.

SIGNS	SYMBOL	SEASON	HOUSE	ENERGY	QUALITY	ELEMENT
Aries	♈	Spring	1	Male - Yang	Cardinal	Fire
Taurus	♉		2	Female - Yin	Fixed	Earth
Gemini	♊		3	Male - Yang	Mutable	Air
Cancer	♋	Summer	4	Female - Yin	Cardinal	Water
Leo	♌		5	Male - Yang	Fixed	Fire
Virgo	♍		6	Female - Yin	Mutable	Earth
Libra	♎	Fall	7	Male - Yang	Cardinal	Air
Scorpio	♏		8	Female - Yin	Fixed	Water
Sagittarius	♐		9	Male - Yang	Mutable	Fire
Capricorn	♑	Winter	10	Female - Yin	Cardinal	Earth
Aquarius	♒		11	Male - Yang	Fixed	Air
Pisces	♓		12	Female - Yin	Mutable	Water

Similar Personality Qualities

Cardinal signs	-Get things done
Fixed signs	-Stay the course
Mutable signs	-Adjust to change

Similar Elements – Similar Temperament

Fire signs	-Inspirational
Earth signs	-Practical
Air signs	-Mental
Water signs	-Emotional

A competent astrologer can see a wealth of information imbedded within this archetypal pattern. Multiple levels of depth, breadth and detail reveal all you need to know about an individual's character, talents, potential and destiny. Every planet, sign, house, and geometric pattern interacts in unique combinations, along with particular placements in each half (top-bottom, left-right), sequences, early and later degrees, exalted and diminished positions, approaching and receding connections, harmonics, secondary progressions, etc. Sun sign astrology in pop culture is ridiculously limited in scope compared to the comprehensive trove of data available in an actual horoscope. The complete Horoscope wheel consists of 12 Houses, 12 Signs, planets plus sun & moon. Each contains unique, direct and subtle patterns forming a graphic personality blueprint. Geometric relationships between planets are key, representing energy flow in harmony or discourse:

Geometric relationships – *The dynamic driver of horoscopes*

  0 degrees *Conjunction* = blended, combined energy, amplified, new cycle
 60 degrees *Sextile* = cooperative balance, supporting energy, opportunity
 90 degrees *Square* = conflicting energy, tension, challenge, blockage
120 degrees *Trine* = positive flow, effortless benefit, harmonious energy
180 degrees *Opposition* = polar energy, dynamic compromise, balance and integrate
360 degrees (Whole Horoscope, 1 complete cycle, beginning and end

Conjunctions, Sextiles, Triangles: good flow, supportive, opportunity
Square and *opposition*: tension, challenges, forced action, change

*Strong personalities driven by difficult aspects (90º/180º); a healthy balance of each is best

The Big 3 Keys of Your Horoscope

Sun Sign	Identity	*Personality*
Moon Sign	Needs	*Inner self*
Rising Sign	Style	*Projected self*

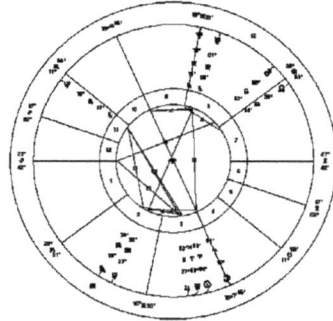

Energy Filters of Divine Spirit

Planets	*What energy expressed*
Signs	*How energy expressed*
Houses	*Where energy expressed*

Numerology applies dualism of numbers possessing both quantitative and qualitative properties. Numbers are not just arbitrary notations but symbols with imbedded universal meanings. Each number is its own archetype; a subtle pattern connected to original forms. Numeric properties were discovered by Greeks, Hebrews, Gnostics, Kabbalists and Egyptians. Numbers correspond to everything, where every *thing* has an energy pattern and a unique vibration. Number symbols act as antennas resonating to a specific frequency, like keys that are shaped patterns matching tumblers in a lock. Everyone born unto the earth is given a name; a stamped marker that sets a template. Names are letter patterns that combine to create a resonant form. Everything on earth is given a name, connecting it with other patterns in the collective ether. Numbers correspond to ratios, proportions, phases of a cycle – parts of a unified circle.

Numbers correlate to colors, sounds, shapes, sizes, sets, places, amounts, bits and bytes of information. Numbers correspond to vibrations, obvious in musical tones – keys on an octave scale, beats, tones and rhythms. Every *thing* is a number and is connected by other numbers on a comprehensive relative scale.

Key Numbers connect to every unique Individual
Always reduce multiple digits to 1 single digit (256 = 2+5+6 = 13, = 1+3, =4)

Primary

| -Life path # | add digits of birth year, month, day |
| -Destiny, purpose # | add digits of letters in your full name (a=1, b=2, c=3. etc.) |

Secondary

-Heart's desire, soul #	add digits of vowels in your name
-Personality #	add digits of consonants in your name
-Personal power, goal #	add life path # to Destiny #

Number Symbology	*Astro-Sign*	*Master numbers*
1 – Independent, purpose, beginnings	Aries	11 – Inspiration, idealist, intuitive
2 – Relationships, harmony, balance	Libra	(Sagittarius)
3 – Creativity, expression, optimism	Leo	22 – master builder, visionary
4 – Stability, building, foundation	Taurus	(Aquarius)
5 – Change, adaptability, freedom	Gemini	33 – Healing, unconditional love
6 – Duty, responsibility, nurture	Cancer	(Pisces)
7 – Analytical, private, sacred	Virgo	
8 – Ambition, power, authority	Capricorn	
9 – Completion, service, transformation	Scorpio	

10 = 1 + 0, = 1, *repeat pattern on a higher octave, cycle of renewal*

I-Ching "translates to "Book of changes", which is divination using numeric patterns of hexagrams arranged in particular, meaningful sequences. It works by tapping into subtle-level consciousness. At the micro level of reality change occurs at the realm where novelty emerges in tiny increments correlated with numerical, contextual patterns. Like numerology, it draws from relational forms but is entirely based on binary code using 2 simple symbols, either a solid or dashed line. These simple dual possibilities represent the alternating push/pull nature of forces in the external environment (Yin/Yang). They manifest as point/counterpoint, yes/no, male/female, light/darkness, expand/contract, up-down, and harmony-discord, all complimentary pairs.

To interpret patterns and generate meaningful insights, the dual symbols are set in patterns of 3. These *Trigrams* are any combination of dashes and solid lines, like 3 coins where each is either a head or tails. There are 8 possible combinations of a Trigram (2^3 power) = 8 different patterns. Think of dominoes with 3 different numbers on each. Next, Trigrams are combined to make further elaborate patterns: 2 Trigrams combined = Hexagram (combination of 6 stacked dash or solid lines). 64 Hexagrams = 32 essences with 2 opposite expressions.

The 8 Metaphorical Elements of I-Ching Emerge from simple Yin/Yang Duality

An I-Ching layout is sort of a periodic table of spirit and vibration where patterns are more important than each elemental part. It represents the fundamental process of reality where 1 emerges into 2, then transitions into multiple variations producing literally everything. Its simplicity is inspiring, taking just 2 basic parts and creating 64 unique yet symmetrical, connected patterns of experiential reality. The 64 patterns are conditions, states of being, dynamic situations, perspectives, phases or modes of action. It's limited to 64 types as a practical range but could easily extend into branches of 128, 256, 512 etc. Each of the 64 produce 6 sub-conditions or behavior responses in an individual.

These combinations of energy patterns produce 384 different pathways; outcomes that are interpreted differently based on situational context. Practitioners use experience and intuition to determine if the issue is internal or coming from the outside world? Is it a problem or an opportunity? Does it require immediate action or is caution advised? Wisdom from a competent reader will determine how to best interpret results, no different than a doctor reviewing symptoms of their patient.

Where astrology uses 12 signs, I-Ching uses 8 metaphorical memes, drawing on the same 4 universal elements recognized by every ancient civilization (air, earth, water, fire) but adds a dual aspect to each, then producing 64 universal conditions.

Tarot is a symbolic system of cards with numbers and images containing layers of subtle information accessible on an inner, psychic level. Pictured cards resonate on an archetypal dimension linking universal themes (*ancient origin of the modern card deck*). Readings begin with a focused question, establishing context to frame the issue at hand. Interpretation is produced through a dynamic process connecting the individual's unique situation, the layout of the cards, and the reader's intuitive response. Factors impacting the interpretation include the sequence of cards drawn, the orientation of laid cards (upright or inverted) and combinations as they relate to each other. The difference between a novice and a skilled practitioner is the degree to which rich information is extracted from each and every card.

78 Cards: 22 Master cards (Major Arcana) + 56 Regular (Minor Arcana)
4 Suits X 10 numeric cards = 40, 4 Suits X 4 Court cards = 16 (56 Total)

4 Suits	Theme	Modern	Element	Mode	Association
Swords	Challenge	Spades	Air	Mind	Reason, clarity
Wands	Ideas	Clubs	Fire	Spirit	Action, passion
Pentacles	Money	Diamonds	Earth	Body	Possessions, security
Cups	Desire	Hearts	Water	Heart	Emotion, creativity

4 Exalted Figures - King, Queen, Knight, Page   * modern version is Ace, King, Queen, Jack

Kings, Queens and Pages correlate with a personality type, timing and area of interest. One card can simultaneously provide insight on *what* you're going to do, *when* you're going to do it, and *whom* you might be involved with. Knights are action cards, indicating change; either a beginning or ending. Each court card correlates to a specific astrology sign (ex. King of wands = Aries)

7 Card Spread – *most common*

Card 1 – Situation
Card 2 – Your current state
Card 3 – Your wish
Card 4 – What will happen
Card 5 – Changes or resistance
Card 6 – The unexpected
Card 7 – Actual outcome

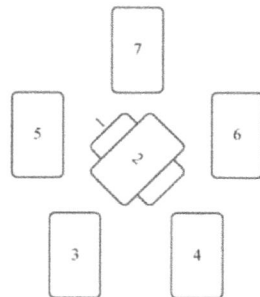

Major Arcana – 22 cards associated with major events in our lives as well as societal issues, humanity, and universal themes. They include the Magician, High Priestess, Empress, Emperor, Hierophant, Lovers, Chariot, Strength, Hermit, Wheel of Fortune, Justice, Hanged Man, Death, Temperance, Devil, Tower, Star, Moon, Sun, Judgement, The World, The Fool.

Minor Arcana – 56 Suited cards 1-10, plus face cards (court cards). They include 40 suited numbered cards associated with mundane, every day issues in one's life, very similar to a modern card deck.

Tarot captures universal experiences of human life, providing insight into one's own unique life journey, at any stage, in any situation. Individual cards contain a wealth of information that correlates to a person's particular area of concern. Ask a specific question and you'll get a specific answer. Interpretation is difficult at first because like all the pseudo sciences, it uses a psychic language requiring translation into normal conversational terms. Symbols, numbers, images, colors, genders, roles, and patterns are rich in symbolic information once you understand the archetypal meaning. It's easy to confuse the Fool as a dummy, the Devil with evil, Lovers with lovers or Death as actual death, instead of understanding the Fool is just taking a blind leap of faith, the Devil is temporary overindulgence, the Lovers represent relationships and choices or that Death just means an ending or loss. Tarot may be taught esoterically but competent application only comes through enduring personal experience and wisdom from accumulated life lessons.

Chess linked to Tarot scheme: King, Queen, Knight, Bishop plus a rook (tower) & 8 pawns

Feng Shui – harmonizes people with their surrounding environment by orienting objects and spaces to optimize the flow of subtle energy (Chi). Living or working quarters are arranged to redirect relationships between positive and negative spaces (solid things and empty space). On a practical level it could be considered a variation of interior decorating where managing furniture and visual displays impacts the mental and emotional state of the those occupying that space. Lighting, visual pathways, doorways, and seating arrangements all make a difference in how we experience being there. Feng shui purposefully orients energy dynamics on a subtle level, supplementing routine changes we make with physical objects and visual elements. The goal is to create a space with positive energy, not unlike being in a room with enthusiastic people versus negative nellies.

Feng Shui's Map of energy flowing in a living space correlates with I-Ching's 8 Trigram essences. It's an Octagon model highlighting 8 relational spaces associated with the following: Fame, money, love, health/family, creativity, knowledge, career, friends

Unity of Pseudo Sciences - *Energy patterns are key, not physical substance.*

Life is essentially driven by energy flow dynamics, not the residual forms we identify with. Whether it's Chi, libido or psychic energy, it's the flow patterns that pseudo-science examines and tracks. In the human body energy patterns resonate at distinct levels (Chakras) as 7 energy nodes. With proper alignment, balance and flow, they line up in a color-coded pattern where resonant energies synchronize. These layered levels of energy are expressed in human behavior and correlated states of being. Manipulation of that subtle energy is the aim of both parascience and pseudo-science, including acupuncture, Feng Shui, psychic healing and many others. All share one key component – man is an energy filter, expressed in unique discernable patterns.

Pseudo sciences: *symbolic systems - universal forms and archetypal themes*

Astrology	Geometric patterns of energy flow through planetary relationships
Numerology	Numeric symbol forms in resonant patterns and unique combinations
I-Ching	2 basic symbols form 8 universal energies expressed in 64 unique patterns
Tarot	Universal archetypal images and resonant numeric patterns

Number Patterns

Astrology – Contains numeric symmetry patterns of 2, 4, 6, 12. The 12 signs/12 houses correlate to universal archetypes in numerology. A 360-degree circle of a horoscope connects symmetrical geometric forms.

Numerology – A small set of original numbers (0-9) lead to an infinite series of derivative numbers, each possessing archetypal meanings. They're imbedded into everything, taking shape in geometric form and manifesting in every situation or condition possible. Prime numbers are special non-divisible values like vowels in the alphabet.

I-Ching – Similar to numerology, the domino-like forms possess meanings based on individual numeric values along with variations created by combinations of base numbers. It's binary language of 2 values (0 & 1) combined to express multiple patterns.

Tarot – All 56 Cards are numbered, correlating to Numerology, including master numbers. Same suit cards that differ by only 1 number can express completely different meanings.
* Dominoes and Dice are variations of I-Ching and obviously correlated to numerology.

Universal Dual Energy Pattern

Astrology	Male/Female signs, sun/moon, mars/venus, 6 sign pairs, inner/outer planets
Numerology	0 and 1, Odd and Even numbers, primes and composites
I-Ching	Yin-Yang, 32 positive-32 negative patterns: perfect symmetry balance of 64
Tarot	Male/female energy: swords/wands vs pentacles/Cups, Major-Minor Arcana

*Pseudo sciences use a balance of left and right brain; structured methodologies interpreted through intuitive wisdom, operating on both a conscious and unconscious level. Horoscopes, number patterns, trigrams and archetypal images all tap into unconscious energy dynamics driving human behavior. Pseudo sciences tap the non-physical realm of subtle energy flow patterns under the surface of all life.

Astrology		*Tarot*	*Mythology*	
Sun	Leo	The Sun	Apollo	Light, Truth
Moon	Cancer	High Priestess	Selene	Moon
Mercury	Gemini/Virgo	The Magician	Hermes	Messenger
Venus	Libra/Taurus	The Empress	Aphrodite	Love
Mars	Aries	The Tower	Ares	War
Jupiter	Sagittarius	Wheel of Fortune	Zeus	King
Saturn	Capricorn	The World	Cronus	Time
Uranus	Aquarius	The Fool	Uranos	Sky
Neptune	Pisces	The Hanged Man	Poseidon	Sea
Pluto	Scorpio	Death	Hades	Underworld

*22 Major Arcana of Tarot correlate to the 12 signs and classic 10 planets of Astrology
**All 3 (Astrology-Tarot-Mythology) share the same universal archetype patterns

It's no wonder science looks down on these disciplines since they don't abide by physical laws. They operate on a metaphysical level that cannot be quantified, dissected or measured objectively. Trying to explain astrology by just the physical aspects of solar energy or planetary gravity is folly. Every pseudo-science deals with subtle energies beyond the physicality of this world. The common denominator is man the filter, where spirit is channeled and expressed from inside out, from a non-physical dimension out into *The World*. Relationships, dramas, lessons learned, opportunities and challenges, are all methodically examined with hints of science-like models but limited to probabilities, not precise actualities.

Practitioners of these disciplines don't receive formal diplomas; they rely on intuition and wisdom borne from real-life seasoning. Like soft sciences or humanities, the primary subject is people, so any attempt to produce consistent measurable results is an exercise in futility. However, this should not be taken as discrediting, since the same could be said about economics, psychology or politics where predicted outcomes are never certain. We could criticize meteorologists who routinely miss the forecast, frequently achieving the accuracy of a coin flip. Heck, quantum theory in physics reduces the ultimate substance of matter and atoms to mere clouds of uncertainty. How ironic that the hardest of sciences, physics, is literally founded upon elusive basic parts which Heisenberg discovered can't be measured definitively, a foundation no better than a house of cards. Since lots of mainstream disciplines don't hold up to the strict standards of science, misguided criticism of pseudoscience practices is a bit like the pot calling the kettle black.

Mystical and Magical practices trace back to ancient periods but still continue today across cultures throughout the world. They're present in various religious traditions which engage in rituals to induce mystical experiences, integrating them into daily life. Perception of mysticism has changed through ages, from a simple experience of becoming one with higher spirit to a catch-all for parapsychology, pseudoscience and religious cultism. It's a somewhat nebulous category but really comes down to personal experience of the spiritual realm; full of mystery and awe, feeling the omnipresence of a divine state, and connecting with a transcendent reality. Some consider it a religious experience while others define it as a personal altered state of consciousness. Common themes include revelation of hidden truths, a sense of initiation, ecstasy, illumination, enlightenment, presence, and transformation. Three basic types of mysticism are *theistic*, *monistic*, and *natural*, parts of which are found in Jewish, Christian and Islamic mysticism. Translating mystical experience and spiritual lessons into language we can understand is generally challenging. Eastern practices use *koans*, which are nonsensical riddles. Jesus used parables. Zen masters teach students using storytelling analogies.

Prior to scientific method, ancient man defaulted to supernatural explanations for just about everything. This created a world of magic, myths and mysticism where anything outside of ordinary experience took on an extraordinary origin. Nature was run by a divine presence. The World was influenced by forces beyond comprehension. Gods with human-like qualities were responsible for every major turn of events. In these conditions a variety of mystical practices emerged, including Alchemy, Shamanism, Witchcraft, Tantric rituals, mantras and mandalas. Ancient religious cults such as the Celtics, Druids, Wiccans, Norse Sufi and Egyptian Kemetic employed similar aspects of mystical-magical practices. Many of these ancient traditions continue today in various cultures.

Though magic was demystified and diminished by the scientific revolution, don't let that fool you into thinking there's no such thing as magic, because there is. It exists as an important aspect of life. It adds wonderment to an otherwise mundane world; the same way color adds life to a black and white image. You'll find it in unexpected places, often at the boundaries of your comfort zone. It pops up periodically in the form of coincidences, pleasant surprises, an unknown helping hand, beating insurmountable odds and finding valued items after giving up the search. Magic is not so much an staged act in front of a captive audience but rather the moment of awe experienced fully in the now where time seems to freeze. It's the eureka moment where you feel an amazing wonderment outside of normal time and space. Though brief and fleeting, its effects linger well after the fact.

Kabbalah is a system of letters, numbers and sounds with roots in Jewish mysticism, explaining the relationship between an infinite, eternal God and his creation of a finite, temporary universe. It portrays a dual aspect of God: a limitless unknowable essence and a manifested world of man. As a hybrid of pseudoscience and religion it includes meditative practices, religious rituals and reading of sacred texts. Kabbalah's broad scope reveals mysteries of God, the Tree of Life and mythic supernatural entities. The foundational model is a Sephirot which connects 10 attributes of God, sustaining the universe and everything in it. Geometrically it's portrayed as 10 centers in 3 columns, resembling a tree with roots. These 10 nodes represent emanations of concealed potential emerging into the mundane world, including specific attributes of wisdom, intelligence, love, fear, beauty, triumph, glory, foundation, upper crown and kingdom.

Sephirot

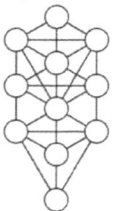	*Yehidah*	Highest plane	Full union with God possible
	Chayyah	Transcendent	Awareness of divine life force
	Neshamah	Higher	Intellect, higher awareness
	Ruach	Middle	Spirit, moral virtues
	Nefesh	Lower	Animal, instinctual

Mythology is a precursor of religion, based on imaginative explanations for the unexplainable, to create order from a mysterious, chaotic world. Myths attempt to reestablish links back to the divine, reconnecting man with the sacred original source of being. Universal themes emerge from collective human experience producing archetypal patterns that continue and repeat in cultures throughout history. These include forbidden actions, cataclysmic events, primal challenges, temptation, hubris, pride, fate, justice, power, beauty and many others. Archetypes driven by innate unconscious psychological forces lead to grand stories, legends and fables of ordinary people doing extraordinary things. Like Shakespeare plays, myths reveal human nature exposed in a world of challenges that showcase limitations, weaknesses, failure and surprisingly, an occasional triumph.

Myths, folklore, fairy tales and parables are narratives rich in symbolism – the metaphorical language of the sacred. They prepare ordinary people for experiences on the human journey of spiritual progress. Such stories chronical the universal search for meaning; how to live in The World and achieve our potential. Those who focus attention on things *out there*, overlook their own *being and becoming*. While legends are based on exceptional human beings, myths incorporate the realm of gods. Most myths feature supernatural heroes, gods of nature and gods with human qualities. They incorporate a wild variety of exaggerated beings and deities including giants, demigods, supernatural figures, ogres, monsters, nymphs, dragons, serpents, and nature spirits. Myths combine truth with imagination, symbolism, nature and intuition in a language not to be taken literally but understood as mystical generalizations of deeper patterns.

Universal mythic themes include duality, creation stories, underworlds, the hero's journey, and gods that direct world events. Duality themes are omnipresent with aspects of life & death, masculine & feminine, good & evil, knights & dragons, conscious & unconscious elements, all imbedded within most fables. Underworlds and dreamworlds are closely connected, with myths and dreams coming from the same place; each transcending thought while operating in symbolic, higher meanings. Creation stories are similar across very different cultures, consistently explaining how initial darkness and nothingness became light and somethingness. In classic mythology it all happened through gods, with creation through procreation, and lots of family drama as an added bonus. Universal themes include: People created and ruled by gods, challenges from a mysterious fearful source, threatening serpent monsters *out there* mirror what's *inside here,* a "Hero's Journey"; challenges conquered, triumphant return, divine power accessed through sacred ritual, apocalyptic events overcome, great Goddess (love, motherhood, earth's fertility), enlightened seer, wise guru, trickster wildcard.

Myth narratives still play a significant role in society. Despite exaggerations and imaginative embellishment, tales of heroes, villains, ordeals and triumphs still ring true to this day. It's no coincidence that roots of myth are deeply embedded in modern culture. In literature, songs and movies, the stuff of myths is still present everywhere. Who hasn't heard of sasquatch, the Loch ness monster, the boogie man, goblins, vampires, zombies, night creatures, etc. Movies, TV and cartoons are loaded with all kinds of superheroes (Hulk, Batman, Superman, Ironman), and the success of multiple superhero movies confirms our yearning for mythic archetypes. Urban legends are mythical story plots dwelling in the collective unconscious. Since human nature doesn't change, neither does our ongoing need for storytelling of adventure, ordeals, triumphs and exploring the great mysteries of The World.

The hero's journey is *the* fundamental ancient theme found in every culture, popularized in countless movies depicting a universal transition we all must abide, each on our own individual journey through life (*see wheel pictograph page 86*). Everyone experiences the process of leaving childhood's safety and security to enter the challenging realm of adulthood, passing through stages, initiations, and facing our own inner demons along the way. Adventure inevitably unfolds as we move forward into the unknown, occasionally faced with metaphorical dragons we must slay in our path, joined by allies who assist at just the right times. Of course, it's a transformation of consciousness as the serpents we face are not real monsters but the fears and insecurities inside projected outside. With insight and wisdom, some eventually discover that heaven and hell are not somewhere else but right here in The World. We might even realize that eternity is not some long period or somewhere else, it's right here and now. The ancient Old Testament of the Bible and today's Star Wars are both rich in mythological themes.

$$ \dagger \quad (\cdot \quad \maltese \quad \text{ॐ} \quad \circledast \quad \Psi \quad \star \quad \circledcirc \quad \text{⛩} $$

Religion is very different from science, which is grounded in absolutes, largely because it's entirely based on invisible spirit beyond this world. While science explains our physical existence, religion informs us of the meaning of our existence, establishes moral and ethical standards, and provides guiding principles of right behavior. The ultimate goal of religion is to connect finite man with the infinite, eternal source, achieved through a transcendent experience. Religion evolved beyond mythology, taking the same connection between man and the heavens while adding a more rational, ordered approach. Both attach human qualities to a higher power and use storytelling to describe man's interaction with a source beyond. Both share narratives of right and wrong living, the consequences of wrongful acts, and recommended steps toward redemption. Religion adds layers of rituals, sermons, holy events and use of prophets or gospels to spread the word, often codified in divine texts.

The *State-Religion* connection was the dominant political structure through history. Kings, Czars, Pharaohs, and other hierarchical leaders required direct support from central church leadership to remain credible and vice-versa (Pope, Cardinal, Mullahs, etc.). Many governmental political systems today are intertwined and codependent with a state religion.

Like metaphysics, religions attempt to explain ultimate causes and higher truths, resulting in creation stories that explains how the physical world materialized. Each prescribes sacrifice and a specific routine for achieving good graces in the afterworld. Emphasis is placed on letting go of the material, mundane world to purify one's soul. The structured frameworks of religion are intended to facilitate guiding man onto the *right* path, following *right* principles, and engaging in prescribed rituals, so as to ultimately reconnect with the divine source – *God*.

"Religion is doing; a man does not merely think his religion or feel it, he lives his religion as much as he is able, otherwise it is not religion but fantasy or philosophy" – George Gurdjieff

MAJOR RELIGIONS

Christianity (West)	Catholic, Eastern Orthodox, Protestant (many divisions)
Islam (Mid East)	Sunni (Orthodox), Shiite (prominence of individual Imams)
Judaism (Mid East)	Orthodox, Conservative, Reform
Hinduism (East)	Saivism, Shaktism, Vaishnavism, Smartism
Buddhism (East)	Theravada (attachment/seclusion), Mahayana (emptiness/non-duality), Zen
Confucianism (East)	Mencius, Xunzi, dong, Zhongshu, Ming Korean, Song, Qing
Taoism (East)	Quanzhen, Zhengyi

Christianity, Islam and Judaism *ALL* originated in the Middle East, believe in a single unitary God, and embraced *the 10 Commandments* delivered by Moses. They all separate man from that unitary God, while their Eastern counterparts don't. They also trace their lineage back to Abraham and in essence, worship the same single God through different lenses. Each of them branches off into different sects creating a virtual kaleidoscope of lenses, all interpreting one God differently.

Christians are the largest religious population in the world, worship one lord and follow the teachings of Jesus, whose birth marks the beginning of the modern calendar. Church structure transitioned during the reformation due to opposition of the traditional Catholic practice limiting interpretation of the bible to only clergy. A Protestant division emerged led by Martin Luther, empowering the common parishioner to read and interpret the word on their own. Denominational branches split further, including Anglican, Methodist, Baptist, Congregational, Pentecostal, Lutheran, Quakers, Unitarian, Seventh Day Adventists, etc. Others pseudo-Protestant branches include Jehovah's Witnesses, Mormons, Christian Scientists, etc. Catholics believe in the concept of purgatory – a middle holding ground between heaven and hell for those not ready for the next stage. Most Christians see Jesus as both God and man, and anticipate his second coming as prophesized in the book of Revelation. Four key books in the New Testament, *the Gospels*, provide a detailed account of Jesus' life and ministry: Matthew, Mark, Luke, John.

10 Commandments Summary: Principles accepted by Christians, Jews & Muslims

1. No other Gods
2. No Idols
3. Don't vain the Lord's name
4. Honor the Sabbath (7th) day
5. Honor your parents
6. Don't murder
7. No adultery
8. Don't steal
9. Don't Lie
10. Don't covet

Jesus Teaching Points
-Love your neighbor as yourself
-Forgive those who have wronged you
-Ask God to forgive your sins
-Don't judge others
-Everyone is redeemable
-When hurt, turn the other cheek
-The way to God is through faith, not law

Islam is the second largest religion, adhering to the teachings of Muhammad and the Quran. Along with Christians, Muslims believe the righteous will be rewarded with heaven (Jannah – paradise) while the unrighteous will be punished in hell (Jahannam). Islamic law (Sharia) covers every aspect of domestic life, including finance, men's and women's roles, and a variety of cultural codes. Where Christians separate church and state, Muslims integrate their religion with the governing system and virtually everything else.

ISLAM 5 PILLARS (obligatory acts)
1. Profession of faith (Shahada)
2. Daily prayer - 5x/day (Salah)
3. Purification tax (2.5%) (Zakat)
4. Fasting month of Ramadan (Sawm)
5. Pilgrimage to Mecca (Hajj)

5 Articles of Faith
1. Belief in a Single God
2. Belief in Angels
3. Belief in Revealed Books
4. Belief in Prophets of God
5. Belief in a Day of Judgement

Judaism is the third religion branching out of the Middle East and has a much smaller following than the other two (.2% of the world population). It's based on a collection of ancient Hebrew scriptures who's founding principle is the Covenant – a promise God made with Abraham and his descendants committing to their protection and offering a promised land as long as they follow the path of God. Jews believe in a single God that is both transcendent (beyond the limits of space and time), and immanent (actively present in the world as a sustaining force).

Judaism Concepts and Principles
1. Monotheism (1 God)
2. God is the only object of worship and praise
3. God communicates through Prophets and prophecy
4. Torah is the immutable word of God
5. Divine reward and punishment of behavior
6. A Messiah will come
7. Ritual without ethics is fruitless
8. The world is a work in progress
9. Evil acts separate one form God
10. Resurrection of the dead

Hinduism is the third largest religion in the world and the oldest at over 3000 years. With more than a billion followers, it's based on a variety of philosophical concepts, mythology, rituals and cosmological systems, all synthesized without a primary founder. Unlike the middle eastern and western religions that separate God and man, Hindus believe God is within each and every one of us and our life purpose is to get in touch with that universal inner essence. Like the Christian Trinity, the Hindu concept of God has three separate aspects: *creator*, *preserver*, and *destroyer*; a dynamic synthesizing mix that produces everything and changes every circumstance. Hindu principles include honesty, patience, forbearance, self-restraint, virtue, compassion, and refraining from injuring other beings.

4 Stages of Life
1. Student
2. Householder
3. Seeker
4. Ascetic

4 Proper Goals of Life
1. Ethics/duties
2. Prosperity/work
3. Desires/passion
4. Liberation/freedom

4 YOGAs (Stages of Evolution)
1. Satya – Golden age
2. Treta – Silver age
3. Dvapara – Bronze age
4. Kali – Iron age

*Brahman is the universal essence – spirit. It manifests locally in a human soul as Atman. Maya is the illusion of things we mistake as reality.

Buddhism is founded on liberation from materialism, desire and worldly attachment, which lead to suffering and pain in an endless cycle of death and rebirth. Liberation is achieved through self-restraint, self-denial, meditation and morally just behavior. The goal is personal transformation and ultimately achieving a state of enlightenment, requiring lots of meditation. Buddhism and Hinduism share similar beliefs, including the concepts of Karma and reincarnation, purging the body of sinful desire, asceticism, liberation from attachments, and attaining a state of Nirvana. Zen Buddhism uses paradoxical anecdotes called Koans, serving as teaching riddles that provokes deeper thought and contemplation. Zen, which means meditation, abhors religious dogma, valuing personal experience above everything. Its simple goal: achieve a state of Nirvana – a blissful, centered harmony, detached from desire and fear.

Buddhism 8-Fold Path
1. Right view
2. Right resolve
3. Right Speech
4. Right Action
5. Right Livelihood
6. Right Effort
7. Right Mindfulness
8. Right Concentration

4 Noble Truths
1. Life is Suffering
2. Cause of Suffering is desire and attachment
3. End of Suffering by overcoming desire and attachment
4. Path that leads to the End of Suffering is an 8 fold path

Taoism (the way) is a combination of religion and philosophy, where one submits to a cosmic spirit of interacting forces (Yin/Yang), to attain balance, health and longevity. This is achieved though exercises and rituals that manipulate the life force *Qi* ("chee"). Interestingly, the TAO Yin/Yang dynamic of opposite clashing forces creating change is similar to the Hindu's trinary aspects of God that creates, preserves and destroys. Taoists practice self-cultivation through various rituals and meditations that synchronize one's inner being with higher spirit. Also called *Daoism*, it shares many principles with Buddhism and prioritizes compassion, frugality and humility.

TAO / Daoism 4 Principles – TAO is the nameless underlying order of the universe.
1. Simplicity, patience, compassion
2. Effortless action, Go with flow
3. Let go, be spontaneous
4. Harmony and balance

*Satori: a brief moment of presence, resonant alignment transcending mental clutter. "be in the zone".

Beyond the major religions covered here there are well over 4000 other religions worldwide – a mix of churches, denominations, religious bodies, faith groups, tribes, etc. A variety of nature religions share the belief that every living thing, including the natural world is the embodiment of divine spiritual essence, not unlike *the force* referenced in Star Wars. This group includes Shaman, Totem, Fetish, various African tribes and Native Americans. Fetish attribute magical powers to inanimate objects. Our modern era adds a category called *New Religion*, which includes the Nation of Islam, Transcendental Meditation, Baha'i Faith, Neo-Paganism, Hare Krishna, Rastafarian, Wicca, and the New Age Movement.

Differences in Religion surface across the globe, especially between the East and West. Eastern religions embrace meditation; western counterparts emphasize prayer. Western perception of reality is linear while eastern religions see life as a cyclic, non-linear experience. The western God is a man-made construct, described as a word, a thought, an image we project, even giving it a human gender (male) as in *God the father* (Patriarchal). Eastern religion relates to a formless God; less structured, less dogmatic and less defined. The western God is *out there*, separate from us, requiring one to reconnect (*our father who art in heaven*). Eastern religions perceive God as already within us so you simply need to stay centered and in sync with the internal divine presence. Western religion is founded on narratives of man falling from grace, engaging in sinful acts causing further separation from God, requiring redemption from sin through faith and atonement to reunite with the divine. Eastern religion focuses on liberation from desire and attachment, losing the self to follow the right path and achieve enlightenment.

Subtle Differences in Tone and Approach of Major Religions

Christianity	Buddhism	Islam
Jesus	Buddha	Mohammed
(Healer)	(Meditator)	(Warrior)
Passive	*Neutral*	*Aggressive*

Differences in religions are generally subtle ones, expressing similar concepts in alternate narratives. For instance, where Islam requires submission to Allah (*God*), the TAO advocates submission to the natural way of things, it's perception of *God*. Yet as central as the concept of **God** is to any religion, it's surprisingly defined differently throughout the world. Of course, any attempt by mortal man to describe an all-powerful omniscient God is bound to be limited, partial, inaccurate and vague. Words cannot capture the essence of anything beyond our comprehension, but still, we must at least try, even if it's just an awkward blind swing, best guess. Unlike precise science, religions are limited to using approximations, clunky language and nebulous descriptions to explain a spiritual experience, leaning heavily on use of parables, analogies and metaphors.

There are plenty of different names for God, including Jehovah, Yahweh, Elohim, Brahma, Allah, the Lord, El Elyon, Bhagavan, Iha, Yu-Huang, and Amaterasu. Imagine how Moses felt when he asked God what his name is and the reply was "I AM WHO I AM", leaving Moses to tell others "I AM sent me. We also describe God by a variety of titles: Savior, Almighty, Holy One, Supreme Being, the Creator, and Holy Ghost. The many facets of God's expressions create the appearance or multiple Gods: Shiva (formless), Brahma (creator), Vishnu (preserver), Shai (destiny), Eshu (fate), Vajrapani, Manjusri, Avalokitesvara (Buddhist deities), or the Christian Trinity of Father, Son and Holy Ghost. God is all things at all times and all situations, so it's no wonder there's so many names, titles and aspects to describe this one, singular omnipresent universal absolute.

Beyond name and titles, there's still a need to better describe just what *God* is. The western concept portrays an all-powerful male deity that created the world, guides it, answers prayers, rewards right behavior and punishes wrongful acts – the "Alpha and Omega" (beginning and end). In the East, Hindu's describe God (*Brahma)* as an eternal, absolute reality beyond forms. The Bible provides a variety of descriptions in the Psalms, such as "a shield around me", "my rock", "my fortress", "our refuge and strength", "a righteous judge", "the world", "robed in majesty", in essence a shield, shepherd and savior. For many, God, like love, transcends words or explanation. God is both seen and experienced a little differently by every religion and every individual.

"God is the tangential point between zero and infinity" – *Alfred Jarry*
"God, to me, it seems, is a verb, not a noun" – *Buckminster Fuller*
"God will be present, whether asked or not" – *Latin Proverb*
"God is a circle whose center is everywhere and circumference nowhere" – *Voltaire*
"God sleeps in the minerals, awakens in plants, walks in the animals, and thinks in man"
                                                                    – *Arthur Young*

Many religions express a dual nature of God. There's the one and only, unitary, single essence but then there's the multifaceted *expressions* of that single essence. Christianity's *Trinity* represents 3 aspects of a single God, confusing some into interpreting it as Gods *plural*. Other religions face similar conundrums. I suspect early creators of mythological Gods simply interpreted life's many aspects as separate deities when they could have just as easily explained them as different expressions of one all powerful, unitary God. No doubt cultural differences have played a part in the evolution of various religions producing local, parochial interpretations of the same God.

Perspectives of God – (Theology is the study of God and "his" relation to the world)

Theism	God creator and participator in The World
Deism	God creator and neutral observer
Pantheist	Many Gods
Atheist	No God

Hinduism mixes 1 God & many Gods: Brahma, Vishnu, Shiva, and sub deities (Krishna-Rama-Shakti)

Religious believers sense the universe was created by God in a grand act of Love. Atheists believe science, natural laws and chance explain our existence. Agnostics just don't know, claiming there's not enough evidence either way. For those who do believe in a God in some form or another, this much is true: God cannot adequately be described in clunky words or imaginary images but can only be experienced as a blissful divine state of being.

Christopher Langan, considered to be the smartest man in the world with an IQ around 200 claims he can prove God's existence using math. Genius or absurdity? You be the judge.

Unity of Religion. Since the ancient times of sun worship, animal sacrifices, mythological gods and nature-based religions, modern versions have evolved and spread through every part of the world with a variety of subtle nuances, approaches and perspectives acknowledging the divine. Despite major differences in systems and dogma, common themes are present in nearly all of them. A place of worship may be called different names (Church, Synagogue, Mosque, Temple, etc.) but their essential practices are remarkably similar.

Common mythological themes and symbolism are imbedded in most religions. The fall of man is a common religious theme. Falling is archetypal – fall to the ground, to earth and nature. Falling is failing. Empires fall. Celebrities fall from grace. Falling is the most common nightmare. Falling from God's grace is a shameful burden man must reconcile. It's a separation from God, requiring reconnection through ritual and right living. Christian's, *man* was kicked out of the Garden of Eden for eating the forbidden fruit of knowledge, establishing a lifelong cycle of sinful acts that must be redeemed through rituals, prayer and forgiveness.

Symbolism is ubiquitous in well-known tales with holy scripts, including snakes and serpents, heaven and hell (underworld), angels, devil and demons, a burning bush, trees, walking on water, fire and brimstone, doves, fish, horns, darkness, candles, lambs, olives, pearls, rain, tree of life, servants, thorns, vines, wells, the void, judgement, redemption, salvation and resurrection (death and rebirth). Personal and collective sacrifice are also universal legacies taken from ancient practices where animals or even people are slaughtered in deference to mythological Gods. Some religious rituals closely resemble sympathetic magic.

Virtually all religions include a creation story that explains how we got here using similar mythological themes. Greek and Roman creation myths involve Gods acting godly, building the world part by part through executive fiat. Some involve serial procreation and incest to get the job done. Genesis uses a 6-day timetable to first create light, then separate the world into land, sea and sky, and finally adds creatures, animals and man. All seem to involve nothing becoming something, ironically similar to the big bang theory that religious zealots claim is preposterous.

Religion and ritual go hand and hand – a universal act of service, submission or commitment to reconnect with the sacred. Rituals are a universal practice often emotionally charged, serving as a fundamental part of any organized religion. They include prayer, worship, fasting, penance, purification, offering, bowing, baptism, lighting candles, incense, drinking (wine), eating (bread), initiation, wearing ceremonial clothing or accessories, and performing acts of sacrifice. Various stages of life are celebrated with ceremonial events including childhood, the transition into adulthood, marriage, and death. Every religion offers unique rituals or beliefs:

Christianity – Mercy, forgiveness, communion, immaculate conception, resurrection, *Trinity*,
Islam – Pilgrimage to Mecca, face covering, Sufi mystics, forbidden images, decorative arabesques
Judaism – The Covenant, God directs all human activity, community-based experience
Buddhism – Reincarnation, Karma, wake up to God within, attain inner peace/bliss, Bodhisattvas
Hinduism – No formal doctrines, animal worship, reincarnation, caste system, arranged marriages,

Holy Events and Observed Periods

Christianity	Good Friday, Easter, Advent, Lent, Palm Sunday, Pentecost, Christmas
Judaism	Rosh Hashanah, Yom Kippur, Sukkot, Hanukkah, Shabbat, Purim, Pesach
Islam	Ramadan, Lailat ul-Qadr, Id al-Fitr, Id ul-Adha, Al-Isra Wal Miraj, Maulid al-Nabi

*Islamic events based on lunar calendar means holidays shift over time and seasons.

In addition to ritualized actions, forms of communication are used with religious connotations. Prayers can become chants, mantras and incantations resonating with higher spiritual energies. Hindus chant the simple three letter word *AUM* which opens up and closes the entire universe in one vibrating breath, interestingly similar to the Christian *AMEN*. Eckankar uses the even shorter *HU* to resonate with spirit, holding it as a continuous tone for much longer than a few seconds. It too has a similar Christian counterpart, *Hallelujah*. Beyond chants and mantras there are practices of speaking in tongues where a different language is spoken with no prior knowledge of it. AUM is symmetric with I Am and Omen. It's the sacred sound of creation containing all sounds. Words are merely fragments of the original whole.

Every religion has its sacred texts; general compilations of writings by inspired men. These include the Bible (Christian), Quran (Islam), Torah and Talmud (Jewish), TAO Te Ching (Daoism), Upanishads and Bhagavad Gita (Hindu), Tipitaka (Buddhism) etc. The *Bible* is a collection of 66 books organized in 2 parts; Old and New Testaments – a compendium of stories, history, poetry, biography letters, love songs – the best-selling "book" of all time. Like Shakespearean plays, it chronicles man's strengths, weakness and ugliest qualities; betrayal, violence, murder, rape, incest, greed, revenge, and a variety of pathologies and general dysfunction – a practical reference for life lessons. Every major religion has their own version: *Bhagavad Gita* is a collection of 700 verses of Hindu scripture, *TAO Te Ching* (meaning "The Way") is a collection of poetry and sayings, *Torah* is the Jewish Bible (basically the first 5 books of the Christian Bible), Quron is the Islamic Bible; the inspired word of God written by Muhammad's scribes, Tripitaka is the Buddhism Bible, a sacred book of discourses of the Buddha.

All of these sacred texts provide rules, laws, recommended behaviors, forbidden actions and general guidance on how to live "right". Guiding principles can be simple narratives or specific rules; 10 commandments (Judaism), Vedas (Hindu), etc. Some are rigid, orthodox doctrines set in stone as uncompromising dogma. In contrast, there are flexible, practical approaches used by some prophets/messengers who teach their gospel in open-ended parables. Since religion is grounded in spiritual guidance it's more useful to convey higher meanings through metaphors, analogies, proverbs, songs and mythological stories. This approach was captured in the 1970's TV series "Kung Fu" with students instructed by masters who rarely answer questions directly but tell stories with coded meanings and subtle insights.

Most religions have a foundational prophet who becomes the very name of the religious movement itself. These key figures in religious history include Jesus, Mohammed, Buddha, Confucius, Krishna, Lao Tzu, and others. They're supplemented by a variety of messengers who play an important role in spreading the word. Each religion has a different name for their messengers, including Priest, Shaman, Rabbi, Imam, Pope, Yogi, Apostle, Disciple, Rishi, Lamas, Oracle, etc. Prophets and their messengers spread the gospel and show the right way to live, often confronted and opposed as they stoke controversy by bring forth a different viewpoint to the status quo.

Duality is another universal pattern found in most religions, emerging as each principle is interpreted differently among messengers of the same religion. These splits are obvious and significant, creating two major branches of every religion: Catholic & Protestant, Sunni & Shia, Theravada & Mahayana, Confucian & TAO, Shaivism & Vaishnavism, etc. This duality pattern is often the difference between fundamental doctrine (orthodox) and flexible application of said doctrine (modern, reform). A simple example is the split in the Christian Bible with its Old & New Testaments; the first conveying a strict, cold, rigid orientation while the latter reflects a more humanistic, accommodating tone.

Religious Duality Present in Every Region of the World

West: *Catholic* – structure and hierarchy   Vs. *Protestant* – faith over deed, humanism
Middle East: *Shia* – Imam directed   Vs. *Sunni* – fundamentalist, orthodox
East: *Confucianism* – practical right living Vs. *Taoism* – mystical, Yin/Yang cycle

*Most religions have an *Orthodox* sect, grounded in original literal texts, fundamental and unchanging with the times. Other branches are more flexible, accommodating to modern life. It's structure vs flexibility, Old-school vs Progressive. Non-compliance can be considered heresy.

Where duality splits a unified message into two sides of the same coin, unifying themes transcend it, rising above the natural tendency in man of interpreting the same thing differently. Universal meanings and messages endure, carried through various religions despite local differences and man-made nuances. These recurring principles include minimizing material attachments, taking personal responsibility, reigning in excessive desires, helping those in need, embracing a higher spirit, following the right path, and treating others the way you want to be treated.

The Golden Rule: A Universal Code in Religion
What you do not want done to yourself, do not do to others – Confucius (500 BC)
Love thy neighbor as thyself – Judaism (Leviticus 19:18)
Do to others what you want them to do to you – Jesus (Matthew 7:12)
Whatever is disagreeable to yourself do not do unto others – Ancient Persia
To do unto all men as you would wish to have unto you – Muhammad (600 AD)
Hurt not others with that which pains yourself – Buddhism (Udanavarga 5:18)
Regard your neighbor's gain as your own gain, and your neighbor's loss as your own – Taoism
One should not behave toward others in a way which is disagreeable to one's own self – Hinduism
A man should wander about treating all creatures as he himself would be treated – Jainism

Universally, religions explain the meaning of life, clarify differences between good and evil, ease the stress of everyday hardship, show us how to find happiness, and address the mystery of what happens after death. Mystical experiences are universal across all cultures. Ironically, however, the unity of spiritual experience is fragmented and separated by religious institutions, limited by interpretations from different people and culture. Man in the middle actually get in his own way and muddies the water through parochial perspectives that are supposedly better than everyone else's. Perhaps the Baha'i Faith captures the unity of religions best by adopting as its core principle the idea that ALL Religions are just alternate expressions of a single God, and ALL the past prophets are unique messengers bringing the same word to different cultures in different languages. *"All the different religions are only so many religious dialects."* – G. C. Lichtenberg

Many Religions...One True Original Source of Spirit

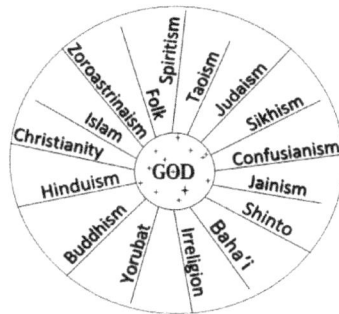

Only Man could filter the pure unity of God into a multifaceted kaleidoscope

THE ARTS

The Arts are a subset of the Humanities, which include History, Literature, Philosophy, and Religion, all covered earlier. As part of the Humanities, art is *the un-science*, open ended and richly subjective. The Arts are man-made expressions in every form possible: painting, architecture, literature, music, performing, cinema, etc. Categories are plentiful and can be further subdivided into dual sets; audio and visual, music making and music playing, written and spoken, works and performance, crafts and productions, film and video, etc.

Often called *the creative arts,* they're a domain of human activity and expression tapping into the transcendent dimension of spirit. They're fueled by the internal inspiration of an artist and put out into The World for others to judge and appreciate. Man in the middle is a natural artist, a filtering agent where spirit passes through to be transformed in various modes; visual, auditory, literary, or physically. Whether the artist is a painter, sculptor, writer, dancer, actor or musician, the process they engage in is virtually identical - a collaborative dynamic between body and soul, spirit and matter, sometimes resulting in stunning indescribable beauty.

Through history art has evolved parallel to civilization's forward progress. Countless periods have come and gone, bringing with them a series of changes in the Arts. Transitions and nuance are constant; from ancient times to early Christian and Byzantine, to Middle Ages, Romanesque, Gothic, Renaissance, Baroque, Rococo, Neoclassical, Romantic, Realism, Impressionism, Modern and Post-Modern periods. Art's evolution produced new categories, branches, variations, techniques, materials, and tools. Each passing phase brought new elements while frequently borrowing from previous influences. The one constant has been the cycle of change between styles, swinging back and forth like a pendulum, though no two periods are ever the same.

Art styles shift in tandem with cultural tastes, always linked to societal states of being – the temporary prevailing zeitgeist. Each influences the other. Fads, music styles, fashion, tastes, attitudes, all come and go in cycles of codependent interactivity between the art world and culture as a whole. Every branch of the arts tends to be swept along with it to some degree in an interactive, connected process. Some innovative artists inject novelty that becomes viral, spearheading an entire new fad. Other times it can be a series of small nuances that accumulate into a tipping point, turning the tide that pushes out the current style. Movements can start at the top with prominent celebrities, outspoken artists and powerful elites, or they can matriculate upwards from ground level developments with newcomers on the street. In either case, small incremental developments either fade away or resonate to produce transformational change.

ERAS	Classical	Renaissance	Impressionist	Modern	Post-Modern
QUALITY	Religious	Humanity	Emotional	Movement	Relativity
ELEMENTS	Idealistic	Realistic	Colorful	Bold strokes	Multimedia
ICONS	*The Parthenon*	*Mona Lisa*	*Ballet Class*	*American Gothic*	*Shuttlecocks*
	Trevi Fountain	*Sistine Chapel*	*Water Lilies*	*Starry Night*	*Campbell Soup*

The arts have blossomed into broad categories, types, branches, and further subbranches with variations on a theme. From the earliest cave drawings and lines in the sand, art has expanded geometrically into a wide variety of applications and venues. The Arts include any activity that uses creative processes to produce something new, often to evoke a response in others. There's an element of art in almost any activity, including hard labor, routine tasks, and science as well.

CATEGORIES OF ART

Visual – architecture, ceramics, conceptual art, drawing/design, sculpture (wood, clay, metal, stone, etc.), origami, modeling, crafts, painting, television/film, photography, animation, video games, comics. *Graphic design* – objects, systems, buildings, cities, spaces. *Decorative* – fashion, interior design, packaging
*Cinema – includes an entire cadre of supporting creative teams: lighting, writing, acting, costumes, props, sound effects, etc.

Literary – prose, drama, poetry, fiction, *Oral:* folktales, myth, legend, story tell.

Performing – dance, song, music, theater, opera, mime, puppetry, improv, circus, magic, comedy, juggling, acrobatics. *In performance art anything goes.*

Verbal – poetry, oratory, singing, comedic speech.

Unique Applications — *Culinary*: cooking, winemaking, deserts. *Apparel*: clothing, accessories, jewelry, fashion. *Applied Art*: design and decorative application to functional objects, packaging (adding aesthetic value). *Miscellaneous*: Graffiti, body art, tattoos, piercings. Martial Arts, athletics, warfare, nursing, politics, etc. *Many of these combine art and science.*

Technology greatly expands the range of applications. Think of all the basic tools for drawing, including pencil, pen, marker, brush, crayon, charcoal, etc. and how computers have enhanced the ability of an artist through mouse, pad, keyboard and powerful software. Engineers and architects must certainly appreciate how their profession transforms into an art form with new electronic tools at their disposal – instantly turning imagination into practical design plans they can now play around with, modify and edit until it looks perfect.

Art never happens in a vacuum; it's always subject to a particular environment and presented in a specific mode (canvas, screen, stage, container, display case, platform, etc.). Where and how it's experienced can be just as important as the art itself. Just as a church serves as sacred space enhancing the spiritual experience, art benefits from a similar management of venue. Examples include performance stages, auditoriums with enhanced lighting and acoustic engineering, art galleries with track lighting plus channeled pathways of observation, theater style home entertainment systems with widescreen 4K high resolution 3D imagery along with surround sound digital stereo, and many others. Art is either enhanced or diminished by the context in which it's experienced. Watching anything that takes time, like a movie or stage performance, requires dedicated uninterrupted attention to maximize the experience, so the venue must accommodate that. Modern blip-culture of quick sound bites and short video clips makes long presentations much more difficult to appreciate with shortened attention spans.

Elements of Art	Principles of Art	Subjects of Art
Line	Balance	People
Shape	Contrast	Objects
Value	Emphasis	Wildlife
Color	Movement	Abstractions
Texture	Pattern	Nature
Space	Rhythm	Animals
Form	Unity	Action

Artists see what others don't see or apply what others may have seen but didn't appreciate. Some tap into a subconscious source and make it conscious, drawing from an inner well to bring it to the surface. They start with an empty slate and appear to create something from nothing. Musical instruments are lifeless, soundless, inanimate objects without an artist. When a maestro walks up to a stationary piano with its 88 static keys, sits down to casually induce a musical masterpiece, the experience can be quite magical. Artists generally look at the world differently. Their unique perceptions match the specific fields they operate in: painters notice shadows everywhere, sculptors notice shapes and surfaces, film directors see the world in rectangular frames, architects see negative spaces and lines, photographers see contrast and lighting everywhere, etc. Artists tend to look at ordinary people differently, including friends and family. Their world is an open-ended pallet of interesting nuance where every little detail scrutinized contains the possibility of something extraordinary. Their inspiration comes from a deep reservoir of potential just waiting to be tapped and released – a place not far from mysticism, magic and enchantment.

Artists may be lonely, but their compensation is the blissful experience of unrestricted creative expression, along with the joy of playful experimentations that bring unexpected discoveries. Nothing is as fulfilling as the act of producing something new into the world that never existed before, using just your own hands and imagination. Anyone who performs a task with an aspect of creativity and flair qualifies as an artist. Just because something seems trivial or menial doesn't mean it can't be conducted in an artful manner. Anyone who directs others to perform a choreographed activity qualifies as well. Obviously, a conductor of a symphony or a choreographer of a dance troupe is considered an artist but why not a football quarterback or a military commander orchestrating Soldiers in the field of combat?

Creativity is a universal quality built into the universe, starting at the quantum level where reality is created moment by moment, each different from the preceding one. It's the secret sauce of life making everything interesting, adding both nuance and mystery to what otherwise would be dull, mundane, predictable. The creative spirit is naturally inside all of us, though some tap into it better than others. A fundamental part of human life is expressing ourselves freely; channeling our inner essence outward, which collectively manifests The World in the process. You as an individual human being were made as a creative act and your life is really just a series of ongoing creative acts. *The World* and everything in it are both the aftermath and the continuance of unlimited creative actions that never cease. The arts are just one specialized field where this process is unleashed wholesale in a wide variety of mediums, platforms, canvases, venues and arenas. It's a subtle force in nature surfacing wherever someone tinkers with a project, edits a great work, tweaks a performance, chips away a sculpture or engages in a renovative act.

Creativity is greatly influenced by the company you share. It's remarkable how much you can create out of thin air when you're working with the right partner or group that stimulates your imagination. Many great comedy teams were successful while paired but upon splitting, the two gradually fade into irrelevance. Some artists are singularly inspired by a special other, their muse, who in the case of some song writers is their sole source of purpose. All sorts of things can influence creative bursts, from vivid dreams to peak experiences. Music sets the mood to get the juices flowing, especially when the right type is played at the right time. Lighting and color scheme can make a difference as well. Just a simple change of scenery can get you into a more creative mode.

Some experiences in the creative fields are universal. All artists eventually get that unpredictable aha moment that sparks a breathtaking original creation, identical to the eureka moment experienced by the scientist or inventor, leading to something profound. But for every breakthrough moment there are much longer periods where blockage and staleness set in, well known to writers, musicians, athletes and most everyone participating in the creative arts arena. Creative block can happen when trying to do too much at once, instead of making incremental progress, or waiting for instant perfection rather than building off of what's *good* or *ok.* Fear of failure doesn't help either. Many creatives consider themselves perfectionists and it's that very attitude that produces self-imposed blockages.

Mental blocks are a sign you're thinking too much or pushing too hard and really need to just let go. Creativity is influenced by attitude, personality, imagination, experience, motivation and the immediate environment. The key to fostering a creative state is embracing and maintaining a playful spirit, letting go of self-imposed constraints, willingness to experiment and not caring about the results. When the conditions are just right you can get into *the zone*; a transcendent place where artist becomes one with the artwork. Like a jazz musician riffing or a Pollack painting frenzy, letting go will get you where you need to be. Just let it all hang out like a free-spirited hippie without a care in the world. Of course, knowing it and accomplishing it may be two very different things.

Factors in creativity	*Influences on creativity*	*Sources of Inspiration*
Learning	Mood	Dream
Experience	Lighting, music	Travel
Motivation	Company	Adversity
Imagination	Environment	Nature
Intelligence	Culture	Idol
Flexibility	Playfulness	Muse
Personality	Time	Meditation

Subjectivity is an implied aspect of the Arts. Any work or performance is neutral by itself, without context. Art is always open ended until perceived by someone whose response and judgement gives it a particular value. Regardless of the artist's intentions, your response is what really matters. Accordingly, ALL art is relative, subjective and dependent on interpretation. Responses are influenced by numerous elements including culture, experience, mood, setting, timing, subconscious factors, etc. Western, Middle Eastern and Asian parts of the world certainly value art differently. Locally, the same is true when considering different values and experiences, such as urban vs rural, educated vs uneducated, high income vs low income, etc. Consider the stereotypical case of an elitist critic fawning over a masterpiece while a layperson just sees nothing but an ordinary canvas someone slapped some paint on. Or the fact that some artists just create for themselves and don't care what anyone else thinks. The Arts are part of the Humanities because they're completely based on the subjective. There's no right answer, it's all interpretive. This gets to the heart of the ultimate question: What makes good art?

Art naturally resides in the transcendent realm, beyond space and time, ever present in the eternal now. We experience it both personally and universally. Our response to it is always an individual experience – no two people see and feel art the same way. Art is best when it connects on a deeper, spiritual level. Some works draw us in inexplicably, capturing our attention and holding it, at times forging a mystical connection so we feel a sense of oneness with the art piece. Trying to share this experience with another is cumbersome, beyond words, largely futile.

The language of art is steeped in metaphors, analogy, and symbolism, which are all elements of the transcendent. It operates on the inner psyche where emotions rule. Art is deeply connected to cultural themes and archetypes, mostly sensed in the subconscious. Much of art is derived from the hidden, invisible dimension of our world only accessible through metaphoric language. Art as an abstract essence exists in an inner dimension, tapped into by poets, mystics and various creatives who translate the unspeakable into images and sounds. Art adds depth to an otherwise shallow superficial surface world.

The Arts are all about expression, a process beginning and ending with man in the middle. The artist is that person in the middle; a filter, prism and transmitter of spirit. Creative expression is processed through a medium in a Man-made/God-inspired dynamic where less is more: less mind, less man and more spirit (*God*). Andre Gide captured it best stating: "*art is a collaboration between God and the artist, and the less the artist does the better*". The art we create is not invented beauty but simply a revelation of the ever-present beauty already permeating the universe; it's always been here, just concealed by our own self-imposed myopia.

All art has a particular style or unique manner in the way an artist does their thing; making, creating, or performing. Every category, branch and field of the Arts has both its own unique associated styles along with shared styles in other fields. Painting is specifically known for abstract, realism, impressionism, surrealism, etc. Architecture uniquely has Gothic, Victorian, Tudor, Colonial, etc. But both also share styles with each other: classical, modern, art deco, post-modern, etc. Every field in the Arts have styles that go in and out of favor over time, often in parallel to other art fields. Many styles diverge or develop hybrids. Naming them is sometimes simply adding a prefix or suffix: Modern, Pre-modern, Post-modern, neo-classical, de-construction, etc.

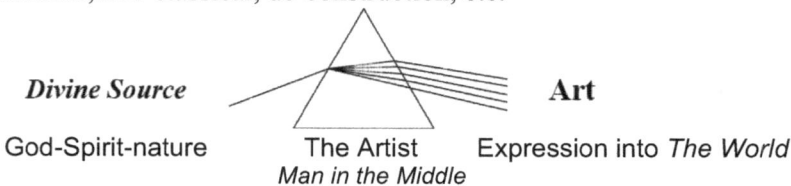

Divine Source   **Art**

God-Spirit-nature   The Artist   Expression into *The World*
                             Man in the Middle

Beauty is an integral aspect of art yet it's very difficult to define. Like art it evokes emotion and connects on a subconscious/unconscious level, similar to music. Philosophy examines beauty in its branch of aesthetics, capturing the nature of art, interpretation, representation, and expression. The ultimate question comes down to what is beautiful? Is beauty a universal quality or merely relative and subjective? How do we account for the difference between individual taste and collective consensus? So much of our response to objects and images is influenced by hidden aspects of our subconscious, as well as moods and states of being, making matters even more nebulous. Being completely relative, beauty will always be a matter of degrees, as in the universal scale of attractiveness: 1-10. From the ugly to the gorgeous, and even *the sublime* (awe inspiring grandeur), what's considered beautiful is relative to everything else.

There's general agreement that simplicity enhances beauty – less is more. Most everyone senses beauty in nature. Beauty is present when there's unity, wholeness and completeness. Beauty emerges in symmetry, proportional relationships, harmony and balance. Absence of these qualities is experienced as ugliness. Monsters and demons generally lack symmetry, possessing disproportional, exaggerated features. Even clowns which are supposed to be fun and friendly can give people an unsettling feeling, actually causing some to reel back in fear, likely due to their lack of proportion or grotesquely exaggerated "fun" features. Ironically, the human face is not completely symmetrical – left and right halves are actually quite different, though hard to notice.

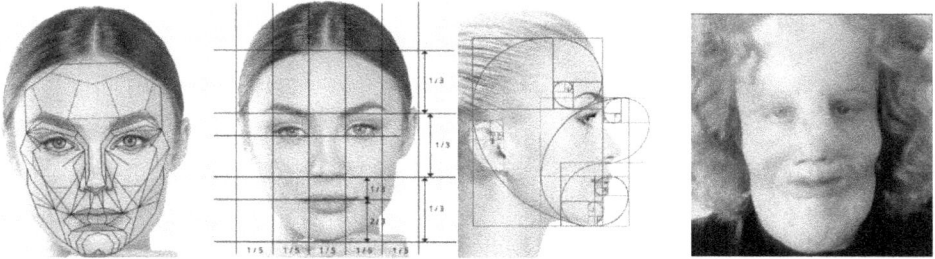

Beauty = Unity, Purity, Symmetry, Proportion, Wholeness
Ugly   = Fragmentation, Separation, Distortion (*Sin is separation from God*)

Everything we judge as attractive or beautiful is entirely subjective, again based on contextual influences: culture, age, personal values, experience, mood, etc. The same individual can look at the same subject and yet have two different responses, depending on context again. I'm sure your tastes today are different than they were years or decades ago. Consider how different the same painting can look by simply putting it different frames. Context always changes. Beauty *is* in the eye of the beholder, but it's an eye that changes with each different circumstance.

*Art reconnects with the divine, reunifying a fragmented, separated external world

Unity of the Arts is found in the common creative processes present in every field or application. All forms of art involve that creative dynamic where an artist produces something new, channeling inner spirit into the outside world. Of course, reception of an artwork depends on context and perception of the audience, whose subjective response is never certain. What's deemed *good* or *beautiful* is always relative, always context dependent. What isn't relative is the universal, invariant creative process and man's role in it. How artists imagine, create, and perform works of art is the common link unifying the Arts in general; an unchanging experience as old as mankind, tapping into a source within and beyond. What does change is the audience's reception and response; how they use, enjoy, and criticize art. Thus, we have two complimentary dual aspects of art; one universal unchanging process, the other variable, subjective unpredictable reaction to it.

The Arts also share fundamental dynamic aspects in every application. On the macro level they involve interrelationships: between the artist and audience, individual and collective culture, external world and internal spirit. This includes the codependent, coevolving relationship between art and culture, where each influences the other as they proceed forward in cycles of taste and styles. Micro level dynamics include the mixture of different elements within a composition, relationships between a work of art and its immediate environment, and the synergy created when parts are arranged in a special way to produce an effect greater than the sum of those original parts. Consider comedy where simple joke telling may seem easy, yet success depends on multiple elements working together – the right material, the right words, timing/rhythm, body language/expressions, appearance and the unique dynamic between performer and audience that's never the same. But what is common to every form of art is the customized creative process where man in the middle uniquely filters spirit to bring novelty into the rigid manifest world.

THE COMPREHENSIVE STRUCTURE OF KNOWLEDGE

The accumulation of knowledge continually proceeds forward much like the evolution of biological life where trial and error tests what works, emergent branches and variations add new forms, and parallel developments progress onward and upward. Over time, the collective compilation takes shape with distinct categories and established patterns making the whole seem intentionally organized. The truth is, it's a painstaking long process of fits and starts, ebbs and flows, with periods of slow change followed by revolutionary shifts. Today we reap the rewards of centuries of trial and error, where great pioneers introduced groundbreaking ideas, standing on the shoulders of earlier pioneers before them. We're blessed with an extensive repository of organized knowledge, left by preceding generations who added to the pile and then passed it forward onto us.

The challenge we face today is making sense of the complex whole, deriving meaning in the age of information overload. With so many branches and perspectives of knowledge we tend to gravitate towards a particular field, defaulting to it as *the* source of information. But we must accept that each discipline of knowledge is both a partial truth and an equally valid source of information. Each seeks answers from different perspectives, entirely appropriate in their specific lane. Problems occur when deviating from one's lane to examine phenomena of a different field using a familiar parochial lens, especially science versus non-science.

Building such an immense, complex structure of knowledge leads to specialization and compartmentalization, complicating cross-talk between different disciplines. It's hard to avoid this given the bureaucratic structure within colleges and institutions of higher learning that reinforce hyper-specialization and separation of fields. Further complicating matters is the use of specialized languages in each discipline, full of jargon and made-up terms that sound foreign to anyone outside the club. There are great differences between the precise terms in hard sciences and the use of metaphors and symbolism in non-science fields.

Our knowledge base is a comprehensive mix of science and nonscience, each revealing a facet of reality through a unique lens. Science examines objects, non-science examines subjects (you). Science is constrained by the measurable, non-science has no such limits. Science is focused *out there*, non-science looks *within*. Science is known through reason; non-science is understood through intuition. Science divides and separates the physical universe while Arts and religion unify it through non-physical dimensions. While science excludes the divine magical aspects of life, they still lurk in the background, much like the star filled heavens masked by daylight. Science is purely objective, with no connection to you or me. But knowledge of reality must include you and your relation to everything else. Science may cover a broad swath of fields, but each is a partial truth – a partial segment of a greater whole extending far beyond physical aspects of the universe.

Non-science examines qualities of a non-physical dimension connecting us to a deeper, hidden level. The Arts operate in this domain, tapping into the sacred to reveal a type of knowledge science is incapable of grasping. Non-science explores deep questions pertaining to man's existence, purpose, morals and ethics. It boldly asks why are we here? What's the right thing to do? What's the meaning of life? Non-science delves into the transcendent realm where knowledge is acquired through direct personal experience and wisdom – the highest level of knowledge. Mystical states and peak experiences cannot be taught or quantified in a laboratory.

Religion operates in the non-science realm, open to the archetypal themes that have informed mankind collectively for thousands of years. Insights are gained through spiritual reflection and ritual practice. Where science explains the how, religion explains the why. Both science and religion have their own structure, formats, methodologies and universal laws. They do however speak foreign languages; one technical and specific, the other symbolic and metaphoric. Religion and mythology are steeped in parables, legends, symbols and archetypes. A religious epiphany is similar to mystical, magical and paranormal experiences in other non-science disciplines, all leading to transpersonal and transformative states.

Art and Science are complimentary sides of the same coin, present in nearly everything. They represent opposing approaches that work fine independently but when combined produce synergistic added value. Collaborative teamwork is optimized when groups include both analytical and creative types. Right brain-Left brain, logic-intuition, both are equally valuable sources of knowledge. Both of these attributes are present in our experiences: sports, crafts, entertainment, business, politics, military, etc. The Arts involve aspects of science to make them complete; architecture needs engineering, cinematography requires camera technology and mathematic principles, music is impacted by instrument design, acoustic laws, mathematic ratios, resonant materials, etc. Successful artists appreciate the science aspects of their craft. And science benefits from intuitive researchers who supplement their methodical approaches with hunches and creative license while investigating new phenomena.

When we say something is an art and a science, we're taking the first step towards reunification of fragmented knowledge. A variety of disciplines embrace both scientific methodology and artistic vision in pursuing success. Any field that involves the dynamics of human behavior qualifies for applying both art and science. Examples would be economics, medicine, teaching, sports, military operations, architecture, communications, martial arts, politics, leadership, design, cinematography, etc. Robert Pirsig's best seller "Zen and the Art of Motorcycle Maintenance", written 50 years ago, captured the essence of art and science present in everyday life. Clausewitz examined warfare as art and science almost 200 years ago, noting that battles are generally won by the side with the most competent staff (science) and creative commander (art). Modern sports teams are most successful when the coach has mastery of field-of-play dynamics, the rules, parameters and playbook (science) and has a knack for mixing in the right plays at the right time (art). Knowledge becomes synergistic when complementary aspects from opposite sides of the knowledge spectrum are combined and integrated.

Science and Philosophy	*The Arts and Religion*	*Art and Physics*
Reductive - Analytic	Soft - Divergent	*Fauvism* - Light
Left Brain - structured	Right Brain – creative	*Cubism* - Space
Quantity	Quality	*Futurist* - Time
Mind – Rational, logical	Heart – Emotional, intuitive	*Surreal* - Relativity
"I think therefore I am" (*Descartes*)	"I feel therefore I am" (*Rousseau*)	*Pollock* - Fields

"Science without religion is lame, religion without science is blind" – Albert Einstein

Science claims elite status with its precision and rigid rules but it's ironically limited by uncertainty due to unpredictable aspects built right into the universe. Physics, the long-standing king of hard science, was karmically humbled when indeterminacy was introduced by quantum mechanics where previous precise measurements suddenly became reduced to best guesses, and absolute realities were awkwardly reduced to approximations. Adding insult to injury, physics was further turned upside down by the introduction of relativity, where space and time abruptly became fluid abstractions with measurements no longer absolute but dependent on each individual observer. Heisenberg's uncertainty principle and the mystery of connected particle action at a distance inconveniently shifted the hard science of physics closer to the mystical realm owned by non-science disciplines.

Non-science embraces the human condition, subjectivity and connection to a non-physical, spiritual dimension. Unlike science, fields in the humanities are limited to approximations and statistical probabilities. Science and non-science disciplines are complimentary and supplementary to each other. Both are more effective when they stay within their respective lanes. But neither is "better". Each offers a different, limited perspective of reality. And both together combine to broaden our knowledge and understanding of the big picture. When all disciplines are synthesized and combined, our collective knowledge gets that much closer to the unity of absolute reality.

The ongoing, ever-evolving spectrum of knowledge resembles a kaleidoscope of data, created in the same way a prism turns pure white light into a diverse color spectrum. Here, man in the middle is the filter, separating a single reality into fragmented bits and pieces. To re-unify the whole of knowledge, we simply need to reverse engineer the process and reconnect our rainbow of colors back into the original white light it came from.

Diversity of Knowledge Disciplines
Expressed through science, partial science, and non-science

SCIENCES – Natural, Hard, Soft, Life, Formal, and Applied sciences.

N A T U R A L   S C I E N C E S – studies of the physical world

HARD SCIENCES – *Physical, quantifiable, measurable*
Physics, Chemistry, Astronomy, Biology

EARTH SCIENCES – **Geography, Geology, Meteorology, Minerology, Hydrology, Oceanology**

LIFE SCIENCES – **Biology, Anatomy, Physiology, Botany, Zoology, Ecology, Microbio, biochem**

N O N - N A T U R A L   S C I E N C E S

FORMAL SCIENCE – Focus on abstract structures and systems
Mathematics, Logic

APPLIED SCIENCES – *Practical application of sciences*
Agriculture, Engineering, Health, Medicine, Nutrition, Management, Military, Electronics, Forensic, Energy, Education, Space, Information

SOFT SCIENCES – *Social realm: study of people/behavior, individual/collective*
Economics, Political Sci, Psychology, Linguistics, Anthropology, Archeology, Sociology, Criminology Law, Behavioral science, International relations

NON-SCIENCES – *Highly subjective, not empirical*

HUMANITIES – Elevated focus on what makes us human vs scientific methodology
History, Philosophy, Religion, Mythology, Literature – language, **Arts** – Visual (Painting, Sculpting, Photography, Film making, etc.) Performing Arts (Theater, Music, Dance, Comedy), fiction & fantasy

PSEUDO-SCIENCE – *Subjective disciplines, archetypal intuitive symbolic systems*
Astrology, Numerology, I-Ching, Tarot, Feng Shui, Palmistry

PARASCIENCE –*Beyond science, unexplainable phenomena*
Alternative medicine, Fringe science, Occult, Paranormal, Parapsychology (ESP, Clairvoyance, Telepathy, Synchronicity, Near death experience)

*Each of these disciplines is a different lens expressing a unique perspective of universal truths. Neither is more "right" than the others, just a specialized focus from a limited vantage point. Each field speaks a unique language that gets lost in translation when directed toward others outside its lane...a sort of unnecessary parochial misguided gibberish.

UNIFYING KNOWLEDGE

Various branches of knowledge separate over time from parochial perspectives, xenophobia, syntax, and esoteric language. It's counterproductive, unnecessarily creating paradoxes difficult to explain. Problems occur when the approach of one field is applied in another where it doesn't work, like explaining physics with mystical, mythical or religious means. The reverse misapplication is equally bad; science trying to quantify spiritual, mystical phenomena. Acts of misapplication foster resistance, inaccuracy and further separation between disciplines, instead of encouraging complimentary ways of examining things. Science, art and religion all pursue the same truth but from distinctly different perspectives and languages.

It's natural for scientists to dismiss religion for lack of empirical evidence but faith-based disciplines and non-science fields operate beyond physical laws. Likewise, religious based attempts to explain science, such as evolution, are also misguided. Practitioners of pseudoscience fall into the same trap, desperately trying to rationalize their domain on a scientific basis when it's not necessary. Just accept the fact that dealing with human beings includes a transcendent level that cannot meet the precise standards of science. Take comfort in the realization that lots of other respectable fields worthy of scientific credibility, like meteorology, economics, political science, etc. are similarly imprecise.

Diverse Multi-lens Perspectives of a Single Reality

SCIENCE	Objective, concise, methodical,	**Objective**
Hard Science	Reductionist, rational,	
Soft Science	Behavioral, social,	
NON-SCIENCE	Subjective, imprecise, wholistic	
Humanities	Experiential, cultural, man as spirit filter	
Religion	Faith based, ritualistic, metaphorical	
The Arts	Transcendent, emotional, creative spirit	
Pseudoscience	Intuitive, probabilistic, psychic, archetypal	
Parascience	Psychic, paradoxical, synchronistic	**Subjective**

Greek Ionians were the first to make serious efforts toward unified knowledge. Centuries later, *Renaissance Man* embraced multidisciplinary awareness over specialization, as did learned men during the enlightenment period. Today there are hints of further unification progress but limited to individual disciplines. Physics continues to pursue its theory of everything though it remains ever illusive. Some sciences have made great progress within but not much without. Others have expanded with further branches and hybrids merely adding layers to the whole. We're getting closer toward integration of multiple fields as a whole, but there's a long way to go, especially between science, art, philosophy and religion. It's challenging when diverse fields use their own unique lenses and language to examine subjects. How do you overcome language barriers to find common ground? We can start by identifying universal laws, themes and properties present in every field. Find the patterns, break the codes and see subtle unities connecting every discipline. Even opposite ends of the knowledge spectrum share common qualities that can be used to bridge the dual nature of each. For example, consider the symmetry between Newton's 3rd law of physics and religion's Golden Rule – basically for every action there is an equal corresponding reaction. Centuries ago, Galileo observed that Bible laws and physics laws are both true. Appearances of contradiction are simply the result of misunderstanding their truths.

Interdisciplinary symmetry is masked by specialization and language barriers, yet hidden clues are there. The unifying language linking sciences are the codified laws. Fields of study are accompanied by set manuals itemizing specific rules, regulations, formulas and laws particular to that discipline. What's overlooked is the similarity many of these laws bear to other associated fields. Interdisciplinary symmetry is graphically portrayed on our TOE matrix model. Common qualities and aspects are seen along columns and rows of the matrix. There's similarity of principle between Chemistry, Anatomy and Economics; dealing with the dynamics of elements in a system. There's similarity between physics, mental processes and politics where interacting forces determine order. Finally, there's subtle symmetry between the metaphysical aspects of quantum mechanics, soul-spirit, and cultural dynamics. Each parallel discipline shares common qualities along the same lines, applying universal principles at different levels of application.

Interdisciplinary collaboration creates synergy, leading to a better, more informed perspective than any individual field can produce by itself. It requires getting out of their comfort zone to cross lines and embrace approaches that seem foreign. When you're naturally a left or right brain person it's not easy to transition to the other side. But successful practitioners of art and science have demonstrated the clear advantage of embracing both ends of the knowledge spectrum. Collaborative groups do this all the time without realizing they're actually engaging in a form of interdisciplinary integration.

Approaches to Synthesizing Knowledge

Transdisciplinary   Take perspective or methods of one discipline and apply them in another
Interdisciplinary   Integrate methods of several disciplines into a coherent whole
Multidisciplinary   Examine a subject from the perspective of several disciplines at the same time
Metadisciplinary   Go beyond parochial limitation of individual disciplines, achieve strategic unity

A Metadisciplinary perspective is key to achieving unity in knowledge, despite society's embrace of specialization. Fields of knowledge continue to branch out with further layers of subdivisions, increasing in complexity and depth. Sciences do it, religions do it, and the Arts constantly churn out a myriad of new styles and variations. Despite the continual expansion of knowledge in depth and breadth, synthesis and integration are still achievable through a strategic perspective and metadisciplinary awareness. It starts by combining each: *reductionist* objective science, *subjective* humanities, *wholistic* perspectives (Eastern philosophies), *approximations* of pseudoscience and the *experiential* awareness of altered states.

Some fields have engaged in this approach for a long time. The medical field is enhanced by its integration of science and non-science. Non-traditional practices are increasingly incorporated into the medical menu of options, including homeopathy, herbal medicine, yoga, chiropracting, osteopathy, etc. Combining art and science has always been an integral part of the healing process. Biochemistry and drugs (science) are supplemented by personal care, empathy, and a good bedside manner (art). A doctor's expensive education is limited without sensitivity, perception, personal touch and a healthy degree of intuition. A good doctor achieves synergy as both artist and scientist, applying knowledge of procedure with a perceptive sense to heal patients.

Knowledge is more than just black and white quantitative data. It's a composite blend of quantity and quality, raw information and relational context. It requires a balance of both critical and creative thought. Not just logic, reason and thinking but imagination and intuition. When knowledge is produced holistically it leads to insight, wisdom, and innovation – the ultimate blend of art and science.

Human interpretation of reality is a lot like the 5 blind men who examined an elephant and reached 5 very different conclusions: For one the leg felt like a tree, for another the tail seemed like a rope, the flapping ear seemed like a fan, the trunk resembled a hose, and the body seemed like a wall. Each partially true from their separate, limited perspective yet embarrassingly off the mark of what an elephant actually is. It's not much different than man's parochial approach, managing collective knowledge as a hodgepodge of separate disciplines, each examining unified reality through their own narrow lens, claiming theirs holds more truth than others. It's the same tendency that produces opposite viewpoints within a single discipline, splitting two sides of a single truth. It happens in religion (Sunni/Shia, Catholic/Protestant, Orthodox/Reform, etc.), Philosophy (Materialism/Idealism), Economics (Fiscal spending/supply side), Politics (conservative/liberal), and numerous fields. It's *Man the Spirit Filter* separating unity, splitting wholeness, dividing-refining-dissecting everything. But with a little reverse engineering and strategic perspective, YOU can easily see the entire connected whole as complimentary parts of a single united spectrum of knowledge.

UNIFIED SPECTRUM OF KNOWLEDGE

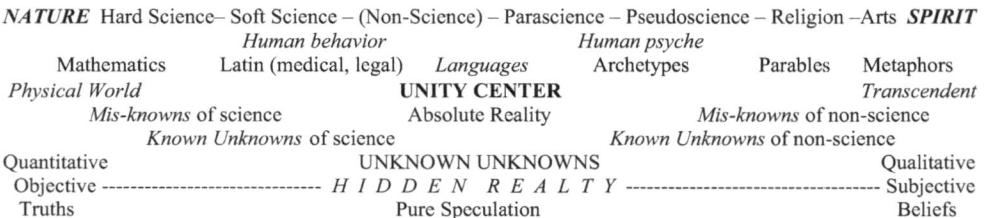

NATURE Hard Science– Soft Science – (Non-Science) – Parascience – Pseudoscience – Religion –Arts *SPIRIT*

	Human behavior		*Human psyche*		
Mathematics	Latin (medical, legal)	*Languages*	Archetypes	Parables	Metaphors
Physical World		**UNITY CENTER**			*Transcendent*
Mis-knowns of science		Absolute Reality		*Mis-knowns* of non-science	
Known Unknowns of science			*Known Unknowns* of non-science		
Quantitative		UNKNOWN UNKNOWNS			Qualitative
Objective -----------------------------		*H I D D E N R E A L T Y* -----------------------------			Subjective
Truths		Pure Speculation			Beliefs

*Hard Science (Black & White) --- Humanities (Shade of Grey) --- Religion (clear) --- Arts (Colors)
**Like a kaleidoscope; a composite filtered reality, unity expressed through diverse connected lenses

9   UNIFYING LAWS AND CODES

A reality where everything is connected and unified must possess an underlying substructure tying it all together. We should expect that all of it is joined in an abstract tapestry of codified patterns, expressed at every level in a variety of languages. They're all around us permeating our environment with great subtly, often hard to see unless you know what to look for. Every spoken language except our own is foreign to us and so too are the coded languages of reality.

LAWS. Codes that describe the way our world works are formatted as laws and principles. They're rule sets laying the framework for orderly, structured processes. Laws are universal principles establishing how and why things take place the way they do. They reflect the invariant, set nature of a system by confirming the interrelated connections of elements within.

The World. Laws are omnipresent in our world, found at every level, both evident and subtle. At the social level obvious examples are laws of restriction established by legislatures and enforced by courts. These are rules defining acceptable behavior. Societal laws of required conduct are present in just about every venue of human activity. Business etiquette, religious doctrine, courting, sportsmanship, table manners, dress codes, etc. These are social laws that apply indirect force to achieve agreed upon behavior.

   Nature is full of subtle laws discovered through trial and error over the centuries. Most have been passed down over time or revealed by spiritual leaders. Universal truths have been chronicled and made available in popular texts. A partial list might include the following:

The Law of Polarity – Everything has an "opposite" complimentary aspect
The Law of Economy – Nature seeks the path of least resistance
The Law of Attraction – Thoughts, Desires, attachments attract like energies
The Law of Abundance –Everything you need is already available, awaiting proper alignment
The Law of Relativity – Everything depends on perspective
The Law of Cause and Effect – Nothing happens by chance (also Karma)
The Law of Correspondence – As above so below

   What all these have in common is a universal rule that holds true everywhere, in every setting. The consistency of these confirms the divine connectedness of all actions and conditions in our world.

Physical Science. Western civilization's broad fields of science catalogue and classify all observable phenomenon to explain how things work. Each identify laws that reflect universal truths in their respective fields. Physics is the big one followed by chemistry, astronomy, geology, earth science, meteorology etc. Most students are familiar with Newton's universal laws of motion:

1: Objects will remain at rest or motion unless an external force is applied
2: Acceleration is related to net force (directly) and mass (inversely). {F=ma}
3: For every action there is an opposite and equal reaction
*The 3rd law applies to many other areas of life beyond physics, truly a universal law of nature

Other well-known laws captured by physics and chemistry include:

Ohm's Law	Relates current, voltage and resistance
Coulomb's Law	Relates force between charged particles and distance
Avogadro's Law	Gases under equal pressure contain same number of molecules
Pascal's Law	Relates pressure applied to fluid
Boyle's Law	Gas volume is inversely related to pressure (constant temperature)
Kepler's Law	Planets follow elliptical orbits with speed and distance correlated

Pervasive laws of science are invariant and apply universally. Principles do as well, such as Occam's Razor, where the simplest explanation is usually the best one. The universal consistency of laws and principles reveals the orderly nature of reality, providing yet another clue to the fundamental connectedness of everything.

Primary Levels. Since laws are present at every level in The World, we should certainly see them at each of the 3 primary levels of our universal matrix. Nested laws capture the essence of life at the physical, biological and social realms. Here is how they're expressed respectively:

1. Physical   - Consistent principles of science describing the natural world (Nature level)
2. Biological   - Sets of universal life experiences (Human level)
3. Social   - Common themes in collective human activity (Societal level)

Physical Level. Most sciences analyze natural phenomenon, each with laws specific to its own narrow area of focus. Since the nature of science is specialization there's a tendency to compartmentalize language, leading to isolation which limits integration with other sciences. Consequently, laws pertaining to each separate field are rarely seen in a holistic context where they would likely reveal similar repeating patterns across diverse disciplines.

What emerges is a hodgepodge of repetitive laws that are describing the same essence from different perspectives. Where hard science looks at phenomenon with absolute concreteness, soft science counterparts might describe similar patterns in terms of subjective, qualitative language. But even among hard sciences the language and perspective of related fields describe similar things differently. I suspect there's a variety of common laws and principles that span the entire scope of sciences falsely perceived as unique to each field. Let's call these universal laws of science ***Micro-principles***. They're singular concepts found in every field of study but codified with esoteric jargon that isolates them. Consider the correlation between water pressure and electrical voltage, each dependent on the medium (pipes vs wire) and the capacity of each medium (pipe width and wire gauge). Flow rate equations for both are identical. Or consider the concept of resistance. Variations of it can be found in various fields as follows:

Physics	Chemistry	Electronics	Meteorology	Biology	Engineering
Inertia	Buffer Solution	Resistors	Head Wind	Immunity	Stress/Strain
Friction	Tensile Strength	Impedance	Pressure Front	Antibiotic	Insulation

*Resistance is just one of many principles that apply universally across diverse fields

Biological Level. Moving up to the second primary level on our universal matrix, we find the emergence of sentient life forms with varying degrees of consciousness navigating through daily challenges, experiencing variations of pain and pleasure along life's journey. In this ongoing learning process living creatures decipher the code of life through trial and error. Laws at the biological level focus on how to survive and thrive in an external world of threats and opportunities.

Man stands atop the animal kingdom, elevated higher through superior tools and the highest degree of consciousness. Learning is cultivated early in childhood and continues through an entire lifetime. Verbal and written communication enable shared experiences and accumulated knowledge. Organized education systems enable rapid, comprehensive learning so individuals can leapfrog previous generations' experiences, achieving great synergy to fuel rapid progress.

Life Lessons are the hard-earned residue of personal experiences, the best teacher you can have. These subtle laws embedded in every pathway make our lives easier or harder depending on if we abide or resist; to challenge headwinds or go with the flow. It's a coded syntax of hacks, and when deciphered allows further advance with less effort. Life lessons are universal; not limited to race, culture, age or sex.

Consider the comprehensive laws of life experienced and shared over centuries via storytelling, handwritten books and electronic documentation. Thousands of songs have also chronicled universal life experiences with more of an emotional emphasis than their literary counterparts. Modern television and movies provide plentiful material, including a trove of enlightening quotes. They capture chaotic ups and downs of life along with extraordinary experiences of others vicariously. Circumstances may vary greatly but the lessons are universal.

We ALL experience the common challenges of competition, discipline, maintenance, stress, overdoing, conflict, risk, illness, work, mistakes, loss, aging, etc. These universal conditions occur in a wide variety of settings; *Life's Arenas*. They come in diverse forms including the home, workplace, school, public forum, athletic platform, performance stage, battlefield, night club, court room, nature, etc. Sports by itself cover a wide variety of its own arenas: stadium, course, ring, pool, mat, range, field, court, track, etc. Nature is full of arenas where man can pursue adventure in a variety of challenging conditions.

Life's bountiful arenas are the places where human theater has played out for centuries. Dramas of ego, control, insecurity, jealousy, desire, and narcissism are actually "normal". Pain, suffering, stress, fear, sacrifice and periodic achievement all surface as common themes. Sublots emerge after attachments and desire lead to disappointment. These recurring dramas are portrayed in television sitcoms, soap operas, reality shows and movies. The trappings of attachment to special interests are in full display on jam packed magazine stands, the mega billion-dollar advertising industry and the repetitive obsessive nature of song lyrics. Though love is the central theme of most music the actual lyrics tend to dwell on misguided habits of mistaking lust, jealousy, and possessiveness for "love".

Myths reflect the universal nature of human experience, despite adding a little exaggeration for effect. Themes become grandiose, creating superhuman drama highlighting our fullest potentials. It's human theater of pride, heroism, fate, hubris, sacrifice, justice, love and beauty. Life's connecting stages culminate with the universal plot of *The Journey*. Mythology is largely about ordinary humans achieving the extraordinary, where great obstacles, challenges and setbacks are overcome. After all, it's the journey that has real value, not the destination. Along the way man learns, adapts and develops into a well-rounded cultured being.

Religion also confirms universal themes of human experience where countless stories of inherent mortal weakness are normal and natural, along with a proclivity to sin. Lessons are conveyed in metaphorical parables teaching right and wrong with role playing and storytelling. Whole sections of the bible are devoted specifically to life lessons. Proverbs is a how-to manual on right living. Common themes include anger, adultery, parenting, envy, pride and humility, perseverance, excess, and careful speech.

Literature has long served as the classic source for chronicling human nature. Great works have survived long time spans given their universal themes still apply today. Many of these classics are the source of commonly referenced quotes that are just as true in modern times. Plots follow the 3 classic types: Man vs Nature, Man vs Himself, and Man vs Man. They fit in exactly with the primary levels on our universal matrix: *The World* with Physical (Nature), Biological (Man himself) and Social (other men).

Shakespeare was the gold standard of classic literature whose tragedies exposed human nature with all its warts and disappointments. Political manipulation was very much routine then and still is today with remarkable similarity. Religious overtones are present as well along with the *seven deadly sins* committed with great regularity. We may have come a long way since classical times but we can't escape our lower nature completely, rearing its ugly self to remind us that these themes continue to plague even the most progressive, "evolved" individuals.

Comedy is all about life lessons. It works best when material is based on the universal human experience. Audiences respond well to content they can relate to and what better topic is there than ordinary life itself. Humor is a release valve for the suppressed acknowledgment of the folly of life. The struggles, disappointments and unfairness of it all is a reality that cannot be denied. But with the right verbal skill, wit and expressive dexterity these annoyances can be revealed as funny divine pranks that we can all embrace as survivors of shared lessons. I may get no respect but as long as you don't either, it ain't so bad.

The age-old question on the meaning of life has a common answer; "To fulfill your life purpose". But that's really about purpose and less about meaning, which comes from life lessons. Meaning is the value we attain as we learn, adapt and grow from experiences in *The World*. Every situation in every arena provides an opportunity to experience reality and draw meaning from it. You are the product and life is the process. The meaning of life is measured in all those differences between who you used to be and who you've become.

Social level lessons are laws we could call **Macro Issues.** Like Micro Principles, they capture universal ways things work at the social level. They're universal experiences emerging from interactions between groups and the greater collective whole. Macro issues result from competing forces generated by opposing interests and group dynamics of economic, political and cultural spheres.

Economic issues involve "fair" distribution of resources. Taxation is a major concern: who is taxed, how much is fair and what system is best to confiscate earnings. Class divisions create disparities leading to questions of how to address the poor and indigent. Should minimum wage levels exist, what should it be and is it worth the tradeoff in loss of jobs? How much should the government meddle with the economy? Given historic cycles of boom-and-bust what degree of preemptive manipulation is prudent?

Interest rates, inflation and economic growth are intertwined and affect individuals differently. How much government tinkers will help some people while hurting others. Should government incentivize certain behaviors such as home purchasing while providing disincentives for others such as vehicle fuel taxes? Every policy tweak made on the economy in some category will directly impact various groups differently; some beneficial, others detrimental. Ongoing disruption within the economic equilibrium leads to dissent and agitation, the very source of macro issues.

Political dynamics generate conflict in society as competing forces constantly via for power. People expect fair representation but rarely receive it. Problems arise when long-tenured politicians become self-serving, corrupted by power, or cater to special interest groups offering kickbacks. Likewise, career bureaucrats become entrenched government employees who tend to put self-preservation ahead of responsive service. Governments are rarely incentivized to operate efficiently, squandering significant amounts of limited resources.

Politicians who possess exceptional speaking skills succeed more often than those who may be more competent, non-polished communicators. Feel good policies trump effective ones. Initiatives implemented today may not be beneficial later as conditions change. Voter apathy and lack of participation in the process doesn't help either.

Government structure is a major factor in producing macro issues. How big should government be, how much control should it have and what recourse do citizens have when it fails to perform? Competing systems exist on a sliding scale with mixed approaches: Democracy-Totalitarian, Capitalist-Socialist, Conservative- Progressive. No nation has a pure form of either – most have a hybrid combination of several. The key issue is degree of control. The necessary evil of required central authority presents a delicate tradeoff inevitably ending with dissatisfying results, then unrest and agitation.

A universal issue in the political arena is the conflict between collective interests and individual freedom. Governments naturally impose limits on individual behavior, but problems arise when too much or too little is allowed. Ironically the same protections offered by a collective whole can become instruments of tyranny when group forces are turned against individuals. Similarly large groups can impose their will on smaller groups and create unfair conditions. Even in pure democracies a group possessing a 51% majority can usurp the rights of a group with 49% of the population. Politics is the act of getting someone else to do what you want; disenfranchisement is an afterthought.

Cultural issues involve competing values among different groups. With greater diversity comes more opportunity for conflicting preferences. Dissention increases when values are deep seeded with little room for compromise. Religion and politics don't naturally mix however they are difficult to keep separate. Strongly held personal beliefs can make coexistence with others difficult when they don't share the same beliefs. Xenophobia is a human tendency to cooperate with similar others while rejecting those who look, sound and think differently than you.

Culture is largely influenced by those who control media and communication systems. Problems arise when the few controllers inject their biases to a majority having different values. Television is a great influencer of thought but it's run by a tiny exclusive minority; those who control the medium control the message. The education system is likewise run by a small minority of controllers who routinely injected political and cultural biases that don't belong. A recurring theme here is how culture is unnaturally steered by small exclusive groups with biases.

Arts and entertainment routinely push thresholds where values and expression challenge standards of acceptance. Governments occasionally intervene to limit or punish artistic license when it is deemed unacceptable in the court of public opinion. It basically comes down to free speech vs standards of decency. Who decides what's acceptable? Is the standard enforced consistently? This issue also applies to any case where government determines individual behavior as deviant from an accepted norm.

Nations that experience imbalances in any of the primary domains become ill, similar to pathologies in an organic living body. Even societies that achieve a decent balance must remain vigilant as conditions constantly change, requiring ongoing adjustment. History is literally the continuing story of nations working through collective dynamics that create *Macro Issues.*

To recap we find unifying laws at each of the primary levels of life on our matrix. We'll label these laws as: Micro principles, Life Lessons, and Macro Issues. Each level expresses the codified nature of reality in its own unique manner and by doing so they confirm the unity of life and the abstract substructure linking everything. These laws apply uniformly on each level but also overlap with other levels in more subtle, indirect ways. Perhaps someday an organized symposium of experts in multiple fields could collectively categorize a comprehensive compendium of universal laws. Such a project could create a single unified primer identifying common laws and their respective application to every major field.

LAWS OF ASSOCIATION (*connectedness among things*)

Whenever any thing or being interacts with any other thing or other being, 3 possible outcomes can occur: *Coexistence, Cooperation* or *Opposition*. The outcome is either Neutral, Positive or Negative. This includes all things, beings, elements or parts interacting in definable situations, from individuals to large groups, from few parts to a very complex greater whole. The key here is the active relationship between interacting elements in any situation.

 The minimum a system can be consists of two independent elements, things or people. Two parts connected always form a polarity and an interdependent relationship emerges. So, when one part acts the other one responds. This dynamic interaction between any two connected things is a persistent pattern resulting in either coexistence, cooperation, or opposition. In *The World* we see this exchange take place in various arenas where two or more people simply get along, work together or compete.

Typical Associative Interactions in the Workplace:
(Neutral, Positive and Negative responses)

Situation:	Coexist	Cooperate	Compete
Meet new coworker	Just acknowledge	Exchange numbers	Jealous reaction
Work same project	No interaction	Teamwork	Withhold information
Pass in hallway	Simple eye contact	High Five	Look away
Boss interview	Neutral feedback	Praise coworker	Criticize coworker

Coexistence — People or things can simply coexist without interfering with others around them. This neutral arrangement is the most common mode of interacting elements and should be considered the normal way of things; the status quo. It includes situations where people ignore, avoid or are simply indifferent to others. What's key here is the absence of influence toward other elements requiring them to change or respond in a free will manner. In the example of people there's no need to change one's disposition after interacting with other neutral people.

Cooperation — People and things can work together aligning their collective efforts to create synergy. This is the most productive arrangement of the three possible responses. It's a win-win situation when elements join together in a unified manner. It's part of teamwork as members trade off some individual freedom and/or identity to serve a greater whole. Integration occurs whenever parts, units or elements combine to work as a subset of a more complex system to achieve synergy, where the whole is greater than the sum of its parts. Markets where people buy and sell goods or share information fall into this category.

 Principles of collaboration emerge in cooperative interactions. Some arrangements simply work better than others. These are captured in universal principles defining best practices of teamwork situations:

Principles of Collaboration

O Leadership – Centralized control enhances group performance
O Communication – Open, free communication + feedback channels enhances group performance
O Resource Allocation – Balance and efficient management of limited resources is essential
O Trust, Confidence, Motivation – Individuals perform better when success is rewarded
O Collaboration – Teams perform better when networking allows cross channel interaction
O Roles and responsibilities – Individuals perform better when they clearly understand expectations
O Goals, Vision – Groups perform better when goals are clearly defined

Cooperative relationships foster integrated effort, producing greater synergy. It's reflected in group dynamics in diverse situations: sports, business, politics, theater, school, government, science, etc. The key is the changed disposition individuals experience to conform and align with a unified whole. Thus, a complimentary relationship emerges where elements interact with other elements to produce a new, synergistic outcome greater than the original separate parts.

Competition – The third relational situation is opposition. Conflict in all its forms emerges when 2 sides clash with contrary goals that are irreconcilable. Competition is a related variation, occurring in the same arenas as cooperation but the process is significantly more dynamic. The scale of conflict can vary from a harmless verbal disagreement to a violent full-scale war. Competitive interactive associations begin with a minimum of 2 elements and have no upper limit in size
    Arenas of various types create the setting where conflict takes place, each defined by particular conditions, rules and boundaries. Typical examples include stadiums, platforms, stages, rings, fields, theaters, courtrooms, courses, society, nature and any place where opposing elements confront one another. Competition, like its counterpart cooperation/integration, also has fundamental principles that determine what works and what doesn't. These rules involve methods, techniques, schemes, maneuvers, approaches, hacks and gimmicks that apply in a wide variety of setting where people compete. For example:

Sports – (Individual Level)
o Consistent Swing (golf, batting, tennis, darts, bowling, free throw, pool stick, pitch, place kick)
o Whole Body Effort (high jump, swim, boxing, fencing, wrestling, weightlifting, pole vault)
o Warrior Spirit (passion & aggression, exceed limits, fearless, win at all costs, ignore pain)
o In Sync (Physical-Mental-Emotional, *In the zone*, instinctive reaction, become one with target)

Sports – (Team Level)
o Put team first, sacrifice self
o Train as a team, become connected single unit
o One leader (coach) unifies the plan, approach and effort
o In Sync with teammates, anticipate actions and respond seamlessly

Personal Success: Getting ahead in *The World*; Society arena, self-help book success principles
o Wealth – Own a business, invest in real estate, limit spending, disciplined saving, avoid debt)
o Love – Communication, acceptance, respect, consideration, trust, compromise, sharing)
o Health & Fitness – Proper diet, exercise, manage stress, progressive load, adequate rest)
o Power – (see "The 48 Laws of Power" by Robert Green) Ex: #35 *"Master the art of timing"*

Games: Same as sports but non-physical. Board games, cards, outdoor activities, and competitions
o Anticipate opponents moves
o Array elements to support each other
o Surprise moves
o Preplan sequences
o Know when to attack and when to defend
o Know the percentages and choose favorable options
o Be flexible and adjust to situational changes

The principles of competition unify elements and processes present when 2 or more things clash. Sports, games and personal success are just a few of the areas unified by conflict dynamics. On a broader scale we can include business, politics and judicial arenas where "competition" begins to resemble warfare. Stakes are higher and the means to an end become more complex. As friendly competition transitions into purposeful fighting these principles transcend to a higher level.

Conflicts we categorize as fighting fall along a spectrum beginning with a simple 1 on 1 clash and expand higher with organized groups fighting other groups. In sports we see controlled physicality also falling on a scale from non-contact figure skating to hard hitting football and rugby. Recent trends reveal greater physicality in many traditionally non-physical sports like soccer and basketball where competitive edges are pushed at every margin. Ironically this is developing at the same time safety is hyper emphasized in school sports and society in general.

Martial arts are an excellent example of controlled physicality in regulated fighting. Each venue has seen an evolution in fighting styles and approaches over time, including Karate, Jujitsu, Aikido, Judo, Hapkido, Kung Fu, Capoeira, Krav Maga, Tae Kwon Do, Tai Chi and others. Many are just variations of hand-to-hand combat. Each share common skills with others, emphasizing variations of approaches. Combinations of punches, kicks, throws, chops, grappling and wrestling are applied in different degrees. Some even incorporate elements of dance and acrobatics! A key principle in all these approaches is to synchronize and leverage the mind-body-spirit to channel maximum energy to an opponent's weak spot. No style is "better" than others but situationally dependent. Bruce Lee famously promoted the value of integrating all styles and then ironically, transcended them all with no style.

Football is a sport sharing great symmetry with actual warfare. There's similarity in terminology, formations and plays resembling battlefield maneuvers. Strategy in this sport continues to evolve. In just the last several decades we've seen the west coast offense, run and shoot, spread offense, wishbone, wild cat, pistol, option and many others. Formations are designed to create situational advantages. Likewise, defenses developed their own array of counter formations. The unifying theme here is ongoing manipulation of group elements to gain positional advantage.

Transcending beyond sports and rising to the highest level of conflict we get whole nations engaged in total war. Laws governing these engagements in conflict are universal, expressed in strategy and tactics. Strategy is the broader, big picture overall plan to achieve an ultimate goal while tactics are lower-level, short-term immediate actions taken to achieve local situational advantage.

Tactics. Actions taken to gain competitive advantage are dictated by the arenas in which fighting occurs. Opposing sides face off in various situations where rules, conditions, boundaries and context provide opportunities to exploit. What works in one setting may be useless in another. And what works now may not work latter, even in the same setting. Relationship between opposing forces and the arenas they engage within become a dynamic ever-changing integrated system – the more responsive participant always has an edge. Yet despite the fluidity of tactical environments there are universal principles that maximize success in any situation:

- o Establish a well-defined goal (desired outcome)
- o Fire and maneuver – pin opponent and move around them
- o Tempo – move and respond faster than the opponent
- o Array resources to concentrate force
- o Weaken opposing forces through attrition and exhaustion
- o Agility – change the approach as conditions change
- o Deception – Confuse the opponent
- o Reserve – Hold a portion of resources for use at the right time

Tactical maneuver can be as simple as a direct assault, relying on sheer force and shock effect. On the other end of the scale is indirect attack, including infiltration or non-contact related disruption. In the middle of the scale is any number of maneuvers used to out flank or encircle the opposing force, thus attacking weakest spots or cutting off communications and supply. Most battles are won by successful coordination of fire and movement that exploit terrain avenues and array friendly forces in positions that maximize relational advantage.

Arenas where conflict takes place are imperfect, uncertain settings that possess an element of the unknown. Unanticipated impediments inevitably arise. Conditions change, surprise developments emerge, and the enemy is rarely as predictable as assumed. Hence the saying "Great plans rarely survive the first engagement". Obviously, a well-prepared plan is essential, but flexibility is equally important.

Conflict isn't limited to physical action. There's a significant degree of psychological warfare that can defeat an enemy before the battle starts. Intimidation, confusion, distraction, bluffs and mental antagonizing can weaken an opponent before and after shots are fired. Small moves done with surprise can effectively throw an opponent off his game with little effort. Well timed actions similarly generate favorable outcomes with minimal resource cost. Terrain and weather changes can be exploited to increase conditional advantage. Adding up a series of small situational advantages becomes a force multiplier that can synergistically defeat a superior opponent.

Intelligence is a force multiplier as it reduces uncertainty and improves responsiveness. Knowing yourself, the opponent and the environment creates a significant advantage. Good Intel should reveal strengths, weaknesses, opportunities and threats. Bad Intel is a prelude to disaster.

Battles are won by offensive actions. Defense is necessary but should be considered a temporary mode, depending on situational conditions. Offense is where initiative is gained and opportunities are exploited. Opponents in defensive postures should be prepared to counterattack at the ideal time. The best defense is a good offense. In sports there is often a fine line between the two modes. Examples of aggressive defense transitioned into offense are the following:

o *Football* defense blitzing the quarterback
o *Basketball* full court press
o *Tennis* attacking the serve
o *Baseball* throw out unsuspecting base runner
o *Boxing* counter puncher
o *Wrestling* reversal

Offense and defense are two sides of the same interdependent coin; either mode can be successful with the right warrior mind set. Offensive actions are optimized when applied deliberately with maximum allowable aggression and persistence. Defensive responses must strive for equal degrees of return aggression and avoid tendencies of wallowing in passivity. Successful defenders maintain an urgency to regain the initiative and resume an offensive posture.

In sports, war or games there are unifying symmetries in the way opponents engage each other regardless of the arena. Like the micro-principles we covered earlier there are similar properties that apply across all competitive settings. The scale of combat begins with simple verbal disagreements or friendly board games, progresses higher to limited, controlled physical engagement, then higher yet with uncontrolled physical aggression (fighting), and finally full-scale war. Combat arenas on either end of the physicality scale share common relationships, universal dynamics and a symmetrical pattern of tactics revealed in the table below:

WAR	FOOTBALL	BOXING	CHESS
Fire & Maneuver	Block and Run	Combo Punch	Pin & Fork
Feint	Pump Fake	Peek a Boo	Check
Shift Forces	QB audible	Switch Southpaw	Castle
Smoke Screen	Hard Count	Bob & Weave	Decoying
Retreat	Prevent Defense	Take a knee	Withdraw piece
Tempo (Ooda Loop)	Blitz	Rapid Jabs	Speed Chess
Sniper	Filed goal	Body punch	Pick off pawns
Delaying defense	Holding	Clinch	Blockade
Out Flank	Screen Pass	Left Hook	Skewer
Ruse	Play Action Pass	Rope a Dope	Zugzwang (force a bad move)
Victory	Touchdown	Knockout	Checkmate

Strategy. Where tactics are the individual actions taken to gain immediate advantage, strategy is the comprehensive collection of planned moves sequenced and orchestrated to support a larger goal. In sports or war, the goal is to beat the opponent. Tactics are used in the plays or battles while strategy is the larger plan supported by those plays and battles to collectively achieve final victory.

Strategy begins with a vision – an imagined concept of how things should be versus how they are now. Once that goal is defined then strategy becomes the plan to get there. It's longer term, broad in scope and applies interdependent elements directed in a linked, synergistic manner. The key is to maximize use of limited resources by channeling effort where it is most effective. Strategy is ultimately about making the right choices between multiple options.

Every competition is a variation of war. Participants may be fooled by the appearance of a friendly setting however opposition always invites cunning agendas. Corporate boardrooms have chewed up plenty of well-intentioned employees dressed neatly and nicely mannered who unwittingly walk into a shark tank. City, state and national politics are similar arenas where the lethality of plotting adversaries can wreak devastating damage on ill-prepared, naïve participants. In politics, strategy is usually the difference between who succeeds and fails regardless of whether their policy ideas are better.

Tactical principles are mirrored by higher-level Strategic principles. Since competition involves universal relationships between opposing elements there must be unifying rules defining them. This is expressed in the 9 principles of war:

o Objective – Clearly defined end goal
o Offensive – Maintain initiative and freedom of action
o Mass – Concentrate to maximize force
o Economy of force – Efficient use of resources
o Maneuver – Position elements for maximum effectiveness
o Unity of Command – Avoid conflicting bosses
o Security – Protect your resources
o Simplicity – Avoid unnecessary complexity and ambiguity

*War or any competition involves *Ends* (objective), *Ways* (strategy) and *Means* (resources)

Subsets of rules and principles apply as well. One is the concept of "center of gravity". It's the key target that makes or breaks the opponent. It could be anything that provides physical strength, freedom of action or will to act. At the national level it's often non-military elements, such as access to oil or public opinion.

Competitive dynamics in warfare were scientifically analyzed by Clausewitz in the 1800's where he identified and categorized its universal aspects. His theory of war comprehensively examines the complex factors and processes involved, revealing connections between military, political and civilian components. Ironically, despite the scientific structured aspect of warfare he also acknowledges its ambiguous uncertain nature making outcomes difficult to predict. On one hand he provides a math like equation; *resistance* = *means* X *will*, and then on the other presents a concept called "the fog of war". This dichotomy reflects the dual nature of any dynamic system where one of the key variables is *man*.

Many centuries before western civilization's scientific analysis of warfare, Sun Tzu developed a comprehensive treatise on it, however his emphasis was on the "Art" of war. This Chinese strategist compiled his own set of universal principles defining what works in battle and war. Mirroring many of the same concepts Clausewitz developed, Sun Tzu provides a more creative flavor in scheming to achieve situational advantage. Weak leaders get stuck in static doctrine; successful commanders exploit creativity and imagination to navigate through fluid situations.

Simply put: adapt to situational conditions. If the enemy is superior – avoid him. If equal, defend. If weaker, attack. If attacked, retreat. If they retreat, attack.

The key takeaway is no strategy is best; it simply depends on the situation. Have an overall plan but be flexible and adapt. Accordingly, masters of strategy possess logical, methodical abilities to plan and structure campaigns while tapping into creative, imaginative capacities to anticipate, adapt and go with the flow as conditions inevitably change. Clausewitz identified a quality he called *Coup d'oeil* where great commanders integrate past experience, observations and imagination to see the interdependent system of parts and orchestrate them as a single whole. Genius for strategy emerges in all arenas: sports, games, business, politics, war.

Strategy and tactics represent laws of association that codify the dynamic relationships between competing forces. No two things confronting each other ever act independently. Each becomes subject to and connect with the other. A whole system emerges where parts are interrelated and affected within the arena. The linked relational action and response is captured in universal laws – coded patterns useful to those aware enough to decipher them.

UNIVERSAL CODES

All language is code and every code is a language. A code is an abstract immaterial pattern linking and unifying things on a particular level. Signs and symbols are basic elements acting as connecting links. Information is the raw material, arranged in bits and bytes forming a patterned tapestry within a larger whole. That whole becomes unified by the signs or symbols joining every part to every other part. Signs are singular one-dimensional markers while symbols possess context-based multiple meanings, often rooted in the subconscious level.

We're literally drowning in a sea of information permeating our environment, but we unwittingly perceive it as mere background noise. Decoding info transforms noise into meaningful data. We're surrounded by *meaningful* information, we just aren't aware of it. What we call learning is really just a form of code breaking.

© § ¶ ○ Ψ † ℀ † ◐ vii ♥ ♫ ♂ ? ♀ ♛ ☼ ☺ ⚘ ∞ ʒ ∫ $ # ☯
Σ ⚵ ♖ ⊘ ✚ ∿⚶♌☌♖ ⊘ ◉ ↰ ☀ ✛Σ ❖ ♟ ⏤ ℀ ⚙ ▤ ⚚ ⛨ ☻ ✿ " ❀ ≈ @
⚷ ♪ ✿ ☄☉ ◆ " ❏ ☊ ⌼ ◁ ☆ ✓ ℝ ¥ ^ % ♪ Ω ≠ ⊠ 𝓜 □ ¶ ® ∫ ! §@ φ□ ¢ ⌘ □

The World. Like universal laws there are also universal codes present at each of the primary levels: Physical, Biological and Social. This translates to the Natural world, human and societal levels. Nature includes the outdoor environment of terrain and weather. Land is codified by its diverse surface features, topography, elevations, depressions, vegetation, streams and lakes. Maps capture this information graphically in composite fashion. Weather patterns include cloud formations, ratios of moisture, strengths of wind, pressure fronts, a heat index and several other components. These attributes are measured through various instruments and then tracked by meteorologists. Towns, cities and states develop in predictable patterns based on math laws, often forming in close proximity to sources of water and traffic ways. Intricate maps replicate these arrangements in precise detail. The physical world we all navigate corresponds to a comprehensive coded information template.

Parsimony is a code offering hints of unseen natural forces. Other phenomena do as well, including tree rings, soil samples, holes, smoke, steam, dunes, heat, etc. Relics and fossils reveal clues to things connected by the passage time. The shape of continents and unique species near coastlines provide clues to grand movements of continental drift. Information cloaked in the seams of nature can be decoded by detectives who know what to look for. These hidden codes contain a message telling us how everything is linked.

Man-made codes are seen in everyday activity including bar codes on products, QR codes, color codes in labeling schemes and numeric classifications used in comparisons. All messaging is coded – most known, some secretive. Cryptography is an entire science devoted to codifying information to secure personal, financial and state sensitive data. Much of the coded information we navigate through on a daily basis subtly blends into every aspect of our lives unnoticed. Social level codes are more noticeable, influencing our behavior in specific situations. Rules of etiquette and conduct apply everywhere in society, beginning with family and extending to large groups and public gatherings.

All professions have their codes: medical, legal, military, engineering, etc. Different venues have their own exclusive languages and syntax requiring a sort of code literacy to fit in. Work, play, sports, entertainment, business and romance all involve specific codes, spoken or unspoken, correlating to favorable or unfavorable behavior. Social codes regulate personal space, hygiene, body language, courtesy, how to use money, wear clothes, eat food or acknowledge others.

PHYSICAL REALITY

Periodic Table. The first primary essence of our universal matrix is the quality of "stuff", the elemental aspect of materiality. Atoms and particles are foundational building blocks of the physical world, each identical in substance but different in number and pattern. It's captured in the Periodic Table; elements sequenced based solely on numbers of protons. Surprisingly great differences in properties emerge as each atom adds more identical particles while changing their shell-cloud shapes.

The table periodic table is a coded matrix, arraying elements with similar patterns and thus similar properties. Periodicity literally means repeating patterns and that's the essence of the coded information imbedded in the table. Size and quantity increase horizontally while shared properties are linked vertically.

There are 94 natural elements plus 24 unstable man-made synthetic ones. They belong to one of 18 families with similar qualities along a vertical column. Each also belongs to a general group (Alkali metals, alkaline earth metals, metals, transitions metals, metalloids, nonmetals, gases). To simplify it further, the table is basically a scale going left to right – from metallic qualities to non-metallic. Metals are characterized as conductive (electricity and heat), malleable, ductile, solid and reactive with acid. Though our environment is largely nonmetallic the actual preponderance of elements are metals – about two thirds of the total.

Periodic trends are the particular patterns of properties grouped within the table. In addition to metallic qualities there are shared tendencies of electronegativity, ionization energy, electron affinity, atomic radii size, and chemical reactivity. Where these properties come into play is at the molecular level where atoms combine to form a plethora of unique new stuff with new particular qualities, often surprisingly different from the original atom's own attributes.

Atoms possess a social trait based on their shell shapes and more importantly, the amount of excess or lacking number of electrons in the outermost shell. Some are very "social" possessing strong affinities toward bonding with that special other atom while those will full shells tend to be anti-social loners. Molecular dynamics begins with attraction, combination and transformation of properties, then progresses to energy transfers as atoms emit or absorb light to balance out the transaction. Each electron jumping between shells requires an energy gain or loss and this is performed in specific incremental amounts (quanta) with light serving as the currency. The process is traceable as each light emission has a particular color signature associated with its energy level, captured by spectroscopy.

The periodic table is simple, symmetric and codified. Three fundamental particles (proton, neutron and electron) fill every nucleus and shell with equal, balancing amounts of each in successively larger wholes. Different properties emerge with larger wholes and added shells. Atoms with similar ratios of particles and shells share similar qualities. There are 7 total shell groupings repeating the same pattern, divinely comparable to a musical octave. All middle atoms possess varying amounts of free electrons which determine their sociability, translating directly to molecular dynamics. The far end group known as noble gases share the quality of completed outer shells and correspondingly have little interest in mixing with other "needy" atoms. These full, satisfied elements resemble nodal points on a sine wave where neither pull or push is present – a sort of equilibrium is achieved.

Materiality is based on creative mixing of identical parts in various amounts. In essence the periodicity embedded in the table of elements serves as a blueprint confirming built in symmetry and connectedness of everything in the material realm. It also reveals the most important factor creating differences in the stuff of life is not the parts themselves but the patterns and combinations they form – a universal code of substance. The elegant symmetry of it can be displayed in multiple formats including these few:

H																	He
Li	Be											B	C	N	O	F	Ne
Na	Mg											Al	Si	P	S	Cl	Ar
K	Ca	Sc	Ti	V	Cr	Mn	Fe	Co	Ni	Cu	Zn	Ga	Ge	As	Se	Br	Kr
Rb	Sr	Y	Zr	Nb	Mo	Tc	Ru	Rh	Pd	Ag	Cd	In	Sn	Sb	Te	I	Xe
Cs	Ba	La	Hf	Ta	W	Re	Os	Ir	Pt	Au	Hg	Tl	Pb	Bi	Po	At	Rn
Fr	Ra	Ac	Rf	Db	Sg	Bh	Hs	Mt	Ds	Rg	Cn	Nh	Fl	Mc	Lv	Ts	Og

	Ce	Pr	Nd	Pm	Sm	Eu	Gd	Tb	Dy	Ho	Er	Tm	Yb	Lu
	Th	Pa	U	Np	Pu	Am	Cm	Bk	Cf	Es	Fm	Md	No	Lr

Periodic Table

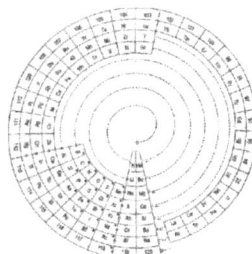

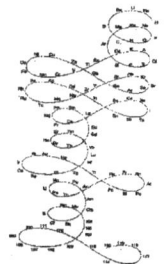

Vibration. Everything following a vibrational pattern is coded information. It's what differentiates and gives meaning to anything expressed as a vibration. It's the syntax of living and mechanical processes governing physical reality. To know anything comprehensively requires awareness of its underlying vibrational essence. The rest is mere façade; exterior window dressing creating illusions of appearance.

Physical substance is a grand illusion. Every object we perceive is made up of fluctuating, flowing wave-like fields that appear solid. This dynamic structure of equilibrium produces a phantom of perception. Absolutely nothing in the universe is "solid", not even the particles we describe inside atoms. Our physical reality consists mostly of empty space and humming, vibrational fields pulsating within.

Among the myriad ubiquitous vibrations permeating our environment there are two fundamental forms essential to our conscious reality: Light and Sound. These wave phenomena connect everything and provide a feedback language informing us of what's happening in our external world. Both are integrally linked to the mechanisms of atomic and molecular dynamics.

Light. Our conscious reality is essentially filtered light. The physical environment is a pervasive macrocosm of light interactions where it passes, reflects, refracts, resonates and interacts dynamically with everything. It can be polarized, blocked, stretched, compressed and absorbed. The entire universe is connected by light – expressed as coded information that when deciphered reveals a great deal. It's the key ingredient of physical reality, both medium and message, and the primary focal point shaping all we experience.

The World we perceive is illuminated and sculpted by filtered light. Highlights and shadows define spaces, variations of intensities add further depth, diversity of colors enrich the experience while all motion and change are complimented by in kind changes in lighting. Every object, space and movement is linked in a feedback loop where light is the conduit. As every atom possesses a specific light signature, so too does every object and process in the physical universe.

Color. Pure "white" light is actually a composite of many parts or wavelengths expressed as colors. White, black and every shade of gray between spans lots of ground but is empty and soulless without the complimentary flavor of colors. As a tiny subset of the larger light spectrum, it's simply a limited band of vibrating waves differing by wavelength, frequency and energy. Color is a continuous sliding scale with 7 nodal points in between, perceive as base colors – resonant points like a musical scale of octaves: Red, Orange, Yellow, Green, Blue, Indigo, Violet. Thousands of secondaries and variations reside between these nodal points – just check a paint store color palette for a small taste of the enormous variety.

The ingredients of color are simple: wavelength and frequency; complimentary sides of a single whole. Like atoms that differ only in number of identical particles, colors differ only in wavelength. Yet both emerge as scales of great expressive difference. And the two are related since atomic structures form unique energy patterns with specific color signatures. It's revealed in spectroscopy where unknown compounds are identified by their color patterns, a byproduct of either reflected or absorbed light. Heat also directly connects atomic dynamics and color. Color spectrum wavelengths directly correspond to differences in energy and heat.

Color is coded information linking elements of our physical environment, possessing survival value for living creatures. Our perception of colors can be lifesaving – informing us of what to eat and what to avoid. A whopping 80% of the neocortex is used to interpret visual information, color being high on the list. Rods and cones in the eye sense luminosity and color respectively. All visual stimuli are interpreted through a mental construct with both conscious and subconscious aspects. Most people can distinguish about a million different colors.

Color codes are present in both our physical and social environments. In physics they reveal the composition of atoms, the direction of celestial bodies (doppler effect), atomic energy transfer, and the composition of distant stars. Red alerts us to potential danger while blue imbues a calming influence. Green at the center of the color octave relates to organic life and the process of photosynthesis. Other colors have specific associations that we respond to both consciously and subconsciously. Culture has created an entire language of color codes assigning specific meanings and values to various things, settings, clothing, status, rituals and beliefs. At the biological level we associate particular colors with moods, feelings, tastes, etc. The common essence with color codes at each level is variation in energy. Vibrations have a resonant quality and as sensitive living beings we harmonize with certain energies based on our own state of being.

COLOR CODE – Meanings and Associations

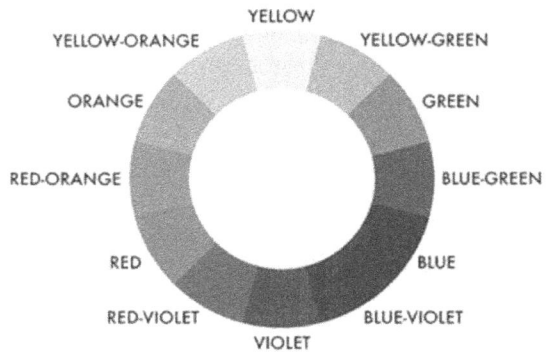

White	Purity, simplicity, goodness
Pink	Feminine, compassion, sweetness
Red	Passion, danger, rage, energy
Orange	Creativity, happiness, vitality
Yellow	Friendly, optimistic, joyful
Gold	Wealth, courage, lustful
Green	Fresh, fertile, natural
Blue	Peaceful, loyalty, security
Indigo	Justice, devotion, wisdom
Violet	Spiritual, luxury, imagination
Purple	Royalty, bravery, fantasy
Grey	Neutral, balance, sadness
Brown	Earthly, stable, reliable
Black	Mystery, power, death

Primary Colors   Red, Blue, Yellow.
Secondary Colors   Green, Orange, Purple
Complimentary Colors   Red-Green, Orange-Blue, Yellow-Purple
Tertiary colors – mixing a primary with a secondary (Blue-Green, Yellow-Orange, Red-Purple)
   *Colored light vs pigment; light colors combined makes white, pigments combined makes black.
  **Color associations differ by culture
***Colors and visible light are just 1 octave of a much broader, vast scale of invisible light

Sound.   Most species communicate using sounds and tones as a primary language, often serving as an alert code to environmental threats. Many incorporate patterns of sound the same way man uses radar and sonar. Both animals and man rely on tonal cues to communicate. Though man supplements it with actual verbal content, the non-verbal aspects can be more meaningful. Sometimes messages are contradictory as content indicates one thing while body language, loudness and tone say something completely different. Context is key. Sound vibration is obviously an essential code that links everything in our physical environment.

Every condition or state of being has a particular sound associated with it, and as states change, sounds change with it. When you're sick, you sound different. When you change moods, you sound different. When your car engine has something wrong with it, a trained mechanic can hear the difference. If you want to know what your kids are up to, just listen to the sounds they're making.

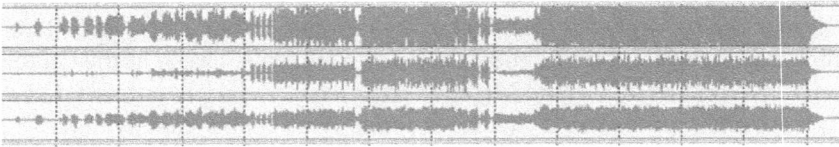

Music. What color is to light; music is to sound. It's a qualitative essence adding flavor to life, transcending an otherwise dull world. Our living universe has a pulsating rhythmic nature with tones and timbres relating everything to specific frequencies. The result is an interdependent mix of harmony and discord, tension and release, and every position in between. This pervasive concert of patterned sound is present at every level but felt uniquely in the realm of living beings. While symmetric vibrations pervade physical reality, music emerges as an elegant spiritual attribute transcending the mundane, lifting it to a higher realm.

Our living universe speaks to us in multiple languages. Music is a temporal language where rhythms and cycles form patterns expressing the mood, setting and tempo of activity. Nature is teeming with rhythmic activity such as wind blowing through a forest, water flowing through stones in a brook, raindrops splashing in a storm or water waves rippling upon a shore. Subtle musical patterns are present at the social level where the ebb and flow of downtown traffic is harmonized with street performers, construction operations, pedestrian movements and the whistle of a traffic cop. Similar patterns are found at the biological level where molecular processes interact within an assemblage of organs that resonate in sync as a whole. Our bodies are vessels similar to musical instruments that resonate to certain frequencies. Health is a function of synchronic resonance where your state of being is centered, symmetrical with musical harmony.

Music is a coded language of tones, pitches, rhythms, tempos, scales and harmonic ratios, all captured by mathematics. Unlike its geometric counterpart of spatial relationships, music is a temporal essence comprised of patterns in time. It begins with a single beat – a quantum moment of creation… out of stillness a sound appears. Then a sequence of further beats and sounds emerge. Like life, there are ups and downs, climax and plateau, repetitions interrupted by change, and mixtures of sequential movements. Life's moments and events correspond with notes, beats and riffs. A breath or heart beat mirrors a note and its rest, and sets the fundamental tempo of a moderate melody. Ratios are determined by nodal points, associated with musical instruments: guitar frets, flute holes, piano pedals, etc. As with life in general, timing is everything. Sometimes the pace is fast, other times very slow. Eventually the tempo becomes just right…until it isn't. Tension is built up only to be ultimately released. Usually an establishing pace defines the "normal" for a particular setting. Even in jazz where spontaneous riffs emerge and break the pace, it eventually settles back to the original "normal" measure.

Melodies and harmony are those sequential flows that just feel right. Like the color spectrum where infinite fractional frequencies exist, there are particular bands that just look nice. Same goes with notes and tones where particular one's sound better than others, especially when played in combinations. It's similar to colors where some are complimentary while others don't mix well. Actually, it's directly related to mathematics where ratios and whole numbers are the key.

Pythagoras discovered centuries ago that strings plucked in particular ratios sounded better. He observed that tones form a band where primary notes separated equally along a repeating scale produced "good" sound. So, our modern music scale emerged having 7 key notes with the 8th being a repeat of the first, called an octave. An octave ratio is exactly twice the original note, with various other fractions in between including 2/3rds called a "fifth", 3/4s called a "fourth", 4/5ths called a "major third" and so on. So, plucking a string half the length of another is one octave while plucking at 2/3rd the length of another is a fifth. The key is that these ratios sound better than any other fractional differences between string lengths. A difference of 1/9th or 2/7ths would produce a discordant sound unpleasing to the ear. There are "half keys" known as sharps and flats that supplement the whole ratios, with complimentary difference as opposed to fractional ones. They're represented by the black keys on a piano.

Math is intimately connected to music, imbedded within frequencies, pitch, tone, wavelength, intervals and harmonic ratios. It's further expressed in the geometric aspects of instruments. The shape of a harp, piano, organ pipes or a French horn reveal the harmonic nature of the sound spectrum. These resemble the curved geometry of a sine wave that mathematically connects music and sound wavelengths. All share the symmetry of uniform ratios that produce harmonious patterns of sound. The same proportionality that makes art aesthetically pleasing is present in the shapes of musical instruments and the tonal qualities they create.

Songs are similar to storytelling, with plots and subplots that climax and return. Most are divided into verses and choruses, sometimes transitioned with an instrumental bridge. A typical format would be verse-verse-chorus-verse. Music supplements and compliments the verbal dialog we call singing, which is simply words spoken in sustained tone where the sound is modulated, usually at a higher pitch than your normal speaking voice. Singing lifts ordinary words from lifeless verbiage to enriched, resonant tones, vibrating in harmony with the accompanying melody. Voices are powered from deep within your chest, filtering through the lungs, throat, mouth, tongue, nasal cavity and lips, producing vibrations that distinguish pitch, frequency and quality of the spoken word. Every singer has a natural pitch within a standard range, somewhere between highs, mids and lows (Soprano, Alto, Tenor, Baritone and Bass), which correspond to tonal differences on the music scale. Orchestras are likewise comprised of instruments playing in a range between highs, mids and lows. Their combination of wind, string and percussion create ranges of pitches and frequencies, lows and highs, steadies and shorts. Both the instruments and the humans that operate them generally spend more time at rest than actually playing.

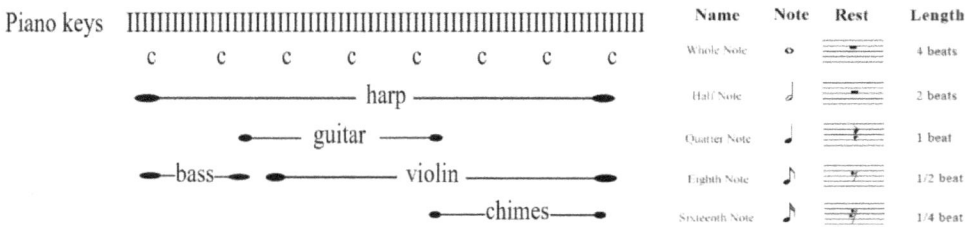

Piano keys		Name	Note	Rest	Length
c c c c c c c c		Whole Note	o	▬	4 beats
harp		Half Note	♩	▬	2 beats
guitar		Quarter Note	♩	𝄽	1 beat
bass violin		Eighth Note	♪	𝄾	1/2 beat
chimes		Sixteenth Note	♪	𝄿	1/4 beat

A human body is a vessel for spirit, filtering it like a musical instrument tuned to specific states of being. Music resonates within man as it mirrors life's harmonic nature. Like good art, music centers us and enriches our being. Songs can capture the experience of passion, challenges, desires, cathartic moments and joyful periods, reflecting the highs and lows of life. The various tempos and rhythmic patterns match the back-and-forth aspects of progress or setbacks in routine activities. Volume and energy vary, sometimes building up in a crescendo only to come back down to earth, just like human drama. Occasionally there's a happy harmony where all seems right and feels good, though temporary and fleeting. A time to celebrate, contemplate, work hard or playfully let it just hang out; all expressed in different forms of music. There's a music style for every personality type: Rock, jazz, metal, classical, rap, folk, etc. Various music instruments (woodwinds, brass, percussion, strings) filter sound uniquely just as personality types filter spirit uniquely. Taking this correlation further, Hermann von Helmholtz saw connections between musical notes and emotional states of being: C –*decisive*, D –*fullness*, E Major –*joy*, E Minor –*grief*, F Major –*peace*, F Minor –*melancholy*.

Light and Sound are *the* essential vibration elements in our physical environment, each with corresponding transcendent dimensions expressed as color and music. These fundamental oscillations connect everything, conveying rich information about what's going on everywhere around us. Life without light or sound would be incomprehensible. Life without their higher octaves, color and music, would be unpalatable. Music and color transcend light and sound, adding a qualitative, spiritual dimension to our lives – a divine coded language of life.

PRIMARY LEVEL CODES – *Math, DNA, Language*

Both Laws and Codes are present everywhere within our universal matrix so naturally they're prominent at each of the three primary levels of our TOE model. The physical level is comprehensively connected by *math*. At the biological level all organic life is connected by the coding of *DNA*. And the social level's connecting code linking collective groups is *language*.

Math is a universal code connecting everything. Pythagoras sensed this over 2000 years ago stating "Universe is number", along with his fellow Greeks worshipping math like a religion, assigning divine attributes to its many transcendent aspects; in essence their *Theory of Everything*. Modern science is now absolutely dependent on math, its multiple fields and diverse applications. New developments in science are complimented by parallel developments in mathematics. Math reveals the elegant simplicity of physical reality, and the realms of numbers, shape, motion, change, space, time and much more. It unifies through formulas, graphs, matrices, tables and geometric figures. Math confirms the connected nature of everything – relatedness and equivalence of it all. Literally nothing is beyond its scope. Static and dynamic processes, quantity and quality, spatial and temporal, visual and abstract, ratios and relationships, and change itself. Every aspect of our interdependent world is definable in mathematic terms. Principle branches are:

Arithmetic	Basic functions of Adding, Subtracting, Dividing and Multiplying. (Multiplication is just repeated addition)
Algebra	Relationship equations – constants/variables ($Y = AX + B$), *Linear transformations* (functions), inputs generate outputs. Quadratic equation ($AX^2 + BX + C = 0$) Describes path of projectile, area of curved surface, etc. (graphically a parabola)
Geometry	Describe shapes using coordinate graph with X &,Y Axis and 4 Quadrants Trigonometry – Relationships of sides and angles of a right triangle
Set Theory	Group elements, connected framework, parts of a whole, permutations. Union, Intersection, Compliment (*And, Or, Not*)
Calculus	Variable change, dynamic motion captured by static function, accelerations and areas. *Integral*-multiplication & addition. Micro small becomes visible, derivatives, differentials, gradients, rates of change. *Differential*-division & subtraction
Statistics	Probabilities and variance, chance and prediction, sequential branches, iterative process, standard deviation. Percentages of aggregates, abstract geometric forms, future portrayed in visual pattern
Topology	Connected objects look different but share identical properties (cup = donut). Objects retain essence after bending, stretching, twisting, deforming – *Self similarity Invariance*
Logic	Rules of inference, propositions and arguments, proofs. Patterns of reasoning, *inductive-deductive*, True-False mode. Algebraic symbols: all S is P, some S is P, no S is P. *Syllogisms*; conclusion-qualifier-conjunctions-implications- negation

*Functional Analysis unifies most of the branches above.

This multitude of mathematic branches serve as surgical tools that probe, dissect and reconnect physical reality, capturing the rich web of diverse interconnected relationships everywhere we look. Various subfields and specialty categories branch out further from each, including Complexity Theory, Game Theory, Combinatorics, etc. What they all have in common is a language framework with particular syntax. Like our verbal language there's an alphabet (numbers), sentence structure (equations), and rules that set order, sequence and combination of operations. A polynomial equation is similar to a verbal statement: $[\ AX^2 + BX + C = 4\ (X + 2)\ ]$. Its *elements* resemble nouns, it's *operations* $(X +\ )$ are like verbs, and $=$ implies a prepositional phrase. Words may describe any action but math formulas and equations do exactly the same thing.

Formulas and equations essentially connect things. In an interrelated universe there must necessarily be an equation or formula that describes anything and everything. Functions connect 2 things in direct relationship. Ratios and constants further describe every particular relationship. Though we perceive differences and disconnectedness in seemingly unrelated things, mathematics confirms they're actually connected, just in varying degrees.

The alphabet of math is only 10 numbers. Humans naturally gravitate to a base 10 number system (ten fingers) but any base will do. Machine language uses a base 2 binary system using just 1's and 0's. It's ideal for computer computation where longer strings of data are easily processed in nanoseconds. DNA code is base 4 or double binary. Since everything in physical reality has a polar opposite, binary code is built into the universe. Binary code is the system used in the I-Ching symbol language where single and double chips arranged in various patterns can represent every situation. Similarly, any visual image can be produced by simply arranging black dots in particular configurations – the same process used in every newspaper image. Any value can be translated into binary code:

#1 = 1   #2 = 10   #3 = 11   #10 = 1010   #100 = 1100100   #101 = 1100101   #763 = 1101000110101100101110

In our base 10 system individual numbers take on qualitative meanings. Oneness and twoness are very different from fiveness or sevenness. The various flavors of different numbers are supplemented by their evenness or oddness. Other special qualities include primes (not divisible or predictable), negatives, infinites and transcendents. Zero is a special number, discovered by ancient Babylonians yet unknown to the more advanced Romans. It's extremely useful as a placeholder, setting a central reference point, and it turns out, nothing is actually a something.

Whole numbers are just the beginning. There are infinite numbers in between we call fractions or decimals. These partial numbers are really just ratios between one whole and the next. Any relationship between two things can be represented by either a ratio or number in the form of a fraction. It's all connected.

Mathematics produced a series of number systems; each subsequent type transcends and includes its predecessor:

Natural	Whole numbers, excluding "0" (1, 2, 3, 4, 5)	*Real*	Combines Rational and Irrational
Integers	Add negative numbers and "0" (-2, -1, 0, 1, 2, 3)	*Complex*	Add imaginary numbers (i, 3i, 4+3-2i)
Rational	Add fractions and decimals (1/2, .735)	*Cardinal*	Only counting numbers (numbers that represent quantities)
Irrational	Can't be expressed as a simple fraction (π , √2, e)		

Math is structured with *laws*. Some are blunt, esoteric and direct, others are subtle and indirect. The obvious ones include rule sets that regulate operations and syntax. Axioms are matter of fact statements that assert the way things are. Proofs are prima facie evidence confirming those same statements. Every branch incorporates rules and structures, assuring consistency, transparency and credibility. The Logic branch epitomizes structure and rules.

Certain abstract math laws possess a mysterious hidden property, operating in a self-directed manner. They hide in the background of nature serving as an invisible force influencing outcomes. These seemingly mystical powers are simply the result of elemental math laws that govern random events the same way gravity causes objects to fall. The Bell Curve is a geometric shape produced by graphically representing the dynamics of an aggregate population of things or events. It's produced when random activities result in a range of possible outcomes where the vast majority fall into a wide middle group while exceptions form smaller and smaller groups. An average will appear in the center while below or above average extend to the far boundaries. Two essential aspects are the mean (average value) and standard deviation (distance from the mean), which describes the most common elements and how far away the exceptions are. Most of the time uniform ranges will be produced (68%, 95% and 99.7%) regardless of how many times a population is sampled or a random number of events are repeated. This consistency in the bell shape pattern is a simple math law that implies an invisible influence.

Benford's law is another case of random events possessing invisible built-in order. It applies whenever a generic sample of numbers are compiled (financial accounting, inventory management). The law simply predicts a larger amount of lower value numbers. For instance, 1, 2, and 3 should show up more than other single-digit numbers (7,8,9) and 11, 12, and 13 should show up more than other double-digit numbers (75, 88,) and so on. It's somewhat counter intuitive as one would expect any number 1-9 to be equally likely to occur. But Benford's law is so consistent in general data sets that the IRS uses it to detect fraudulent returns.

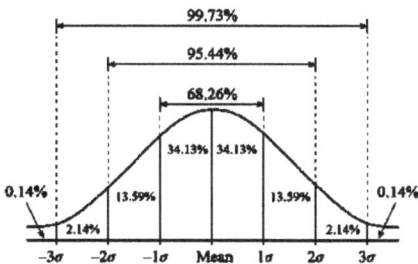

Normal Distribution - *Standard Deviation*   Benford's Law - *1st Digit Distribution*

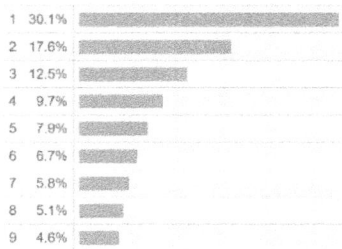

Power Laws create situations where small change in one thing leads to big change in another. This occurs in many non-linear systems where things seem random until order mysteriously and unexpectedly emerges. A tipping point forms from tiny, incremental change adding up without notice until suddenly an avalanche happens. Similar emergent behavior occurs in growth progressions, including business expansion, living organisms, economic growth, athletic performance, city development, etc. Complexity science focuses on these situations where self-organization appears out of nowhere. Math power laws act as invisible regulators on ordinary activity, creating situations that seem to take on a life of their own. Consider the geometric expansion of possible chess moves: 1st move = 20 possible choices, 2nd move = 400 choices, 3rd move = 20,000 possible choices, etc.

Other power laws express the connectedness of aggregate ratios. For instance, as cities increase in size there will be fewer of that size (smaller distribution). The difference in city sizes is an inverse ratio to how many there are. It's similar to Pareto's Law that states 20% of a group cause 80% of the effects. So, a group of 10 people will have 2 people doing 80% of the work.

Cellular automaton is a programable system where basic rules create induce unpredictable life-like activity. A grid is set with random cells that act according to programmed regulations, telling them what to do based on neighboring cells. Each "random" cell responds to its own circumstances while neighboring cells do the same until the sequential interaction creates emergent motion patterns that resemble living entities. This process is a variation of power laws where a sequence of small, random change surprisingly leads to emergent order.

Another aspect of the invisible hand directing nature is present in feedback loops. In a closed system where neighboring elements affect each other, change becomes self-regulating. We see this in population dynamics such as predator-prey relationships where imbalance between the two creates a counterbalance. Whenever one half of the connected pair grows too fast it becomes a victim of its own success, causing the ratio to flip back. Similar relationships are mathematically expressed in a logistic map where growth creates a negative feedback loop limiting the very progress that caused it. Fractals are formed in this manner with surprisingly ordered patterns suddenly emerging out of what was previously utter chaos.

The law of large numbers produces its own invisible hand when dealing with repetitious events. Flipping a coin a few times can easily result in more heads than tails, appearing to break the 50-50 chance rule but it's just an illusion. Flipping a coin thousands of times will magically weed out short term anomalies and all but guarantee a 50-50 outcome. It's a statistical fact that given large numbers all odds become a certainty. Insurance companies and economists understand this as a function of aggregates. Balancing out the odds over time is also known as regression to the mean. Anyone who's experienced a winning or losing streak knows this all too well; the inevitability that your luck will soon change.

If you have any doubt about power laws creating statistical enigmas, consider this: Given a room of 50 random strangers the odds of 2 people sharing the same birthday are almost 100%. It's completely counter-intuitive considering there are 365 possible birthdays, but there's that invisible force again!

Principles of mathematics form the abstract structural foundation in a universe where everything is interconnected; equations reveal unity, ratios establish relationships, graphics illustrate symmetry and repeating patterns confirm order. Whether it's a thing, living being or process, there are math properties that govern virtually every aspect of it. If you can't see the mathematic principle involved in something, you're just not looking close enough.

Math connects and unifies everything through statements of *equivalence*. Of all the symbols used to describe things, an equal sign "=" is preeminent. Equations are firm proclamations that two things are connected. Formulas symbolically define connected relationships between elements. Equivalence is also expressed in an algorithm; a recipe for processes – connecting inputs to outputs. Of all the equations in math, a special few stand out as fundamental in physical reality:

Newton's law of motion – Connects force and motion ($F = MA$)
Special Relativity – Connects matter and energy ($E = MC^2$)
Pythagorean theorem – Connects sides of a right triangle ($a^2 + b^2 = c^2$) (Also Sine-Cosine angles)
Newtonian Gravity – Connects forces between objects ($F = G\, m1m2/r^2$)
Logarithm – Connects variables using proportional ratios of 10 (LOG xy = LOG x + LOG y)
Calculus – Connects rates of change to a function (df/dt – Lim/h>0 = f(t+h)-f(t) / h
Maxwell/Faraday Equation – Connects electric charges with magnetic fields ($\nabla \times E = aB/at$)
Chaos Theory – Connects process changes through time (logistic map) ($X_{t+1} = kx_t\,(1 - x_t)$)
Schrodinger Equation – Connects a wave function with quantum energy ($i\hbar\, a/at\, \Psi = H\Psi$)
General Relativity – Connects gravity with curved space-time ($R_{uv} – 1/2Rg_{uv} + \lambda g_{uv} = 8\pi G/c^4\, T_{uv}$)
Euler's Identity – Connects fundamental numbers and constants ($e^{i\pi} + 1 = 0$)

While these equations connect things directly, *constants* connect everything on a relational basis; preset values built into the fabric of our universe. They provide stability, consistency and invariance to the operating system of physical reality. Changing any would change the fundamental ways things work, making our universe unrecognizable, just as parsimony regulates the size and shape of living creatures; consider how different life would be if the force of gravity or the size of molecules were even slightly changed. Constants possess specific values in a delicate, interdependent relationship with each other and could not exist otherwise.

Universal Physical Constants

Speed of Light   C – Physical speed limit of the universe (300,000 M/S approx.)
Plank's Constant   h – Fundamental quantum of energy ($6.626\,070\,10^{-34}$ Js)
Gravity Constant   G – Proportional force between objects (6.674×10^{-11})$m^3\, Kg^{-1}\, S^{-2}$)
Elementary Charge   e – Charge carried by an electron ($1.602 \times 10^{-19}C$)
Avogadro Constant   L – Ratio of particles to mass ($6.022\,140\,76 \times 10\,23$mol)
Vacuum Permittivity Eo – Capacitance of free space (8.854 187 8128 (13)x10 -12 F M-1)
Faraday Constant   F – Electric charge carried by 1 Mol (96 485.3329 sA/Mol)

Transcendental Constants: enigmatic infinite mathematic relational properties
PHI – Golden Ratio producing diving proportion (1.61803....)
π – Universal ratio between circle's curve and straight diameter (3.14159....)
e – Euler's Constant present in all exponential growth functions (2.71828....)

Diverse formulas and constants tie everything together in mathematical relationships. Polynomial equations with a variable element are easily solvable since fixed constants and established relationships provide a framework limiting the correct answer. Quadratic equations in the form of $Y = AX^2 + BX + C$ are used to figure out so many everyday applications including calculating areas, speeds of an object or business profits. Calculus is useful when situations involve multiple variables and coefficients. Each approach represents a varying degree of relationship and connectedness between any two things.

Every facet of life is subject to these connections at every level, in every situation. Beyond the purely physical level, math lurks beneath the surface of biological and social realms. D'Arcy Thompson pioneered the connection of math to biology in his groundbreaking "On Growth and Form". Others have discovered numerous applications in social activities including finance, transportation, sports, communications, gardening, cooking and even art.

Mathematics comes into play wherever aggregates are involved. Collective activity takes on a life of its own, becoming a formulaic function subject to computational analysis. This also applies to repetitive activity as biological and social behavior follow cycles whose patterns are defined by math equations. The human body itself is a system of systems governed by orderly regulations and limits, each subject to mathematic laws and relationships. Bilateral and radial symmetry reflect the built-in parsimony and harmonic ratios that maximize use of space and energy. The same math ruling the microcosm of our bodies corresponds directly to order in the macrocosm.

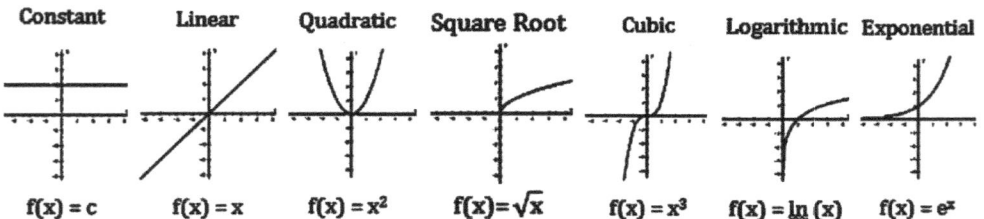

Constant	Linear	Quadratic	Square Root	Cubic	Logarithmic	Exponential
$f(x) = c$	$f(x) = x$	$f(x) = x^2$	$f(x) = \sqrt{x}$	$f(x) = x^3$	$f(x) = \ln(x)$	$f(x) = e^x$

Math functions interconnect relationships present in every thing or process everywhere

The connectedness of things captured in equations is easier to comprehend when displayed graphically. Tables, diagrams, coordinates and graphs provide visual forms of equivalence. Geometry serves this purpose with formulas for every possible polygon, connecting shapes, sides, angles, areas and volumes. Venn diagrams illustrate equivalence via set theory using interconnected circles. Screen images and printed pictures are simply composites of orderly dots arranged in unique mathematic patterns. From the black and white dots used in every newspaper picture to the colored pixels of a TV screen, every possible image can be constructed using a binary formula of 1's and 0's. In essence math is both the form and substance of everything we see.

Statistics incorporate a concept where the degree of connectedness between any two things is measured using a coefficient of correlation. Everything connected in physical reality is linked by degrees – a coefficient between itself and every other possible thing. So, in addition to direct equivalence revealed by formulas with an equal sign there are many other relationships of partial equivalence falling somewhere between complete and minimal, 1 and 0. Connectedness is therefore not a matter of *if* but to *what* degree.

Pushing the boundaries of dimensions can create dilemmas involving the infinity concept. Paradoxes emerge due to differences in properties and language that apply in one dimension but not another. A dramatic example of this is *squaring the circle*. Straight and curved lines may share a common essence but live in different realms. The equations that describe each are significantly dissimilar. It's no coincidence that comparing the straight-line radius of a circle to a curved line circumference creates an irrational number (π). This is the intersection of transcendent dimensions; comparing each is like speaking in a foreign language. Infinity is similarly hard to describe and grasp though hints of it are seen in fractal images of never-ending repeating patterns retaining their elegant symmetry through every perspective. When Galileo noted the universe is written in the language of mathematics, he was speaking of the physical realm. Infinity transcends that. Comparing the finite to the infinite in mathematic terms creates paradoxes resulting in absurdity.

Math definitively codifies the relational essence of our physical universe in a precise language connecting and unifying all things and processes in physical reality. Though elegant in its simplicity, symmetry and minimalism, variations can be complex, abstract and counter-intuitive. Every day, new discoveries are made as fields of science expand, requiring corresponding progressions in math – a sympatico relationship. As new words and phrases are added to our language lexicon annually, so too does mathematics expand and evolve. In a universe of interconnected patterns and ubiquitous symmetry math serves as the perfect cypher, making the abstract definable and the invisible now visible, capturing it all in a single, comprehensive language.

DNA. All biological "living" creatures are connected by DNA. Every organic living thing past and present is interlinked by the same coded molecules present in anything we say is "alive". This includes millions of different species permeating the earth today, plus the hundreds of millions of life forms that preceded us.

DNA is the barcode that contains instructions for building a living being. It's not just a symbolic code but actual genetic material; nucleotide bases formed in long strings where sequence is key. Information is embedded in the combinations and sequences of genetic material which are passed on through replication and translation. With the exception of red blood cells (a specialized oxygen transporter) every other cell in the body contains a complete copy of the DNA code.

DNA is a short alphabet with only 4 letters – A, C, G, T. These nucleotides are formed in 3 letter words called codons, ordered into 64 different combinations (4^3). Curiously there are 20 codons (amino acids) that are used while many of the other possible combinations are redundant versions of the 20. Those 20 codons are set into different sequences that can be thought of as sentences. So, there's 4 letters, 64 words and an unlimited number of sentences whose sequence and combinations comprise the rich information shaping living beings.

DNA is a double helix pattern where 2 parallel strands with rungs between them resemble a ladder. The rungs consist of the 4 letter nucleotides restricted to specific paired placement. A & C, G & T are complimentary pairs and must connect in that matter. It's a chemical template that makes self-replication easy. The ladder splits down the middle and each split pair can be made whole again by filling in a new complimentary opposite letter.

Information transfer is the essence of DNA. Instead of using symbols found in written code the information is embedded in the chemical, genetic material itself. Molecules with their unique qualities serve as raw data and the patterns they form provide meaning. Abstract information is stored and shaped by combinations of material units and copied repeatedly. To facilitate this process something needs to pass the copied information on for translation and execution of the embedded message. That key ingredient is RNA.

RNA is the complimentary cousin of DNA and probably evolved first. It's an information messenger copying the DNA and transporting it to a ribosome. RNA is versatile in that it too can store information while also capable of performing its own functions. In addition to coding, decoding and regulating genes, it a catalyst for building proteins. RNA resembles DNA however it bends itself into various shapes that align with specific functions, much like a template or key. RNA is a like a jack of all trades, essential to the info exchange and message translation.

DNA code is primarily focused on building and directing proteins, the founding molecule of organic life. Proteins are the base material of bone, muscle, blood, nerve and tissue. They are mobile enzymes needed for most of life's physical processes and are produced in every cell. Proteins form long, complex chains and come in 3 types: Fibrous, Globular and Membrane. They're essentially just a series of amino acids formed into unique end products.

Genes are the sentences formed by combining 3 letter codons in the DNA code. They become the message derived from raw data ordered in meaningful patterns. Genes act as instructions for directing growth and function: a recipe for development and a program for organic construction of a whole being. They're the basic unit of heredity containing all the specific traits that make each of us unique. Each one has a specific task to perform in the cell. Most work sequentially and operate like parts in an assembly line.

Genes are asleep or dormant 90% of the time. When active some genes regulate others, including both building and destruction…the intentional teardown of other cells! Think of temporarily useful parts on an assembly line which are discarded after use. Genes are like the foreman directing and regulating the production process of proteins. Sequence is paramount to determining what, where and how functions are performed. The entire coding-transfer process is sequence dependent.

DNA is the code data, RNA is the copier and messenger, and Ribosomes are the translator. The communication process proceeds as follows:

```
DNA – TAC  GGA  GAT  CAT   – Includes Start and Stop Codons
mRNA – AUG  CCU  CUA  GUA   – Replaces T with U (Uracil)
tRNA – TAC  GGA  GAU  CAU   – Anti Codon (corresponds to original DNA lettering)
```
Amino acid Met   Pro   Leu   Val   – Produced acids (*abbreviated*)

So, DNA is copied with complimentary lettered pairs, including the substitution of T with U and transported to the Ribosome for translation. Then proteins are produced by forming long chains of these amino acids. The process varies in lengths, combinations and sequences but each is the result of an individual Gene message. A typical sequence of a Gene message would look as follows:

```
AUG    CTG    ATA     GTT    CTA   UAA
Start  Acid 2 Acid 3 Acid 4 Acid 5 STOP
```

Note that certain amino acid codons also function as regulating syntax that indicate when a sequence starts and stops. Here AUG and UAA signal the beginning and end point of the DNA sequence. Interestingly the STOP codon is not an amino acid, just a syntax functioning molecule. This communication process consists of genetic material serving as both language and action substance. When arranged in a multitude of combinations and sequence patterns they translate to specific messaging, directing and production.

Like binary code DNA is extremely simple and redundant. What makes the 4-letter code so powerful is its extrapolation into myriad combinations and lengthy chains that can be readily copied and transported. With the assistance of RNA, the chemical substances can perform a wide variety of functions that build and regulate organic material. When operations are conducted in orchestrated sequences, the results can lead to complex development. A variety of cells are directed to specialize in a coordinated manner that produces an integrated living organism. It all starts with simple step-by-step instructions that direct shaping, sculpting and further specialization. Cells are directed to divide and replicate on cue. A simple communication process that is repeated millions of times:

```
    DNA         RNA      Ribosome    Protein
Message > Messenger > Cypher > Function
```

DNA is the primary code of biological life; a recipe for making a living being. At the human level it is the basis of what we are. Other codes within us are derivative from it. Additionally, body language and facial gestures are codes as well. So too are moods, talents, mannerisms, body types and personality traits. Like DNA, the Briggs-Myers personality type model also uses 4 elements that combine to describe a wide variety of personalities. These coded expressions are traceable back to our embedded DNA patterns.

DNA code contains the record of our past, of humanity's past and the very origin of life on our planet. It connects us to our ancestors and every generation that preceded us. DNA links us to every other life form that ever existed on earth – from the single celled rudimentary genetic material to the odd collection of various species that occupy every niche. DNA connects us to every animal, plant and virus;

to every land, sea and subterranean organism. It branches out in every direction to connect numerous species further branching off, diverging into variation after variation, similar yet different. It traces an enormous tree of life linking every living organism as a distant member of the same family.

Close up cropped view of broad branching web of organic life connected by DNA

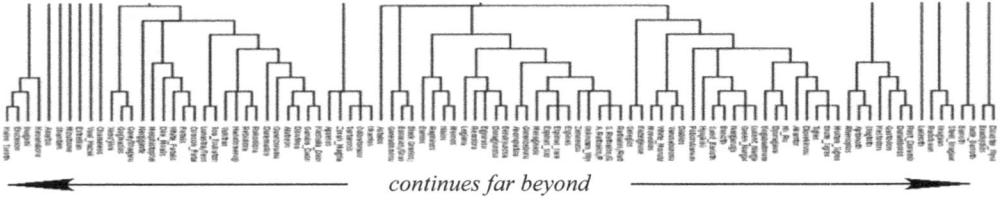

◄───────────────── *continues far beyond* ─────────────────►

DNA code possesses a key attribute built into life: a balanced mix of stability and flexibility. First, it maintains order through a succinct hi-fidelity information transfer leveraging self-replication. This ensures continuity and sustainability. However, to succeed in an ever-changing physical environment beset with ongoing threats, the process must include a built-in mechanism to adapt. That function is made possible through sex and mutation. Each introduces variation that allows for new traits to emerge for testing in the field. Those that are successful become part of the new improved package and are incorporated into the DNA code. The rest are disposed of in a very insensitive trial and error process where unsuccessful traits die off with the host prototype.

Life itself is a mirror of this constant/variable balancing act. Water is the molecule of life with perfect balance between solid substance and fluid adaptable form. It possesses a combined structure and flow quality ideally responsive to life's environmental demands. DNA mirrors this same attribute to ensure adaptability, thus enhancing resilience and survival aptitude for countless different species.

Language. Information is present everywhere, surrounding us at every level in every possible format. It's either rich with meaning, mere background noise or anything in between. Information codified into language at the social level is the key ingredient of civilization – a foundational principle. Information exchange is a catalyst creating synergy where collective outputs are greater than inputs. What we call language is simply formatted information codified into user-friendly syntax.

Language is essential to social functioning, serving as both the glue and lubricant that keeps independent elements working together. At our level those elements are people, individuals or groups, small and large. Language enables integration and cooperation; key requirements for development and progress. It facilitates coordinated effort, expression of intent, sharing experiences and storage of new information. Language connects and unifies.

Mathematics expresses information via numbers and equations. DNA expresses and transfers information via chemical patterns. Language expresses information via sounds and symbols, verbal and written. Types of languages include computer language, text, braille, morse code, phonetic, sign language, and much more. The common thread in ALL codes or language is the symbolic transfer of information.

Verbal spoken language evolved from basic guttural sounds to highly articulate, distinctive pronunciation of complex words. Oral speaking enables instant transmission of thoughts and feelings to others. One can quickly share useful info, suggest something be done, ask a question, confirm a status, express a feeling or just engage in small talk. Where spoken language in everyday situations is useful, a formal speech to a large group can be powerful. Innovative groundbreaking ideas can be instantly conveyed on a broad scale.

Verbiage is only half of the content. Much of the message is masked in non-verbal forms, supplementing and enriching the complete message. Body language can reveal more meaning than the actual words themselves. Kinesics focuses on body movements, posture, facial expressions, hand gestures, blink rate, etc. connected with the verbal message. A lot can be said in just a simple handshake. Tone and volume impact the message as well. Voice quality, pitch, resonance, inflection and rhythm make a huge difference in how the message is received. Within each person's unique voice signature is a trove of coded information offering clues about their personality, intention and general state of being.

Non-verbal communication is influenced by visual cues attached to a messenger. Tattoos, dress, jewelry, make-up and most importantly, eye contact affect the message. Without saying a word, you can express confidence, apprehension, insecurity, anger, disbelief or general nervousness. Physical space and comfort zone vary greatly person to person – another aspect of the dynamic. These collective visual signs either reinforce the verbal message or contradict it. Same with auditory cues in your voice including tone, inflection, and resonance.

Like all codes, the languages formed at the social level have their particular syntax, grammar and structural patterns with rules and regulations. Phonetic aspects come into play in verbal modes while punctuation, spelling and sequence formats apply in written applications. As with chemicals and music tones, written language also has its own lexicon of elements (symbols, letters, etc.) and products (words, sentences). Where math equations express a constant and its function, language sentences express nouns and verbs (subjects and actions). Math refines the value of constants with notations of + or -, squares or square roots, fractions and ratios; language refines subjects using pronouns, adjectives, tense, gender and plurals. Sheet music follows similar format patterns with notes and melodies.

$$\{ \, . \, , \, ; \, : \, ! \, ? \, " \, (\,) - \} + - \, / = X \, \% \, \# > \int \neq \Sigma \, \pi \approx Y^2 \, \flat \, \flat \, \flat \, \flat \, \flat$$

Letter combinations into words resemble the periodic table where identical particles make up everything based on their unique groupings. Mixing either letters or whole words creates meaning largely based on neighboring elements. It's the same principle with DNA since sequence, combination and context are key. Special letters we call vowels act like harmonic nodes similar to prime numbers, noble gases, primary colors and music octaves. Temporal reality is built into the code using word tenses that relate to past, present and future. Punctuation supports this regulatory formatting with symbols that start, pause, delay and stop sentences.

Language collectively connects human life the same way math connects the physical world. Story telling can include fiction, fantasy, mystery, tall tales, myths, fables, and legends. Speech formatted for pure entertainment can include drama, poetry, comedy or the elevated mode of singing – a sustained, continuous analog sound transcending the ordinary digital mode of monotone words. And the boring but practical conveyance of facts can be expressed in a lecture or documentary.

Language further accommodates universal human experiences using a variety of particular functional devices. These include analogy, hyperbole, juxtaposition, idiom, euphemism, oxymoron, simile, metaphor, metonymy, and synecdoche. Just for fun you can use a pun. Or make up an onomatopoeia, which is a word that resembles the sound it represents. Maybe even a paradox, which is a statement that contradicts itself: "deep down inside he is a shallow person", or "this statement is false". These various formats help capture the diversity of human experience and settings, just as math equations incorporate variables to capture change. Other variables present in story telling include the plot, characters, theme and points of view. The counterpart of a mathematic constant would be the story's subject.

Language connects all living creatures in their collective social environments. Beyond humans the rest of the animal kingdom is basically limited to verbal rudimentary sounds. Some species communicate using patterned movement such as the dance of bees while others communicate via chemical pheromones like army ants. Consider how many creatures rely on grunts, howls, hisses, screeches, roars and cackles for expression? Birds sing and convey a variety of messages from affection to fear to hunger. Dogs bark, growl, bear teeth and wag their tales. Dolphins and whales communicate through ultrasonics in water while bats do the same in the air. Early man certainly expressed himself in similar primitive fashion. Human words are essentially made from 5 basic sounds – the vowels A, E, I. O, U. All other consonant letters direct and shape those sounds in various combinations (F, sh, G, th, br, etc.).

Onomatopoeia – Word resembles its sound: sizzle, clap, buzz, thud, pop, squeak, meow

Evolution of language progressed as neighboring tribes comingled and merged into larger organizations. Over time small groups expanded into clans, towns and cities, so language became unified and improved. Progress accelerated as humans possessed the right combination of vocal cords, larynx (voice box), tongue and lips and developed greater mental processing capacity. While gains were made locally languages diverged globally as groups migrated to faraway places, resulting in great distinctions and variations. Differences in climate, culture and demographics led to the development of disparate languages and dialects. Linguistics studies these variations, analyzing phonetics (sounds), syntax (rules structure) and semantics (meanings). Different dialects are treated like unique species. Memes are to culture what genes are to organic life.

Today there are roughly 7100 living languages. We can trace our language lineage to 6 major branches: Indo European, Sino-Tibetan, Afro-Asiatic, Niger-Congo, Austronesian, and Trans-New Guinea, mostly corresponding to continents. The 21st Century's most popular "native" languages are Mandarin, Spanish, English and Hindi, primarily based on population size. The most popular "total" languages spoken are English, Mandarin (close 2nd), Hindi, Spanish (again close 4th). English still tops the list due its popularity as a second language, and both English and Spanish remain high as a legacy of colonialism and conquistadors.

Language, like DNA, contains artifacts connected to ancient origins. Root words and meanings get passed onto successive generations like traits in genes. Western civilization is connected with various languages sharing common words and phrases based on Latin. Medical, legal and mathematics are replete with Latin words. Some take on entirely new meanings once they're incorporated into new cultures and settings. Parts of languages get appropriated by other larger ones while many more just die and become extinct.

Culture and language are interrelated and codependent. Values of a society are directly reflected in its language as the coded expression of collective being – the zeitgeist of the whole. Language with its imbedded phrases, sayings, cliches, titles and nicknames are inseparable from the social setting influences it. Conversely, language can also separate individuals when it's esoteric and incomprehensible to those who are not fluent. Words become mere jargon to the illiterate. Words and phrases used in medicine, law, science, math, military, metaphysics, auto mechanics, etc. are essentially foreign languages. *Conlang* are fabricated languages created for fictional worlds (normally science fiction) using established logical syntax but laced with a made-up lexicon of logical yet incomprehensible babble.

Good language connects. Clear, understandable words with brevity and directness achieve high level communication. Contrary to "experts" in specialized fields, the use of "big" words and excessive jargon is actually low-level communication. Getting into the weeds will lose the listener. Simple is better. Less is more. When it comes to the written or spoken word, high level communication occurs when the message is clearly understood immediately. Swear words are actually high-level language. The highest level of communication involves images.

Any language not understood separates us. The same goes for communication in general where messages are misinterpreted or simply not understood. Each of the various states of being carries with it a particular parochial language. Anger, love, depression, joy, sadness and fear all bear corresponding expressions that require a certain degree of empathy to relate to. Their expression is not necessarily understood without appropriate listening skills. Love resembles an esoteric language with its own lexicon and associated mannerisms that the uninitiated soul may not fully grasp. "Love language" includes affirmations, acts of service, gift giving, physical touch, etc. Other subtle esoteric languages reside along the seams of culture in less than obvious ways. Everything in society is interconnected to a tapestry of language codes at every level. One's success depends greatly on possessing multiple literacies beyond the established main language. It also helps to be skilled in the lost art of focused listening.

A universal cause of strife in relationships is a disconnect in communication. Differences in values, perceptions, education, age, moods and a multitude of biases can create barriers despite having decent language competence. Differences in perspective can get in the way as well where two people who may share a common goal but still disagree because the language they use comes from different vantage points. The key to good communication is speaking the same language and providing immediate feedback. Without the later part of the process, messages are bound to be misinterpreted. Good feedback is anything that reaffirms the sender is understood. It can be as simple as a nodding head or smile. A popular example of a barrier to communication is the meme "Men are from Mars and Women are from Venus". Obviously, men and women share the established language but often differ on the subtle esoteric language involved in courtship. Bernard Shaw referenced a similar disconnect with his famous quote "The United States and Great Britain are two countries separated by a common language."

Modern civilization's progress accelerated by leveraging verbal and written communication modes that enabled the expression of abstract thoughts into symbolic form, enabling duplication and sharing on a broad scale. The printing press was a major catalyst for mass proliferation of new information, further enhanced by the telegraph, radio, television and a world-wide internet. Coded information facilitated storage and mass dissemination, leading to a unified, integrated societal mind capable of unlimited progression. Each successive generation passes forward greater amounts of information and knowledge, building upon its predecessor with technology leap frogging upon itself. Coded information is the catalyst and glue linking it all together. Society, culture and language working in unison, inseparable and essential to the rapid evolution of mankind.

Life Energy Codes. Some people naturally brim with high energy; every moment driven by action with a seemingly bottomless well of biological fuel. This dynamic spirit can take the form of physical, mental or emotional energy or a combination of all three. Where does this force come from? Why does it vary so much person to person? The answer is likely related to what makes personality patterns so different.

Energy states in people are manifested in body language, behavior patterns and the residue of their activity. These states of being we all possess fluctuate through time. Our daily rhythms follow an orderly cycle with peaks and valleys that are surprisingly predictable. Body temperature and brain waves mirror each other as energy levels shift. Breakfast, lunch and dinner are part of the equation along with rest periods and outside influences.

Longer periods of time mirror similar cyclic patterns as well. Our energy level goes through peaks and valleys over weeks, months and seasons. Physical, mental and emotional aspects can rise or fall for extended periods of time. Professional athletes routinely alternate between slumps and success streaks. Writers get blocked and then out of nowhere get on a prolific roll. Some people can get depressed for weeks, months, even years and later snap out of it to move forward and achieve great things. States of being resemble a vibrational energy essence distinguishing a living being's disposition like a fingerprint. Each state resonates with particular situations and settings in a codified manner.

Living energy patterns within inevitably manifest outwardly, expressed in the things we possess, desire and act upon. It comes through in our moods, tastes, attitudes, mannerisms, facial expressions, level of effort and degree of passion. It also surfaces in our laughter, speech, play and dance. Our unique energy signature communicates with the outside world in so many subtle ways that it's definable as a linked pattern. Every relationship we share reflects our energy pattern at the time, but transitions as states change. Relationships themselves take on a unique energy pattern intertwined with the states of each involved. Likewise, all groups possess a collective energy pattern that attracts members who resonate with it.

Individual and group consciousness are identifiable by the people, things and actions connected with them. Who you are is defined by where you spend your time, energy and attention. People naturally gravitate toward certain activities influenced by their pattern of consciousness. Each of us occupies a position along a scale of conscious being, expressed in the things we do, the people we associate with and the attachments we cling to. What televisions shows do you watch? What kind of friends do you have? All these things consistently form a pattern that identifies your vibrational state of being. Each of us is associated with clusters of others who share similar levels of consciousness. Society as a whole is a combination of diverse strata of people with varying states of being. The masses tend to occupy the middle range while smaller segments occupy levels above and below it (bell curve again). People generally don't associate with those in other groups unless they themselves are in a transitioning mode.

Environments attract certain people depending on their current state of being. Towns, cities, states and countries have an associated resonance about them that fits for some people but not for others. Same goes for your own home. Sometimes changing your residence results in a significant change in perspective. Athletes often elevate their output by simply switching teams. An environment itself can change as well by introducing different mixes of people who then change the overall energy pattern and reset the resonate frequency of it. Nothing happens by coincidence. Everything affects everything else through interconnected relationships, each having an associated pattern of living energy.

Weather affects our physical and spiritual energy level; damp overcast days lower our spirits, sunny days energize us. Changes in weather correspond to changes in our moods. Daily, weekly and seasonal changes all contribute to our general state of being. Various climates create particular environments that are good for some but not for others. Again, every situation has an associated vibration resonating with certain types of different people.

Our time in history has its own unique energy pattern or zeitgeist, as does each passing generation with unique collective experiences distinguishing it from other periods. During America's 20th century certain decades were given nicknames, like the roaring 20's, the flying 40's, the fabulous 50's, psychedelic 60's, the disco 70's, the awesome 80's and others. Every period in history has its own ambience, embodying a unique collective state of being. And there are always certain individuals who thrive in those periods as they resonate with the period's zeitgeist.

Living energy expressions take on many forms, but it's the vibrational pattern that connects and unifies them all. Moods, weather, tastes, attitudes, body language and spiritual energy all share a symmetry that can be sensed and felt yet difficult to quantify. Every emotion has an associated color, facial expression and attitude. Music styles, melodies and pitch likewise associate with particular states of being. Hard rock and easy listening clearly fit specific states. Weather conditions mirror the situational temperament of man, from calm conditions to stormy outbursts and all the dispositions in between. To the spiritually illiterate the connection between all these states of being may be hard to see but the code isn't hard to decipher. Whether it's sensed intuitively or experienced directly like in drug induced states, coded living energy patterns are corelated expressions of resonating vibration.

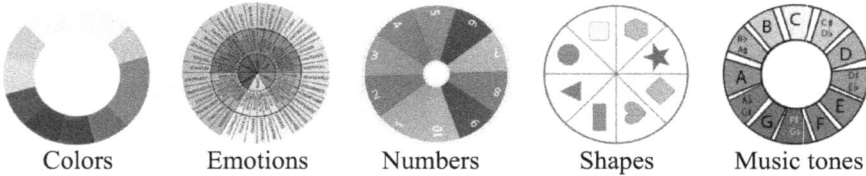

Colors   Emotions   Numbers   Shapes   Music tones

Pattern Codes. The simplest, most universal code of our entire reality is pattern language. Everything is connected to everything else through a variety of different patterns, including spatial, temporal, numeric, process, resonance and more. The meaning of anything is entirely context based, linked by relationships of form, position, scale and orientation. Patterns are a language describing anything and everything. YOU are a compilation of patterns; physically, emotionally, mentally, spiritually, behaviorally. Your identity is derived from patterns you're born with plus others you choose to embrace. What makes you and anything else unique are the specific patterns chosen out of infinite possibilities. The stuff we're made of is very limited; however the combinations they form are virtually unlimited, and the only difference between you and me are the patterns of those limited parts.

Patterns are ultimately about relationships or how things compare to other things. Form and order are relational concepts, along with arrangement, association, structure, correlation and connection. These concepts tie everything together – an essential ingredient in any TOE. And we must not limit this to a static perspective because change is omnipresent, affecting all things all the time. Relationships connect everything in space and time, so pattern code applies to processes, cycles, vibrations, and dynamics in general. Process is pattern in time, expressed in music, organic growth, and dynamics of behavior.

Within the infinite variety of patterns are recurring themes of sub-patterns that emerge everywhere we look. A prime example is polarity and duality. Everything has an equal opposing version of itself, a polar opposite, expressing the Yin/Yang essence built into our universe; equal and opposite, complimentary, each containing the seeds of the other. It's a self-similarity quality where 1 whole expresses 2 sides. It produces opposition, competition, tension, and conflict; creative or destructive. Polarity manifests digitally; either/or, Yes/No, without middle ground. Analog is a scale or spectrum with partial values in between opposite extremes.

Analog Polarity Scale   *Digital Polarity*

●— YES — mostly yes — maybe — mostly no — NO —●   **Yes**   **No**

DNA coding activates sequences using yes/no, on/off chemical switches, similar to computer logic

Duality is a higher octave of polarity. It's a single whole with two transcendent aspects, where the original part retains its essence but changes in kind (body and soul). Black and white are polar opposites and similar in kind yet both are transcended by colors, producing a dual relationship. Duality is also an imbedded feature of our universe where every physical thing exists in a dual framework; matter transcended by spirit. Duality is a subset of transcendence, the first being 1 whole with dual aspects and the latter being 1 whole transcended by a greater whole, above and beyond it. A child grows up to be a fully developed adult, two stages of one person – the later version transcends and includes the former version. Polarity, duality and transcendence are omnipresent recurring patterns in everything, easy to spot when you know what to look for.

CODES OF FEW ELEMENTS BUT MANY COMBINATIONS

BINARY

Morse Code	– 2 Elements (dots and dashes)	—● ●—● ● ●●● —●● ●— ●●—
Computer Code	– 2 Elements (1's and 0's)	10011000111101011011100001010101101
I-Ching Code	– 2 Elements (single/double dash)	—— —— —— —— —— —— —— ——

QUAD CODE

DNA – 4 Elements (ATGC)   ATTTCTACTAGAATACCGTATTCGAT

Briggs/Myers Personality Code (IE TF SN JP)   IFSJ ETSP ITNJ EFNP IFNJ ETNJ EFSP

7 Node Codes: Music, Colors, Chakras, Periodic Table (Atoms - 3 particle code in 7 interval scale octaves)
10 Element number code – produces infinite combination of numbers   0 1 2 3 4 5 6 7 8 9
26 Element Alphabet mixed combine to form a million words: abcdefghijklmnopqrstuvwxyz
    DNA Codon patterns = 64   I-Ching Hexagram patterns = 64   Chess board squares = 64

Language	Letter	Word	Sentence	Chapter	Book
DNA	Nucleotide	Codon	Gene	Chromosome	Genome
Music	Note	Chord	Melody	Song	Album
Matter	Particle	Atom	Molecule	Compound	Mixture

***Vibration* Code** is a fundamental pattern built into the universe. It's a temporal essence expressed in cycles and waves or anything repeating with regularity. These Patterns in time range from the micro to the macro, long and short, fast and slow.

Everything vibrates, beginning at atomic levels and continuing through living beings and societal organizations. Rhythm and tempo are present in every activity and event. Rhythmic vibration is symmetrical in music, light, plant growth, tides, breathing, heartbeats and biorhythms. Life is a process with tempo, pace, beats, and timing. Verbal language is simply patterns of vibration. Music instruments only produce vibrations, with strings, winds or percussion, just variations of the ancients' Lyre, Horn and Drum.

Vibrations are simply patterns in time with varying degrees of temporal symmetry. Expressed as cycles of periodic repetition, they're both linear and non-linear, differing only by degree of regularity. Rock music maintains regularity using a 2nd and 4th beat via a drummer while jazz is less regular with occasional riffs changing tempo but always settling back to the original. Harmonic vibration is symmetrical to equilibrium. Harmonic vibration in water produces geometric patterns of symmetrical mandalas. Synchronicity and resonance emerge when vibrations unify and harmonize. You may experience this personally as Jungian synchronicity in those moments where inner self and higher self-resonate as one.

Resonance is a unifying vibrational pattern producing emergent forms. It's a powerful creative force that concentrates energy. Resonant vibrations become temporary structures which in essence make something out of nothing. Repetition of anything has an inertial effect leading to cosmic habits that transform subtle possibilities into stable actualities. It's a force multiplier as single vibrations align and amplify. Material objects feel solid because they're composed of millions of vibrating atoms held together as a single resonating energy structure.

Relationships are the root of patterns, including aspects of arrangement, proximity, connectedness, context, meaning, form, composition, structure, association, design, ratio and order. Nothing can be anything without context or a relationship to another thing. We measure relationships using space and time. These dimensions are quantifiable and reduceable to mathematical numbers and ratios. Any whole can break down into subcategories, each with its own numeric values (size, weight, shape, attributes, etc.). No matter how complex, the sum total of parts can be expressed as a series of numbers, in much the same way bar codes describe virtually every consumer product made. Since all relationships can be quantified numerically, they can therefore be portrayed graphically. Algebraic equations translate to visual forms captured on a graph, both 2D and 3D. While unity transforms into polarity, duality, and transcendence, all three are expressions of relationship and pattern. *Relationship* is the highest-level essence of reality.

SYMMETRY PATTERNS

Everywhere we look symmetry reflects the hidden unity within nature and our physical environment. It reveals self-similarity, likeness and connectedness. It's also a unifying quality with attributes of proportion, harmony and beauty. The concept of symmetry is derived from geometry where changing the disposition of a form doesn't change its appearance. Different types include translational, rotational, reflection and glide, which are just various ways to change the orientation of a thing without affecting shape or form, either linear, planar or spatial (1D, 2D or 3D). Translational symmetry is a shift of an object in a consistent, repeating distance or interval. Humans and animals possess both radial and bilateral symmetry, along with some exceptions of asymmetry (left-right handedness, left-right facial balance).

Geometry is the most obvious expression of symmetry, limited to spatial orientation, but there are other less obvious forms: temporal, sound, color, etc. Beyond basic shape, other qualities of symmetry include sameness in composition, arrangement, regularity, uniformity, consistency, proportionality and harmony. Sameness implies connectedness, self-similarity, invariance and unity. It's found in repeating patterns of identical parts and harmonious proportions, which is not about shape but rather, relationship.

Buckminster Fuller connected shapes and form with geometric energy, linking atomic and molecular patterns to man-made structures. His geodesic dome mirrors organic cells and organelles, providing maximum strength with minimum material.

Symmetry can be temporal or sequential, where repeated moments or events produce a cyclic pattern of sameness, like the recurring tonic key in a melody. Time exhibits both symmetry and asymmetry since some things can be done in reverse sequence while others are constrained by the forward one-way arrow of time, like breaking an egg. All things are connected by some degree of symmetry, with rare cases of asymmetry, which is simply lower by degree. Polarity is actually a partial symmetry since 2 elements are equal and opposite, both same in kind but with exactly reversed expression. Black and white are more similar than any shade of grey in between. Love and hate are more similar than apathy or lack of interest. Each are two sides of the same coin; connected equal opposites.

Symmetry is revealed in languages, especially math, but others include physical laws, metaphysical principles, metaphors and analogies. Laws are simply universal truths serving as axioms that reveal order, regularity and congruence – all products of a symmetry bound universe. As above so below, opposites attract, no pain no gain, karma, etc. Analogies are an abstract language connecting seemingly disparate phenomena. Metaphors achieve the same with a purposeful degree of ambiguity. The language of poetry is rich with symmetry, combining harmonic, melodic timing plus colorful metaphors and rhyming words. Poet John Keats saw symmetry in the connection between beauty and truth. Analogies and metaphors serve as connecting unions capturing similarity on an abstract level. Languages of symmetry range from the directly equal ($E = mc^2$) to the nebulous associations expressed in the divine language of poetry, each connecting and unifying in their own particular way.

Symmetry is easy to see in art where the whole includes elements in mirror image arrangement, such as portrayed in Davinci's "The Last Supper". Symmetry in design is expressed as repetition of features, shapes, and ornaments, applied in architecture, furniture, and decorative art. The ancients embraced symmetry as homage to the divine, obvious in Egyptian pyramids, Greek temples, and Roman architecture. It was a constant theme continued through succeeding architectural styles, such as the variations in Greek columns: Doric, Ionic, and Corinthian. Virtually every past culture incorporated symmetry in its art, architecture and ornaments, including Arabic design patterns, Chinese lattice works, Islamic tiles and even modern Penrose tile patterns that radiate without repeating.

Symmetry is literally built into the universe in every nook and cranny. It begins at the atomic level with particle symmetry (particle/anti-particle), molecule lattices, chemical reactions, electricity and magnetic fields, and extends through the macro level with stars, planets, systems and galaxies. Crystals are structures of repeating patterns in a 3-D lattice. They have 32 possible shapes, all symmetrical. Topology is the ambiguous symmetry possessed by an object after a major distortion of its original shape. For instance, a coffee cup and donut are both objects with a central hole, so they share a self-similar quality but with a diminished degree of unity.

Symmetry can be found everywhere, even within a chaotic system where hidden forces eventually produce emergent order through the presence of strange attractors. Complexity occurs when sensitivity to small initial changes produces major upheaval later on (butterfly effect). This chaotic relationship doesn't eliminate symmetry, it just means there's a lower degree of it. Ironically, a totally chaotic system possesses a significant symmetry since it looks the same from every point. Contrast that with its opposite, a sphere – a perfect form with unlimited symmetry and maximum efficiency.

Symmetry is a universal pattern, ubiquitous in math laws and principles, universal constants and physical laws of nature. It's more than just similar shapes in space; it's an omnipresent essence of sameness imbedded into the universe, in a wide variety of expressions. It's reflected in unifying laws, truth, beauty, self-similarity, proportionality, repetitious patterns and harmonic vibrations; clues to a unified interconnected higher reality.

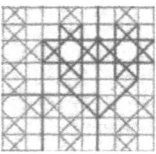 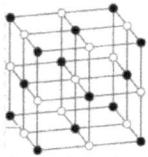   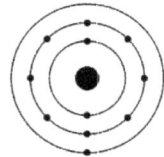

Islamic Tiles   Crystal lattice   3D lattice   Triskelion   Sodium Atom

Nature Symmetry is apparent everywhere we look, rooted in the environment, organisms and creatures of all types. It begins with the material world in elements, minerals, crystals, compounds and manifests into the entire organic realm with plants, animals, fungus, bacteria, etc. The entire animal kingdom is teeming with symmetry in mammals, fish, birds, insects, reptiles, and in every kingdom, phylum, class, order, family, genus and species. Symmetry is ubiquitous in all life.

The universe operates in the most economical manner possible, always doing the most with the least, particularly in material and energy. This translates into the law of parsimony, where counterbalanced forces regulate the shape and sizes of natural objects, especially living organisms. Organic forms are etched, chiseled and shaped by internal and external forces interacting to produce symmetrical products born out of an equilibrium balance. This interplay between competing forces is how nature designs all living forms. From the geometric shapes of viruses to the elliptical frame of fish, nature always maximizes space and minimizes material.

Environmental forces shape organic forms in the most efficient manner possible

Parsimony explains why vertical moving objects are cylindrical, why objects moving through water are curvy shaped and why large bodies like planets and stars subject to great forces of gravity are spherically shaped. It obeys Curie's law, which states bodies tend to take shape using minimum surface material and energy. Nature operates as a vast integrated shaping filter of organic forms, guided by invisible forces made visible through geometric patterns in living creatures. Most buildings are squared in response to gravity, compression and tension forces. Fish and airborne projectiles are elongated, curved cylinders. The largest objects in the universe are always spherical. None of it is arbitrary. Hexagonal symmetry is a perfect balance between the material efficiency of a circle and the structural integrity of a rigid boundary, used repeatedly in nature: snowflake, honeycomb, turtle shell, crystal, carbon molecule and much more. There's a physical bias towards spherical shapes at every level: a sphere is *the* optimal form, providing maximum surface area and internal volume while using the least material.

| Dendrite | Soil cracks | Leaf | Tributary | Roots |

Dissipative structures — branching as nature transfers energy along the most efficient path

Growth is an essential aspect of organic life, manifesting as directed energy constrained by physical laws and mathematic properties. Nature uses two approaches to achieve growth: *Organic* via cell division and replication (mitosis), and *Inorganic* via accretion that adds identical parts to the existing structure (crystal lattice). Organic forms predominantly follow a pentagon shape (star fish) while inorganic forms default to a hexagon shape (snowflake). Life processes are captured by differential equations describing patterns of growth, development and decay, which proceed in intervals of either steady increments or geometric change.

Proportional progression exhibits self-similarity based on the golden ratio (PHI). It follows the Fibonacci sequence resembling a logarithmic spiral – the universal algorithm for organic growth.

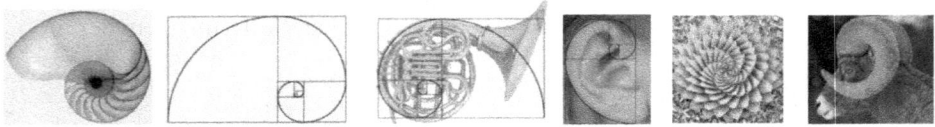

Platonic solids are the most common form within organic life molecules.

As anything grows it changes in several dimensions simultaneously. But outside forces act upon dimensions differently, often in a non-linear manner. Small changes in surface area can cause a much large increase in volume, (arithmetic growth vs geometric growth), so size is universally constrained by the difference between a being's surface area and volume – physical constraints based on simple math relationships. Nature is resource efficient, always choosing the optimal balance between shape, form and size, resulting in the least energy and materials.

Growth frequently follows a spiral pattern expressed as radial energy in motion; the geometry of self-similarity. They can extend vertically with a fixed radius (DNA, rope, spring) or laterally with an expanding proportional difference (shell, flower, galaxy). Either way it retains shape and maintains a fixed center. Radially expanding spirals are a form of acceleration in nature (organic growth, molecules, horns, inner and outer ear, weather patterns – tornado, hurricane, dust devil, etc.). Vertebrate embryos grow out as expanding spiral patterns. It's a universal process pattern enabling continuous growth without changing shape or center of gravity.

Flowers patterns are radial, limited by specific mathematical increments. Petals are often spaced in spiral patterns following the Fibonacci sequence in sections of 3,5,8,13,21,34, etc. nature's design to ensure adequate space between each petal to maximize surfaces facing sunlight. Plants and all other living forms are less *things* and more *processes*; more verb than noun, which becomes apparent when observed over longer timeframes. They're really just pattern forms of moving energy.

Some things exhibit asymmetry, due to advantages of functionality. Crystals and molecules can possess chirality, which is a handedness or one way directionality. All amino acids in living organisms are left-handed. Nucleic acids are always right-handed. Other compounds are either left or right-handed. Our brains have a handedness given one side is logical, the other creative. But asymmetry is the exception in nature, not the rule. Omnipresent forces in balance are the overwhelming default, revealed by a wide variety of living organisms shaped and sculpted to geometric perfection. Pervasive patterns of symmetry in and throughout nature provide beautiful visual clues to the hidden connections unifying all life, and express purposeful cosmic dynamics driven by efficiency and resourcefulness.

"The great book of nature can only be read by those who know the language in which it was written…mathematics" – *Galileo*

Human Symmetry is simply an extension of nature, and as the saying goes, "God made man in his own image". YOU, then are a living example of symmetry, containing the same built-in attributes of mathematic principles and harmonic ratios present everywhere else in nature. Bilateral symmetry in a human body is obvious, with mirror duplication of every main part on both the outside and inside (skeleton, ribs, limbs, eyes, ears, etc.). Man is living geometry – exhibiting radial symmetry (hand – fingers, joints), bilateral symmetry and a bone structure fitting into 3 golden rectangles. Leonardo Davinci first captured the divine patterns engineered within our bodies, applying arithmetic, geometric and harmonic proportion. Davinci's Vitruvian Man transposes the heavenly circle and earthly square connecting the human body as a whole. Each relate to nature's elegant proportions of 5/8th and 3/4th, exactly mirroring music's diapente 5th and diatessaron 4th. Like most organic forms, we can find plenty of nature's mathematic ratios imbedded within, including the Fibonacci sequence, Golden ratio, spiral patterns (DNA), etc.

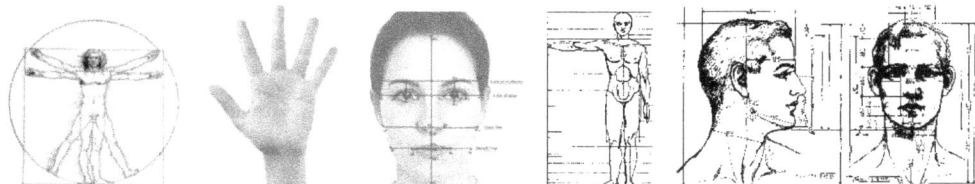

Symmetry in Life, symmetry in Man, symmetry in YOU

Symmetry is an integral part of Man's being. It's the difference between who's considered attractive (facial symmetry) and who's a healthy, athletic specimen (body proportions). On a higher level, peace and harmony are cherished states of being. Sacred books and their Gurus preach the wisdom of embracing the golden mean, seeking balance in everything. Enlightened man becomes one with nature, with feet on the ground and head in the sky, balancing earthly needs on the physical plane with spiritual needs in the heavens above. Polarity and duality are plentiful as well. Muscular man and curvy wo-man mirror the symmetrical duality of human beings in the exact intersection between matter and spirit. Athletic humans demonstrate poetry in motion pushing biomechanical engineering to the limits. Some of the finest art is not on a canvas but displayed in arenas where athletes dive, jump, swim, run, slide, dance, weightlift, swing a racquet, or catch a fast-moving ball, with grace, precision and breathtaking elegance.

Repetition of proportional ratios: neck to hip + knee to ankle = hip to knee *squared.*

Mathematic Symmetry is a universal theme of numerical self-similarity, expressed wherever you see a square root, inverse, fraction or something squared. The square of a value (X^2) is simply something times itself; not two of itself as in X x 2 but times its own self. Nowhere is this more significant than in the equation $E = MC^2$ where the incredible speed of light is multiplied by itself creating an enormous number. An interesting pattern with squares is their common association with acceleration. Square roots are just the opposites of squares, both connecting a thing with itself. An inverse is an upside-down version of self-similarity, connecting complimentary flips. When one value increases the other decreases in exact proportion. Fractions, decimals or angles are actually forms of ratios between a thing and itself. They could be thought of as degrees of wholeness in a thing.

Self-similarity in ratios and relationships allows us to leverage logarithmic powers where exponents capture an enormous range of numbers with just a few representative symbols. Large values like a Googol can be expressed as 10 to the 100th instead of 1 followed by 100 zeros. Or consider a Googolplex, expressed as a Googol to the Googol power – 1 followed by a Googol of zeros… way too many to show here. Long decimal numbers can also be captured in the same practical manner. What makes logarithms so practical is the orderly ratios between all things and between a thing and itself. 1 to 10 is an identical ratio as 10 to 100 or 100 to 1000, just higher octaves; symmetrically equal ratios with 10^2, 10^3, 10^4, etc.

Logarithms operate with exponents and inverses to express self-similarity as multiples or divisions of the same number. For example [$b^X = n$] just means Log X as repetitions of B. So if $2^3 = 8$ then LOG 3 is simply 2X2X2, self-similarity. Transitioning from addition to multiplication is a transcendent shift, an acceleration of addition. Phi is where addition and multiplication intersect in a single process, producing the golden ratio found everywhere in nature.

Self-similarity symbols in math: X^2 $\dfrac{1}{x}$ $\sqrt{x}$ $\log X$ Φ

Every number regardless of size is just a derivative of some value between 0 and 1. This is why an old-fashioned slide rule can perform complex calculations with very large numbers. It makes sense when you see that a number like 1 billion is simply 1×10^9. In a similar way, every circle regardless of size is identical with every other circle, differing only in ratio. Relationships of self-similarity are everywhere in mathematics, including squares (X^2), cubes (X^3), Square roots, cube roots, inverses, integrals, derivatives, and much more.

The orderly ratios we recognized in nature illustrate math's connection to parsimony and the physical constraints shaping organic forms, where survival of a living creature is optimized by achieving the right balance among competing forces. Self-similarity attributes of the Fibonacci sequence emerge with geometric growth based entirely on adding increments of the previous size of a thing. Expansion is not simply additive but proportionally additive. (1,2,3,5,8,13,21 etc.). The same principle is present in the golden ratio that produces such an aesthetically pleasing shape. The formula is simple: take any two-line segments where the ratio of both together is identical to the ratio of the longer to the shorter. Self-similarity ratios look and sound better, probably because they imply unity. This pattern is present in plants and spirals. The Fibonacci ratio is pure mathematical symmetry, producing growth through the simple addition of identical parts in geometric progression. It's built into a logarithmic spiral directing curving patterns that vary by dimension while retaining its exact original shape.

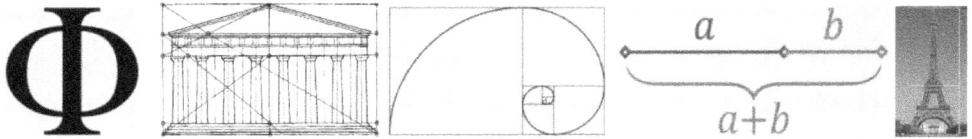

Self-similar ratios are the secret sauce in music where only particular intervals sound good, especially the octave (2:1), the fifth (3:2) and the fourth (4:3). Similar harmonic ratios emerge at the cosmic level where planetary orbits follow Bode's law – a harmonic scale resembling a music octave. Kepler's laws also capture the symmetric, orderly connected relationships between planet orbits, size and speed. Many of the equations of physics involving squares and square roots apply to acceleration, transformation, expansion and contraction, etc. They all seem complex but are really dealing with symmetrical self-similar change.

Symmetry in our physical reality starts with the foundational structure of dimensions. Deconstructing a symmetrical 3D cube results in the following dimensions: 0 – Point, 1 – Line, 2 – Plane, 3 – Solid. These dimensions associate with Position, Distance, Area, and Volume. They're all separate yet connected by a 90-degree intersection. Euclidean geometry defines these relationships in a flat-space manner. Cartesian coordinates graphically represent shapes and objects using a 3-axis grid, each 90 degrees apart. After curved space was considered, a 4th dimension was necessary to accommodate the time axis. Minkowski introduced 4D space-time coordinates as (x,y,z,t). Time is transcendent to the others therefore its 90-degree difference is problematic to visualize. Later, Riemann created mathematics to address a 4D space-time construct sympatico with Einstein's theory of relativity. Symmetry becomes more abstract, and a little funky once non-linear math and transcendent dimensions are introduced.

Mathematic relationships connect things, either directly with an equal sign (=) or indirectly with a congruence sign (3 lines). Symmetry is a little bit of both, differing only by degree, such as the difference between constants, invariants, squares, roots, inverses and fractions or decimals, which are self-similar parts related to a single whole. Ratio connects 2 elements while proportion connects 2 ratios. All three versions of Golden ratio, section and rectangle, plus the logarithmic spiral are equivalent (Sq Rt of 5 + 1 / 2 = 1.618….). The Fibonacci ratio expressed as either $2/5^{th}$ of 360 degrees or Sq root of 5 - I /2 equals the fractal divergence of 2/5, 3/8, 5/13, 8/21, etc. Euler's identity links 5 key constants in mathematics in one elegant formula: $e^{i\pi} + 1 = 0$. It's a statement of pure symmetry; all exponential functions are proportional to their own derivative, and that proportionality is 1.

Mathematic Medians and Progressions

Arithmetic $B = A + C / 2$,   Arithmetic progression = 2,3,4,5
Harmonic $B = 2AC / A + C$]   Harmonic progression = ½, 1/3, ¼, 1/5 1
Geometric $B^2 = AC$

Math is the science of connected patterns, expressed in symbolic language with numbers, equations, graphs, matrices, ratios, angles, etc. Most equations include an = sign which implies sameness. This is pure symmetry since it's not just similar but *equal*, one and the same…a unity. A matrix is a visual depiction of connected parts, a graphic version of an equation with its own equivalence. Geometry is the ultimate visual form of symmetry, portraying connected relationships in graphic images. Self-similarity is the very foundation of mathematics, conveying the inescapable omnipresence of unity in our physical reality.

Music Symmetry manifests as mathematic ratios, proportional relationships, and patterns of frequency, pitch, beat and amplitude. Harmony is based on whole intervals similar to quantum mechanics where intervals only exist in whole numbers. Though harmony is somewhat subjective, it's universally optimized in specific math ratios: 1, 4/3rd, 3/2nd, and 2, musically known as a 4th, 5th and octave, which sound the most pleasing to hear. An octave is the foundational pattern; recurrence of the same tone, which is self-similar yet different. Like a cycle beginning and ending with return to the starting point, very much like an upward spiral pattern returning to the beginning but a little higher than before.

Music with too few beats is dull, too many beats is chaotic. Harmony is achieved through balance among different elements – avoiding extremes between monotone sound and discordant noise. Music symmetrically corresponds with life experiences, with cycles of ups and downs expressed as rhythmic patterns of tension build up and release. Every event in life has its own tempo, volume and pitch, whether it's a celebration or catastrophe. Rhythm is a combination of sameness and newness, no different from life's routine activities punctuated by novelty and nuance. Music, melodies and songs exhibit a mini-life cycle with rising crescendos and falling decrescendos, always returning to the original baseline, reestablishing wholeness. Rhythm is harmony in time, paced by beats per second, with a full note divided into four quarter notes. Beats, measures, notes, rests, tempo, etc. are all relational ratios in time.

Piano keys are a perfect representation of the music scale, with a 13-note range of alternating white and black keys (8 whole notes and 5 semitones), grouped in 3's and 2's, corresponding to the Fibonacci series. Symmetry is obvious with alternating patterns of white and black keys, ordered in recurring sets of 7 that harmonize in progressive octaves across. Semitone sharps and flats balance out neighboring keys in mathematic ratios. Notes and chords combine in complimentary sets. The entire scale and groups within correspond arithmetically and geometrically in patterns that can be expressed graphically, in static form (graph) or motion (sinewave). Music symmetry is both spatial (Chords of notes played simultaneously) and temporal (melodic notes played sequentially). Songs are symmetric with poetry; verbalized melodies unified in rhymes. Written language connects minds; music language connects hearts.

Shapes of instruments mirror mathematic ratios and geometric forms (harp, organ pipes, French horn, piano, violin, trombone, etc.). Most rounded instruments correspond to exponential curves shaped by math equations with square functions ($y=x^2$). Algebraic and quadratic equations depicted on graphs are visual templates matching the shape of musical instruments. Leibnitz intuitively sensed the divine connection stating "Music is sounding mathematics"

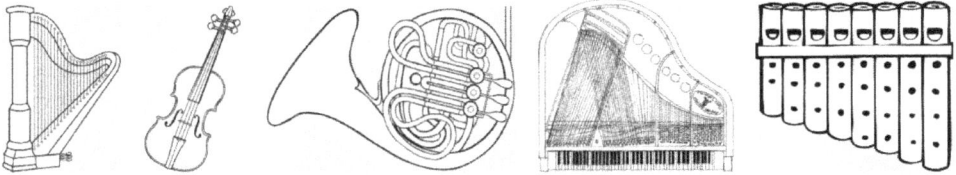

Hierarchy Symmetry reflects self-similarity as top and bottom elements are comparable. A special kind of hierarchy called a *Holon* possesses a higher degree of symmetry where each whole becomes just another part of a greater whole. A perfect example is the page you're reading with letters, words, sentences, paragraphs; each whole being a part of the next level's whole. This principle is imbedded into the physical universe with particles, atoms, molecules, compounds, etc. Every *thing* is a holon in its own particular hierarchy, including You. Holon patterns are not obvious at first because our minds generally operate in a reductionist manner, focusing first on parts, while intuition works oppositely, grasping the whole first through a top-down perspective.

Fractals are a special type of holon where every whole-part is identical at every level. It's a pattern that never ends, continuing on to infinity without deviation. They're scale invariant patterns, which means you see the same shape and forms no matter how high or low you look (self-similarity on steroids). While fractals may look extremely complex and unpredictable, they're produced by simple mathematic rules. The logistic map captures fractal patterns in graphic form, generated by simple equations. Recurrent relation connects inputs with outputs, creating a feedback loop that builds self-similar infinite patterns. Fractals and spirals are two sides of the same symmetry coin, expanding in opposite directions with perfect proportionality. The two patterns bridge duality, connecting the finite with the infinite. Like fractals, spirals are scale invariant as their shape, pattern and proportion remain identical at every level.

Holograms take the holon principle to a transcendent level where self-similarity is built into each and every part. In a hologram each part IS a whole, not just part of one. Metaphorically, *Man the microcosm* is a hologram of the Universe, possessing the very same qualities present everywhere above and below. You too are just such a hologram with perfect self-similarity built in, as every cell in your body contains the same DNA biological blueprint of You – *every* cell is a whole.

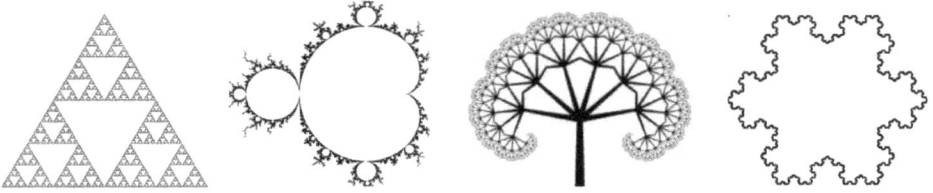

Process Symmetry is a temporal quality obvious in music but present through life in general, exhibiting a flowing harmonic pattern of self-similarity. Organic growth follows this rhythmic process pattern code, with ebb and flow of expansion and contraction, abundance and scarcity, growth and decay. Beyond the obvious patterns of radial and helical growth in plants the entire external environment is brimming with fluid dynamics, air flows, swarms, eddies, currents, and a myriad of whirling vortex motions. Turbulence may produce chaotic movements, but order is still present as patterns of symmetry inevitably emerge. Similar processes take place in the desert where sand dunes are shaped and formed by hidden forces, partially revealed as they dance across the sandy surface in symmetrical patterns. Non-linear chaotic systems can't be understood by simply looking at the parts. Only the whole provides context necessary to see its emergent properties of order.

Equilibrium is a symmetrical state of being, omnipresent in the universe, beginning with immense stars whose spherical shape express the balance between external compression of gravity and internal pressures of expansion. That same balancing act is present everywhere, just mostly hidden. Equilibrium dominates the landscape, transforming pervasive forces of process into balances of structure. Yet it's all just a temporary state; sooner or later asymmetric forces intervene, creating imbalance leading to change. As elegant as symmetry is, the world would be a boring place without its asymmetric counterpart. The same dynamic interplay of forces is at work in human behavior and social discourse. Culture is held together by a delicate balance between individual freedom and state directed laws. Even ideas and concepts are shaped by competing forces that produce temporary consensus. It's thesis, antithesis and synthesis; point, counterpoint and middle-ground where the result is a position maintained by equilibrium.

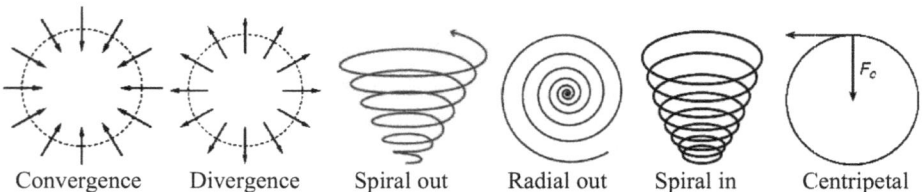

| Convergence | Divergence | Spiral out | Radial out | Spiral in | Centripetal |

Contraction-expansion, tension-release, grow-recede, order-chaos:
Life's ongoing oscillation between pressure and release, balance and imbalance

The physical world, nature, man and society are all shaped by ongoing interacting forces competing between extremes of constraint and freedom. Temporary equilibrium states emerge along a spectrum of opposing forces. It's the same process that determines what form water will assume; solid, liquid or gas (Ice-water-steam). Each form is just a temporary state dependent on which forces prevail at the time, constraint or freedom. Forces in general possess a symmetrical polarity that counteracts itself. Fusion and fission are inversely opposite yet symmetrical, one integrates, the other separates. Same thing with reflection and refraction, the first separating, the other integrating. Or consider entropy and synergy; one divergent, the other convergent. These opposite forces of order and chaos are the yin/yang of physical reality, each a symmetrical mirror of the other.

Symmetry is an omnipresent attribute of a unified connected universe, built right into the software and hardware of everything within it. Mostly hidden beneath the surface and between the seams, yet visible once you know what to look for. Symmetry is a universal pattern code manifested in just about everything, including physics (particle supersymmetry), math (geometry, algebraic equations, logarithms), music (octaves, harmonics), nature (organic forms, parsimony, dissipative structures, Man (radial & bilateral symmetry), Color (complimentary), Process (polar forces, equilibrium), Hermetic laws, universal principles, Polarity, Holons, Fractals and ubiquitous Spiral patterns. Harmony, beauty, peace and proportionality are just rewards when we experience the perfect balance between heaven and earth, matter and spirit; two sides of the same symmetric coin.

Divine Codes speak the language of duality, connecting the mundane physical plane with a higher spiritual realm. Our experience in The World is one of dual status where the interaction between matter and spirit shapes our perception in ways often difficult to define. This dichotomy is bridged and transcended through divine language. The word *whole* is associated with *holy*, implying linkage between wholeness and a divine origin. Shalom means peace and wholeness. Kabbalah reveals man as a split being whose primary task in life is to restore wholeness. Most religions seek reunion with God; bridging the separation between man dwelling in an earthly arena full of sin with the original divine source, restoring wholeness.

Metaphysics is couched in language describing the link between matter and spirit. Pseudo sciences similarly operate in coded, mythic languages. Astrology, Tarot, Numerology, I-Ching all speak in archetypal symbolic code. The Arts draw heavily upon divine imagery penetrating the inner psyche through metaphors and archetypes. Hermetic Code connects everything in universal principles; *As above, so below*, along with the connecting law of Karma. Religious chants, mantras and speaking in tongues produce coded language of divine origin. Synchronicity is a subtle code unexpectedly revealing an elusive interconnection between dual realms.

Laws of nature are expressions of divine pattern code, correlating universal truths and principles to manifestations on the physical plane. Hermetic Law serves as a bridge with 7 universal principles: Mind, Correspondence, Vibration, Polarity,

Rhythm, Cause-Effect, and Gender (+/-). The perennial philosophy of the East also embraces this linkage, providing the basis for the TAO. It connects the relationship between thoughts, feelings, and ideas with the actions that express them. As within, so without. Western religion covers similar ground, translating laws of nature into recommended guidance, such as the golden rule of do unto others…

| SPIRIT | Curves | Analog | Flowing | Quality | Sing | Imagination | Intuition | O | ～～～ |
| MATTER | Lines | Digital | Static | Quantity | Talk | Memory | Thought | □ | --------- |

Greek philosophers applied systematic order to divine codes, producing TOEs based on mathematics, music and geometry. ***Music of the Spheres*** was their divine model connecting celestial motions with musical form. This metaphysical concept intertwined the movement of planets with harmonic relationships, tracking dynamic interactions expressed as energy tones with aspects of pitch, angle, ratio, sound and shape. The planets move and interact with divine harmony producing elegant music imperceptible to the human ear. Both musical harmonics and aesthetic beauty in geometry are equally linked through ratio and symmetry. Like poetry, they speak to us psychically on a deeper, intuitive level.

Plato's TOE incorporated 5 universal essences (earth, air, fire, water, cosmos) and linked them with the 5 perfect geometric solids (tetrahedron, cube, octahedron, dodecahedron, and icosahedron). These unique forms are special having identical faces on every side plus meeting symmetrically at every dimensional angle. These ***Platonic Solids*** were perfect geometric shapes considered divine forms from a higher plane, possessing identical lengths, faces, and corner angles (a surprising number of sea creature forms mirror the Platonic Solids). Kepler incorporated this metaphysical concept and linked it to the cosmic realm, intuitively connected planetary motions with harmonic proportions, sensing direct symmetry between music, geometry and planetary bodies. While seeing polygon geometry as frozen music he concluded that each planet possessed its own distinct melody.

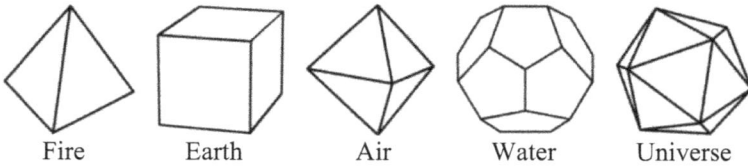

| Fire | Earth | Air | Water | Universe |

Kepler's 1916 Musical Harmonics of planetary motion

Planetary Pairs Symmetry

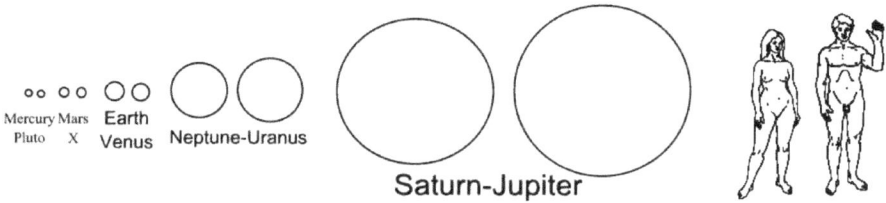

Mercury Mars   Earth
Pluto   X   Venus   Neptune-Uranus

Saturn-Jupiter

*Every planet has a matching twin. *Very likely the asteroid belt was planet "X" (the Mars twin)*

Bode's law reveals geometric relationships in distances between the sun and the planets and further connects musically given the harmonic ratios involved. It predicts a planet where the asteroid belt is. Like Kepler, Bode saw a musical connection in the pattern of intervals between each planet. The entire solar system resembles a mammoth harmonic scale.

Planetary orbit ratios: Earth-Mars 2:1, Earth-Venus 3:1 (musical 5th + 1 Octave)

The Golden Ratio is the divine proportion identified by the Greeks as the most pleasant esthetic geometric relationship. Also called the golden mean, golden section or golden rectangle, it's the divine connection where smaller and larger parts share the same ratio as the larger part to the whole. It's imbedded in nature and organic life, including You. When expressed in algebraic form it produces an infinite number. (1 + Sq root of 5, divided by 2). The musical major 6th is an emotionally pleasing harmony having a ratio of 8:5, which is the golden ratio.

Sacred Geometry is a graphic pattern language Greeks conceived as a Theory of Everything based on a universe generated by emergent divine forms of geometric patterns. Ancients similarly perceived a world founded by archetypal shapes and structures omnipresent in nature and living organisms. They applied these principles into religious structures, temples and holy artifacts. This metaphysical paradigm links hidden universal natural forces to their manifestations in the physical world. Plato saw it in geometry, Pythagoras saw it in music, Kepler saw it in cosmic orbits, Durer in art, D'Arcy Thomson in organic forms and Davinci in man. It's present everywhere, imbedded within everything. The influence of this ancient philosophy pervades Egyptian, Greek and Roman architecture, and its residue extends into medieval European cathedrals, Islamic and Hindu art, Buddhist mandalas, Indian temples and Chinese spiritual traditions.

Sacred geometry is essentially a graphic portrayal revealing archetypal clues of invisible universal forces that produced and created everything – a divine process of making something from nothing. It derives every geometric form from the same divine numbers of 0 and 1. This concept is a metaphysical deep dive into patterns, rules, laws and principles, transforming essences into variety and complexity. Sacred geometry contains layers of hidden meanings, expressed through lenses of number, shape and relationship. It's the source of nature's harmony, manifested in

315   EVERYTHING FITS

mathematic properties, musical melodies and artistic beauty, all built right into the
infrastructure of the physical universe, including YOU.

In the beginning there was nothing, represented by an empty circle possessing
unity as a whole. Then, in an act of self-similarity the circle reflects upon itself,
generating an image of 2 circles overlapping. The middle joining section is the
center of creation, resembling a cosmic womb from which every possible
geometric shape originates from. From the circular center comes a line, then a
triangle, square, pentagram and all the rest. Each variation is linked to their
corresponding numbers: 0, 1, 2, 3, 4… This process is symmetrical with organic
birth: 0 represents a seed or a cell, which then divides through mitosis to produce
two identical cells, then 3,4, 5… The Vesica Pisces is the shape of divine
emergence present in nature and man (eyes, ears, nostrils, vagina, pineal gland,
etc.). It's the starting point for all creation, mirrored in geometry as the origin of all
shapes and forms.

The Flower of Life symbolizes the unity and interconnectedness of everything.
It's found in cultures and religions around the world dating back to ancient
manuscripts, temples and works of art. 19 interlocking circles within one larger
whole circle represent the cycle of creation, unity and oneness. It's a powerful
symbol in sacred geometry.

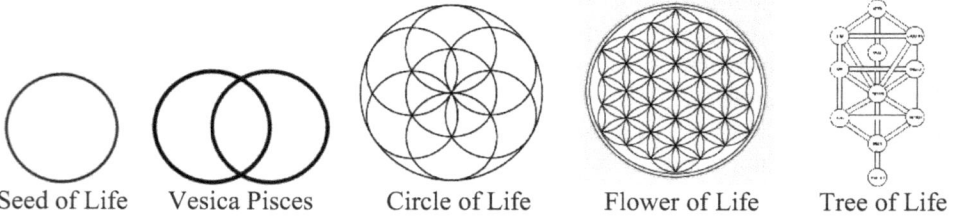

Seed of Life   Vesica Pisces   Circle of Life   Flower of Life   Tree of Life

*Zero in mathematics is equal to a point in geometry.
**1 is unity, 2 (line) is the door way to the many (3,4,5….)
***A circle represents nothing and everything simultaneously. Every circle is identical. Circle
accommodates all other shapes within. It's the only object that won't fall into its own hole.
Radius to circumference will not equate. Universe means "one turn", a completed circle.

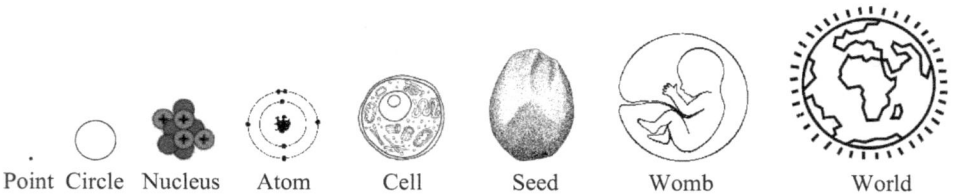

Point Circle Nucleus   Atom   Cell   Seed   Womb   World

◯ = unmanifested reality, the infinite, *spirit*   ☐ = manifested reality, the finite, *matter*

circle	line	triangle	square	pentagram	hexagram	heptagon	octagon	nonagon	dodecagon
1	2	3	4	5	6	7	8	9	10
360	--	180	360	540	720	900	1080	1260	1440

Number Codes. Numbers and shapes are related cousins – building blocks of sacred geometry. Every number has a quantitative and qualitative aspect. This dual characteristic is the basis of numerology, and the divine essence of the I-Ching.

1 & 2 as parents, 3-9 as children, 10-100 as cousins and 101 --- ∞ as distant relatives. 10 returns unity at a higher level, higher octave; one complete cycle.

1 – Monad – Oneness. Unity, wholeness, beginnings.
2 – Dyad – Twoness. Difference, separation, polarity. With time twoness produces vibration and opposition Polar opposites create tension, seeking reunion and wholeness.
3 – Triad – Threeness. Relationship, arrangement, pattern. Self-sufficient structure. Establishes positional context: beginning-middle-end, good-better-best, birth-life-death, past-present-future. The basis of scales after polarity creates separation and differences (by degree between opposite poles).
4 – Quad – Fourness. Firmness, stable, solid, mother earth, matter. Represents spirit manifested into dense, rigid structures on the physical plane.
5 – Pent – Fiveness, within organic life, quintessence. Ubiquitous in nature, flower pedals, 5-point star, fingers, toes, facial symmetry. Pentagram fractal properties, radial symmetry in organisms.
6 – Hex. Sixness, dimensional measurement: Time (60 seconds, 60 minutes, 24 hours, 12 months. Space (360-degree complete circle, 30, 60, 90, 120, 180-degree polygons). Base number for mystical 12 Disciples, 12 tribes, 12 mythic Gods, 12 zodiac signs and houses, 12-man jury.
7 – Septa. Seveness. Mystical essence, religious icon, 7 colors, 7 music notes, 7 Chakras, 7 octave Piano keyboard, 7 Octave Periodic table, 7 World Wonders, Lucky number 7, and the sabbath day. Not found within nature's structures, unlike 5&6. (2 triangles, 1 point up other point down: diamond)
8 – Oct – Eightness. Power, strength, will, self-renewal. 8 seeks balance/alignment. 8 is in I-Ching (8X8), DNA Codons, Checker board (8X8), 8-Fold path enlightenment. 8 sideways is infinity (∞)
9 – Nona – Nineness. Completion. We spend 9 months in a womb and get the whole 9 yards. Aztecs Mayans & Indians identified 9 cosmic levels. Every polygon (circle to triangle, square, pentagon, to 10-sided decagon) is based on a factor of 9 (180, 360, 540, 720, 900, 1080, 1260, 1440). While 1 & 7 are whole, 9 is complete.
10 – Deca – Ten. Fulfillment, new beginning. 10 is a higher octave of 1, both unities. Culmination of the Cosmic creative process, journey from 1 – 9 ending with reunion; return to origin. Anything times 10 is just 1 at a higher octave. Similar to a spiral as it returns to the beginning, but just a little higher.

*Each number possess unique attributes acting like a charged field, operating as keys that fit particular patterns, or a tuning fork resonating to certain vibrations. From atomic micro level up through macro societal level, everything associates with a specific numeric pattern.

Special Numbers. All numbers are not created equal. Consider ***Prime numbers***, special in that they cannot be divided by any other number, nor are they predictable. Except for 2 they are all odd, so beginning with 2, then 3,5,7 but not 9, which is divisible by 3. As the sequence progresses, fewer and fewer remain indivisible. ***Palindromes*** are numbers that look the same when reversed; 747, 969, 1001, etc. **Doubles** seem to carry more weight than other 2-digit numbers (33, 77, 88, etc.). **Triples** are considered angel numbers (111, 222, 333, etc.). Whenever they appear in daily life they indicate synchronicity.

π (3.1417...) The relationship between radius and circumference of a circle is a never-ending value, uniquely mysterious and divine. Infinite numbers, irrational numbers, complex numbers, repeating numbers all carry with them a mysterious quality of a higher realm. π reveals the intersection of duality in mathematic code.

Euler's number: 2.71828… is an infinite number and an exponential constant imbedded in various physical, organic and economic phenomenon, mysteriously popping up in unexpected places everywhere.

Φ **(1.61803**…) Golden ratio produced when a+b/a = a/b, in nature, organic life.

13 carries an archetypal resonance of bad luck, whether it's superstitiously buried in the psyche or an unconscious self-fulfilling prophecy. *666* is associated with the beast, interpreted from the bible. *7* is both "lucky" and a mystical, spiritual number in religion. Numerous sets of *7* are referred to in the Bible, including hundreds of references to the number itself. Religion, mythology, Kabbalah and pseudoscience all make reference to mystical numbers *3, 7* and *12.* Note that 3 and 4 represents spirit and matter, *7 = 3* + 4 and *12 = 3* X 4. *11, 22, 33* are special, spiritual numbers in numerology. Music scale of *7* Nodal tones and 5 semitones = **12**

1440 and *2160* are resonant numbers that intersect in time, space and geometry. There are 1440 minutes in a day, 1440 degrees in an octahedron (platonic solid) and 1440 years in the earth's equinox season (2 deacons). There's also 144 inches in a square foot. Similarly, there are 2160 degrees in a hexahedron, 21600 minutes of arc in a circle, 2160 years in the equinox season (3 deacons), and 2160 years in each *astrological age* (currently we're in the age of Aquarius). The moon's diameter is also 2160 miles.

108 is a mysteriously common number in the cosmos: Sun diameter is 108 earths. Distance between sun and earth is 108 sun diameters. Moon's radius is 1080 miles. Diameter of the earth plus moon is 10080 miles. Diameter of mars is 1080 x 4 miles. Diameter of Jupiter is 10800 x 8 miles. Diameter of Saturn is 108000Km. Saturn's orbit is 10800 days. Distance from sun to Venus is 108 million Km.

0 AND 1. 0 represents nothing, 1 represents everything. 1 is unique as it's neither composite or prime. Anything multiplied by 0 gets absorbed into 0. Anything multiplied by 1 remains whole, intact and unchanged. Both 0 and 1 symbolize completeness in their own dimension. Both 0 and 1 are symmetrical. Logarithms derive everything between 0 and 1, all decimals, fractions and parts of a whole. Any number beyond 1 is simply a derivative or exponential power of a value between 0 and 1. 1-10, 11-100, etc. are octaves of 0-1; just a higher scale of it. Wholeness is 1. 100% is 1. Binary logic covers everything with 0's and 1's. Euler's divine identity equation culminates with 0 and 1. The Fibonacci series begins with nothing and everything (0 and 1). A circle's center is 0 while its circumference is 1 whole. Angles are like fractions – simply parts within the whole, values between 0 and 1.

*Monad unity "1" contains every number: 1111111111 X 111111111 = 12345678987654321

Euler's Identity
$(e^{i\pi} + 1 = 0)$

*Every gauge on your car dashboard represents a value between 0 and 1

PATTERN AS THE HIGHEST-LEVEL LANGUAGE

Laws and codes covered here are languages describing relationships; how The World works and how things within it are connected. They're all pattern languages focused on particular expressions of physical reality. But the highest language of all is simply pattern itself; the ultimate Rosetta stone of language. All other codes are simply subsets of pattern code. I suspect Plato intuitively sensed pattern language as the ultimate code when he described pure essence as a non-physical archetypal form, one that serves as an immaterial blueprint for manifesting physical forms. If we're essentially spiritual beings, then our original essence is formless and shapeless, like water, which takes the shape and form of everything it embodies, all differing by pattern. What makes each of us different from others are the unique manifested patterns of our spirit, expressed physically, emotionally, and mentally as habits, thoughts, desires... a composite personality blueprint.

Pattern is more essential than substance or materiality. Buildings may crumble but blueprints are eternal. Gourmet meals come and go but recipes endure. Images on any screen are simply illusions. Movies and videos consist merely of frame by frame still images creating the illusion of motion. Print photos are merely patterns of dots organized to present the illusion of a recognizable image. Atoms are all made of the same stuff yet simple differences in arrangement create an immense multitude of unique qualities and attributes. Take just one element, Carbon, whose atoms arranged differently produce coal, fullerite, graphite and diamonds.

Smell is determined by geometric shapes and frequencies of molecules. DNA molecular shapes play a role in organic processing; protein shape is key. What's the difference between a pile of lifeless organic molecules and a living being? The answer is order, form, arrangement, structure; essentially just pattern itself. Organic molecules contain essentially the same parts as inorganic ones; they merely form much longer chains in particular combinations. Organic molecule patterns create specialized functions as they fold, coil, corkscrew, chain, thread and expand as pleated sheets. Thousands of different colors are produced by simply stretching the same light wave. All forms of light including X-rays, microwaves, radio waves, gamma rays, etc. are the same, just stretched to different lengths.

Symmetry is a universal pattern representing more than just shapes in space; it's an omnipresent quality of similarity. Supersymmetry may be a better description, beyond the limited principle in physics connecting particles but extending to our entire reality of ubiquitous patterns connecting everything in an elegant, inter-dependent unity. It's manifested in attributes of self-similarity, polarity (two things opposite and equal) and duality (one thing with two aspects). Time is to music, what space is to geometry; both reflections of duality in harmonic relationships. Similar to these forms of connectedness are the transcendent; things whose nature transitions to a different scale. All 3 are connected, but where polarity and duality differ by degree and type, transcendents go above and beyond the original. Symmetries may be created equal, but some are more equal than others.

Supersymmetry Pattern Code: Polarity, Duality, Transcendence

POLARITY			DUALITY		TRANSCENDENT	
Yes/No	On/Off	Odd/Even	Constrained form	*Flexible order*	Quantity	*Quality*
Male/Female	+/-	Yin/Yang	Structured matter	*Flowing energy*	Knowledge	*Wisdom*
Big/Small	Fast/Slow	Near/Far	Crystalline lattice	*Fluid dynamics*	Strength	*Power*
Hard/Soft	Win/Lose	Stop/Go	Geo symmetry	*Musical harmony*	Black-White	*Color*
Open/Closes	Cause/Effect	Left/Right	Digital	*Analog*	Fact	*Faith*

Patterns distinguish information from noise, encoded by numbers, shapes, geometric forms, cycles-waves-vibrations; all expressed in spatial and temporal relationships. Ultimately, patterns connect and unify everything, personified by mathematic equations and abstract laws of nature. Math is *the* science of patterns and connections, revealed in numbers (arithmetic), shapes (geometry), motion and change (calculus), relationships (algebra), reasoning (logic), probabilities (statistics), etc. Math captures patterns that are real or imagined, static or dynamic, quantitative or qualitative, spatial or sequential.

Patterns can emerge out of nothing through a resonance or repeated action. We see this at the most fundamental level of physics where vibrating energy manifests as physical forms. But this process is ubiquitous at every level, including your own habits and routines. The images on your TV or computer screen are produced by rapid repetitions of single frames that appear as a stable picture – a resonant form. This process of repeating vibration is what produces variations in anything along a spectrum, including colors, sounds, shapes and periods of time. Every scale that contains elements will include emergent nodes within that stand out, different from the rest. These resonant points are a pattern code that produces distinct contrast among what otherwise would be a generic scale of sameness. Think of all the people you know and then count those special few you would consider as close friends; they are the nodal, resonant elements among the many. Or think of all the events in your life and then list the few that were life changing; those are nodal points. Everything at every level is subject to this node pattern code.

Some patterns *only* exist in the aggregate, collective level. There's no such thing as temperature in a single molecule, nor is there any sense of direction in a single point or position. An economy is a collective concept, meaningful only in large quantities of people. Single behaviors repeated continuously become habits – resonant patterns of real form. The Bell curve emerges out of nowhere, only taking shape as more elements are added over time. All attributes of matter (tensile, brittle, magnetic, stickiness, density, roughness, etc.) are collective qualities that don't exist at the individual atomic level. Substance is trivial, pattern is everything.

SPECTRUM OF UNIVERSAL SCALE CODES WITH NODES

* **Divine Octaves** (*7 Music Notes, 7 Primary Colors, 7 Chakras, 7 Atomic Levels*)

A E I O U Y	Vowels as Nodes
BCD FGH JKLMN PQRST VWX Z	Consonants
2,3 5 7 11 13 17 19 23 29…	Primes as Nodes
4 6 8,9,10 12 14,15,16 18, 20,21,22 24,25,26,27,28	Composite Numbers

Tetra Cube Octo Dodeca Icosa (*Abbr*)	Platonic Solids as 3D Nodes
Triangle Square Pentagon Hexagon Circle	Primary Shapes as 2D Nodes
	Ordinary Polygons

C C C C C	Octaves as Nodes
AB DEFGAB DEFGAB DEFGAB DEFGAB DEFG etc.	Music Notes
Octave Fifth Fourth Major Third Minor Third	Harmonic Intervals
He Ne Ar Kr Xe	Inert Gases as Nodes
HLiBeBCNOFNaMgAlSiPSClKCaScTiVCrMnFeCoNi etc.	Atomic Elements
Play School Career Marriage Family Retire	Stages as Nodes
Birth Child Teen Young adult Middle age Senior Death	Life
Plague WWI Depression A Bomb Moon 9-11 Covid	Major events as Nodes
H I S T O R Y ---------- *T I M E L I N E*	Periods and eras
Disgust Sad Anger Fear Surprise Happy	Primary Emotion Nodes
Shame Jealous hesitant lonely playful eager proud loving	Derivative mix of emotions
R O Y G B I V	Primary Color Nodes
IIIIIIIIIIIIIIIIIIIIIIIIIIIIIIIIIIIIIIIIIIIIIIIIIII	Shades of every color
Reds Oranges Yellows Greens Blues Indigos Violets	
Mars Jupiter Saturn Uranus Neptune Pluto Exoplanets	(Planets beyond Earth)
red orange yellow green blue indigo violet	
Each planet matches the color spectrum in perfect sequence order	
Root - Sacral - Solar Plexus - Heart - Throat - Third Eye - Crown	Chakras – Energy Nodes
Each Chakra matches the color spectrum in perfect sequence order	

    Perhaps the most fundamental pattern of physical reality is the dynamic interplay between fluid change and static form, an ongoing dance of tension and release, order and chaos, constraint and freedom. Living organisms naturally gravitate towards the middle ground between these two extremes, a *Goldilocks* sweet spot with just the right mix of constraint and freedom; the same pattern in life-supporting water having both order and fluidity; freedom to move beyond the fixity of ice but with order not present in chaotic steam. It's a universal pattern found in everything and situation – dynamic elements of both structure and flow.

Heisenberg observed that atoms are less things and more events, implying all matter and materiality consists of fields and process, not the solid stuff we perceive. Everything is a temporary pattern in time, a resonant structure intact for the moment until inevitable forces of change create enough disorder to distort, alter or disintegrate it – nothing is exempt, including You. Flux is the natural state of things everywhere, everywhen. Despite our misguided perceptions of stable things in this illusive temporary moment here and now, every *thing* is merely a fleeting part within a larger grand composite; part of an ever flowing everchanging dynamic universe permeated by infinite fluid patterns of pattern.

"Patterns don't lie; and at the center of every pattern is you." – Daniel Mangena

The ubiquitous QR digital code is a graphical information pattern linking YOU to a remote website somewhere within the vast world wide web. Scan this one with your high-tech smart phone and enter *The Matrix*

10   THE UNITY OF LIFE

What is Life? In essence it's *everything*. For any living being it's the only thing. Without Life nothing would exist. Yet for all its ultimate importance who can really say what it is? The best anyone can do is describe its qualities but in the end, it remains elusive, beyond definitive explanation. This much is certain: Life cannot be explained by classic laws of reductionist science.

Defining life naturally falls within the bounds of biology, where conventional dogma requires a set of criteria – qualities present in living things. Given the elusive nature of life, it's no wonder the criteria from "experts" in the field varies. Most lists include some or all of the following attributes that define life: *Adaptive, metabolism, internal organization, stability (*homeostasis*), growth, reproduction, functionality and information storage.*

Some of these qualities by themselves are found in various "non-living" things but most are present in "living" things. They vary by degree and importance depending on the life form. Additional qualities appear as complexity increases in an organism. Whole living systems evolve with emergent properties not found at rudimentary levels. Biologist James Miller proposed a comprehensive theory of living systems identifying 8 component levels and 20 fundamental processes. Essentially, they cover storage of info and energy, processing it to capture meaning, and then replicating and reproducing. In this framework we see the qualities of life vary according to its level of complexity.

Living organisms are considered open systems because they must interact with the environment to draw in life-sustaining energy. They possess a variety of physical substances including some very common elements. Hydrogen, Oxygen, Carbon, and Nitrogen are the 4 primary elements found in almost all organic life along with other less common ones. Quantities and combinations very widely. So, we see a variety of qualities and functions in beings and living systems but there really is no definitive criteria that applies to all forms. Where does life begin? Where does it end? Even biologists struggle with this question, full of ambiguity.

Science should be able to determine clear boundaries for life. On the lower end of the scale, it's generally accepted that the smallest unit of life is a cell. The next level below that is a virus: a self-contained membrane with its own DNA but no functionality. Biologists define a virus as non-living – a parasite that can't reproduce by itself. At best we can define it as pseudo-life. Below it are simple components of life: organic molecules separate and non-integrated.

On the higher end of the scale, we have man and "non-living" extensions beyond human flesh. This includes self-sustaining machines, institutions, and social organizations functioning as complete integrated "non-living" systems. Cities, states and nations as a whole come into play. Like viruses on the lower end of the scale, these higher-level forms could at least be considered pseudo-life. Looking at both ends of our broad scale of life there is no distinct threshold, only vague boundaries. *A virus can join with a dead cell and bring it back to life.*

Focusing in the center areas of the life scale you would think delineations would become more obvious. They aren't. When does an individual life begin? When does it end? At the human level there's plenty of controversy over whether a baby is alive upon conception, after a beating heart, or after its first breath at birth. There's equal uncertainty on the other end of the spectrum; individual death. When is a person actually dead? When the heart stops? When breathing stops for extended periods? Is a brain-dead human with a beating heart really alive? How should we define a person pronounced dead who is later revived? Documented cases reveal people "brought back to life" many hours after their hearts stopped beating. What about the life or non-life of amphibians that actually freeze for more than half a year only to thaw back to life. Was it dead? If not, are other dormant forms of life and "non-life" temporarily dead? Or is it just one more case of pseudo-life? What about bringing back an extinct species through DNA tech?

Looking at a living organism internally adds further ambiguity. Human beings who are fully "alive" are actually a mix of living and dead components. Your skin contains an outer lining of mostly dead cells. Hair and nails are alive on one end and dead on the other. Teeth and bone share hard surfaces of mostly dead cells. In fact, over 100 billion cells die in our bodies every day as a routine process of continual maintenance and replacement. In nature we find similar examples of the living-dead mix within organic life. Besides animals with dead horns, antlers, feathers and thick dead skin layers, ponder the fact that trees are primarily just dead pulp. What about a person whose lost several limbs and must be assisted by machinery? Or pushing the line further how should we define a person with mechanical internal parts, including an artificial heart? These inconvenient realities further blur the line between life and non-life.

There's a measurable degree of aliveness among similar beings, including us. Clearly there's a difference between a young thriving, vigorous individual and an old, decrepit senior barely able to move. Consider the contrast between two healthy individuals where one is active, productive, continually learning and developing while the other lays on a couch lazily binge-watching hours of tv in a sedentary non-active life. Certainly, we can observe disparities of aliveness based on a few obvious attributes; responsiveness, resilience, adaptability, and vigor.

Such numerous gray areas suggest life is an intangible essence manifested across a continuum with no clearly defined thresholds. At best we can delineate general ranges along a spectrum of changing degrees. This applies to both the quality and forms of life.

A Spectrum of Life Intensity (conscious state)

Coma - Vegetative - NonRem - Dream Sleep – Drowsy - Awake - Active – Alert - Manic
Awake by itself has a wide range of degrees of "aliveness"

This Spectrum may be limited to organic life, however, ambiguity increases as we expand our perspective of life's continuum to include forms beyond just the organic range. On the lower extreme living attributes begin at the physical level of nature. On the higher end we find emergent complex systems – the social level of collective humanity. In between lies the biological level of life we relate to.

Spectrum of Life and Pseudo-Living Forms

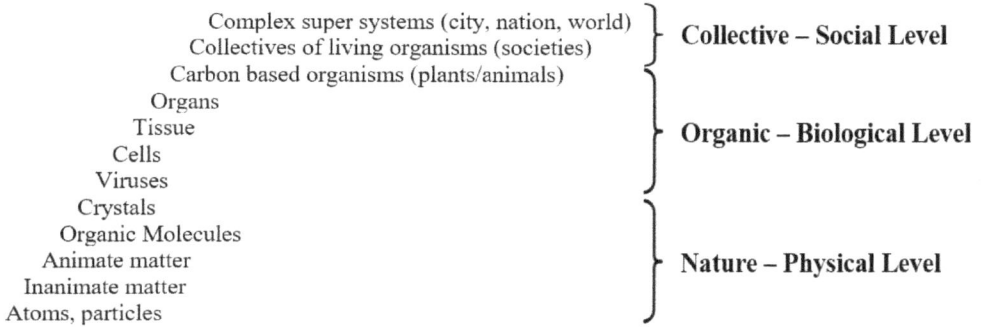

Complex super systems (city, nation, world) Collectives of living organisms (societies)	Collective – Social Level
Carbon based organisms (plants/animals) Organs Tissue Cells Viruses	Organic – Biological Level
Crystals Organic Molecules Animate matter Inanimate matter Atoms, particles	Nature – Physical Level

Pseudolife.   Anything not classified as organic life is considered either dead, non-living or a form of pseudo-life, depending on your perspective and definition. To be accurate and inclusive it's necessary to go beyond the limits of typical human perception. Our narrow sense of time, both long and short term, obscures our view of life and reality. Speeding up a cosmic movie projector where eons pass quickly, brings to life glaciers, mountain ranges, changing lakeshores, moving sand dunes, shifting rock formations, etc. The whole surface of earth suddenly emerges as a dynamic metabolizing life form ebbing and flowing in rhythmic motion resembling a living organism. Plants are seen twisting and twirling in spiral dances reaching higher towards life-sustaining sunlight. Turbulent weather patterns now rapidly churn, moving about and all around in a very lively grand performance, expressing a moody temperament complimenting the planet's surface. Reality is significantly transformed when viewed on this time scale and tempo.

Similarly slowing down the movie would expose life hidden at the atomic level where particles swirl, shift and jump around in dizzying rhythmic patterns. Molecules could be seen possessing a very social aptitude for joining other groups, frequently changing partners and dancing in, out and around other social circles. Thousands of aligned particles could be seen vibrating in symmetrical orders, resembling a choreographed cabaret dance. A cross-section of various substances at different temperatures and pressures would reveal a diverse assortment of particles, shaped structures and unique orders dynamically oscillating while demonstrating the metabolism of an interconnected complex organism. Limits of human time perception obscure both of these levels of life from our awareness.

Other ambiguities arise with cases of artificial limbs and parts; components morphing into extensions of an organism. Nature is full of examples including a Hermit crab shell, caterpillar cocoon, bird's nest and various shells that encase life's diverse forms. Wouldn't the same apply to man-made shells such as clothes, vehicles and home shelters? The only difference between body hair and clothing is material: one being organic substance, the other inorganic. How is the snail carrying his home shell around any different that humans living in a mobile camper, other than material? From an evolutionary, adaptive perspective consider how different homes look in various climates (Wood frame, concrete, mud hut, igloo, cave, etc.). Life always finds a way, creating various forms, shapes and patterns to meet needs, regardless of the substances used.

Pseudo-life is omnipresent in The World where extensions of man permeate the environment. Civilization is saturated with all kinds of machines and systems that are an integral part of modern life. Engines and mechanical beasts behave in ways resembling living organisms as they take in fuel, breath air, perform work, belch exhaust, run down and eventually die. Factories with networks of machines and integrated systems encased in an outer shell function eerily similar to living beings, taking inputs of energy, information and raw material, process them, produce finished products and discard waste material. Is this not a form of inorganic life?

Pseudo-life emerges out of various collective communities. In nature we see ants, bees, birds and fish operate with a group dynamic that resembles a higher-level order – a transcendent collective organism. Ants form complex societies where individual members specialize and perform unique roles supporting the whole. Beehives as well. The same collective phenomenon manifests with human life where sports teams, military units and whole nations take on qualities of a synchronized living organism. Viewing a city overhead reveals a bustling interaction of people who look like ants flowing in, out and all around, performing a variety of essential functions that keep the whole living system operating in a balance of homeostasis. The same applies to earth itself as one single giant superorganism where cities, states and nations resemble the cells, organs and internal systems of a living being.

Ambiguity is inescapable in the arena of life, present at every level and every state of being. At the micro level we observed atomic particles and molecules behaving socially and dynamically in a sea of hyper-fast integrated activity. In that same realm, various structures take on rudimentary qualities of life – especially crystals. Here we observe diverse patterns of repeating alignments containing order, necessary for reversing entropy to establishing progressive form. Like organic life, crystals can grow and reproduce. They may not have all the ingredients of complex beings, but they exhibit the same order present in more complex counterparts.

Pseudo-life appears in other phenomenon in nature where subtle qualities of life are displayed. Consider the behavior of a raging fire, a menacing tornado or a lively thunderstorm, each exhibiting bouts of temperamental tantrums in bursts of directed energy that builds, thrives and then dies off. Fires are especially interesting as they require a spark to be born, ingest fuel to grow and develop, consume all in its path, breath oxygen, cough up smoke, and express various moods – from calm and collected to threatening rage. Man-made engines behave similarly as they too consume fuel ravenously, suck in streams of oxygen, expel exhaust, start in fits and spurts, often gasp with convulsions, only to inevitably run down and wear out.

Life consists of a variety of ambiguous forms everywhere we look, bringing us back to our original paramount question; just what exactly is life? Conventional consensus seems suspect at best, claiming life is only organic within specific boundaries. It's a needless self-imposed constrained position, limiting life to a narrow midrange as a privileged placement within a broader spectrum of potentiality, failing to explain the presence of vague boundaries at every turn. A good place to start therefore is defining exactly what *organic* life is.

Organic Life.

The primary substances of organic life are easy to identify; the first few elements of the periodic table. *Hydrogen, Oxygen and Carbon* are three of the top four most abundant elements in the universe and along with Nitrogen are key ingredients of all organic life. Three of these four (H, O, N) make up the bulk of earths biosphere, including all bodies of water and the entire atmosphere. *Carbon* provides the glue integrating them all in "living" organisms. Excluding Nitrogen, three key elements (H, O, C) make up 99% of organic molecules, the building blocks of life. These elements are highly social, purposefully partnering up to form thousands of macromolecules supporting biological life.

Organic life is considered carbon-based because of the integrating nature of that key element: Carbon. It's unique as a molecule super template – enticing up to four neighboring elements to join with it including other carbon atoms. With its 4 strong covalent bonds it encourages long chains in a variety of lengths, combinations and shapes. These extended structures are flexible yet durable, so they can twist and skew in unique patterns with particular attributes. Carbon and Hydrogen love to join (Hydrocarbons), repeatedly in millions of different organic compounds.

The building blocks of organic life operate at two successive levels: Atomic and molecular. The same four key elements (H, O, C, N) make up 96% of the human body. They naturally socialize, integrate and form into organic molecules. At the next higher stage, particular sets of organic material emerge, essential to building beings. They include *proteins, lipids, carbohydrates* and *nucleic acids*. Each performs a necessary function in living organisms:

Proteins	-Building material
Carbohydrates	-Provide energy (sugar forms)
Lipids	-Store energy and shield areas (fats and oils)
Nucleic Acids	-Genetic material

These are the prime ingredients for producing organic life's foundational unit: *the cell*. Proteins are "jacks of all trades" that make up the bulk of stuff in cells and living organisms: tissues, organs, muscles, tendons, etc. Additionally, they serve as enzymes and catalysts which are integral actors in cellular processes. Exactly twenty amino acids combine to make thousands of different proteins. Diverse types vary by combination and shape, which determines specific functions. They manifest in a variety of molecular patterns crystalized in 3 dimensional lattices, producing unique structures that twist, swivel, bind and curl up.

Proteins, Carbs, and Lipids serve as the hardware of organic life. Nucleic Acids which form genetic material provide the transcendent attribute of software – the information code that tells organic parts what to do. Genes made from these acids are a hybrid since they're both substance and information, like dollars and coins which are both symbolic and *physical* codes. DNA is the chemical code linking all organic life regardless of how rudimentary or complex it may be. It's the blueprint and recipe for making any living organism.

Where carbon is *the element* integrating organic life, water is *the molecule* integrating organic life. It's no coincidence our planet's surface composed of two thirds water is mirrored by the same ratio within our bodies. Life began in water and all organic forms are linked by that special molecule. External water dynamics (weather, storms, rivers, falls, irrigation, evaporation) are likewise mirrored by internal dynamics within living organisms. Life as we know it cannot exist without an abundance of water. Every great civilization emerged near large bodies of water and to this day all major metropolitan centers are next to rivers, lakes or coastlines.

Water retains a balanced mid-point between structure and flow – a sort of happy medium between stable solid and chaotic gas. Its arch shaped molecule creates an attracting polarity and a highly social quality. This polar force holds its liquid state when most other molecules would transition to a gas. High polarity also encourages the formation of lattice patterns leading to viscosity and surface tension, supporting the shape and rigidity of cells. Besides being a universal solvent for substances, it possesses the rare attribute of becoming less dense after freezing solid. It's easily absorbed into substances which aids living processes, as it does with plant roots. All the key molecules of life dissolve in it and are transported by it.

Water molecules possess a high specific heat, making it a temperature stabilizer. It takes a lot of energy to change the temperature of water, which is why oceans on earth's surface minimize temperature changes on land. This stabilizing quality is key in homeostasis where body temperature must be maintained in a narrow range to survive. It's the primary ingredient in cells where it makes up the same two thirds ratio as our bodies and the earth. Cellular activity is greatly facilitated by water where adjustments of concentration influence transportation, production, exchange and discharge actions. Metabolic processes in cells and complex organisms are emersed in this living liquid. Organic life is not just carbon based; it's equally water based.

Special Qualities of Water - *Colorless, Odorless, Tasteless*

States	Only natural substance existing on earth in all 3 states: Solid-liquid-gas
Specific heat	Higher specific heat than any common substance (has cooling effect)
Compressibility	Volume doesn't change much with added pressure
Surface tension	Can hold up dense object without sinking
Adhesion	Sticks to dissimilar materials
Capillary action	Flows in narrow spaces without use of gravity (plants move it upward)
Solvent	Dissolves more substances than any other liquid

Spectrum of Water
pH balance scale
Middle sweet spot

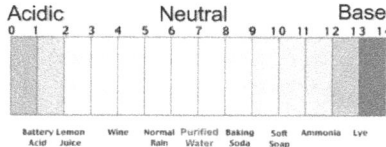

Acidic   Neutral   Base
0 1 2 3 4 5 6 7 8 9 10 11 12 13 14

Battery Lemon   Wine   Normal Purified Baking   Soft Ammonia Lye
Acid Juice   Rain Water Soda Soap

2/3rds of your body is water
75% of your brain is water
80% of your blood is water

**pH States based on Ions: Acid-Neutral-Base – varies by extra or deficient electrons

The Cell. Man is a microcosm of the universe. Cells are a microcosm of Man. Both are self-contained independent systems enclosing an inner identity mirroring an external world outside. A cell is the most basic unit of life in the organic realm, similar to its counterpart, the Atom, in the lower physical level. And like atoms there are a variety of types and attributes with each one.

Despite its tiny size, cells contain a remarkable set of unique components and perform a surprising number of sophisticated functions. Many of these are carried out in larger, more complex organisms but the rudimentary order of them all originates inside the microscopic world of cells:

COMPONENT	FUNCTIONS
Membrane	Protective Shell
Nucleus	Control Center
Ribosome	Material Assembly Line
Mitochondria	Energy Production
Cytoskeleton	Fiber Structure
Vacuoles	Regulators, Maintenance
Lysosomes	Digestion
Golgi Apparatus	Material Storage
Endoplasmic Reticulum	Transport Tubes
Peroxisomes	Catalyst Enzymes
Protoplasm	Inner Cell Fluid
Cilia – Flagella	Mobility Hairs
Organelles	Inner Organs

These components and function work in an integrated fashion. Parts support the whole cooperatively and with purpose. Cells operate similarly to complex organisms where the essential activity is intake of information and energy, then processing it for growth and action, and finally discarding waste. It's a dynamic that putts cells on the border line between physical and biological realms. Nature's trick is converting physical energy (most of it from the sun) into organic energy to support living systems. It accomplishes this though integrated chemical processes much like the mechanical and industrial ones at the societal level.

Cells perform all the necessary tasks that a complex factory requires. They have a manger (Nucleus), directing foreman (DNA), communication network (Chemical messengers), raw materials (Proteins), regulators and controllers (Enzymes), power source (Mitochondria), warehouse (Golgi Apparatus), roof/walls (Membrane) and of course assembly lines (Organelles). They maintain balance so the whole may persist (homeostasis), ensuring stability in temperature, pressure, resources and general metabolism through various filters, enzymes, regulators and feedback loops. And they reproduce independently through simple division to make identical copies. Where crystals grow and multiply through accretion of single geometric templates, cells duplicate each internal part and split in half to create two new independent twins. There are exceptions including sex variation (meiosis) and cell specialization but the process of cell division occurring in our bodies (well over a trillion daily) are exact replicas.

Cells are produced in a variety of shapes, though sphere and cuboidal are most common. They can vary in size by up to a power of 10. Shape, size and content are determined by function and purpose. Different types are needed for bone, tissue, cartilage, blood, nerves, fat, muscle, etc. Human brain cells are rigidly set after the first few months from birth, never to be replaced again. Red Blood cells are the only type without a nucleus, probably because it has such an important specialized function of carrying oxygen throughout the system. Platelets are designed to clot up as needed when injury damage requires it. Other cells are specialized to perform each and every possible function supporting a complex living organism.

Organic life's sequence flows as follows: Atoms socialize into molecular groupings, molecules socialize into cells, and diverse-specialized cells integrate into higher level organs. Collective socialization leads to emergent higher-level complexity. Along the way cells cluster into larger parts such as tissue, tubes, liners, webs, fibers, and various substances and fluids. Cells then become the very components and ingredients of the organs they produce. At every level of increasing complexity along this chain of parts and wholes there's a corresponding emergence of new attributes and functions:

CELL	ORGAN	SYSTEM
Nucleus	Brain	Nervous
Secretion	Heart	Circulatory
Absorption	Stomach, Intestines	Digestive
Chemical messengers	Glands, Hormones	Endocrine
Diffusion	Lungs, Gills	Respiratory
Membrane	Skin	Protection

The highest level of organic life is an organism itself. At this level we see further emergent functions and components but more importantly there's a higher degree of integrated complexity emphasizing the systems nature of the whole. More parts, more subsystems, more interaction and more coordinated integrated operating dynamics become the attributes of a living system. All the internal action is connected, coordinated and complimentary within itself. Like human group dynamics, the members work together and create synergy – a whole greater than the sum of its parts. As a complex system its ability to maintain balance, order and stability (homeostasis) requires continual dynamic interaction of each subsystem. Survival depends on it.

Living beings eventually develop more sophisticated capacities to address increasing internal system demands, along with external environmental stresses and threats. Senses become more elaborate, mobility increases and capacity for intelligent response expands. As size of an organism increases, so too does its requirement to acquire, process and store energy. Hence lungs pull in gallons of air, stomachs hold pounds of food and hearts pump oxygen-rich blood through elaborate arterial channels to every cell in the organism. Limbs powered by muscular mechanics grow in size and dexterous fingers develop to maximize capabilities. Eyes and ears develop high sensitivity to environmental signals. And each and every innovative new subsystem is synchronized in a highly coordinated, integrated whole.

Organic life builds upon the physical realm of matter and molecules adding layer after layer of higher complexity. At each successive level novel attributes emerge providing increased functionality and capability. These new developments are driven by environmental demands, including basic physical laws of nature. The organic realm is essentially based on a combination of special molecular ingredients and biomechanical processes. While we've looked at internal processes at the micro level, equal scrutiny must be applied externally, at the macro level. The dynamics of organic life are codependent on *both* internal and external processes that work in unison to advance the progress of higher-level forms and greater complexity. That progress is driven by the evolutionary forces of nature.

EVOLUTION

Evolution is a fundamental aspect of this elusive thing called life. It produces organic success via progress and persistence. It's a pervasive force woven into the fabric of everything around us. Once you know what to look for, you'll see it everywhere. In our personal lives we learn, grow, and adapt – in both small incremental changes and in major stages. We see commercial products continually evolve, from the Model T car to a Tesla, from outdoor clothesline to digital dryer, from abacus to supercomputer. Likewise, we've seen *processes* evolve, from campfire to instant microwave, from planting corn seeds to robotic farming, from hand crafted woodwork to modern assembly lines. Organic evolution seems different because it involves invisible, self-directed change as opposed to our human directed change we're all familiar with.

Evolution has been mischaracterized through history. Creation myths from early civilizations persist to this day as remnants of ignorance and imagination substituting for real science; feeble attempts to explain a complex world without facts. Consider the fairy tale of Noah's ark, which would need over 1,000,000 species plus tens of thousands of micro-organisms necessary to sustain life.

Evolution theory is evidence based; scientifically supported with a fossil record, genetic record and logical patterns of geographic divergence. The natural selection process can be demonstrated in real time with a petri dish, bacteria and an injection of anti-biotics; most are quickly killed off but a few survive to become immune or better fit. Diseases and recurring epidemics are evolutionary forces acting in real time. Evolution is already demonstrated at the physical and chemical level; atoms evolved into molecules, then organic molecules, and eventually cells.

War mirrors nature's survival pressures as each combatant side is compelled to develop innovative weapons and tactics to endure. Necessity is the mother of invention and there is no greater necessity than survival itself. World War II produced a wide variety of extraordinary new weapons, methodologies, accessories and technologies in just a few years. D Day alone introduced numerous innovative landing craft, mine sweepers, gliders, Bangalore Torpedos, artificial harbors, and a variety of specialty machines so odd they were called Hobart's funnies. All wars involve a back-and-forth arms race, such as the continual one-upmanship of tanks where each side continually adds thicker armor and bigger guns.

Life *is* evolution. It includes learning, growing, developing and *transforming*. The first 3 are differences in degree while the 4th is a difference in kind. This transcendent transformational attribute of life is key. It's the power behind a seed becoming a plant, an acorn becoming a tree, and an egg becoming a chicken. Transformation is synergy – the whole becomes more than the parts. It's unpredictable and the very essence of progressive life.

Evolution is a subtle process of transferring information from one form to another. During that process energy is exchanged with invisible directionality. To put in component terms, we can think of evolution in the following basic equation: *Purpose + Info + Energy x Process = Life*. At the organic level this consists of cells with DNA replicating, specializing, then organizing into complex wholes. The entire process seems complex with a variety of mechanisms and physical principles involved during every stage. To understand evolution, one must appreciate the interactions enabling life's ability to change, adapt and progress by both incremental degree and transformational stages.

Dynamics of Evolution – Change is a fundamental, ordinary aspect of reality yet various forms of it confuse people. Small incremental, linear change is familiar but when there's a pattern of stops and starts, zigs and zags, crawling along and then suddenly a giant leap… it can seem a bit perplexing. Organic evolution is such a process. Certain aspects of it are predictable, others are not. Basic principles of physics and chemistry are obvious while other subtle dynamics of integrated parts co-evolving may be difficult to discern. Complexity, ambiguity and principles of collective emergence make evolution seem complicated. Surprisingly, it's not.

Evolution is difficult to grasp for those who don't understand the time frames involved. It's not hundreds or even thousands of years but millions and billions. The odds of something happening once in hundred years is very remote for us. But those odds become greater and greater with longer time spans. Once you deal in millions and billions of years, the likelihood of something happening become a virtual certainty. "Random" chance in the short run becomes possible in the mid-term and then probably in the long-term. With unlimited time spans everything becomes possible and probable.

The key to understanding organic life and its evolution is seeing the mechanism of change at play. Like atoms at the physical level, much of it is recombining simple elements to create new attributes with emergent synergy. It's surprising how simply mixing parts can have a dramatic effect. Consider how Hydrogen and Oxygen, (2 very combustible elements) feed a fire separately but when combined make water – the very thing that puts out a fire. Organic substances operate in similar fashion. Recombination of parts is an essential mechanism in evolution.

Organic parts are mixed up both intentionally and accidentally. The first is achieved through sexual variation, the later involves mutation. DNA, the code directing cell activity, is a chemical recipe for making a living thing. It's made up of gene sequences that determine how something will turn out. When two beings give birth to a new being, genes from each are passed on with the new set containing parts from both. This simple mechanism ensures variation, a key component in adaptation and survival.

Genes are replicated millions of times, in millions of cells and passed on through millions of further duplications. Periodically errors are made where the gene passed on is either damaged, changed in a bad way or changed in a good way. These occasional mutations are normally unimportant but over time a few significant ones lead to a better fit in the new set. The same random chance that brings conflict or opportunity in our daily lives also plays a part in basic evolutionary processes. Recombination of DNA through built-in variation and occasional mutation provides diversity and opportunity for improvement. The culminating step is choosing which changes are beneficial to new life.

Natural selection is a simple, sort of dumb mechanism that basically weeds out stuff – parts or wholes that aren't as good as others. Think of a filter that limits what gets through. You could pick a setting allowing only small, lightweight, colored, circular objects while everything else is blocked. Natural selection is identical to this where its filter settings choose life forms and attributes – and the settings change all the time. Sometimes big and strong is preferred, other times small and swift is better. Nature and the general environment of pitfalls and predators act as settings on the filter of life, choosing which types will survive and which ones will die off. It's a fairly simple problem-solving mechanism but one that provides a subtle directionality to evolution.

Preceding natural selection is internal selection – the mixing of genes. Unlike Mitosis where cells duplicate identically, Meiosis is a differentiated cell that produces gender variation. This enables two parent organisms to join and combine their genes to create a new mixed product. Selection here is based on simple math ratios along with an added twist of dominant and recessive genes which influence the outcome. Built-in variation is a key aspect of genetics but other mechanisms play a role. All cells contain the entire DNA code yet most use only a very small part of it. Surprisingly, genes are dormant most of the time. Active genes tell other what to do and when to do it. They regulate in an on-off manner, much like computer circuits triggered by on-off transistor switches.

Cells behavior is based on a few instructions within the extensive code, plus they respond to the behavior of neighboring cells. DNA is both recipe and algorithm, directing action in both sequential and parallel dynamics, so cells carry out specialized tasks at a designated time and in an integrated manner with neighboring partners. Consider the unbelievably sophisticated recipe that directs a single fertilized human egg to differentiate into over 260 different types of cells that duplicate, specialize and form a complete being with trillions of cells.

Evolutionary processes combine internal and external mechanisms working both independently and jointly with each other. Operations involve filters, valves, funnels, templates, catalysts, repressors and regulators. Internally they're directed by DNA while externally they're guided by environmental forces. In both cases the ultimate driver is necessity. Survival is need based. Life succeeds and persists by meeting needs – rudimentary at first, then refinements come later. Conscious human beings meet needs by problem solving and engineering. Nature is no different. Evolution is primarily a very long, ongoing process of solving problems through simple trial and error, then choosing the most optimal results.

Nature is an extremely creative, determined innovator. Its various end products are efficient, effective and often elegant. Case in point are the clever solutions used to create different types of animal limbs, combining attributes of both strength and dexterity to meet the needs of a physically demanding world. Versatility is achieved through cunning engineering such as skeletons with arched structures, ball and socket joints, and articulated limbs. Successful solutions are reused and applied across species with minor modifications as needed in each regulating habitat. Fish fin structures transition to mammal arms and legs. Gils transition to lungs, membranes become skin, shells transition to bones, and a nucleus evolves into to a brain inside a head. Innovations begin in primitive form, develop in layers and eventually emerge into complex variations of the original.

Nature is an impressive problem solver, especially when you consider how simple and low tech its approach is. It just tries every possible option through millions of separate trial and error tests until something works. This crude, dumb, pedestrian methodology produced remarkable innovations including our acidic stomach that doesn't eat its own liner, an umbilical cord not rejected by the new born within, antibodies that destroy foreign invaders without harming the host, testicles housed outside the body to maintain acceptable temperature, blinking eyes to achieve moisture balance, skin membranes filtering moisture out while shielding harmful external particulates from entering, etc. The list could go on and on.

The transition from plants to animals required solutions to a new set of problems. Food acquisition beyond photosynthesis tied to the fixed position of plants necessitated mobility, along with digestive systems and a variety of other needs for animals. So, nature developed a skeletal-muscular framework with a stomach, sensory organs, circulatory system, nervous system, and respiration. Then as life forms occupied a variety of environments with vastly different conditions, animals adapted and differentiated to survive. Some were successful, most were not. Engineering standards in nature are quite high: meet the need or die.

Evolution proceeds in a push-pull manner where internal-micro level processes continually push organic change from bottom-up, while the external environment pulls organisms in directions that favor their survival. This internal-external dynamic is then integrated with parallel forces pulling species into divergent pathways leading to further differentiation. Think of a sports team changing its members and then being put into a different league. The point is there are always multiple interacting forces at play adding complexity to the process – difficult to track. Further complicating the process are variations in the pace of change where modifications may take long, extensive time periods only to be followed by sudden, transformative change, generally from environmental crisis.

Laws of nature, physics and chemistry limit the end products of evolution, which debunks notions of evolution being just a random process. Chance plays a part in everything however it is always constrained by forces, filters and limits of physical reality. The deck is stacked, the dice are rigged. In organic life this is revealed in the rather exclusive, narrow forms and shapes chosen out of many possibilities. Nature insists on efficiency and conservation of energy. Seeds and eggs are spherical because they optimize space using the least amount of material. Symmetric patterns are ubiquitous for similar reasons, based on simple math ratios that reveal parsimony (balancing internal-external forces to produce optimal shapes/forms). Spiral patterns are omnipresent, including horns, shells, fruits and flowers – it's the universal pattern of proportional growth. Other common, efficient forms are hexagonal honeycombs and snowflakes, tetrahedron minerals and bubbles, and spherical everything. Parsimony limits the infinite possibility of forms to just a select few. There's actually no random chance behind the shape of fish in water, birds in the air, or human beings on earth. It's the same way nature limits man's engineering solutions to just a few forms and shapes in the design of planes, boats, submarines and racing cars.

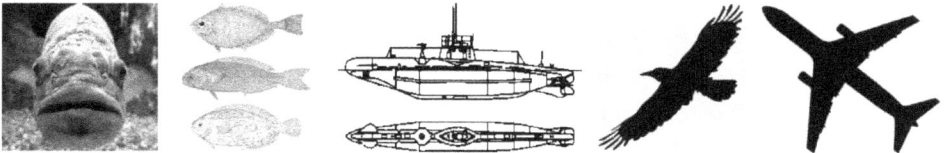

Physical laws explain the recurring patterns in organic structures that meet survival needs. Limbs, vertebrae, skulls, hands and fingers all follow similar configurations that balance strength and agility with external constraints. Height, girth and weight are constrained within a narrow range of possibilities. When growth of an organism increases surface area by *addition*, its volume increases *exponentially*, making temperature regulation difficult. Similarly, size restricts speed so optimal ratios are chosen based on predatorial threats. Due to math power laws, size also relates to species populations – bigger sized creatures have proportionally smaller populations.

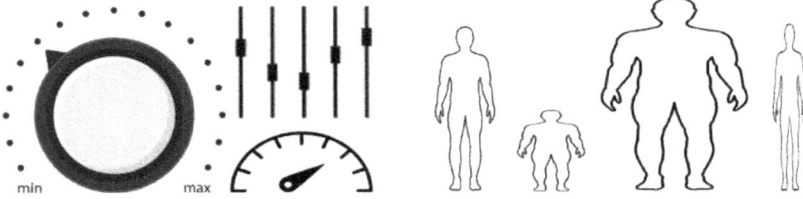

Under different external force conditions you would be shaped/molded accordingly

Nature's constraining forces limit engineering solutions to conserve resources and optimize functionality. From the micro (cells) to the macro (beasts), systems are designed by nature to do the most with the least. We see this in recurring patterns of coiled tubes, layered surfaces and arched structures to maximize space, surface area and weight bearing strength respectively. Other recurring forms include branching patterns (water tributaries, roots, leaves, lightening, arteries and veins), crystal forms (minerals and bacteria), and cantilevered bone systems (limbs, feet, hands). The human body is a microcosm of nature's limiting selection process that sculpted a structure with 208 levered bones, 650 skeletal muscles, a ribcage, vertebrae with discs and cartilage, cranium, pelvis, and pivot and ball-socket joints.

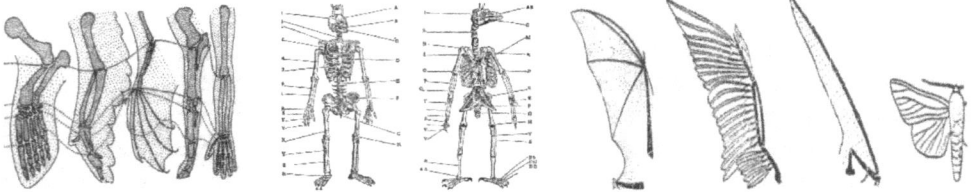

Limb engineering   *Skeleton structure*   *Wing engineering structure*
Seal-Bird-Bat-Horse-Human   Human   Bird   Butterfly   Pterodactyl   Bird   Bat

Evolution never happens in isolation. Every life form is connected to every other life form and its environment. Like dominoes, every event causes a chain reaction – an equal and opposite reaction symmetrical with newtons 3rd law. Ecosystems are interconnected webs of life sensitive to ripples in the joining fabric. Geology, ecology and climatology all play a part in this dynamic environment along with interlinked species connected by adjacent territories, predator-prey relationships and host-parasite partners. This lateral connectedness is amplified by vertical linkages between minerals, microorganisms, plants and animals. Given this integrated wholistic network of organic dynamics, life's process of change is more accurately described as Co-evolution. All parts and wholes abide and adjust together. Those that don't fail or die.

Unpredictable properties surface in the course of evolution that can' be explained by reductionist science, let alone ancient man's fabricated myths, fables and religion. The process of combining parts and integrating them into complex new wholes creates novel conditions leading to changes in both degree and kind. It's a similar emergent process seen in transitions between atoms to molecules, cells, organs and organisms. Increases in complexity often bring transformational change that induce synergy – the new whole becomes more than the sum of its parts. Dynamics and mechanisms come into play here. Social interaction among parts impacts the behavior of each individual part. Sequential actions are no longer simply linear but include non-linear networking interactions with parallel paths and feedback loops. Groups and clusters behave in ways that singles or a few don't. More is different. This emergent process follows a basic pattern: parts grow in number, then specialize, then integrate and become a unique whole. Next that whole and other wholes like it grow in number, specialize and integrate to become an even larger, more complex whole. Rinse and repeat, all the way from atomic particles to a fully developed human being.

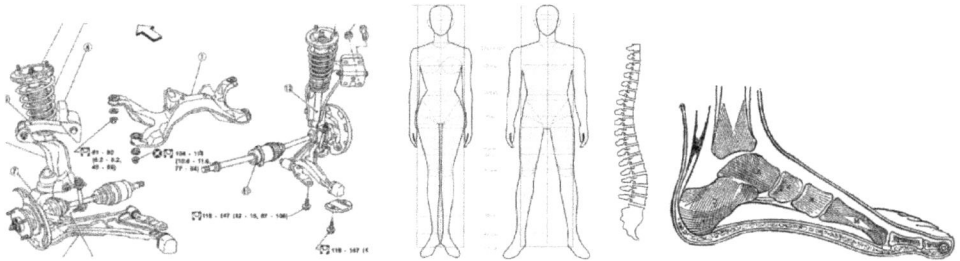

Human designed car suspension vs. nature's elegant engineering: form follows function

As life forms attain greater complexity, they become fully anchored in holism where integrated patterns and processes supersede static structures. Transcendent attributes arise where cells, tissues and organs synchronize, behaving as one purposeful system responsive to collective needs. This social quality makes the organized whole more adaptive – a value-added bonus lending itself to survival in an ever-changing external environment. New dynamics emerge in the realm of holism, supplementing linear processes with non-linear ones, incorporating networking versus hierarchy, and integrative relationships versus simple organic parts. Somewhere along the way individual parts sacrificed independence for service to a larger whole: a win-win for both. Not much different than a member of an organization who willingly complies with constraining rules and regulations in order to reap the benefits from group membership.

Complex life forms gravitate toward a state of dynamic flux – occupying a perfect balance between rigid order and fluid agility. Chaotic environments incentivize this. Static structures can't adapt. Disordered assemblages won't survive either. A happy medium is optimal. Again, it's like water occupying the middle ground between rigid ice and chaotic steam, containing the right degree of order and flexibility. Ironically, organisms attain stability through adaptive agility. Persistence is achieved through flexibility. Since environments are both fixed in certain ways and subject to periodic change, the most successful disposition for a living system is to maintain ongoing balance between order and fluidity, with the capability of favoring either extreme as needed.

Holism principles are fundamental in living systems and complex life forms. Systems are present at every level of reality, from physical to biological to social. Man-made ones are ubiquitous: machines, vehicles, factories, organizations and societies. They possess inputs-throughputs-outputs, they have connected parts contained by a boundary, and they're managed by a controller. Systems take in energy, material and information, process it and produce something. Here again, parts give way to the whole, relationships take precedence over things, and structures are less important than processes. A key mechanism ensuring stability is the feedback loop. Positive feedback reinforces successful action but when unchecked it can lead to excesses. Negative feedback limits action or confirms failure. Both work together to maintain balance in the system to enable forward progress. In a human body feedback regulators include sweating when hot, shivering when cold, breathing harder when active, yawning when tired, and a faster heartbeat when scared. Like a thermostat regulating room temperature, living organisms possess a multitude of similar regulators that ensure homeostasis by narrowing tolerable ranges of metabolic response.

Living systems possess the highest degree of holism, with nested sub-systems and integrated networks. These supersystems grow and develop with combinations of gradual and transformational change. As open systems they coevolve with the environment, demanding continual input of energy. Living systems generally take in energy and information, convert and store both, synthesize them for productive use, move around, enact defense mechanisms, reproduce, then output materials and information. Most of these functions are present in man-made systems as well. Both are complex nested wholes – layered subsystems functioning as synchronized networks responding as a unified team supporting the collective one. On the surface, complex superorganisms present an orderly structured pattern, yet internally they operate as an ongoing dynamic metabolic flux taking in energy, replacing parts, changing form and adapting to environmental pressures.

A human body is just such a complex living system constantly taking in and passing pounds of food, volumes of air and gallons of water to maintain a steady state organism with subsystems regenerating life sustaining nutrients and fluids, while hundreds of billions of cells die and get replaced every day. The whole is a temporary illusion of form whose essence is a dynamic process flowing through space and time. Consider a car made up of thousands of integrated individual parts working together within synchronized subsystems operated by a controlling driver. The vehicle appears as an enduring, unchanging single structure but its parts are replaced routinely, systems are upgraded periodically, and its particular model is transformed over several generations of progressions. Its pattern persists despite ongoing internal transformative change. Evolutionary processes transformed the embryonic model T Ford into today's high-tech, high-powered modern automobile.

The complexity of living systems brings them into a realm of chaos; a condition of ordered flux – another middle ground state between extremes. Chaotic systems exhibit fluidity and ambiguity while maintaining subtle degrees of order. Wobbling tops moves in such a manner. Every new spin leaves a similar pattern of circularity but always a little different than the one before. Centripetal force acts an "attractor" – a limiting force pulling the top toward a center point. Chaotic systems maintain a sense of order though attractors that pull or push in a limiting, order producing pattern. Complex organisms are pulled and pushed by environmental attractors orienting them away from stable states. Likewise, organic bodies including yours are always changing with external conditions, moderate or extreme.

Chaotic systems are unpredictable despite having some degree of order. High sensitivity to small change ensures this. No one can say with certainty when an avalanche will occur even though tiny additions seem harmless. Watching a cooling glass of water suddenly change its state to ice seems equally peculiar. Small incremental changes reaching a tipping point results in transformative change; emergence produces built-in ambiguity in chaotic systems. The butterfly effect illustrates this phenomenon where a seemingly small action results in enormous impact after a continual series of bit by bit, step by step, tiny actions.

Complex living systems thrive and survive through whole-being agility. In a world where change is constant, external pressures are demanding, and unpredictable events create periodic chaos, it's only natural that evolution would place higher value on an organism mirroring those very qualities. Living systems possess a fluctuating balance between stability and fluidity, and will shift towards either one when needed. Advanced adaptive capabilities emerge over time as both subsystems and processes continued to progress towards higher complexity. Non-linear networking improves responsiveness, along with regulating feedback mechanisms and synergistic synchronization of individual parts. None of this fits the realm of physical science where reductionist scrutiny cannot capture transcendent realities. Life evolves along lines of unpredictable emergence where novel attributes and spontaneous order are the norm. Life itself is beyond science which at best can only ponder its multifaceted aspects, but never its comprehensive character as a creative, progressive force in the universe.

PRINCIPLES OF EVOLUTION

   o Mix parts and choose best results
   o Parts and wholes change together
   o Pace of change fluctuates between slow crawl and sudden leaps (environmentally driven)
   o Not "Random Chance" as external forces limit fewer possibilities (Nature stacks the deck)
   o Species variation uses serial and parallel process, branching, divergence and convergence
   o ALL living things Co-evolve in an integrative dynamically linked manner
   o Emergent evolutionary change is unpredictable; can't apply scientific reductionism
   o Forward progress is non-linear, zigs and zags, involves complexity and layered wholes
   o Life survives thru stable agility; balanced middle state best adapts to changing environment forces

Science occasionally stumbles upon important insights while probing with blunt instruments around the periphery. A case in point is *Cellular automata*; a computer-based simulation testing principles found in organic beings. In this video game of life, graphic tiles are displayed and programmed with simple rules they must follow. Tiles can flip, twist, roll, and move in a variety of ways based on neighbor tiles. Their social response to others creates behaviors that animate each tile. Making simple social rules results in otherwise inanimate tiles acting in very life-like manners. Adding a few rules here and there can suddenly create complex, chaotic behaviors. Individual tiles can coagulate with others and then oscillate, blink, multiply, glide, zig zag and move in elegant patterns. Predictable rule sets occasionally produce anomalies. Even the simplest rules can create complex movements, beautiful symmetry patterns and fractal forms. The game of life lends insight into fundamental principles of evolution where complex beings can trace their roots back to simple rules of social behavior originating at the basic physical level of atoms and molecules.

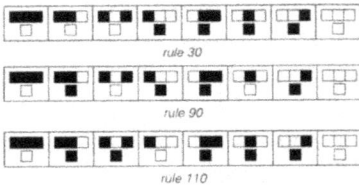

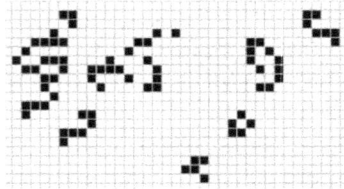

Simple rule sets create unique patterns   Process rules create moving creatures

LIFE'S CHOSEN PATH

Billions of years ago when the first atoms began socializing with each other to form partnered molecules… life became inevitable. Not a miracle, not a random chance occurrence, not some divine intervention – life had to happen. The social qualities of matter combined with physical properties of energy made the emergence of life a certainty. All that remained was guessing the path it would take on its way towards omnipresent manifestation.

 If early conditions in the universe were duplicated and life had to make the long journey again it would certainly be different, however many things would end up surprisingly similar. Some life forms that made it this time wouldn't, while other that didn't, would. The dynamic process of coevolution and environmental chaos would alter the natural selection process, change some rhythms of the timetable, and favor different organisms here and there; resulting in an altered outcome. But the general forms of creatures and patterns of living organisms would be surprisingly similar. As long as physical properties and general environmental conditions remain the same evolution will produce similar products.

 In any timeline the early progress of life is excruciatingly long. Producing the first organic substance resembling a cell took about ¾ of a billion years. One billion years later photosynthesis emerged. It then took another billion years of cell progression to form a nucleus. And then yet another billion years later cells coagulated to form multicellular rudimentary organisms. So, after about 4 billion years of laying the groundwork life was ready to take off. Plants and animals developed rather quickly given the time scale, increasing in complexity after several hundred million years.

The journey was intriguing and unpredictable with unexpected turns and detours. Some organic forms progressed along a steady narrow pathway only to come to an abrupt dead end. Others weaved in and around pathways that branched off in different directions, each offering unknown payoffs and perils. Where paths widened and payoffs increased, new life forms appeared. Wider, deeper routes opened more branches and variations. Graphically visualizing the comprehensive path life actually took would be impossibly complex but we can certainly capture the general chosen pattern of divergence:

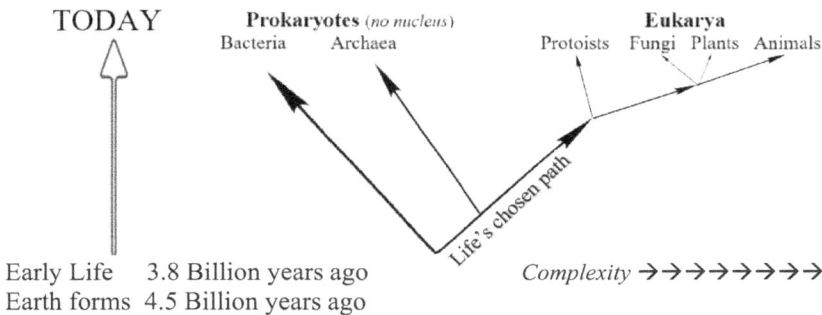

This path emerged after countless turns, twists, dead-ends, and do-overs as life pursued success, each turn adding hundreds of further branches with their own twists and turns.

Forms of life diversified over time from environmental conditions, coevolution, natural selection and a variety of factors favoring or limiting specific types. Organic complexity varied early on as fully developed cells branched off from rudimentary types that didn't even contain a nucleus. Later on, further branching divided single cells from multicellular forms. Mutation, genetic drift and adaptive radiation in both parallel and divergent paths spread life in all directions, carrying with it common forms and mechanisms proven successful. Then complexity took a big leap forward with the development of multicellular plants and animals. From there diversity exploded in all forms, including 350,000 species of plants alone.

Plants had no mobility, so they achieved stability using a fixed structure specializing into 3 parts: Roots for water/nutrients, a stem for transport, and leaves for capturing sunlight and CO_2. Plants did have limited mobility in one direction – upward. Anchored plants grew towards the original source of all life, reaching for the sun, the same way conscious vertically erect man reaches for the heavens. As coevolution proceeded, plants partnered with insects to pollinate in a rudimentary form of sex.

Animals progressed beyond plants, trading fixity for mobility. Roots or branches became arms and legs. Mobility enabled greater opportunity to seek food but now required development of digestive systems. Self-sufficiency required a central nervous system, eventually developing into two parts: a sympathetic system for fueling organs and muscles plus a parasympathetic system for maintenance functions. As complexity increased, nervous systems developed a complex brain to run the whole organism.

Evolutionary Sequence of Species Development – *increasing complexity*

Mammals *dogs, cats, cows, horses rabbits, chimps, humans*
Birds *(subset of reptiles) owl, eagle, swan, crane, vulture*
Reptiles *lizards, snakes, turtles, crocodiles, dinosaurs*
Amphibians *salamanders, frogs, toads, caecilians*
Fish *urchins, starfish, sharks, cod, perch, eels, sturgeon*
Bugs *fleas, flies, bees, ants, lice, termites,*
Fungi *sponges, molds, mushrooms, yeasts*
Plants *flowering, conifers, ferns, mosses, algae*
Bacteria *cocci, bacilli, spirilla, vibrios*

Mammals achieved the greatest complexity of the animal kingdom with significant improvements offering advantages over other life forms. Birth would now occur within the mother, transitioning away from external shelled eggs and into the warm protection of a womb. Similarly, mammals would adopt an internal skeleton to replace outer shells used by other life forms. These "vertebrates" would include some type of spinal cord. Along with birds, mammals became warm-blooded creatures – a development that increased adaptiveness to a wider variety of environmental conditions. The tradeoff for this advance would be a greater need for energy intake to maintain proper body temperature. Lastly, the most important transition to mammals was emergence of a larger brain size, paving the way for early primates and the ascension of man.

Major kingdoms of organic life break down much further into detailed classifications. Groupings are based on features, types and functionality. The complexity and variety of life forms requires a hierarchal categorizing scheme with multiple layers. Every living thing falls within the following schema:

TAXONOMY OF LIFE	*HUMANS* (Example)	*GENERAL (Many)*
Domain	Eukarya *(Cell type)*	
Kingdom	Animalia	
Phylum	Chordata	
Class	Mammalia	
Order	Primates	
Family	Hominidae	
Genus	Homo	
Species	Sapien	*SPECIFIC (few)*

The comprehensive classification of life reveals both diversity and unity. Evolution's path may split, branch and diverge into a million different directions but every single living organism is connected by a common ancestor. Our grand taxonomy of life confirms the connectedness of diverse organisms. Further break downs and classifications of life provide clues on how environmental pressures shaped and formed bodies of every creature at all levels. A brief look at any kingdom illustrates a filtering process where life's path through variation and selection branches into multiple choices, each successful in their own unique way.

ANIMAL CLASSIFICATION

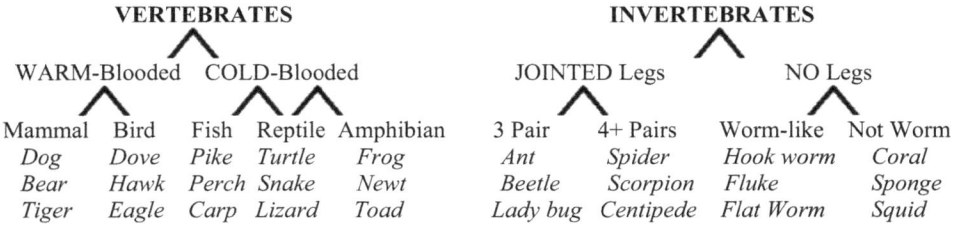

VERTEBRATES					INVERTEBRATES			
WARM-Blooded		COLD-Blooded			JOINTED Legs		NO Legs	
Mammal	Bird	Fish	Reptile	Amphibian	3 Pair	4+ Pairs	Worm-like	Not Worm
Dog	Dove	Pike	Turtle	Frog	Ant	Spider	Hook worm	Coral
Bear	Hawk	Perch	Snake	Newt	Beetle	Scorpion	Fluke	Sponge
Tiger	Eagle	Carp	Lizard	Toad	Lady bug	Centipede	Flat Worm	Squid

This is just a summary of the millions of living species tucked away within sublevels and branches of the hierarchy. Ponder the sheer immensity of every living form on earth: thousands of mammal species, birds, reptiles, amphibians, and tens of thousands of fish species…and these are just the vertebrates. There are well over a million invertebrate species, mostly insects and bugs. And these are just the species we know of. Many others go uncounted – hidden in remote habitats, too small to detect or residing within the bowels of other life forms. Now consider *billions* of prior species, came and went, never to be known or seen again.

Tracing life back to its origins involves a journey spanning billions of years, beginning about 500,000 years after the earth formed. The period of time where rudimentary life developed excruciatingly slow covers most of earth's history. This Precambrian era lasted about 4 billion years. During this long, long time span simple organic substance emerged, followed by basic bacterial cells and then fully nucleated cells. By the end of this enormous 4000 millennium era organic life developed into multicellular organisms.

Life Radiates and Diversifies   *Life Always Branches Out*

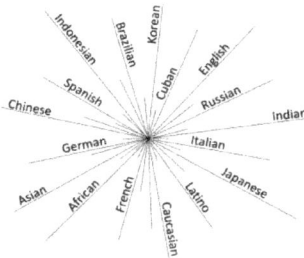

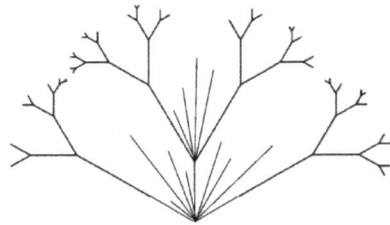

Race-Ethnicity-Culture all blend and vary on life's path. Life splits, divides, multiplies…

Finally, after all that time we get to the last half billion years of earth with the Cambrian explosion of life, where the timetable of evolution begins to dramatically accelerate. Early ocean life in the form of microbes and sponges gradually increase in complexity, leading to vertebrates, fish, amphibians and reptiles. Life next transitioned from sea to land, paving the way for plants and simple animals. Eventually dinosaurs appeared along with their airborne cousins that would later transition into hundreds of bird species. After the most recent mass extinction about 65 million years ago 75% of all living creatures perished, including dinosaurs, paving the way for the rise of mammals and early primates. Modern man wouldn't arrive on the scene until very late in life's long journey – just a mere 2 million years ago. There are 4 Major geological periods (Precambrian, Paleozoic, Mesozoic, Cenozoic) each containing separate periods that denote significant developments in organic evolution, including extinctions.

Crunched Time Scale   *Actual Time Scale*

4.5 Billion Years	Earth formed	
3.8 Billion	First Cells formed	4.5 Billion Years
2.7 Billion	Cells develop a nucleus	
630 Million	First multicell living substance – Sponge	
540 Million	Cambrian explosion of emergent life	
500 Million	First plants appear	
470 Million	First Vertebrates – Fish	
370 Million	First Amphibians	
320 Million	First Reptiles	
300 Million	First Dinosaurs	
250 Million	Permian Mass Extinction (96% species gone)	
200 Million	First Mammals	
150 Million	First Birds (flying Dinosaurs)	
135 Million	Americas split from Europe & Africa	
65 Million	Last Mass Extinction (Dinosaurs gone)	
50 Million	First Primates	630 Million Years
4 Million	Upright Hominids	
200 Thousand	Modern Man arrives	Today

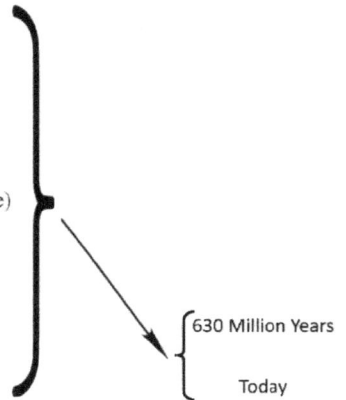

Organic life has quite a storied past. It's been a long slog through billions of years, periods of mass extinction and patterns of ever-increasing complexity. Life took a remarkably long time to just get out of the gate and then prodded along, crossing certain thresholds to become unstoppable in its accelerating progress. It all began in the oceans but eventually expanded beyond, permeating the lands and atmosphere. Yet each and every creature that made it this far is connected to a common ancestor in a long unbroken chain of organic substance. To this day land mammals like us are filled with blood that matches the salinity of the ocean, along with our sweat and tears. Every transitional period along the journey merely passed the baton onward to a new paradigm offering novel products yet always carrying with it original hereditary cargo.

We can deduct from this history of organic evolution that given enough time, life will find a way. Simple matter will organize into greater complexity and manifest into a variety of forms – the most successful of which will live on while the others go away never to be seen again. Four and a half billion years of evolution resulted in the cumulative progressing of slow gradual change supplemented by unpredictable punctuated shifts and dramatic change; no longer by degree but changes in kind. These transcendent leaps combined in novel ways that produced emergent new-improved forms. The methodical dynamic progress from atomic particles to organic molecules to complex beings involved many hundreds of step-by-step incremental advances.

Key Organic Evolutionary Stages:

1. Organic Molecules form with Amino acids and nucleotides
2. Self-replication emerges with RNA chains copying each other
3. Enclosed cell forms after membrane appears via bubbles containing RNA molecules
4. Cell Division emerges from growth and internal pressure
5. Proteins made by pseudo-coded RNA
6. DNA forms info template – RNA links it with substances
7. Photosynthesis learned method of converting sunlight into sugar energy
8. Cell mobility via hair-like cilia enables movement for food
9. Sex mechanism mixes DNA between cells
10. Nucleated cell – pseudo control center
11. Complex cell forms after symbiosis incorporates energy producing mitochondria
12. Multicell groups form via social interaction in cooperative bond
13. Plants form from collective integration of multicell groups
14. Primitive animals form with segmented bodies and further integration of parts
15. Central nervous system develops as rudimentary basis for sensing organs
16. Internal skeletons transition from external shells promoting growth and mobility
17. Co-evolution integrates plants and animals – encourages communal cooperation
18. Warm blooded animals increase adaptability in variety of habitats.
19. Brains grow and increase capabilities of animals
20. Humans develop individual and collective consciousness

EVOLUTION DECODED

The process of evolution is widely and needlessly misunderstood. First, reductionist science trains an analytical thinker to break things down to focus on parts, unwittingly overlooking emergent synergy in wholes that transcend the parts. Second, perception of time limited by a relatively short human life span along with a natural bias to here and now obscures the required comprehension of geological time frames involving millions and billions of years. Third, the subject is simplistically taught as a series of random events, failing to account for physical laws that limit possibilities and provide directionality. Fortunately, a mountain of physical evidence reveals the process, including a comprehensive fossil record, linked genetic record and layered geological residue – all serving as a physical memory catalog. Each reinforces the other providing obvious clues, such as the consistent pattern of deeper sediment layers possessing increasingly simpler life forms frozen in incremental time periods.

Evolution's multifaceted and multidimensional aspects must be perceived in a unified manner to appreciate its dynamic variety of mechanisms working together. Both replication and variation are essential, the later including both mutation and sex. Once differences are produced natural selection filters the results to maximize and optimize. Successful developments are repeated and passed on, modified to fit changing needs. Physical laws of nature direct the process as environmental forces shape and limit life forms. This parsimony applies to all life types, changing and adapting in sync, producing a coevolutionary dynamic – *all change is integrative*.

The process is long and drawn out but periodically takes an unexpected great leap, leading to emergent innovation. Punctuated, transcendent novelty occurs when numerous specialized parts coagulate and collectively form new, integrated wholes. As complexity increases, sophisticated systems soon develop, possessing greater capabilities. Organic open systems operate in a dynamic balance between stability and flexibility to achieve greater agility needed for adapting in ever changing environments. A chaotic internal balancing act puts organic life in the sweet spot between structure and flow.

Evolution is an imbedded aspect of physical reality where natural forces induce dynamic processes of change, producing original, unpredictable products that persist or perish in variable conditions. The mechanics and principles involved are repeated over and over, producing universal patterns in its wake:

Life's Recurring Patterns

o Builds from the bottom-up
o Assembles in chains
o Organizes information and energy
o Exploits variation
o Co-evolves with environments and other life

o Recycles materials
o Leverages water and solar energy
o Opportunistic
o Relentless persistence

A key attribute on that list goes relates to the component of evolution not covered yet: *Purpose*. This is the final element in our evolution equation, which already includes information, energy and process. It directly relates to definitions of the meaning of life that focus on aspects of purpose: Attainment, fulfillment, progress, learning, or reaching potential. It's the enigmatic ingredient adding a special quality to life that science cannot grasp. What drives the process of evolution? It sure isn't some quantitative unit to be measured or defined but it is easy to see and obvious in its disposition. The results speak for themselves. Evolution is a driven process whose purposeful spirit is formidable.

Evolution Equation: Evolution = (Information + Energy + Purpose) x *Process*

Relentless Progress of Life. Nothing pushes forward with such relentless force quite like the focused continual ongoing diligence of life. As a force of nature, it's persistent, resilient, and downright incorrigible. It cannot be stopped and it won't take no for an answer. Life knows only one mode: ferociously pressing onward: up and down, in and out, all around. It thinks outside the box and pushes beyond all boundaries. It breaks through all limits, takes short cuts whenever possible, but can also patiently wait in a dormant state until future opportunities arise where it can press forward again without missing a beat. Simply put – life will not be denied.

Five mass extinctions have occurred over millions of years and yet life persists, moving forward wherever it can. If a nuclear holocaust rained upon the earth... somewhere life would survive and continue on. Like the genie let out of the bottle, life cannot be recalled. It grows and spreads like a wildfire that must run its course. But where a fire runs out, life goes on. It branches out, runs into some dead ends, moves around those and extends into multiple new branches. Life began in the seas but the vast oceans just weren't enough. Inevitably it spread onto land – all land and all places covering the entirety of the earth's surface including beneath, in the soil and sediment, and above in the skies too. Life won't be deterred. Consider how quickly nature recovered after Mount St Helens blew a cubic mile of dirt, rock and ash, killing scores of humans and destroying entire forests. If not convinced, just compare Hiroshima then…and now. Life *always* goes on.

Consider how unlikely you are to be alive. Every human being in your ancestral chain back in time had to be born, survive and have children, repeating the process thousands of times through history. Each birth required 1 sperm out of about 100 million to reach and connect with 1 egg. What are the odds your great, great, great, grandparents survived the common afflictions of the past: famines, disease, war, fatal sicknesses, and simple accidents that became life threatening due to lack of basic medicine? Yes, you are a highly improbably anomaly in the big scheme of things but even when the odds are greatly stacked against it, life WILL find a way.

Life knows only 1 mode: persistence, which leads to inevitably. Life goes everywhere and tries everything. It radiates in all directions as it diverges, converges, branches, zigzags and loops around. It infiltrates and saturates, finding its way into every nook and cranny. No hiding place is safe. Like water flowing in a stream, it meanders, bends, winds, soaks into the ground and when it falls over a cliff it just resumes flowing on a different plane. Life is constantly probing out, reaching further and renewing. Trees illustrate this pattern as the tops radiate out with branches and leaves to embrace the atmosphere while the roots reach out below to absorb water and nutrients. It's like a newborn baby pushing out head first, emerging into a new world of possibilities. From the beginning of conception right up to the birthing process, life is moving on and out and won't be stopped.

Life is extremely adventuresome as it seeks new environments and tries pathways never explored before. Harsh conditions just mean more adaptations to fit in. Deserts may seem lifeless, but they aren't. Frigid antarctica couldn't possibly sustain life but it does. Same with the pitch-black dark depths of oceans where organisms are present 8000 feet below. Life is always on the move – vines climb, birds fly, creatures burrow into soil, and snakes slither along. Mobility is built into life starting with cilia on cells, then fish fins and tails, bird wings, pollen spores, sticky prickers, and legs in all shapes and numbers.

What distinguishes life from other natural forces is its degree of ruthlessness. It will try anything and do everything to survive and persist. No Marquis of Queensbury rules here. No implied fairness. It's often kill or be killed. When it comes to getting what it wants life is fanatical. It will do whatever is necessary to survive. Human combat sports are child's play in comparison. About the only thing remotely resembling life's will to win in the field of human endeavor is war. The closest thing to it in sports is the do or die playoff bracket where each team wins or gets eliminated. Nature plays the real tournament of life where species compete on a daily basis just to have the privilege of getting to tomorrow. The urgent demands of this tournament led to developments in offense and defense in every life form.

Evolution's brute creativity produced a wide variety of defense mechanisms: claws, fangs, shells, horns, needles, smells, and sprays. Instinctive survival behaviors emerged such as hissing sounds, herding defense, hiding, changing color and faster running speed. Some life forms simply regenerate as in the case with severed worms or a torn salamander limb. An often-overlooked survival tactic is partnership and alliances. Cooperation and conflict are both dynamics of survival built into the tournament of life and either will be exploited without judgement.

Life is indifferent when it comes to getting from here to there. It will fight, cooperate, exchange, change course or plow straight ahead. It if has to kill, it will without remorse. There is only one judgement it makes: does this work? Success is all that matters. Take the case of a parasite that feeds off its host. If it can get away with it, good. If it ends up killing the host and thus itself, well, not so good. Success has a way of persisting while bad approaches disappear.

Assimilation is a simple strategy employed by life, even more successful than Star Trek's *Borg* which unapologetically will borrow, copy, steal, upgrade or reuse anything that might help it move forward. This approach produces complex organisms at the leading edge of a long chain of incremental modifications. Sometimes these final products carry with them old vestiges of parts that were once useful but now just come along for the ride. In humans these left over remnants include wisdom teeth, body hair, sinuses, male nipples, tonsils, coccyx, appendix, and others; all left behind as life pressed upward and onward.

Some pathways included parasites riding on someone else's effort. Life has no shame and will leech onto others whenever it serves its own purpose. Viruses routinely invade a host, use its resources to reproduce and then move on to invade another. And what is reproduction other than life's selfish method of exploiting organisms to pass on ancestral stuff before they die. As comedian Jerry Seinfeld noted, "make no mistake, our children are here to replace us". Evolution indifferently churns out organism after organism to pass the baton forward for better products later on. Hey thanks for the assist…now go away and die. *Sheesh*.

Life's greatest accomplishment as an agent of assimilation is its engineering feats. As a world class problem solver evolution has produced astonishingly efficient mechanisms of survival and self-improvement. From the early days of cell membranes, cilia and nuclei, organisms developed complex systems of mobility, defense, fuel conversion, telecommunication, sensory devices, and computers. You know these more commonly as stomachs, eyes, ears, noses, vocal cords, skin, muscles, skeletons and brains. The sheer variety and generational modifications of these engineered systems through diverse species is breathtaking. Environmental necessity required organic material to be formed into cantilevered levers, skeletal frames, tubular networks and hydraulic systems, cunningly designed by nature to be effective and efficient. Life routinely exploits physical laws to maximize energy, space and resources, seen through built-in symmetry tied to mathematic codes, including Fibonacci, the golden mean, power laws and geometric principles.

Death – Topping life's relentless qualities is its sheer ruthlessness and nowhere is this clearer than in its most powerful innovation: *death*. Everything life creates dies. It's built-in to the very process and for good reason – it ironically provides survival value. That's right, death is necessary for life to survive and thrive. Death is the flip side of birth, both essential to life. Death is about sacrifice – trading something now for something better later. It opens the way for new life. This is why natural forest fires are intentionally allowed to run their course. It's a convenient solution to overpopulation of anything. Death clears away outdated ideas and obsolete products. Slow dying cities, run-down urban centers and ghost towns are not exempt. Out with the old, in with the new. Every form, every product and everyone is expendable. Painful in the short run, positive in the long run.

All life essentially feeds off death. We're all parasites in this respect, eating food from living plants and killed animals. Even plants are parasites of the earth. And all life on earth is parasitically dependent on the sun. One long integrated chain of living organisms sustained by the consumption or death of other life forms. This can be seen in real time with a browning apple where thousands of microorganisms decompose it, continuing the parasitic process that sustains all future life. Or consider the manner in which humans infested the entire earth in the blink of an eye on the geological scale of time, spreading rapidly across the globe like an out-of-control cancer. It's just the way life pervades and assimilates everything in its path.

Still, there's great value death brings to the table as it facilitates the filtering process of eliminating weak products to make room for better ones. Predatory animals attack the weakest in the heard, leaving the strongest to live on. Similarly, natural selection filters out the least fit organisms in an ongoing incremental process collectively improving all life. Sex was developed to enhance this dynamic by providing more products to choose from. Once again death facilitates life as a tradeoff mechanism where simple duplication is replaced with organisms that create variation and then die. Nature's selection process enables consideration of more maybe's to choose from, where only a few become yes's while the vast majority become no's. It's a simple, literally dumb mechanism: pass = live, don't pass = death. This mechanism while dumb is incredibly powerful and effective. All life that ever lived has faced this test. The results: 99.99% of all species that ever lived failed the test, are dead and gone, never to be seen again.

Without death there could be no life as we know it. It's integral to living beings that couldn't function without it. Right now, cells are dying by the thousands inside you in a very routine manner. Many of your cells are programmed to die – a form of intentional suicide. It's built into the DNA recipe for constructing a living being. A human hand is shaped by forming a general one-part whole and then directing cells in the middle to die, differentiating space between to form fingers. Sculpting tissue by instructing cells to commit suicide is life's way of exploiting sacrificial parts to benefit the whole. Likewise, whole organisms, including humans, sacrifice by producing children, supporting them through maturity, and then dying later, conserving life's resources for a new improved generation of offspring.

Life and death are not opposites – Life transcends death. Birth and death are opposites. Both birth and death are subsets of life, parts of a greater whole. Birth and death are easy; living and dying are hard. On the spectrum of birth and death there's a degree of ambiguity where the former starts and the latter ends. Does a human's organic life begin upon conception, at first heartbeat, birth, first gasp for air, or cutting the umbilical cord? Does a human die when breathing stops, the heart stops, brain wave activity ceases or some other arbitrary point? Does a human being maintain its identity after 7 years when every cell in the body has died and been replaced? How can a person be alive in one moment and dead in the next when their internal parts are identical? Our conventional definition of life is just a limited subset of a broader scope, beyond organic attributes – one that transcends birth, death, and everything in between.

Between birth and death is a spectrum of living and dying. The two are complimentary opposites that go hand in hand. Ironically, living is dying. Like a battery that must run down to function, living beings begin dying the very moment they become alive. Every subsequent day after birth is one day closer to death. Yin and Yang, both part of the other.

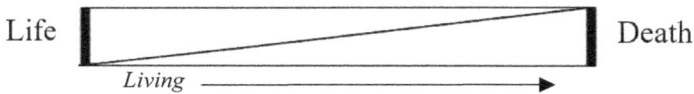

Life [   ] Death
Living ⟶

Fear of death is universal, but the act itself is more routine in our lives than we realize. Every day we go into deep sleep and later wake up anew – a sort of pseudo-resurrection. Each day we incrementally transition into a new form of ourselves, imperceptible until long periods of time pass. Similarly, our whole lives can be seen as a series of stages, each with attributes of an older, discarded version of ourselves that made way for a new improved one. Death and dying are a natural part of life and even when it involves great pain there's often comfort and release at the end. The final transition from life to death can be a blissful state, commonly bringing a fulfilling sense of warmth, weightlessness and visions of bright, pure light. This fits with death's nature as a transformation agent. Like the tadpole becoming a frog or the caterpillar becoming a butterfly, death possesses transitional value everywhere it manifests. Evolutionary progress, transformative change, transcendence and ultimately life itself is connected and unified by death at every level.

OMNIPRESENCE OF LIFE

Evolution's relentless path through physical reality culminates in a comprehensive web of life integrating everything. The appearance of lifeless, inanimate objects are optical illusions. All is alive to different degrees. All is connected and integrated. Life is an eternal process, weaving through ecological networks interconnecting all organisms to each other and their shared environments. Within this wholistic interrelationship we find principles of symbiosis, complementarity and coevolution operating everywhere. System dynamics dominate the action with networks, webs and woven tapestries. All together you get a unified superorganism with countless independent specialized parts working in sync as a massive collective whole, including "dead" matter, organic life and everything in between.

Ecology is a comprehensive environmental playground of living species responding to external natural pressures and internal competing forces. Climate and weather are an integral part of this connection as well as geology and oceanology. The integrated biosphere is a concentrated organic layer covering the earth's surface, extending up several thousand feet in the atmosphere as well as several thousand feet deep into oceans. Life in all its varieties and parts is synchronized in harmony like an orchestral symphony, including those in conflict relationships. Ecosystems operate in a continued state of dynamic flux, maintaining *balanced instability* – chaos within limited boundaries regulated by stabilizing feedback loops. Predator-prey relationships fluctuate yet stay within limits as each population symbiotically depends on the other. Host-parasite relationships operate with similar regulated connectedness. Food chains vary from plant-bearing fruit to animals eating each other, each linking all species together in one continuous feeding cycle. Interdependent species adjust their biorhythms, migration patterns, hibernation cycles, etc. in sync with the greater whole. Coral reefs serve as a perfect example of interconnected habitats with a wide variety of interlinked marine life, including sponges, oysters, clams, crabs, sea stars, urchins and numerous species of fish, all interdependent as a living whole. Symbiosis is a connecting glue unifying all life.

Every organism is linked by an unbroken chain through eons of time back to an original ancestor. Species of all types have coevolved in parallel with all other life while progressing and developing in a serial sequence of individual paths. These interlinked webs through time include *environmental coevolution*, each changing and affecting the other in mutual complementarity. Life's long journey brings with it an imbedded physical memory contained within everything it touches, linking the here and now to every past. While organisms code this right into their DNA, the physical environment possesses similar pseudo-memory linking things back in time. A rock lifted higher contains greater potential energy than before; a difference of stored energy that's equivalent to stored information as physical memory. In this same manner memory is stored in sedimentary layers, imbedded fossils, sculpted landscapes and especially every organism alive today serving as a living legacy.

Protocell→Bacteria→Eukaryote→Sponge→Plant→Fish→Amphibian→Reptile→Mammal→**YOU**
An unbroken, incredibly long linking chain connecting the original ancestor cell to YOU

All organic life shares the same resources, recycling everything in the process. One organism's waste is another's feast. Sometimes an entire being serves as a complete meal for an even larger being. The ingredients of life are universal, varying only by degree. Ecosystems are busy feeding frenzies where energy packaged in a variety of forms is transferred from the environment to and between all species. Remember, it all came from the sun with both its life sustaining energy and the material source of elements on earth. As Carl Sagan famously proclaimed "we are all made of star stuff". Ingredients haven't changed; they just assume new forms and get further recycled.

Water is a primary ingredient of organic life. We're a microcosm of the greater whole, mirroring the composition of earth with identical proportions of water along with blood salinity and sweat matching that of the oceans. Water is essential to metabolic processes, without which most life would perish. Though hidden, it saturates our atmosphere, only becoming visible after increasing density bursts out as rain. Weather is essentially about fluctuating degrees of moisture, dynamically changing in response to temperature and pressure differences, all connected by water in the air. The entire biosphere is linked by cycles of water passing through living systems and their habitats.

Parsimony reinforces the symbiotic framework connecting all living organisms to external environmental conditions. Bodies are formed by physical forces which limit dimensions, shapes and parameters. Fish are shaped to navigate in water; birds are engineered to effortlessly glide through air, and mammals are shaped to maintain proper body temperature and survive against predators. Man, the intelligent engineer, follows the same rules as nature, designing the shape of boats, planes and vehicles to abide by natural constraints. Life's dynamic coevolutionary processes are comprehensively integrated with everything else – manifested in ubiquitous symmetry patterns expressed in a variety of life forms, confirming the connection between organic processes and physical properties of nature.

Seed	Stem	Flower	Fruit		
				Minerals	*Grow*
				Vegetables	*Grow, Sense*
				Animals	*Grow, Sense, Move*
Point	*Line*	*Plane*	*Volume*		

Life's hierarchical scale begins first with particle interactions and atomic-molecular social dynamics, then continues with organic interaction of bacteria, microorganisms, and scores of tiny creatures, eventually developing into the higher complexity of plants, fish and animals. The lower end of the scale has smaller parts but very large populations. Microbes dominate the lower rung of organic life, collectively possessing the largest mass of living matter on the entire planet. There are trillions of microbes in every person. You have more bacteria cells in your body than human ones! A clean glass of water has about 10 million bacteria in it...yummy! They permeate the biosphere and are essential to all life, including our own. Human digestion requires symbiotic assistance of microbes in the stomach and intestines. The micro level of life also includes dirt, dust, soil, and the variety of tiny creatures hidden from sight. Dirt serves as a general suppository of dead organisms, sewage and waste. More than half of all life on earth is hidden in soil, home to a host of creatures burrowing in, tunneling and sheltering, including beetles, bugs, ants, spiders, worms, slugs and many others.

Man in the middle enters life's scale at the midpoint, above plants and animals but below the social realm. Higher up the scale life integrated into complex wholistic systems, transcending organic substance. Living beings coalesce into organized communities taking on a life of their own. In the same manner that ants form colonies, humans progressed collectively upwards to form families, towns, cities, states, nations and planetary communities. Organically this process is mirrored by cells-organs-organisms-living systems, ecosystems and the entire biosphere. At the top of this nested series of parts and wholes is the earth itself; a living superorganism or GAIA, a concept introduced by James Lovelock in 1970.

Earth exhibits a discernable metabolism of activity including shifting tectonic plates, receding valleys, churning rapids, violent volcanos, vibrating quakes, floods, fires and constant turbulence of changing weather – mimicking the mood swings typical in human drama. Metabolic dynamics are present in land, air and sea, with erosions, shifting fault lines, currents, waves and raging storms. Each of those big three are interconnected and inseparable as weather always impacts air, land and water together. The planet's skin has hair in the form of grass, bushes and trees. Gases (CO_2) taken in by this surface tissue transform into breathable air, symbiotically supporting animals permeating the biosphere who in turn release CO_2 right back. Ancient civilizations sensed mother earth as living sustainer of all life, yet earth itself is just a nested whole in a larger cosmic system. Her moon plays a key role in life with a lunar cycle regulating organic behavior and of course our sun is the ultimate source and sustainer of *all* life. Greece's Anaximander perceived the entire universe as a living organism as far back as 600 BC.

Expanded Scope of Life. Our limited perspective of life comes from years of programming and natural xenophobic bias. Since we're "alive" we assume other life must resemble us. We've been long programmed that life consists of beings like ourselves along with lesser animals, plants and various creatures. Curiously, we occasionally describe inanimate objects with metaphorical qualities of living beings, like a ship or car we treat and describe like a pet. We can't help but notice how a car engine comes to life, purrs like a kitten idling, roars in acceleration and coughs, gasps and dies when starved for air or fuel. Certainly, weather displays volatile mood swings and tantrums not unlike those of human beings. Communities exhibit a collective personality shifting in moody ways that fit the zeitgeist. Much of the inorganic "non-living" things around us express distinct aspects of life.

Our spectrum of life reveals ambiguity present on both ends, the beginning and end of a life. On the micro level uncertainty is the rule as to when something definitively becomes alive or dies. Similar indeterminacy is present at the macro level as life extends beyond just the biological. A more comprehensive spectrum has organic aspects in the middle but then extends out both ways to include non-organic life at the lower and higher ends, physical and societal respectively.

Societal	– Communities & Organizations	(Social Life)	linked by *language*
Biological	– Carbon based Organisms	(Organic life)	linked by *DNA*
Physical	– Substances & Things	(Non-Organic Life)	linked by *Periodic Table*

* Each successive level is a phase change like water transitioning from solid to liquid to gas

Note the transitional overlap between each level. At the bottom, particles, atoms and molecules form organic compounds that somehow develop into organic life. Higher up, organic beings coalesce to form groups entities. Transitional intermixing of domains occurs between each level. Man in the middle integrates physical, non-organic parts into his very being. This interior mix includes artificial limbs, pins, plates, stiches, implants, pacemaker, etc. External surface components are merged as well such as contact lenses, hearing aids, braces, lotions, colostomy bags, chemical patches, etc. Both internal and surface attachments are only the beginning. There's a variety of non-organic stuff serving as symbiotic extensions of us, such as clothes, tools, and vehicles. Man's collective extensions include 1.5 billion cars, 2 plus billion homes and 5 plus billion smart phones. These extensions mirror nature's own. A human's house is not much different than a snail's shell.

On the upper end of the spectrum, biological life transitions into societies with emergent properties of life on a higher level. Living members that coagulate into groups create collective attributes similar to ant colonies, bee hives, schools of fish, etc. The whole exhibits a group mind transcending the individual parts or members. Human collectives grow in complexity, forming super systems with specialized functions and all manner of group dynamics. Consider a city as a living system with internal networks of pipes, wires, channels, processing centers, transportation webs, continual throughput of resources, and waste materials. Life at this level resembles a living superorganism.

The transition between organic and societal life has its share of mixed forms as well. People are incorporated as specialized parts integrated into factories, businesses, utilities, agencies, armies, etc. Societies and sub-communities interact and respond symbiotically as connected entities. Each human cell sacrifices freedom and independence to serve and support larger organizations, which in turn provide reciprocal benefits to individual members.

Extended Life Spectrum

E	Societal	City, State, Nation, World community
F	*Mix*	Business, Factory, Collective, Army **neuralink*
I	Biological	*Man*, living beings, Organic Substance
	Mix	Implant parts, Machines, Vehicles, Homes, **cyborg*
L	Physical	Non-organic stuff, raw material and things

A comprehensive definition of life must consider phenomena beyond the limited biological domain. Organic life is really just a middle point on a spectrum extending and transcending outward. On the micro level there's individual birth and death, each with uncertain delineations. At the macro level there's physical and societal attributes that transition with equal vagueness.

Comprehensive Transitional Stages of Life:

Physical forms > Organic forms > Social forms

GAIA
Global Community   SOCIAL FORMS
Nation
State
City
Town/Village
Family/Clan
Married couple
---------------------- *Arbitrary dividing line*
"Organic Life"
---------------------- *Arbitrary dividing line*
Virus – Non-nucleated   ORGANIC FORMS
Proteins with side chains
Functional Compounds
Polymer Chains
Methanes
Crystals
---------------------- *Possible domain dividing line*
Macromolecules
Atoms
Particles   PHYSICAL FORMS

Crystals introduce pseudo-living qualities, with simple geometric patterns of repeating lattice structure. Early reproduction scheme via replication & accretion. Pieces accumulate and break off like rudimentary genes. Cell division mirrors this process: tubules pieced together and extended – emergent order.

The underlying spirit of life permeates each level and domain, imbedding its presence into everything. Like water, it takes the form and shape of different patterns that fit every new set of conditions. Life at every level is shaped and regulated by parsimony, where forces, dimensions, limits and emergent properties interact to create unique, symmetric forms. That's why a fish at the organic level translates to a boat at the societal level. Life is a directed universal flowing essence – a purposeful process emerging through successive domains. We observe it second hand, through derivative temporary forms and structures left behind in its wake. It integrates and connects everything. Substances, things, organisms, communities of organisms and the entire environment are linked in a comprehensive metabolizing network – parts in each domain coevolve in unison. Life is a process of becoming, progressing, developing, and evolving. It proceeds by doing. Life is a verb.

Life's omnipresent nature seems masked in a cloud of invisibility due to limits of human perception. Our sense of reality is confined to a narrow range of speeds, sizes and rates of change. Man *is* in the middle, especially in matters of scale. At the micro level, particles, atoms and molecules are moving, dancing and metabolizing at extremely fast rates. Life at the atomic level is beyond our reality. Ordinary objects surrounding us look like dead, static non-living things yet they're comprised of millions of vibrating parts, linked together in chains of pulsating, throbbing rhythms that would make a disco dance floor look static. Likewise, the hidden reality of microbes saturating our planet is masked well beyond our awareness, despite the fact they're present everywhere and outnumber all other life forms by far, both in number and sheer weight.

Human perception of speed is too narrow to grasp motions of plants, trees, mountain formations and glaciers, all moving outside our awareness. Similar limitations prevent us from grasping societal evolution where the progress from small towns and villages into fully developed complex cities only becomes obvious when viewed in high speed time-lapse. We simply can't relate to trees that live for hundreds of years, canyons that form over eons, and geological developments occurring over millions or billions of years. We can't even relate to our very own ancestors, who directly link us to members of our own species thousands of years ago. Nor can we grasp metabolic processes on the cosmic scale involving stars, nebulae and galaxies. Our confined position in the middle of all scales creates bias to life forms in a narrow range, oblivious to those on every other level or scale.

Fortunately, we can overcome limitations of perception to see recurring patterns life creates; the residue and signature left behind as it progresses through each domain. Organisms, living systems and man-made structures possess common components and configurations in different domains. Note the similarity between a human body (veins, intestines, nerves, openings, valves, tubes, tendons, skeleton) a car (wires, hoses, fluids, rods, bolts, frame, vents, filters, springs) and a house (wires, pipes, cables, outlets, studs, siding, ducts, doors). Or the recurring pattern of cells and shells: membrane, fur, skin, hive, nest, igloo, cave, cocoon, tent, seed, womb, submarine, or honeycomb. All the engineered products that evolution designed are repeated through different species with recurring common patterns of limbs, appendages, joints, wings, defense mechanisms, etc. Ancient people had a better sense of this omnipresent spirit of life than modern man. We're isolated by science separating nature out of the equation, compounding limitations of built-in scale bias. Long ago Aristotle observed "Nature makes so gradual a transition from the inanimate that the boundaries between these kingdoms is indistinct and doubtful". Life has no boundaries other than arbitrary ones we create. Everything has some degree of life in it, even inanimate objects and "dead" things, all of which are vibrating robustly inside. Notice how we label living things as *organic life*…implying "organic" is a caveat with a counterpart reality of *inorganic life*.

Societal Life. Hierarchies or holons are just symbiotic patterns of parts and wholes. All evolved higher levels always contain elements of lower levels. Where organism parts consist of cells, societies' parts are made of people, who cluster and coagulate into a variety of forms resembling pseudo-organisms. Group dynamics at the societal level are just like the social attributes at the atomic level, molecular level and organic cellular level. Similar specialization of parts takes place as people separate into divisions of labor with particular roles and functions. Organizations begin to resemble organisms as living systems with integrated component parts working in sync to form a more complex, greater whole. This pattern repeats all the way up, from families to cities to nations.

Organizations possess both hierarchical and network order. Sustainment is achieved by throughput of people over successive time periods who fill roles, earn titles and perform required tasks. Lines of communication connect human "parts" regulated by a managing authority, usually a single leader but sometimes a layer of management. Organizational guidance is directed much like DNA from a nucleus, instructing what tasks should get done and when. And like a living organism whose cells are replaced continually, organizations have people come and go; title holders are replaced regularly while operations just go on as if nothing changed. People are less important than the actual positions they fill, simply serving as temporary place holders until the next shift or permanent replacement.

Social organizations are no different than other living systems requiring some input of energy and information; they process it to produce a product (tangible or intangible). Businesses are an obvious example where management of input and output determine success, which is just the social equivalent of survival. Nations follow the same scheme with greater complexity, including entire economic systems processing inputs and outputs. Governments serve as the regulating function while culture and national identity provide purpose to the whole. The same is true of cities, just on a smaller scale.

If people are equivalent of cells in a society or nation, then cities are the primary organs that support the overall superorganism. First, necessity incentivizes specialization, then network coordination occurs, and soon complexity follows. A growing town gradually transforms into a bustling integrated city where primary functions arise supporting and maintaining the living system. Hubs are established linking transportation, communication, manufacturing, maintenance, repair, utilities, security, and the sustainment of ongoing requirements of fuel (energy) and materials (food, resources). This complex but synchronized system is supported by a constant turnover of humans, who pass through daily, yearly and over centuries while the system lives on. The city itself is continually evolving as structures get replaced, modified or supplemented. Growth starts out radially, then sprouts up vertically, pushing up tall buildings and modern skyscrapers. City evolution reflects both political and cultural change, working together to direct ongoing progress. Long term shifts in values are revealed in relics or leftover structures. The tallest buildings indicate societal priorities: Ancients glorified churches, then civilizations emphasized government, and today, modern societies exalt big banks.

Tokyo 1945   Tokyo today

Within cities and states, social dynamics mirror metabolic processes of living beings. Economics involves material processing to sustain organized systems. Businesses are the internal organs, producing specialized products and services. End products resemble species, changing or adapting based on environmental consumer demand – some survive, others die out. Niches are open, filled and closed, only to be reopened later as consumer environments' conditions change. Business survival is achieved through a combination of competition, cooperation and neutral exchange, just like organic species in nature. Business structures have hierarchal positions filled by humans, just like all other social organizations including political, religious, military, and cultural. Constant throughput is normal – people replaced by other people while the whole's structure lives on.

Other types of social organizations follow the same pattern of living systems. Each societal group interacts with others in symbiotic relationships whether competitively or cooperatively. Collective groups interact with an integrated dynamic environment, leading to coevolution amongst all. Some groups adapt better than others, leading to greater success and longer-term survival. Those that grow exhibit expansion patterns that resemble species branching in organic life, such as a franchise chain or spin off enterprise. Businesses often "infest" a geographic area rapidly while slowing down or stopping growth in less favorable areas…just like species in nature. Society is like a biosphere for social groups, each connected by external conditions and relationships among other social species. This interrelated, interconnected network system coevolves with everything within, including structures (building materials), transportation (vehicles), communication (telecom, language itself), education (knowledge), healthcare (medicine) and every human-interest area from sports to work to entertainment. Nothing ever changes in isolation, it's all integratively connected.

Human Body compared to Modern City

Brain	Government
Body	Infrastructure
Veins/arteries	Water pipes
Nervous Sys	Electric Grid
Stomach	Processing Plants
Muscles	Machines
Skeleton	Street grid
Blood	Money
Capillaries	Banks
Food	Fuel
Excrement	Sewage
Sweat pours	Hydrants
Memory	Library
Glands	Agencies
Adrenaline	Fire dept
Immune Sys	Police
Circulatory Sys	Transportation system
Blood Vessels	Cars, ships, trains

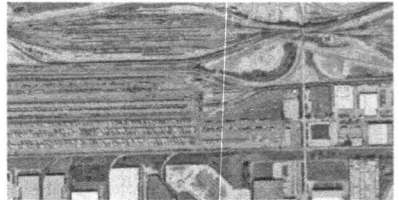

The Spectacular Ascension of Man. After billions of years developing stable life forms, evolution produced upright walking hominids with special qualities taking living progress to a whole new level. Our ancestors split from the ape lineage about 5 million years ago, developed into homo sapiens about 250,000 years ago and very recently emerged as modern humans about 35,000 years ago. At this point, man's cultural aspects begin to transcend biological evolution. Humans created basic tools, forged functional weapons and engaged in organized hunting to accumulate and store food. Still primitive, they relied on sticks, stones, and animal skins, communicated with grunts and gestures, but became the first beings to adapt the environment to their own needs...not the other way around.

Evolution endowed man with special advantages including upright bipedal posture, dexterous hands and feet, binocular vision, opposable thumbs, verbal capacity and an enlarged brain, tripling in size within a few million years. The early brain was basically autonomic, then progressed to a second surrounding layer expanding capacity and functions, and then that too was transcended by a third layer on top to form the complex version of today. This modern, organically produced brain evolved beyond autonomic functions, now possessing capacity for abstract creative thought, advanced language (naming things/representing ideas), imagination, and self-conscious awareness. For the first time after billions of years of evolution, life was now aware of itself.

Gibbon Human Chimp Gorilla Orangutan   *Long continual progressive ascension of Man*

Man's brain is quite an achievement with its trillions of cells and billions of neurons, each with over 1000 synapses firing up to 100 times per second. This organic network system of chemical relays is one of the most complex things in the universe, giving human beings an incredible advantage over all other life forms. With such an advanced super computer man is able to shape the environment and create an entire new domain of evolution at the societal level. Suddenly, individual brains could synchronize into a collective network with all other brains.

Human Brain Attributes

Triune – (3 Part Brain): evolved in successive stages

Reptilian	Instinct, compulsive, ritualistic
Paleo Mammalian	Survival: eat, fight, flight, sex (basic emotions)
Cerebral Neocortex	Imagination, reason. Free will, override instinct/emotions

Dual Halves – Accommodate duality of life (art & science aspects of nature)

Left Brain	words, numbers, abstract thought, logic. **Linear**. (*doing*)
Right Brain	art, feeling, laughter, love, *creative*. **Non-linear**. (*being*)

*Brain and mind co-evolve. Experiences change both brain and mind in tandem. Repeated experiences strengthen connections. *100 billion brain cells mirror 100 billion stars in our galaxy.*

Finally, inevitably, the stage is set for modern humans to take over the entire biosphere, spreading throughout the vast surface of the planet filling every niche. About 10,000 years ago man transitioned from hunter-gatherer clans into clustered organized centers of rudimentary civilization. Food was then processed and stored, animals were domesticated, tools were improved to produce pottery, woodwork, metal smelting, and then agricultural organization became the sustaining system of mini societies. Collective wholes of humans engaging in specialized tasks created synergy not present before, setting the stage for further advancement of the species.

Human global migration →

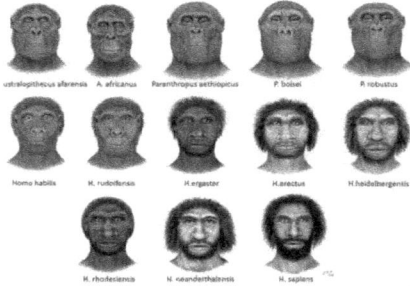

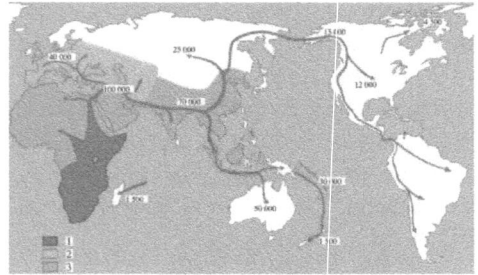

Evolution proceeded into this new domain in exactly the same manner as all previous stages: human species branched out spreading everywhere possible, saturating the surface of the planet, developed connected networks and formed complex living systems. A network is basically a group of elements linked by nodes, which is how humans coalesced into towns and cities. Early civilization marked a transition from pastoral groups into feudal communities and concentrated central towns. Hierarchies were established providing stable authority through political systems and religious orders. Class structures formed with the preponderance of individuals (peasants) on the lower rung, then middle-class traders and merchants above them, and finally, privileged landowners and ruling elites at the top. Commoners paid tribute (tax) in exchange for the promise of protection from above. Order is maintained through the rule of law, supplemented by religious edicts setting moral values. A major drawback of organized societies with concentrated populations is the periodic rapid spread of germs and disease.

Around 6000 years ago the first major civilization developed in Mesopotamia (Iran-Iraq area) followed soon after by Egyptians. Societal progress accelerated as organized structures took shape while specialization and collaboration continued further. As governments became more centralized, layers of bureaucracy grew and professional armies were conscripted. Agricultural production increased dramatically, facilitated by publicly traded products in markets with currency-based transactions. Collective know-how steadily increased leading to advances in every area of civilized life. A great leap came about 2000 years later with Greek civilization where human intellect was esteemed and nurtured, producing a comprehensive expansion in all fields of knowledge. Science, philosophy, arts, astronomy, engineering, literature and medicine were taken to a whole new level. Inventions and improved methods flourished in their modern society, accelerating the rise of western civilization. Romans followed, took their lead and built upon it, adding major contributions in all areas, especially engineering and the arts.

After this period of several thousand years of punctuated human advancement, the pace slowed dramatically, inching along sluggishly during the dark ages which lasted about a thousand years. But as always, life picked right up again and lurched forward to an emergent renaissance, where classic times of Greco-Roman society were revisited, revitalized and embraced. This period was followed by a subsequent similar renaissance called the enlightenment, catalyzing further expansion in knowledge and science. Conditions were now favorable for renewed acceleration of human societal evolution in a non-stop series of transformative developments that would change human life dramatically.

Acceleration of Evolution. The historic surge of civilization and scientific technological progress built up unstoppable momentum. A few hundred years ago a tipping point was reached, unleashing a series of advancements culminating in an industrial revolution. With machines doing the work of many humans, production exploded with endless supplies of goods while entire systems ran around-the-clock as manufacturing centers. Steel, being ten times stronger than stone, enabled construction of monstrous sized buildings. Engines powered all types of vehicles leading to faster, further and cheaper transportation. Industrialization of societies drew more people into concentrated urban centers, fostered increased specialization and transformed civilization into complex integrated networks.

Mechanization then transitioned into electronics which supplemented and superseded machines. Engines and punch-cards gave way to motors, dynamos and circuitry. Transistors expanded capacity as it miniaturized circuits, leading to electronic components getting embedded into every aspect of the societal Technosphere. Vast improvement of everyday products was followed by great leaps in telecommunication and information networking. Computers emerged as an industry requirement – first big mainframes, then smaller practical versions making way into consumer markets. Several hundred years of industrial based societies now merged into an information-based tech landscape.

***GRADUAL* SOCIAL EVOLUTION** ***PUNCTUAL* SOCIAL EVOLUTION**
    (*Societal Improvements*)   (*Key Inventions*)

Standard of Living	Wheel	Light Bulb	Battery
Mortality Rate	Compass	Radio	Transistor
Individual Opportunity	Gunpowder	Telephone	Television
Human Rights	Printing Press	Electric Motor	Jet Engine
Access to Information	Telegraph	Gas Engine	Computer
Choices and Options	Penicillin	Refrigeration	Internet
Healthcare	Concrete	Air Conditioning	Smart Phone

Modern civilization evolved into a complex supersystem with continual advances permeating every area and category. In just a short time humans progressed from stone caves to huts, cabins, houses and massive buildings. Man's world quickly shrunk with transitions from walking to horse and buggy, cars, trains, planes, jets and even spacecraft. Knowledge exploded in every field of science, engineering and medicine with continual breakthroughs and expanded branches. Manufacturing proliferated into everything: plastics, petroleum products, chemicals, synthetics, textiles, and metals. Progress built upon itself in an accelerating manner where future developments and transformational change became a certainty, without knowing what may come next.

During evolution's continual push onward and upward there are always vestigial remnants left in its wake. In the societal domain we see this in Third World nations lagging behind along with pockets of primitive clans still abiding in nature's various undeveloped, isolated regions. Similar dissociation is found in the persistent presence of human pathologies of modern society, still dealing with lingering problems of drug addiction, crime, poverty, government waste and abuse, human trafficking, and inequality. This disparity in evolution's progress is natural at every level and shouldn't surprise anyone. Its spirit is hyper forward oriented, pushing relentlessly ahead while leaving all other concerns as secondary.

Moving onward without hesitation, the rapid advance of technology combined with both industrial capacity and information proliferation to reach a tipping point of evolutionary germination; a stage of dramatic transformation where life afterwards bears little or no resemblance to anything prior. During this phase shift comparable to water freezing or a plant blossoming, civilization transforms into a highly integrated network of information where individual minds linked across the planet synchronize as a collective integrated whole. The relatively recent establishment of the world-wide internet web marks a watershed event in world history no less significant than organic evolution's greatest product – the human brain. Suddenly a global mind emerges, sustained by a collective web of human being as individual cells, each with a variety of backgrounds, cultures and ideas sharing information in a vast collaborative network of accumulated knowledge.

Information itself becomes the primary currency of progress as things become digitized, formatted and processed. Bits and bytes packaged in a variety of mediums facilitate its transfer, beyond old schoolbooks, magazines and journals and onto magnetic tape, discs, video, and fiberoptics. Everything is interlinked in real time via webs of both hardwired cable and wireless signal. Miniaturization continues with microchips while storage compacity doubles, then redoubles, along with processing speeds increasing exponentially. This technological quickening represents a novel evolution of information transfer: replacing chemical DNA with manmade electronic codes. Genes give way to memes. Information is exchanged as ideas, images, plans, blueprints, signs, symbols, algorithms and video. Biological transfer of info is now superseded by societal forms. Where sex was the principal biological means of differentiating genes, now a great social nexus of linked webs with varieties of information products enables super-differentiation of ideas and creative novelty. And like DNA, the writings of a contributing author survive and continue on long after the original creator dies, adding to the cumulated knowledge of mankind in an unbroken chain built by a lengthy line of dead pioneer ancestors.

   Mankind is quietly transitioning into a new paradigm where everyone, everywhere, is connected here and now in real time while simultaneously linked digitally to every other person in recorded history. Thoughts and feelings are shared instantly on countless social media platforms including YouTube, Twitter, Instagram, Facebook, Tik-Tok, along with other mediums: texting, messenger, email, etc. Unlike old-school phone calls these options enable mass communication, linking individuals to a large audience covering great distances in real time. This new tech biosphere brings with it positive and negative aspects. The benefits are obvious while side effects are subtle. People now default to group think, trading off intimacy for distant connections and filter out opposing perspectives leaving echo chambers of like-minded drones. Isolation and polarization increase. Individual adaptability and resilience to adversity diminishes. An Orwellian dystopia might set in if these side effects persist and spread. One remedy to this concern is to just maintain a healthy balance between individual autonomy and collective submission.

   As this new info-tech world order takes permanent shape, every person, thing and process coevolves with it. Collective linkage incentivizes economic globalization and cultural inter-connectedness. Mobile phones, Go-Pro cams, live feeds and streaming video produce a constant inflow of information into our global network. The *internet of things* links places and spaces where humans abide, including smart cars, smart homes and the smart office. Earth's surface becomes literally saturated with cables, fiberoptics and wireless signals of increasing bandwidth. This info-tech social biosphere takes on a life of its own and seems to be driven by a pilotless operator.

   Humans are currently adapting and coevolving with the emerging tech-nexus environment. During the previous phase of societal evolution man integrated physical supplements like artificial limbs, hearing aids, braces, pacemakers, etc. Now mankind itself is being transformed as tech merges with human biology – man and machine become one. Smart phones are already a required personal attachment as some people feel physically ill when separated from them. Google glass is a hybrid component example of man and machine merging. Neuralink directly connects human brains to computers with chip implants greatly accelerating mental processing. These smart devices link human beings directly into the global mind, making all individuals part of a greater collective not unlike the Borg. We're already physically connecting human brains into the matrix. Once directly linked the transformation of man may reach the point of no return. Further integration will come from self-thinking artificial intelligence; catalysts for automated systems accelerating transformation into a new human-tech paradigm.

   Machines and life-like robots will eventually surpass man's limitations, taking everything to a whole new level. Manufacturing is increasingly populated by non-human bots performing all manner of labor functions mimicking human actions: drawing, walking, jumping, emoting, playing games, creating art works and doing heavy lifting. Eventually these non-human bots will look much less mechanical and more human-like androids. Combine this with AI Chatbot, digitizing human behavior into bots, AI self-learning, quantum computing, deep-learning and direct chip implants…and the result is comprehensive transformation of human life, with no going back. AI is already solving problems in days that previously took teams of experts years. Technological acceleration is quickly approaching a singularity; a hypothetical point where it becomes uncontrollable and irreversible, potentially culminating in an automated dystopian future.

A world inhabited by cyborgs and hybrid humans may seem alien as portrayed in science fiction movies, but that reality is already here. It's hard to notice because the human-machine balance is currently far skewed to the human side but as the ratio flips and automated, robotic sentient machines predominate, the transition will become obvious. We already engage in biological engineering with a completely mapped out human Genome and companies like CRISPER actively engaging in alteration of human DNA. While offering the promise of extending life spans by reducing replication errors of the aging process, it opens up a pandora's box of unimaginable consequences by manipulating human beings' internal components. Or consider the unintended consequences of bringing back extinct species through application of the same technology.

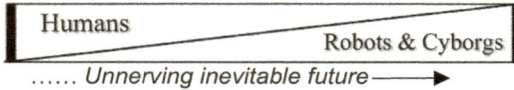

```
Humans
                    Robots & Cyborgs
…… Unnerving inevitable future ——→
```

Meanwhile, a whole generation of young people are embracing virtual reality, many favoring it over actual reality. The combination of these developments and their coevolutionary impact on human beings is starting to change societal behavior in significant ways. A new stage of life has arrived and it's taking all of us into a future we can only speculate about. This much is certain: Evolution is relentlessly proceeding forward with or without our consent.

VIRTUAL REALITY	*AUGMENTED REALITY*	*MERGED REALITY*
Digital environment	Real world with digital overlay	Real and virtual interlinked

Isolated synthetic experience	*Real but digitally enhanced*	*Mix of real and virtual world*

LIFE REDEFINED

So, we come back to our original question: what is life? We can acknowledge popular definitions as generally valid but seriously limited. Consensus based life qualities are merely symptoms of life, not the actual essence. Traits like adaptive ability, metabolism, internal organization, stability, growth, reproduction are fine to use as associations with "living" things, but they're more effects than cause. Life transcends all of them as an elusive flowing progressive essence imbedded within everything and it only knows one mode: continual progress forward and upward. Life's an eternal process. It's the action, not the stuff. It's the processor, not the product. It's a verb that produces nouns – not the nouns themselves. Life interacts and connects everything, both internally within organisms and externally beyond organisms, integrating every part with much larger wholes.

Evolution is an expression of life's progressive force – a purposeful drive towards greater complexity and higher consciousness. Man is merely the leading edge of this force; a recent stage in a very long linked sequence of transformative progress through emergent domains. Life relentlessly presses forward spreading everywhere through space and time – expanding radially, branching out endlessly and never holding back. It transforms everything it passes through, bringing the entire environment of things along with its integrated coevolutionary dynamic. Continual change is the norm, both gradual and punctual. Living systems are just balanced fluidity riding the sweet spot between rigid structure and unstable chaos.

The process of life is a persistent on-going state of flux where energy is transferred, forms emerge, and transient structures take shape. The process combines fixed forms with variable states, stability and flow, being and becoming. Things and beings all come and go, are born and die, pass through temporary states, transition through blurred lines between living and "non-living". Spirit goes on eternally, with no beginning and no end…it just always was. Science can't explain it. Life is beyond physical explanation, transcending birth, death and all the traits we ascribe to it. Life's emergent properties cannot be predicted; uncertainty is built right into it – a feature, not a bug. Ponder how a tiny acorn transforms into a giant redwood tree. Or a microscopic sperm cell joining a microscopic egg somehow produces a complex, integrated, conscious human being. Life is the most important essence of reality yet remains a wonderous veiled elusive enigma.

Evidence of life is present everywhere once you know what to look for, though hidden by scales of size and speed. Life literally saturates the entire planet's surface, including the air above and ground below. It pervades and proliferates everything, integrating the entire biosphere. It's the glue connecting a vast comprehensive tapestry of organisms, living systems and encompassing environment – all of which dynamically coevolve. It's expressed in a series of parts and wholes linked together to form ever larger, comprehensive wholes, like a long chain of Russian dolls within dolls. At the highest rung earth appears as a unified superorganism which one could consider just another part of larger wholes that form our solar system, galaxy and the entire universe itself, teeming with life on planets in every corner.

Drake Equation $N = R^* \times f_p \times n_e \times f_l \times f_i \times f_c \times L$

*Estimated number of advanced civilizations in our galaxy
(Stars with planets, habitable ones, probability of life, intelligent life, developed civilizations, duration)

If life is present everywhere and in everything then we must conclude that everything is alive. Certainly, there's great variation in degree but not in essence. A great spectrum of life would obviously contain a wide variety of forms, conditions, attributes and states of being – naturally exhibiting a complex ambiguous nature. It would reflect a sliding scale of qualities meshing through different domains, just as it does when passing through physical, biological and societal levels. Man's attachment and familiarity with the middle, biological level obscures awareness of life expressed in realms above and below. Our narrow, limited perspective is understandable given how life generates unrecognizable forms with different rates of metabolism, different levels of complexity and different states of development than our own. Ambiguity is present all along the way with vague delineations of birth and death, living and "non-living", order and chaos. Once you accept that everything is alive, it all makes perfect sense. Ambiguities disappear and arbitrary classifications dissolve to reveal a continuous unified spectrum. Differences suddenly reduce to simple degrees. Man is no longer separated here from nature out there. Chimp's possessing 99% of Man's DNA become extended family. Even cats sharing 90% of Man's DNA become distant relatives. Divergent organic species of all shapes and types join the same extended family. Now just include the entire living environment and every life form that ever existed through history, and you end up with one continuous, interconnected unified whole.

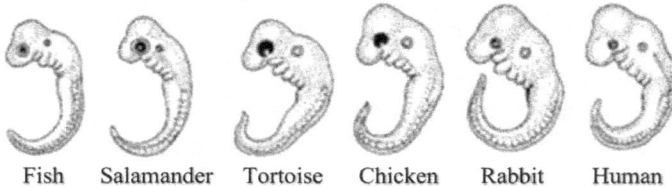

Fish   Salamander   Tortoise   Chicken   Rabbit   Human

Embryos = nucleus of organic life; a continuous repeated pattern of spirit unfolding into life. An evolutionary pattern persisting for millions of years, repeated every second of every day.

This illusive, ambiguous omnipresent essence is inescapably difficult to define however it does leave a trail with clues. Life is a pattern language decoded by tracing the emergent temporary forms left behind. Within each living structure are repeating themes shedding light on life's progressive spirit that produced them. Life's a flowing purposeful process progressing forward through mechanisms, physical systems, biological organisms, man-made machines, factories, societal networks, and collective communities. Living beings and living systems are resonant patterns in time and space, persisting as delicately balanced wholes of material and energy. Internal dynamics operate with constant churning of parts – processing stuff in a metabolic flux while maintaining a coherent wholistic pattern of stability. An organism is really just material residue: a by-product of processed energy. You are no different. Throughput of stuff goes in and out, cells die and renew, energy transfers continuously, yet the shape and form persist in what appears to be a stable structure – an ongoing illusion of balanced tension and flux.

Life's pattern language reveals parsimonious interconnections between evolutionary forces and environmental constraints, expressed in branching patterns, engineering solutions, and organic geometry. Recurring themes provide clues to the methods life exploits to press forward and upward. Universal patterns appear everywhere in various mechanisms of progress shaped by nature's physical laws subtly hiding the spirit while it leaves a comprehensive trail of residue everywhere it flows. YOU are part of that residue; a temporary form serving as a tiny part of a much larger, interconnected whole, doing your small part until passing the baton along to continue the ongoing journey forward and upward.

    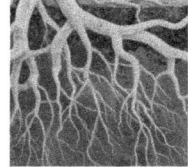

Rivers   Leaf   Dendrites   Lung   Roots

Ubiquitous branching patterns express permeating spirit of life in beings, plants, veins, arteries, kidneys, flowers, intestines, trees, tributaries, brains, fractals…

Omnipresent patterns of living hierarchies reveal the interconnected nature of life's forms saturating the entire organic biosphere of codependent species and societal community supersystems where individual human beings serve as interdependent cells – all integrated into a single unified spectrum. Man-the-microcosm is just one component of that spectrum, occupying a privileged middle sweet spot. Our existence is merely a metabolic wave…a temporary pattern, more fluid than form. An illusion of stability in an ever-changing dynamic state. You too are merely an abstract momentary mirage masquerading as stable structure; an illusion maintained by persistent patterns of relational form generated by transient parts within. You and every other one of life's versatile forms lie in the wake of a divine progressive spirit passing through, speaking to us indirectly in coded pattern language, masking the underlying elusive essence of life that comprehensively unifies everything, everywhere, everywhen.

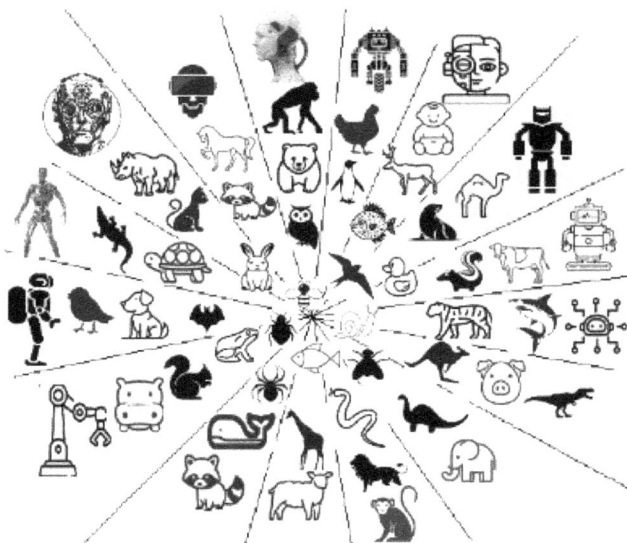

11   THE UNIFYING INFRASTRUCTURE OF PHYSICAL REALITY

So just what are the key ingredients of the physical universe, how do they interact, and what are the laws that govern how The World *out there* works? These are the fundamental questions long pursued by science, particularly physics. Ancient philosophers defaulted to mythology, mysticism and magical explanations until Greek scientists applied rational methodology. In time they were superseded by classical physicists during the renaissance and enlightenment periods centuries later, who established a structured framework with classical mechanics, thermodynamics and the essential laws of nature. But just when science was considered settled the modern era introduced radical new concepts, turning classical physics on its head. Relativity and Quantum theory completely displaced previous notions of what was normal, adding a significant element of abstraction and high a degree of absurdity to how things really work in the physical universe. Finally, the post-modern era shifted attention to the pursuit of physics' grand Theory of Everything; one that would unite the fundamental forces of nature.

The physical infrastructure of the universe basically comes down to a few key ingredients dynamically interacting in space and time. Material substance interactions involving compounds and mixtures are analyzed in chemistry while their component parts are broken down further in physics at the atomic level. Energy transfer and conversion is the essential process at work via conduction, absorption, convection, mechanical, thermal, electrical, or radiation. Power is generated and transferred through electrical dynamics involving currents, voltage, resistance and capacitance. Forces drive motion and all action-reaction events. Objects are moved, modified, heated and transformed through mechanical dynamics that abide by physical laws and universal constants. Standard equations capture the interconnected relationships of all the parts and processes.

UNIVERSAL PHYSICAL CONSTANTS

Speed of Light	C
Plank's Constant	h
Gravity	G
Elementary Charge	e
Vacuum Permittivity	$\varepsilon 0$
Fine structure	α

KEY INGREDIENTS OF PHYSICAL REALITY

Our physical domain consists of material substances and what we perceive to be solid objects. Stuff in this realm varies from sparce airy gases to liquids to gelatins to dense matter. Even solid things vary greatly from lightweight Styrofoam to hard wood to rock to steel. Substance is a quality of materiality deriving its attributes from degrees of structure and rigidity. But these diverse forms are primarily surface features and not the principal ingredient within physical reality.

Energy is the primary ingredient in all things. Matter is merely a derivative patterned form of structured energy. The base unit of all matter is the particle – a singular material thing with mass and spin. Particles are in essence energy knots which interact with other particles to form all manner of elements, compounds and mixtures. With few exceptions, all particles carry charges, inducing a social nature to mingle and attach to others in kind. The fact that particles are naturally charged reflects their inner essence derived from pure energy.

Energy can assume numerous forms, changing its disposition routinely and continually. When in a temporary fixed state, it's considered potential while its flowing state is kinetic. Each of these conditions have multiple forms. Potential energy includes nuclear, gravitational, chemical, and springs. Kinetic involves motion and includes radiation, mechanical, sound, heat and electrical. A significant portion of kinetic energy is derived from vibration. While energy continually changes form, its total quantity remains the same. Every transformative act in physical reality merely mixes parts while the amount of energy never changes.

Energy manifests *structurally* as matter and *dynamically* as force. The four fundamental forces in physical reality are the Strong force (nuclear), Weak force (particle decay), Electromagnetic (radiation) and Gravity. The strength of these varies greatly, especially between the strong force and gravity, a difference of 10 to the 38th power (1 followed by 38 zeros). These foundational forces translate into all the push-pull, tension-compression aspects of mechanical dynamics.

Electricity and its compliment magnetism are *the* essential forces at work in our world, including all chemical interaction. They're derivative attributes of particles formed from knotted energy. Electricity is simply a collective flow of electrons – charged particles that attract and repel. Magnetism is a structured form of electricity where force is projected in a static field instead of a flow. Both are two sides of the same energy coin and play a paramount role in powering modern civilization. Technological achievements in transportation, communication, industry and general mechanization are entirely dependent on them. Without electricity-magnetism civilization as we know it would collapse.

VARIOUS FORMS OF ENERGY

Mechanical	Nuclear		
Chemical	Gravity		
Thermal	Elastic		
Electric	Light		

POTENTIAL *vs KINETIC*

Gravity	*Radiant*
Elastic	*Thermal*
Chemical	*Electric*
Nuclear	*Magnetic*
Mechanical--------	*Mechanical*

Atoms are the first self-contained system in the physical universe with an integral center nucleus and surrounding electron cloud. It's the first complete whole unit – an entry level where wholes emerge along a great spectrum of ever greater wholes. The nucleus containing protons is where self-identity first emerges, establishing unique differences among a family of 92 natural elements, each with special personalities. All physical substances are manifested out of simple patterns of identical particles arranged in unique orders.

Chemistry tracks interaction among particles, atoms and molecules where energy is transferred and matter changes form. Differences in charges, shape and fullness incentivize sociability. Chemical reactions are a driving force behind change in the physical world. Energy is transferred; parts are exchanged. Nothing is ever created or destroyed at this level, just rearranged into different order.

Light is fundamental. Light is *the* key ingredient, preceding matter and everything else in the physical universe. It has no mass and no charge because it is pure energy and non-material. Every *thing* is a derivative of light. Every physical thing, every substance, every "solid" object is a form of structured light. $E=mc^2$ confirms this. Not only is matter composed of light energy, but it is directly connected to the speed of light; one essence with multiple manifested forms.

Space and time are the omnipresent background substructure of physical reality possessing a pseudo-physical quality, always present yet ambiguously intangible. Space-Time is a dimensional constraint of physicality, limiting the speed and position of everything within it. Matter-Energy may be the primary ingredient of physical reality, but it is complimentarily connected to and limited by Space-Time. Everything begins with energy manifesting in both a flowing form (kinetic and radiative) and in a structured form (potential and materiality), always interacting within the dimensional confines of space and time.

PSEUDO-PHYSICAL INGREDIENTS

When we think of physical reality, we associate it with material substance and anything tangible. If it's physical, you should be able to hold or feel it. However, anything and everything made of substance has a dual nature. That solid hard rock you pick up is actually composed of tiny particles held together through interlocking forces at the atomic level. And all those tiny component parts are simply energy fields knotted up. So, the physical realm really consists of varying degrees of matter taking form along a spectrum of intangible energy, forces and structured objects. Every "solid" thing we sense physically is actually composed of pseudo-physical ingredients.

Energy and forces are foundational yet possess pseudo-physical attributes. Where material objects exhibit static fixity, energy is in perpetual flux, transferable and naturally fluid. Energy is not really a thing but more of a property of a thing. It can be measured, stored and affect matter. It's defined as the capacity to do work. Energy is in everything, drives all action, and all changes via temporary conditions through dynamics that produce heat, light and motion. It can neither be created or destroyed, only transferred or transformed.

Forces can be felt but not grasped. We feel the effects of forces but they remain intangible; immaterial without substance, direct or indirect. And like energy, forces come in a variety of forms: Inertial, centripetal/centrifugal, tension/compression, momentum/angular momentum, buoyancy, push-pull of electric/magnetic, etc. Structural forces apply stress to objects via shear, tension, torsion (bending) and compression. The effects of these forces can alter an object's shape, size, heat and motion, including changing its speed and direction; either constant or accelerating. Internal and external forces determine the shape, size or disposition of everything:

4 FUNDAMENTAL FORCES		*CONTACT vs NON-CONTACT*	
Electromagnetic	(*Infinite reach*)	Applied	Gravity
Gravity Weakest	(*Infinite reach*)	Normal	Magnetic
Nuclear Strong	(Nucleus)	Friction	Electric
Nuclear Weak	(Particle decay)	Tension/Spring	
		Air Resistance	

Fields are the foundational ingredient of physical reality - the root essence that makes up all matter, substance, things, objects, etc. and they're simply vibrating resonant patterns of force. What you think of as "empty" space is literally saturated with various fields, though most are too weak to detect. Fields are measurable quantities that can carry energy and momentum, just like a particle. Their physicality differs by magnitude, speed, direction; captured mathematically as a vector. They're the very essence of quantum mechanics where concepts of solid things break down to nebulous clouds of probabilities.

Particles already possess pseudo-physical qualities so the presence of fields just make them even more ambiguous. At the quantum level everything is essentially vibrating patterns of energy. What makes one particle different from another is its unique resonant pattern. So, the building blocks of material substance are basically vibrating fields that interact and exhibit particle dynamics, including spin, charge, interference, and anti-matter particle counterparts.

Modern physics achieved an integral breakdown of particles and their subparts, known as the Standard Model. Protons, neutrons and electrons considered primary in atoms now break down into smaller constituent parts called quarks and leptons, each with variations differing by mass and charge. As we go further down the rabbit hole of physical substance, parts just keep getting smaller *and* more nebulous. Einstein's funky relativity at the macro scale is no match for the abstract elusive quirkiness of tiny vibrating fields at the micro level. This is the basis for a foundation of solid substance we perceive from our middle level in between it all.

Fundamental Parts – Key Ingredients of Pseudo-Physical Reality

FERMIONS – Particles of Matter (12 total)
1. Electron, muon, tau
2. Electron neutrino, muon neutrino, tau neutrino
3. Up quark, charm quark, top quark
4. Down quark, strange quark, bottom quark

BOSONS – Particles of Force (12 total)
1. Graviton – gravity or spacetime warping
2. Photon – electromagnetism
3. Gluons (8) – strong nuclear force
4. W & Z Bosons – weak nuclear force
5. Higgs boson

Light is essentially a vibrating energy field, one with both electric and magnetic components. It's a radiation wave with frequency and intensity (wavelength and amplitude) serving as a universal conduit of energy transfer. Atomic level dynamics involving particle interactions and electron shell displacements are all connected by emissions or absorptions of light. While light is an energy vibration in constant motion, it can be reflected, refracted, filtered, interfered (constructive or destructive) and polarized. Everything in the physical universe is a derivative of light. It's the first cause, spirit manifest and the first act referenced by religion in "let there be light".

Light is incredibly elusive. Its dual nature allows it to act as both a particle and a wave. As a photon it has no mass or charge. It operates in a timeless realm. It's never at rest and knows only one speed; the maximum possible speed. The moment it's produced it's instantly gone… thousands of miles away in a split second. The moment a photon makes contact with matter it's instantly annihilated. It manifests in numerous varieties (x rays, radio waves, microwaves, infrared, gamma rays, ultraviolet) "visible" and invisible, pure white light and hundreds of millions of different colors (humans can distinguish several million of them). Sun light is pure white, only looking yellow or orange when filtered through our atmosphere. All electromagnetic radiation is light, varying only by intensity/amplitude and wavelength, the latter of which determines type and color. Light can transform into matter, transitioning from constant motion to static field. Particles are simply resonant field structures bound together in a concentrated aggregate we call matter.

Misconceptions about Light. When we say a small band of light is visible, we're stating a partial truth. The surprising reality is that ALL light is invisible! The light we do see in that tiny band of frequencies is merely a derivative form of light, a secondary after affect, not the real thing. What we actually see are things light illuminates. We can see a light bulb but not the light between the bulb and our eyes. We're fooled by sunrays which are actually atmospheric particles illuminated by light, not the light itself. We can't even see color unless pure white light has been filtered, such a prism does. But all light is invisible. Think about it; if you can't see a speeding bullet traveling around 1500 *feet* per second, how in the hell can you see light traveling at 186,000 *miles* per second? Light travels about a million times faster than a bullet. That's not just beyond a blink of an eye…it's beyond comprehension. So, what we do see are all the things that light illuminates or reflects, not the original light itself...believe it or not.

Saying the speed of light is 186,000 mps is a misleading statement. This is not actually the speed of light; it's the speed of space. Light travels at various speeds, depending on the medium it's goes through. In water its speed is about half as fast as going through space. 186,000 mps or 300,000 kms is the fastest speed anything can travel because space itself has a capacity attribute called permittivity. It's a limiting quality similar to the resistance in a wire that regulates electricity going through it. Space is not nothing; it's definitely something. That's why it's gravitationally warped by large masses and resists light passing through it.

Portraying photons traveling through space is a misconception. Light travels as a wave. Anything that produces waves does so by disturbing a medium. Water waves are simply energy moving horizontally while the water itself merely bobs up and down all along the wave path. Moving water is an illusion; water doesn't travel horizontally with the energy wave. Wind waves moving through a wheat field create the same illusion, looking like the wheat is moving horizontally, but we know wheat strands are all tethered to the ground. Or consider cracking a whip where a wave travels from your hand to the end while the whip itself doesn't go anywhere. It's the same thing with light in space. Light is produced by induction; charged particles disturbing space produce a ripple, and since space has very little resistance, the waves it generates move incredibly fast. Energy passes through space in much the same way energy travels horizontally across water. There are no photons or particles moving, just an energy field traversing a medium which vibrates in a sinewave pattern. But once light (electromagnetic radiation) makes contact with something, its energy field ossifies into a particle; a massless, chargeless photon. It's a process similar to water hitting a surface producing splash-droplets. Surprisingly, none of this is captured in any modern physics book.

DYNAMIC ASPECTS OF PHYSICAL REALITY

Physics at the micro level is about how stuff interacts with other stuff. It's a how-to subject on the way our universe works, and at its core is dynamics – how objects move, interact and change. Change can be cyclical or secular. Some is mere surface change, some is internal agitation and transition, and the rest is full transformation. Energy once again is the central factor, both driver and conduit. Work is the measure of what gets done after parts are exchanged – stuff is moved and states are changed. Power is a measure of the speed of said work. Dynamics involve the translation of forces into action and reaction, cause and effect.

At a fundamental level everything begins with atomic dynamics consisting of charged particles attracting, repelling and shifting positions. Electrons which whirl around each nucleus in a whizzing cloud of unpredictability interact based on their internal position and composition. All the action occurs in the outermost shell where vacant positions desire to be filled, while single lone particles are eager to mingle. The social quality of atoms could be compared to dress behavior where a person's clothes either attract the opposite sex, turn them off or present a neutral indifference. Atoms are satisfied when their outer shells are full. However, satisfied, content atoms are less sociable and down-right boring. Those with a lone extra electron or lone vacancy space tend to be the most exciting and active.

Atomic dynamics are driven by polarization, the same way a magnet with charged opposite ends attracts or repels. Atoms are naturally balanced with positive protons matched by an equal number of negative electrons; however, they can be ionized where the number is imbalanced, creating a temporary polarization. They can be further polarized when electrons in many atoms fall into alignment, creating a magnetic field. It's similar to polarized light becoming a laser. Polarization is complemented by agitation, which is the temporary excited state of an atom after absorbing excess energy. This sets up a continual dynamic of charge and discharge – an atomic level sexual interaction of build-up, excitement and release.

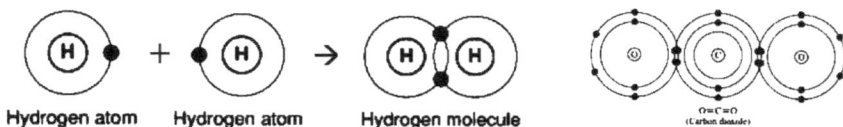

Hydrogen atom   Hydrogen atom   Hydrogen molecule   O=C=O (Carbon dioxide)

Exchange or sharing of electrons always includes energy transfers equal to the strength of each transaction. Atoms are like mini storage cells containing energy. Electrons in higher shells carry more energy and emit light when dropping to lower ones. Different shells with different energy levels correspond to specific light frequencies and unique color signature. Atomic interaction is traceable by the light show produced. When electrons change shells or engage in lateral social exchange, their atoms absorb or emit light energy in whole units called quanta. Strangely, electrons are never between shells – they are either here or there but never in between. And the energy exchange never involves fractions, only whole units.

Chemistry transcends atomic action, enabling a higher stage of physical wholes – molecules. Mixing and combining different atoms creates a new level of attributes in substances, often very different from the component parts. Chemical reactions are the primary dynamic in a micro-world where stuff gets mixed, transformed and reintegrated. Like energy, there is conservation of materials where parts may be changed or exchanged but the final product is always equal to the original components. The sequence is fairly basic: Products → Process → Resultant. Processed stuff is either mildly transitioned or completely transformed, and the attributes often bear no resemblance to the original parts. Consider how adding a neutral substance like sodium to metal and dropping it into a bowl of water produces an explosive reaction yet mixing the same sodium to volatile chlorine results in common table salt.

Chemical reactions are simply transactions involving electrons; always those in the outer shell. The key mechanism used in the process is bonding, either covalent or ionic. In the former electrons are shared and the latter they're exchanged. Bonds are stronger when closer, requiring more energy to break. Bonds can be single, double or triple, making the union of two elements very tight. Social behavior at this level is quite similar to atomic action, however with molecules shape adds a new dimension. Dynamic interaction is influenced by the shape and disposition of each molecule. Chemistry thus involves a subtle pattern language playing a significant part in how substances mingle. Still, the primary driver of both atomic and chemical dynamics is electromagnetic force, the very essence of light. And like atomic reactions, chemical exchange transactions often include discharge of heat and light, plus other reactants (gas, odor, precipitate). But no matter what's exchanged, the final products always equal the original parts' total mass.

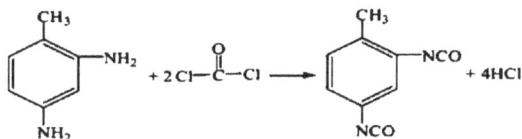

Energy and force come in two primary modes: fields and flows. One's a static structure, the other a stream of current or radiation. All charged particles project a field with lines of force penetrating adjacent space. All *accelerated* charged particles produce radiation as electromagnetic waves. These waves are invisible light, varying by amplitude and frequency. Where atoms resemble a static energy cell, light is a mobile one. Since electricity and magnetism are two forms of the same stuff, so too are fields and radiation. Both fields and matter are structured energy forms, one *penetrates* space, the other *displaces* it.

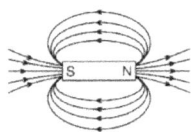

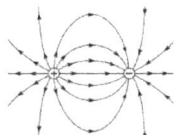

| Magnetic | Charged Particles | Gravitational | Warped Space |

Most forms of energy are linked to vibration. Light, sound, heat, alternating current, etc. all derive from vibrating motion. Energy in rhythmic form agitates or flows. Heat is simply vibrating particles whose collective average is measured as temperature. Note that a single particle has no temperature. Heat is also what differentiates states of matter. The primary difference between ice, water and steam is heat – the vibrational energy within. Light is also differentiated by vibrational energy, a factor of frequency and amplitude. The *only* difference between all the colors we see is frequency of vibration. Same goes for all forms of radiation including gamma rays, microwaves, radio waves and visible light.

Vibration in sync leads to resonance; persistent patterns of force that create fields. A spinning particle becomes a resonant field pattern, creating polarity and static force. Matter is really just a pattern of resonant energy where interlocking vibrating forces establish a tension field, creating the illusion of something solid. At the micro-level, physics is primarily about field dynamics involving the interaction of static and moving force patterns. Charged particles induce electric fields while moving current induces magnetic fields.

Heat is simply agitated particles reacting to the addition of energy, whether mechanical, chemical or electrical. Thermodynamics examines heat-energy interactions between objects; transfer of energy through pressure and vibration that

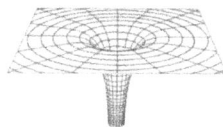

changes states of matter. Its key principle is the 2nd law of entropy, where disorder always increases over time in a closed system. This entropic process implies a one-way direction of time and a diminished dead-end state. Fortunately, negentropy is a counter principle of increasing order in an open system, evidenced by organic life.

Fluid dynamics focuses on gases and liquids – flowing phenomenon versus static ones. Fluid actions are more difficult to capture given their ambiguous nature and lack of structure. The same concepts in thermodynamics apply but must accommodate a variety of unique conditions. Fluids can exhibit differences in steadiness, uniformity, compressibility, rotation and aspects of turbulence. Patterns and disposition may vary but it's still about energy transfer and interaction.

Classical mechanics analyzes the dynamics of objects in motion. Newton's 3 laws of motion summarize the relationship between forces acting on objects, whether at rest or moving. Vector dynamics add the element of angles when applying force, incorporating geometry in space, where velocity, acceleration and direction combine spatial and temporal aspects to describe basic motion. Classical mechanics also covers energy transfer using gears and levers which introduce mathematic ratios to relationship between force and object, along with all sorts of push-pull interactions including pressure, tension, torque, compression, shear, elastic-spring action, friction and equilibrium force-counter force relationships.

Newtonian physics accurately describes how things work at our level (man in the middle), but doesn't apply at micro and macro levels. The higher macro-level of super-fast speed and massive cosmic bodies follow Einstein's relativity, superseding classical rules. And at the small level of atomic particles, quantum mechanics takes over. Weird stuff like time dilation, action at a distance, quantum duality, may accurately describe physical reality but have no practical application at the human level of perception where speed and distances are "normal".

MICRO LEVEL	*MAN in the MIDDLE*	*MACRO LEVEL*
Quantum *(Small)*	**Newton-Classical** *("Normal")*	**Relativity** *(Large/Fast)*
Quantized whole energy units	Fractional amounts of energy	Light as universal constant
Particles as wave clouds	Solid particles	Mass and time dilation
Heisenberg uncertainty	Absolute time and motion	Relative frame of reference
Non-local Action at a distance	Separate independent particles	Mass warps space
Ambiguous fields	Straight smooth lines	Curved, non-linear

*Quantum mechanics simply means the particles we perceive are actually fields; clouds of motion that only act as single particles once they interact with something, either an instrument of detection or just our own observation. It's like a roulette wheel where the ball is in motion around every possible slot until it finally falls into 1 and only 1 position. The quantum cloud is also a wave of motion that once interfered with collapse into a single position – becoming a defined particle…transitioning from many possibles to 1 actual. Hence the physics term *"collapse of the wave function"*.

PROCESS DYNAMICS

Change is *the* essence of life. Reality is never static but rather a continual stream of eternal moments. It's an ongoing condition of constant flux, ebb and flow – continuous novelty. Every moment is different from the previous and the following. No two moments are ever the same. That's the TAO. It's also the very center of *being*. It's the primordial pulse within the substructure of the universe where each quantum moment creates and sustains reality, like the heartbeat of a living organism. What we call life is simply the flowing stream of sequential, progressive action. Duality begins here with the eternal beat of *now* in the middle, which is both preceded and followed by a stream of other *nows* that establish an active dynamic process. This is the basis of all change.

Process is the temporal action of changing states. Every process has a beginning, middle and end. It's the *doing* dimension transcendent from the *being* state. It's measured by difference – the comparison of before and after conditions. Process involves a variety of interactions: Internal and external, temporary and ongoing, singular and multiple elements. In physical reality, process always includes transfer or transformation of energy in one form or another.

Change is a natural attribute and built-in feature of the universe; a foundational essence that affects everything. This structural aspect manifests in different ways and types. Change occurs along a spectrum of degrees, from the very slow gradual to the rapid revolutionary type. Change can occur as a minor surface nuance or a complete transformation. One is difficult to notice, the other quite dramatic. It's graphically expressed in motion. An object moving at a constant speed is 1st degree change. The same object moving with a constant acceleration exhibits a 2nd degree change. And if that object were to accelerate at an increasing rate (acceleration of acceleration) it would experience a 3rd degree change. These can be depicted as a flat line, an inclined line and a curved line respectively. Also, the first two involve linear change while the latter one is non-linear (geometric). In addition to speed, direction is also involved in an object's degree of change. Any change in direction is an acceleration, a significant change in and of itself.

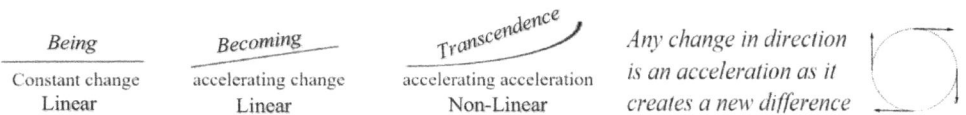

Being	*Becoming*	*Transcendence*	*Any change in direction*
Constant change	accelerating change	accelerating acceleration	*is an acceleration as it*
Linear	Linear	Non-Linear	*creates a new difference*

Within the spectrum of change are different types based on continuity. Change can be variable, incremental, discontinuous and punctuated. Variable change can be as simple as different rates of change within a single process. Incremental change is a process occurring in stages, with equal or unequal parts. Discontinuous change is a process that goes in spurts. Punctuated change involves a steady, gradual process suddenly interrupted by a major spike. Process dynamics include a mix of ongoing differences in both rate and degree of change.

Spectrum of Change Mechanics

Simple – Subtle – Gradual – Discontinuous – Incremental – Punctual – Transformational

Repetition of an action is a symmetric type of change qualifying as a cycle. *Circular motion* and *vibration* are two variations of the same essence. A turning circle viewed from the front displays an obvious simple cycle yet viewing it from the side reveals a vibrational version exhibiting harmonic motion. The same turning circle moving horizontally creates a universal vibration known as a sine wave. So one motion produces 3 different but connected expressions of change.

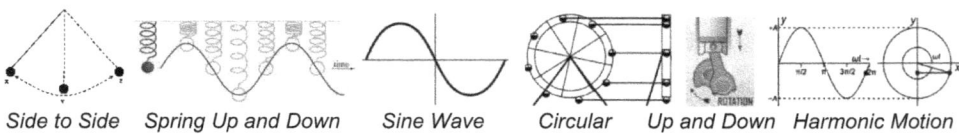

Side to Side   Spring Up and Down   Sine Wave   Circular   Up and Down Harmonic Motion

*Circular motion = Cycle = Vibration = Wave = Pendulum = Spring = Piston = Harmonic motion
This dynamic applies to charge/discharge, growth/contraction, pressure/release, compression/radiation

A related dynamic of change is angular momentum. This involves an object moving in a forward direction *and* moving around a central point. An orbiting spaceship or a tethered ball on a rope illustrate this process where the force pushing straight ahead is equal to the force pulling down to the center. Speed is constant but direction is continually changing, thus accelerating. Once speed exceeds or falls below the force pulling down, the orbit will either spin out or spiral in. This produces a growth or decay dynamic expressed as expansion or contraction, spiraling inward or outward, a universal process on the spectrum of change.

Process is about the dynamic interaction of forces, either in conflict, harmony or neutral. Physical dynamics include attraction/repulsion, synergy/entropy, build-up/release, resonance/discord, creative/destructive, and a variety of combinations. All processes result in either equilibrium or disequilibrium (balance or imbalance). The universe and all its contents are in a perpetual state of shifting in and out of equilibrium. It's an interconnected web of ebb and flow where constant change leaves the parts in and out of balance.

Equilibrium is a state of temporary stability. It can be a place of rest where energy fills a niche or a tension where equal and opposite forces balance each other out. Nature abhors a vacuum, which is another way of saying it doesn't like disequilibrium. Energy seeks the path of least resistance, so gaps are filled, weak walls are broken, and opportunities are taken. And because everything is connected, one action filling a gap opens another. Instability contains the very seed that leads to future stability and vice versa. It's the polarity principle where a Yin-Yang dynamic of opposites represents the interaction of polar forces in nature; each half contains the potential of the other. It's an active principle, not a passive one. Each polar opposite attracts to unite with the other. We see this in electric charges, molecular compounds and sex. It's an equilibrium/disequilibrium dynamic dance with cycles of tension and release, graphically portrayed in the sine wave: repeated vibrations of up and down but always coming full circle.

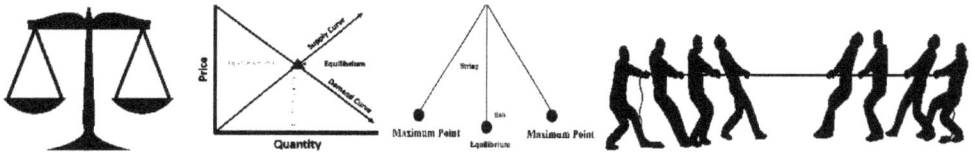

Interaction of forces generally follows a patten of dialectics where two elements clash to produce a 3rd element, summarized as Thesis-Antithesis-Synthesis. All creativity and novelty follow this pattern where something new is proceeded by something prior that was modified, tweaked or updated. Change itself contains this structure since all events have a beginning, middle and end. All things or "products" come from a sequence of input-process-output. The results follow our spectrum of change ranging from minor alteration to complete transformation. The clash between forces can balance into stability or push through the breaking point and radically change or destroy the original order (equilibrium dynamics).

In threshold transitions, the process of change is gradual until a critical point is reached where it suddenly becomes dramatic. A wide variety of phenomenon change slowly, often unnoticeably and then surprise – BIG CHANGE! Think of an avalanche where just the slightest addition collapses the whole. At the social level this can be seen in a variety of black swan events, like stock market crashes, viral memes, business disrupters, fads and revolutions. Your own life follows this pattern as stages suddenly transition after periodic life altering events. The essential principle here is how slow change precedes a large unexpected one.

Process dynamics involve a variety of *mechanisms* that influence the rate and form of change. These regulators include filters, valves, inhibitors, catalysts, diverters, and resisters. They're present at all levels (physical, biological, social) and direct the way change takes place. Regulators fall along a process spectrum ranging from slowing things down to neutral to speeding things up. They can work separately, in combination, in series or parallel. Process dynamics are fundamentally influenced by regulating mechanisms.

A Spectrum of Process

Flow – Shift – Divert – Filter – Resist – Inhibit – Agitate – Intensify - Catalyze
*Processes vary based on regulating mechanisms that do all of the above

Collective dynamics are quantitative phenomenon where unexpected attributes emerge simply by increasing the number of something: "more is different". This manifests as a social quality as clustering creates emergent behavior not present in individual parts. Consider an atom which by itself doesn't exhibit distinguishable qualities alone but gradually takes on specific attributes in greater numbers. A metal bar made from millions of iron atoms possesses unique traits: tensile strength, conductivity, malleability, heat capacity, etc. None of these traits exist in single or few atoms but they emerge mystically when those same atoms are clustered in large groupings. Other social qualities appear when elements interact with other like elements. All that's required are pre-set rules that govern how each responds to particular neighboring ones. Adam's Smith's economic invisible hand is based on this dynamic where individuals looking out for their own selfish interest collectively create a stabilizing force balancing a capitalist market system. This same collective order property is elegantly illustrated in virtual tile dynamics where inanimate objects exhibit living behavior based on simple programing. Once basic instructions are set and the action begins, individual tiles move, mix, giggle, glide, vibrate and dance. Social properties emerge and the result is interacting patterns resembling organic life. This further demonstrates how novel unexpected behaviors arise at the collective level that don't exist at the individual parts level.

*Digital creepy crawlers come to life, move around, zig and zag based on simple rule sets

Power laws are mathematic principles that govern proportional change. Since everything is mathematically connected, interactions in one element invariably affect all others. Ratios between things and their changing states can be greater than 1, equal to, or less than 1 ($> = <$). Differences can be arithmetic, geometric or logarithmic, including inverse relationships and squares. Change can take place in Log normal proportions, represented by skewed curves with greater initial difference followed by gradually slower, smaller change at the end, and vice-versa. Scaling laws capture all kinds of relationships in nature and social realms where bigger leads to fewer and smaller leads to many. City sizes are an example where scaling limits create patterns of many small ones, some medium and few large ones in orderly predictable ratios. The bell curve graphically depicts abstract process law where the proportion of average items to exceptional ones is a predictable mathematic pattern. Repeat any action thousands of times and the results will *always* follow a predictable bell-shaped pattern. This mystical property of change creates certainty out of seemingly random activity. Since mathematics is *the* code of physical reality it's only natural that it governs process dynamics.

Systems theory exhibits similar principles governing change covered above. Because systems involve cohesive groups of interrelated interconnected parts there will be collective dynamics, regulator mechanisms and dialectic interactions, all subject to mathematic properties of process change. Systems theory is synonymous with concepts of complexity and chaos, where small change and apparent instability unexpectedly lead to big change and actual stability. It's based on threshold transition mechanisms where crossing a critical point results in dramatic action. A key change mechanism is the feedback loop, which either reinforces the initial state or diminishes it. An iterative back and forth cycle can lead to a buildup (growth or expansion) or a limiting constraint that either reduces or stabilizes conditions (homeostasis in living beings). Complexity and chaos conditions display the appearance of disorganization and randomness but mystically transition into a stable system due to order generating rules. Systems share conditions of dynamic interaction, interdependent parts and tensions sensitive to small change. The resulting self-organizing process is an unpredictable emergent property.

The Greek Heraclitus sensed change as the natural principle of a universe in perpetual flux, the only real constant of life, stating "No man ever steps in the same river twice". The I-Ching "Book of Changes" embraces the process nature of life where every day presents a different set of conditions; a perpetual sequence of situation-response-new situation. Science does it's very best to capture the nature of change in a variety of fields and mathematic codes. The quadratic equation comes pretty close as it connects a constant, linear change and geometric change $(ax^2 + bx + c)$, although calculus goes further capturing rates of change and accelerations. Chemistry does its part by revealing the process of order transitioning into disorder by simply changing temperature (adding or subtracting energy), which leads to dramatic sudden change of state (ice to water to steam). And while unpredictable transformations occur suddenly, other change proceeds painstakingly slowly, as when a flowing river whittles down a rock over hundreds of years, not out of power but simple persistence. *Change* is the ongoing eternal attribute applied along a spectrum of tempos, increments and filtering mechanisms, while *process* is *the* fundamental dynamic driving the physical infrastructure of our living universe.

PHYSICAL REALITY UNIFIED

Post-modern Physics emerged mid-twentieth century focused on the ultimate quest of unifying everything into a single universal equation. Various domains of physics share similar symmetrical aspects suggesting a common thread built-in beneath the substructure. Magnetism, gravity, nuclear forces, electricity, etc. must be linked at some level. Einstein himself pursued a unifying principle unsuccessfully for decades while many others have continued the same pursuit ever since. Ultimately the end result should be a single equation that puts it all together.

A Theory of Everything (TOE) is the grand prize *goal* of post-modern physics. What began with Einstein's illusive attempt to unify the 4 fundamental forces of nature has progressed through decades of ongoing refinement to come up with a practical framework. Theories of gravity, supersymmetry, strings etc. have been hypothesized with multiple hidden dimensions, branes, hypothetical particles and bizarre concepts. A brief summary of this progression would include:

Quantum Gravity	– Linking General Relativity with Quantum Mechanics
QED	– Linking Electromagnetism-Special Relativity-Quantum Mechanics
Electroweak Theory	– Linking the above with the Weak Force
QCD	– Linking the Weak Force with the Strong Force
GUT	– Linking QCD with Electroweak Theory
TOE	– Linking ALL OF THE ABOVE (GUT & Quantum Gravity)

Unification in physics includes a quest to find the *Fundamental Particle*. While exploring deeper into the rabbit hole, science dissected the Atom only to find smaller, interlinked particles with associated anti-particles, leading to still more complimentary counterparts. Fermions, Bosons, Quarks, Leptons, etc. with up, down, top, bottom, etc. and variations in spin may seem like a confusing jargon of puzzle pieces yet they form a simple, balanced substructure of symmetry and unity. The standard model correlates both particles and forces in a manner that offers promise that one day we'll know exactly how every part of the whole is linked.

Grand Unification requires discovering both a fundamental particle and a fundamental force from which all other particles and forces are derived. The quest remains elusive yet clarity, precision and elegance have increased during the arduous journey. I sense the search for one equation that would account for everything is an exercise in futility because physical reality exhibits scale variance. In other words, what happens at the micro-level doesn't apply at the macro-level. Physicists attempting to unify Quantum Mechanics with Relativity are running into the same paradox of squaring the circle. There's a reason why a round circumference doesn't mathematically connect with a straight diameter without generating irrational numbers. We'll get into this in more detail further in.

Other obstacles to achieving a physics TOE are the ambiguous nature of dark energy and dark matter, undiscovered particles, Heisenberg uncertainty and the elusive weak force of gravity. It seems the more physics digs deeper into the rabbit hole the more unforeseen nuances emerge gumming up the works; approaching a distant horizon that never gets any closer. But this is just a half-empty cup perspective. Physics has indeed made much progress unifying a variety of essential ingredients that make up our physical infrastructure. Despite its reductionist approach and elitist hard science perspective (the same one that ridiculed ancient superstition and magic), it ironically comes full circle in its quest to discover deeper physical realities by revealing mystical aspects first intuited by early man.

Out of separation, dissection, and isolation, concepts of wholeness eventually surfaced, unifying a wide variety of seemingly dissimilar phenomenon. Modern and Post-modern physics have taken classically separate Matter, Energy, Space and Time and unified them. Distinct, isolated aspects of physics have gradually converged as more concepts, premises and theories came together resulting in compelling unification of the entire field. When it comes to the key ingredients of physical reality, what was once two separate individual things are now single interconnected sides of the same coin.

Matter and Energy. The special theory of relativity connects Matter and Energy simply and elegantly with Einstein's famous equation: $E=mc^2$. This is a powerful statement. First it denotes that matter can be converted to energy. Second, it denotes the reverse: that energy can be converted into matter. Most importantly it implies that matter *is* energy and energy *is* matter – complete unification.

An overlooked, subtle implication from this equation is that light is an integral component in the connection between matter-energy *and* space-time. Specifically, it's the speed of light that's key. All electromagnetic radiation is subject to the universal speed limit of light (C) because space possesses a kind of resistance called permittivity and permeability. And since speed is a function of distance and time, $E=mc^2$ links and unifies Matter, Energy, Space and Time. It's an important connection you won't see in any physics book.

Space and Time. The general theory of relativity connects Space and Time as an inseparable 4-dimensional framework, one where previously perceived isolated vantage points now become linked events in space-time. Position-motion are no longer absolute but dependent on all other positions-motions in one interconnected system. Matter generates gravity as a force by curving space-time around it. Space-time reciprocates by directing the motion of matter moving through it.

General relativity also unified forces of *gravity* with *acceleration*. There is no difference between standing still on a planet with gravity and riding an elevator up at a rate whose force is similar to the pull of gravity. And time slows down equally with either high gravity or an equivalent high speed of motion. Both involve compression of space-time which further unifies each aspect of physical reality.

Electricity and Magnetism. Electrons produce charges that manifest in both static fields and moving current. One essence, two forms: Magnetism and Electricity are two sides of one coin. Charge tells fields how to look while fields tell charges how to move, just like the relationship between matter and space. Faraday linked both forms, demonstrating how moving magnets generate electric current while flowing electricity generates magnetic fields. The result:

Electromagnetism, two complimentary forms of a single essence. This omnipresent force is present in atomic and chemical interaction, cellular processes of living organisms and the machinery that drives civilization. Light and all other wave lengths of radiation are electromagnetic. Like the strong nuclear force, light is also significantly stronger than gravity, by 10^{36} power!

Quantum Unification. Reductionist probing into physical reality paradoxically revealed an interconnected framework where parts are never separate and whose appearances are merely alternate versions of a single essence. Even light, one of the most fundamental ingredients of physical reality can express itself as both particle and wave; two forms, one essence. All matter composed of particles may change, interact and transform but in the end every single part remains eternally connected to every other part, regardless of distance and time. Quantum fields abide as unified wholes and remain so until interfering human perception creates the illusion of separateness – isolated things trapped in space and time.

Particle Unification. Atom, which means indivisible in Greek and once believed to be the foundational unit of matter has been sliced, diced and ripple cut into further microlevel parts by modern physics. Deep dives into reductionism have revealed a wholistic harmony of symmetrically connected subparts. Despite a host of quirky names for each elemental particle, they form a symmetrical united family of parts and twin counterparts, further linked by complimentary forces. Consistent symmetry and partnership indicate an interlocking wholeness among everything at the most fundamental micro-quantum level, captured in the Standard Model.

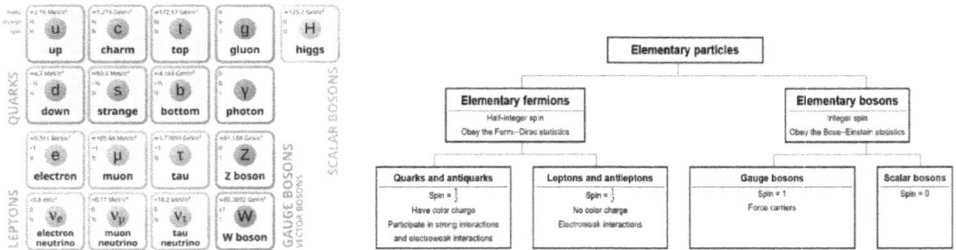

Fundamental Forces Unification. The 4 Forces of physics possessing both similarities and differences have long been the focus of relentless unification quests. The Strong Force and Weak Force operate exclusively at short range while Electromagnetism and Gravity extend through much longer range; literally infinite. Force strengths all vary greatly by magnitudes of scale, yet symmetries are present that imply inherent association. Equations for gravity and electrostatic charge are identical. Attributes among each of the forces vary primarily in strength and range. Their dissimilarity seems oddly similar to the way matter assumes different forms despite its obvious uniform internal essence. So far three of the four have been unified while linking gravity remains elusive.

Comprehensive Unification. The infrastructure of physical reality is essentially unified at both the cosmic scale and microscopic level. Matter and Energy ARE one. Space and Time ARE one. Magnetism and Electricity ARE one. Light is connected to ALL of them. Each exhibit dualism – a single essence expressed in two forms. Fields and Radiation likewise are two forms of a single essence. Same for Gravity and Acceleration along with the Particle/Wave duality. But what really makes for a Grand Unification is the interconnectedness of ALL these components.

Matter-Energy is inseparable from Space-Time. Large masses warp Space-Time while Space-Time directs the path large masses move through. Light is the substructure knotted up to form atomic elements. Light is the conduit that transforms atomic fields within Matter-Energy into radiation through Space-Time. Light *IS* both Magnetism and Electricity. Light follows the curved path of Space-Time created by a large mass of Matter-Energy. Light *IS* both Particle and Wave.

Gravity *IS* Space-Time curved by Matter-Energy. Gravity *IS* acceleration. Greater mass increases Gravity and reduces Space-Time. Greater acceleration reduces Space-Time. Light speed completely transcends Space-Time just as a Black Hole singularity's gravity collapse completely transcends Space-Time.

Grand Unification's Theory of Everything may be on life support but there are still compelling theories that qualify as potential TOE's. String theory is the leading candidate, requiring 10 dimensions (3X3 plus time) and seems to mathematically fit most of the requirements necessary for unification. The concept is eloquent as it breaks everything down to tiny vibrating strings, serving as the fundamental essence making up material elements. This is consistent with the field nature of forces and matter, along with dynamics of resonance which transform immaterial wave patterns into substantive structure. It's also congruent with pattern language which differentiates all reality by relationships of parts and the unique patterns they form.

The long journey towards unity in physics was a culmination of pioneers standing on the shoulders of predecessors to move one step closer to wholeness. Newton connected the mechanics of moving objects with gravity, Maxwell and Faraday connected magnetism and electricity, and Einstein connected energy, matter, space and space. But that progress produced an enigmatic split conclusion with relativity on one end and quantum reality on the other. And don't expect this dichotomy to be bridged anytime soon because it may turn out that physical reality is not scale-invariant; what applies at one level may be different at another. This will become self-evident as we dig a little deeper.

Conclusion. The reductionist approach used in physics ironically led to wholistic unifying of disparate, contrary concepts. The tunnel vision, focused microanalysis isolating physical phenomenon pushed everything to the limit, which unexpectedly flipped the script; Yin became Yang, micro emerged into macro. It turns out the parts were never separate but eternally interlinked – interconnected flowing threads of one reality. Matter in all its forms, Energy in all its forms; both expressed as one essence with many faces. Space and Time each illusive and abstract, emerge as intertwined dimensions of one mystical substructure.

Quantum revelations exposed illusions of substance previously appearing solid, distinct and certain, now just temporary snapshots, momentary selections out of infinite possibilities. Super-positioned wave clouds now take precedent over previously assumed particle forms. Our perception of separated particles misleads us from true reality where entanglement connects everything…where parts and wholes are revealed as inseparable whole-parts. All forces, substances and relationships intermingle as one unified system of systems observed locally, separately and futilely through human eyes limited by self-induced constraints. We're left with the inescapable conclusion that the fundamental ingredients of physical reality are not particles but rather abstract fields, not solid objects but temporary energy patterns perpetually fluctuating through space and time.

NON-PHYSICAL REALITY – *Metaphysics = beyond the physical*

The physical realm comprehensively analyzed by science is merely a partial reality, an incomplete picture of everything; only half of the whole. It has a complementary non-physical counterpart realm, consisting of everything non-material, non-substantive. It includes all those aspects of life that are invisible, intangible, abstract, ethereal and sublime. It consists of anything we can't touch, feel or sense. The entire mental domain of thoughts, imagination, intuition and anything our minds can conjure up abide in this realm. And there's no need to look *up above* or *out there* because the non-physical realm is right here. Both the physical realm and its non-physical counterpart are always here, occupying the same space and time simultaneously. One just transcends and includes the other.

The non-physical domain is where ancient mystics, mythologists and religious prophets sensed a higher reality beyond. It's the source of their magic, archetypes and gods. Pseudoscience and Parascience operate in this domain. It's where modern religion's spiritual source originates from. If there's a higher power governing man and The World, here is where you'll find it. If there is such a thing as one's soul, it too will be found here. The non-physical domain is spiritual by nature and the place where duality originates from.

We must resist the natural tendency to view this non-physical domain using the same lens applied towards physical reality. It won't work. This is unseen territory with a foreign language. While rational thought works well in the physical realm, intuition works best here. There's no empirical evidence or logical basis for comprehending the ethereal nature of the non-physical realm so other means must be employed. To perceive anything beyond our normal perspective we must look with a different set of eyes.

Reality is really a composite of two coexisting realms (physical and non-physical) both unified, inseparable, and intertwined. Yet human experience perceives separation, disconnection and muddled divergence. There's the material plane we're bound within and the illusive higher dimension permeating it, hidden from immediate perception. This higher realm is difficult to explain. It's formless, timeless, transcendent and beyond simple explanation. It's the region where nothing exists yet everything emerged from. You could call it pure consciousness, spirit, primordial ether or just simply "God". Some call it non-Dual reality. Think of it as a unified undifferentiated plenum of potential – the original source of everything. It's everywhere and nowhere. No words can adequately describe it.

Our familiar physical universe mirrors the less familiar non-physical, higher domain. Everything physical is preceded by an original non-physical source. From a detached perspective we see this in basic content of the physical universe with its virtual emptiness. Invisible dark matter and dark energy overshadow objects and dense matter. Atoms are 99% empty, just like everything else we perceive as solid substance. The ancients sensed this principle, proclaiming "As above, so below". Higher and lower realms symmetrically relate and share a common essence. One original whole expressed in two realms: formless and formed.

THE INGREDIENTS OF NON-PHYSICAL REALITY

Every *thing* in our physical realm originates from a non-physical source. What we think of as "solid" forms have a non-physical originating template, like a blueprint of a building. All material substance in our domain is merely residue mirroring non-physical forms that existed prior. Since matter is merely fields of structured energy, it follows there must be similar counterparts of non-physical energy in a higher domain. We can think of this subtle higher-level energy as the etheric original source of the various forms of energy in our domain, including electrical, mechanical, atomic and of course, light. Energy already barely qualifies as physical to begin with, possessing a more pseudo-physical nature compared to dense matter. It's just one more transcendent step transitioning from the non-physical realm.

Subtle energy in this etheric non-physical dimension is the source of *prana*, the vital principle permeating life at all levels. It's no different than the Chinese *Qi* or *Chi*, pertaining to the life force referenced in yoga and martial arts as the body's bio-energetic field. It's also the same subtle energy at play in pseudoscience disciplines that track behavior and human drama, along with parascience investigations of phenomena beyond traditional scientific methodology. It's the source of psychic energy patterns submerged deep within our unconscious being, collectively expressed in archetypes. Undoubtedly, it's the same mysterious energy referenced in Star Wars as *The Force*. Subtle energy is simply a non-physical version of the stuff we're familiar with, except it can't be seen, stored or measured conventionally, but it can be experienced.

Since energy in our physical domain has a subtle non-physical counterpart, matter must also have a mirroring substantive form in the non-physical domain.

That intangible stuff is *Consciousness*. Explaining consciousness is like describing the nature of love – both are illusive essences with a variety of expressions and context. It's as futile as defining the TAO, an indescribable reality simply beyond words. Yet consciousness is the primary stuff of both physical and non-physical realms. It's both the medium and the message, content and context. It's everywhere and nowhere. Both realms are defined by it. Its presence assumes various forms and states. In its purest form it has no form. Consciousness is both the ethereal spirit of the higher domain, and the imbedded fields present in lower domains. A comprehensive spectrum of consciousness spans all levels, varying in degrees from higher spirit to dense constrained materiality bound by space and time.

Water is a close analogy for consciousness – imbedded in things at different degrees and in different forms. The earth's surface is comprised mostly of water, in large ocean bodies, flowing rivers, lakes, small ponds and even little puddles. It's present at the inner depths of land as well as in the atmosphere above – the very air that we breathe. It's the main ingredient in living beings that populate the earth. Water is *the* omnipresent primary substance of life at different degrees. Same with consciousness which is the primary substance of reality and is present in *everything* at different degrees. It's *of* the non-physical realm yet it's *in* our physical realm. It's difficult to pin down as it takes the form of everything on the material plane to become that thing, same as water taking the shape of every container it fills,

Adding further layers of confusion is the human filter, where consciousness becomes self-aware, individuates and creates man-centered reality. Now we have identity – me, myself and I, *here and now*, and everything else *out there*. The same stuff built into everything now appears separate and apart from me. Consider a rubber band twisting into 2 loops: we now have a singular whole appearing as separate parts. From this relational illusion *The World* is created. Perception emerges as a filter between parts of the same whole. Attachment, desire and fixation towards external things further reinforces the illusion of separation.

To further this illusion, as individuals begin to cluster, multiple perspectives emerge. I becomes we, leading to us and them. So, after consciousness becomes self-aware it transitions to collective awareness where few, some or many share perspectives that others don't. What was originally a single reality now appears different among various individuals and groups. Two people can look, listen, touch, taste or feel the very same thing and perceive two different experiences, both valid. This reinforces perception of things and objects *out there*. As conscious awareness expands, more elements are connected in the equation. Multiple viewpoints arise with a new reality based entirely on a relational perspective. True reality requires consideration extending beyond me, myself and I to include perspectives of other individuals, groups, and the context either operates within.

Conscious participation in the physical domain involves experiential processing of information. Each and every moment of being brings with it a personal internal experience. Thoughts, feelings and sensations are attached to actions, interactions and events. Experience is the currency of consciousness while the payoff is knowledge and eventual wisdom. Learning is ongoing, involving pattern recognition and breaking codes. Intelligence is a metric of that interaction: how much an individual can process, interpret, remember and make correct conclusions. Intelligence is a derivative product of a universe with built-in mental attributes.

Living beings are composite mixtures of consciousness experienced through multiple layers, levels, degrees and states. Every person resonates to particular levels of awareness, some higher, some lower, most somewhere in the middle. Within a broad spectrum of consciousness is a human band loosely following the bell curve as individuals vary between low-middle-high relative levels. The majority of any population will occupy the middle levels while much smaller segments occupy the low and high end. People naturally resonate to each particular level however during life's journey with movement up and down the spectrum, depending on progress or regression. Life stages are associated with specific levels corresponding to situations and relationships. Over time, your consciousness increases until a phase change occurs, lifting you up to a higher level different from the previous stage – an ongoing ratcheting up process forming a stair case pattern. Consciousness is experienced as various states of being. There's the daily spectrum of wakefulness where attentive awareness shifts between three primary levels: Conscious, Subconscious and Unconscious. These levels are directly associated with wakeful attention, dreaming and sleeping. Of course, there's a sliding scale between each and the transition is generally subtle. Different people can vary greatly within the same states. Consider how "awake" some individuals are compared to others who seem to be working on autopilot. Or note the different intensity of dream experiences between people. Consciousness is a slippery slope where states of being change frequently, differ greatly among individuals and intermix with various groups.

Mental attributes built-into the infrastructure of physical reality are derivative of consciousness. It's the ordering principle giving rise to intention and thought, as well as the emergence of relational context providing meaning in every situation. There's a direct connection between consciousness, mind, order, relationship and information. It's the non-physical substrate that emerges to create patterns and forms everywhere, setting the stage for our perception of reality *out there*. Human experience is entirely based on perception of forms and relationships, all products of consciousness serving as both a pseudo-substance in them and a perceptual filter sensing them. Relationships are patterns that define. They provide information about the world out there and how everything fits in it. That information is a derivative attribute of consciousness, manifesting as an ordering principle. Our awareness consists of processing information in various forms. It's a pattern code whose highest language form is an image. Relationships among different things are merely information patterns that can be expressed in bits and bytes (binary language), easily expressed in images composed of arranged dots and gaps. Whether we see things directly with our eyes or see digital representations (photo-video), the information is virtually identical. Non-physical information patterns precede form and order, yielding a pattern language manifested as images, dimensions, relationships, constructs, figures and abstract codes. It's the source of the mental quality pervading every *thing* and *non-thing*, mirrored by a universe with a mind of its own along with built-in physical memory.

Consciousness is like resonant spirit metaphorically similar to pure white light with its millions of embedded colors, coming in infinite wavelengths or intensities. Both are filtered locally by us, manifesting into particular forms based on our limited perspectives. As resonant spirit, consciousness is a a self-reinforcing entity, leading to a localized self-aware identity. Then mind emerges to establish local order and further reinforces a state of *"I"*. As a resonant pattern of self-aware identity, *I think, I act, I am.* Mind's ordering process now produces perceived separation between I and everything else. It's a process of universal consciousness forming a localized pattern, just like a water droplet on an ocean wave crest. It's difficult to grasp the true nature of non-physical consciousness with its formless, empty ethereal essence, which can't be explained by reductionist science. It's the intangible fundamental non-physical "substance" within everything, manifesting in every conceivable form, pattern and relational condition. Residue of consciousness can be found in the pervasive mental attributes of matter-energy dynamics, the mind of man, and the ordering principle of physics. It's the source of intention, purpose, identity and self-awareness, which by extension through man, enables the universe as a whole to become self-aware.

SPECTRUMS OF CONSCIOUSNESS
States of Awareness
coma – vegetative state – non-rem sleep – dream sleep – drowsy – awake – alert – manic

The Great Chain of Being
matter – body – mind – soul – spirit (*Modern*)
inanimate objects – plants – animals – people – angels – God (*Classic*)

Collective States of Consciousness (*Societies*)
instinctive - animistic - egocentric - absolutist - multiplistic - relativist - systems - holistic
survival　security　power　order　success community synergy holism

*ALL things and beings vary by degrees of consciousness, both substantively and in states of being. Collective societies are aggregates of individuals sharing a mix of spectrum traits.

DYNAMICS OF DUAL REALMS

Everything in our material plane emerged from the non-dual higher original source. Physical reality as we experience it is generated moment by moment from this source in a continuous quantum oscillation. The higher-level plenum represents the infinite – a realm of all possibilities. It mirrors the quantum wave aspect in physics where all possible positions of a particle suddenly collapse into singleness – a distinct point in time and space. So too does everything emerge into the lower physical plane, a moment-by-moment process of many possibilities becoming specific actualities. Think of a long silence suddenly interrupted by a single note; a disturbance in the pure stillness now transformed into manifest sound. Then a series of notes produce a pattern of form. Thus, nothing becomes something.

The highest realm of reality represented by etheric spirit emerges into the lower realm through an ever-flowing process essence. Quantum moments transform potential nothing into actual something. Process dynamics stir, rustle and agitate the undifferentiated plenum into subtle orders. Patterns resonate into templates of abstract form. Relationships emerge, appearing to separate what was once undifferentiated. All is in nebulous flux; moving, meandering and permeating. This higher spirit-like non-physical domain pervades everything, imbedding an etheric non-physical essence into every nook and cranny of the physical universe. It is *the* source of what eventually manifests as time, change and process dynamics.

Hindus intuitively described this process as the dance of Shiva – a sort of cosmic interplay between static and dynamic energy flow through which everything is created, transformed and destroyed. Chinese divination captures the sentiment using the I Ching (*Book of Changes*). Human experience is just one level of this ongoing cycle of birth-life-death and the creative interplay among the old and new, past and present, original form and transformation.

Duality originates through a process of undifferentiated spirit unfolding into patterns of forms. This unfoldment produces greater degrees of constraint, order and rudimentary substance. It's not unlike invisible electromagnetic radiation which resonates into structured matter, where all objects on our material plane are essential composed of crystalized light. Duality is primarily a process of emergence and interaction both within and between two conjunct realms. The result is a unified whole progressing into multiple variations – from the monad emerges several, then many. In physics it's similar to a photon transforming into particles, particles pairing into atoms, then repeated pairings produce 118 unique atomic elements with widely different attributes. From one original whole…*Everything* emerges.

Describing the essence of higher realms is difficult if not futile, comparable with listening to a foreign language. Comprehension requires a heavy dose of metaphors and analogies along with channeled intuition. The higher non-physical realm has no boundaries, no beginning or end, free from the constraints of our familiar physical domain. A mystical transformation takes place where the higher realm manifests into the lower, dense realm of form and structure. Like pure white light magically emerging as a color kaleidoscope when passing through a prism. Sound as well, exhibiting the ground state of empty silence until suddenly interrupted by invisible wave impulses to form a variety of patterned notes, even melodic music. The quintessential metaphor of duality is water, the very essence of organic life, where the ocean represents an undifferentiated plenum – a unified whole...silent, still and calm. Energy flows through it, surface tension builds, waves emerge and rip tides form; from pleasant ripples to tidal waves turbulent conditions, appearing out of formless original unity – a transcendent process where nothing becomes something.

The transformation from the absolute, undifferentiated plenum into our domain of physicality progresses through a scale of increasing concentration of essence. Free flowing, unincumbered spirit descends into gradient dimensions of resonating constraint. *Nothing* gradually becomes *something;* order and structure emerge into patterns of form. Degrees of constraint increase along the descent into the physical plane where stuff becomes material order, bound by space and time, differentiates, enfolds, and descends into lower constrained levels that manifest into everything. Out of the primordial plenum, self-similar relationships emerge, then patterns form and order arises with individuated attributes and dimensionality. The original substance of this non-physical domain is primarily a subtle, spiritual form of energy enriched with information. The transition is gradual, but its transformation is dramatic. Like the state change of gas into liquid or liquid into solid, the phase change between realms is equally mystical.

Interconnectedness is easier to see in the physical realm than between it and the non-physical. Unmanifested spirit of the higher non-physical domain permeates the lower, physical realm with great subtlety. It's always present, lurking in the background as stillness – the pause between action and non-action. It's the silence when nothing is spoken or heard – the gaps between music notes. It's imbedded within the empty space surrounding everything. The physical realm consists entirely of derivative patterns from the higher realm, manifested as temporary resonant forms taking physical shape. As living beings, we experience both realms simultaneously, one through actions and change, the other through stillness and being. The duality of these realms is built into each of us. Dual realms are not separated but rather complimentary; two sides of the same coin.

Every quantum moment involves realm 1 emerging into realm 2; the non-physical transitions to the physical. The possible becomes actual. Life happens. We don't see it because every new moment seems just like the one before, as does the next one and the next. Here and now is not just a timeless, eternal moment but an ongoing string of nearly identical here and now moments. Thus, realm 1 permeates realm 2, moment by passing moment. It's an invisible eternal process regenerating our reality. Over time, physical objects are transformed, living organisms grow and die, history is made, and change in all types and varieties happens.

Light presides at the seam between the physical and non-physical realms, timeless and massless yet abides the fabric of space. Once a photon makes contact with matter it's immediately annihilated. Light may be *in* this world but it's certainly not *of* this world. The same maybe said for the concept of God. If the attributes of one's God are all powerful all omniscient supremacy it cannot be limited by the physical realm. Higher level qualities precede the lower levels, interpenetrate them but are not bound by them.

Physical and Non-Physical Realms Connect and Overlap

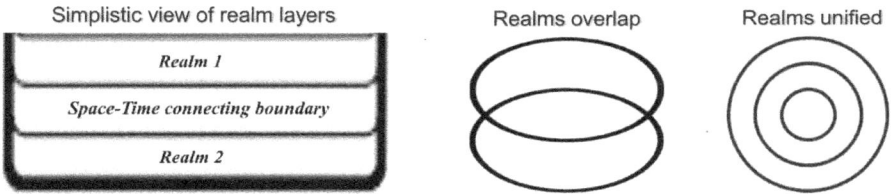

Space-time links dual realms, serving as a transitional boundary seam where light operates

The universe is an information processing system with mental aspects imbedded in everything including atomic dynamics. Matter and energy interactions define physical reality while information-relational patterns define non-physical reality. Information present in matter is transferred through physical actions. Living beings mirror this action through chemical processes. DNA is *the* information conduit.

Information is fundamental, matter-energy is secondary. The pattern of an atom is a template, preceded by a form necessary for its existence. Particle behavior exhibits preference of action and intentionality – a sort of pseudo intelligence. Atoms and molecule dynamics exhibit the same intentional interactions, mirroring human social behavior. This mind-info attribute is present everywhere,

The simplest, most basic piece of information is the bit. It represents 1 of 2 possible states: either this or that, yes or no. Everything can be defined as a string of bits. Binary language is based on this principle. A more complex thing just needs a longer string of bits to define it. On this basis the only difference between a rock and a complex machine is the length of a bit sequence used to describe it.

Binary Code   Morse Code   ACSII Encoding   Simple Image of Dots

Images are raw information patterns, easily represented by a binary pattern of 1's and 0's. Black and white photos are just patterns of dots. This relational essence becomes the fundamental substructure of non-physical information-based reality. The transcendent linkage connecting all of this information is pattern language. All meaningful context is derived through relationships between things, encoded in the various patterns they form. Physical relationships originate with immaterial patterns of information built into the non-physical domain. Like a template, this foundational matrix translates into physical objects, things, structures and general materiality with attributes based more on patterns and relationships than actual substance. The 92 periodic table elements contain identical parts but exhibit radically different qualities due to different relational internal arrangement.

Non-physical reality originates from nothing yet somehow manifests into something. The process begins with a special kind of order: self-relation. It unfolds the same way the TAO begins with absolute unity, where a wholeness of 1 becomes 2, then 3, then many, all starting with self-relation. Using our rubber band analogy, take a single, whole loop, twist it around once to form two connected loops, each related to the other and unified, but now possessing pseudo-separateness. Twist it again and you get three, then four, then many. Similarly, spirit unfolds to find itself in multifaceted forms creating diversity from unity. A simple, rudimentary function multiplied numerous times eventually leads to complexity. For a comprehensive explanation of the process see "Unified Realty Theory" by Steven Kaufman (2002).

Similar transcendent order emerges in living systems where organic elements develop into progressive hierarchies without predictable, logical direction. Parts combine in connected networks forming unforeseeable new wholes. They then join other wholes in a higher-level network, making them the parts of an even greater whole. Thus, every element in the hierarchy chain is a whole-part, relative to what's above or below it. Holon structure is a fundamental order in physical reality, resembling a series of encased Matryoshka dolls. This text has a holonic pattern of letters-words-sentences-paragraphs-chapters-etc., a series of greater parts and wholes. The universe progresses as particles-atoms-molecules-oganisms-families-cities-states-nations-worlds-solar systems-galaxies. Transcendent order results in higher level wholes not predictable from original parts. New attributes simply materialize without explanation. This is how abstract immaterial dynamics emerge out of our two transcendent realms, which themselves form a holon pattern to begin with, serving as the original source of all other holons and hierarchies.

World	Chapter	Compound
Nation	Paragraph	Molecule
State	Sentence	Atom
City	Word	Proton
Town	Letter	Quark

The infrastructure of physical reality is an interconnected fabric combining physical and non-physical realms in a mirrored tapestry where the higher *whole* unfolds into the *manifest many*. Every part is connected to the same whole, and every whole is part of a larger whole. This super-fractal pattern operates like a holographic image where the complete original whole is contained in every part. Physicist David Bohm developed a holographic model describing two realms as implicate and explicate orders, one preceding the other and manifesting as a result of our conscious participation. Human perception and experiences create specific realities out of infinite possibilities, similar to collapsing the quantum wave cloud. Reality is created moment by moment as the cosmic plenum teeming with energy and potentials is crystalized into man-made physical reality, driven by intention, desire and attachment. Everything already exists, here and now, in a higher domain until conscious unfoldment – to be revealed or discovered but never invented.

Realm 1 is comprised of infinite forms and patterns whose immaterial essence is non-physical subtle energy. It's the source of Realm 2 physicality where fields emerge and induce order. Resonant patterns evolve into relational structures and dimensions. These invisible forms are connected in fields of dynamic equilibrium – a balance of forces in a domain of self-relation. Light mirrors this dynamic as it combines electric and magnetic components in harmonic cycles, both aspects possessing pseudo-physicality with dynamic results.

All actions and events are essentially exchanges of energy and information in a variety of forms, primarily through field dynamics. Fields and resonant patterns are omnipresent, original forms described by ancient Greeks. Forms or patterns repeated over time acquire a persistent resonant field attribute. Archetypes are created in this manner along with a variety of pseudo-physical and non-physical phenomena. Waves progressing as cycles with particular frequencies become resonant patterns of order. Gyroscopes are merely circular moving rings but they generate real force resistant to change from simple spinning motion. Electricity flowing through circular coiled wires generates magnetic force in a similar manner. Tuning forks vibrating in sync create a force greater than each original source.

Vibration is a universal dynamic present in the eternal dance between energy and matter, freedom and constraint. Waves, cycles, and vibration combine fixity with flow, form and change. Everything has a beginning and end, which can be considered a single cycle or opposite ends of a single vibration. The universe itself is one big cycle of boom and bust, with lots of mini cycles within it, from the birth and death of stars to the coming and going of every life form. Vibration is built-in physically and manifests everywhere: growth and decline, ebb and flow, up and down, day and night, breath and exhale, life and death, and the ongoing heartbeat of living beings and bustling cities. This universal dynamic is induced by the moment-to-moment transposition of higher realm 1 into lower physical realm 2.

Thoughts become resonant fields that create order, as do emotions. Persistent thoughts or strong desires lead to intentional action. Thoughts and emotions have a subtle energy pattern with a life of their own. Obsession, fixation or attachment become powerful forces manifesting into directed change. Repetition of simple patterns transforms anything and everything. As within, so without.

Every material object is proceeded by an immaterial pattern. This applies to organic, living matter as a subtle energy field mirrors the physical body of living beings. Rupert Sheldrake proposed a theory of morphic resonance where physical and nervous systems of organic life correlate to vibration patterns. Morphic fields penetrate bodies of living organisms, including plants or animal, via vibrational resonance, influencing healing and states of being. Symbiotically they mirror brain waves and changing mental states. The seven Chakras likewise represent nodes of energy in similar resonant fields. They overlap and interpenetrate to form one unified whole with nodal points, like the octaves of a music scale.

 Spirals, spheres and geometric shapes are omnipresent artifacts of original forms in realm 1. Harmonic motion, chaotic turbulence and process dynamics in general also originate from the flux and flow of higher levels. As above so below, each realm is a mirrored complimentary pattern of the other, one subtle/non-physical, the other material/physical. Energy in all its physical forms (mechanical, chemical, thermal, electrical, nuclear) must mirror subtle immaterial counterparts in realm 1. The fundamental forces in physics (strong and weak nuclear, electromagnetic and gravitation) must do so as well. Everything in our physical universe is derived from non-physical original forms from a higher immaterial domain. And notice how our universe of mostly empty space is mirrored on our lower physical level by mostly empty matter.

Both the physical and non-physical realms are gradations that mirror each other. We can approximately project what the higher realm is like based on the patterns in the lower one. Our physical domain is a gradation of constraint, with energy and gaseous matter on the higher end and dense solid matter on the lower end. It's likely that the non-physical realm would also have varying degrees on non-physical ingredients, including subtle energy, information and consciousness. At the highest level is complete formless emptiness, beyond description. We can think of this level as realm 0, the absolute original plenum from which everything descended out of. It's the seat of pure spirit, *being* and the power of stillness. Non-physical realm 1 and physical realm 2 are its lower manifestations. We don't know what connects realm 0 to everything else, but we can surmise how realm 1 and 2 connect - *space.* What we think of as empty space is not nothing; it's a pseudo-physical membrane connecting the non-physical and physical realms. It's the linking boundary between the tangible and intangible. It's where light operates with no mass, no charge and no time. And it's the epicenter where all quantum action takes place.

Non-Physical Realm Precedes the Physical

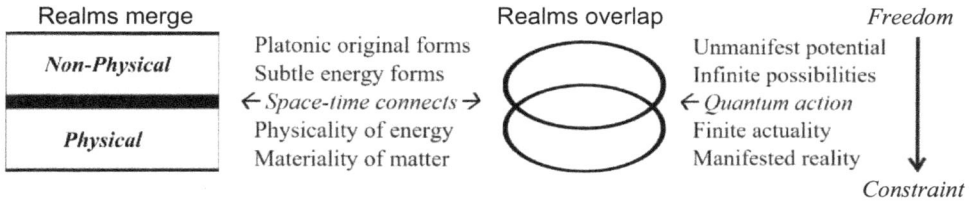

Realms merge		Realms overlap	Freedom
Non-Physical	Platonic original forms	Unmanifest potential	
	Subtle energy forms	Infinite possibilities	
	← *Space-time connects* →	← *Quantum action*	
Physical	Physicality of energy	Finite actuality	
	Materiality of matter	Manifested reality	

Constraint

Is it any wonder that quantum mechanics produces bizarre, abstract realities that bear little resemblance to our everyday classical physics? It's what happens when you connect extreme boundary layers crossing different dimensions to produce emergent new attributes. Space is a transition layer where limitless freedom from constraint in the higher realm intersects with the dense limiting attributes of the physical domain. Infinite possibilities exist in the higher realm. As time passes moment by quantum moment, higher realm 1 intersects with lower realm 2, freezing the free-flowing cloud of possibilities into a single actual. Remember our roulette wheel spinning eternally but stopping on a particular number every moment. This is how our physical reality is created, moment by quantum moment. There's the infinite higher realm, the buffer zone of space in the middle, and the finite lower realm. But don't think of these vertically; they overlap each other. The higher non-physical realm is already here within our physical domain. Heaven was never *up there*; it was always *right here*.

Boundary layers perfectly explain common paradoxes of quantum mechanics. Physicists must have been annoyed to discover built-in uncertainty of quantum action and certainly perplexed by the unpredictable dual nature of light, acting as both wave and particle in different situations. And surely it was a hard pill to swallow when particles exhibited connectivity regardless of vast distances apart. It all becomes completely understandable in a framework of connected layered realms. It equally applies to the connection between human consciousness and manifesting localized individual personal experiences out of infinite possibilities.

Space and time are a unified connected dimensional medium. While space is the *boundary layer* between higher and lower realms, time is the *process layer* linking both realms. Our physical reality may seem stable, continuous and permanent, but it's created moment by quantum moment like a film projector with hundreds of individual frames moving so fast it creates the illusion of continuous motion. Our experience of time is produced by the limitless higher realm intersecting with the limited lower physical domain, creating a moment-by-moment difference. This is why you can portray anything moving as a series of individual increments. It's the same with analog music converted to digital bits. The reality you experience is quite literally a complete and total illusion! But don't worry, we can't tell the difference anyway, just as analog and digital music sound virtually identical.

As self-aware beings we can appreciate the process of evolution, our history of progress towards greater complexity and higher consciousness. What we don't see is the preceding opposite process called involution where the higher non-physical realm descended lower, in apparent regression. Think of mature adults who produce an infant child that enters the world ignorant, inexperienced, but full of potential – awaiting a journey of discovery progressing upward, learning, growing, evolving to eventually reach adulthood and complete the cycle. Involution and evolution are similar – one phase precedes and enables the other. Philosopher Arthur Young developed an intriguing model of this process, described in his work "The Reflexive Universe" (1976). His model captures the descent of spirit into dense matter and its rebounding ascent up, culminating with fully conscious man. There are 3 basic levels and a lower fulcrum. The process flows down from Light to Particles to Atoms to Matter (molecules), then up to Plants to Animals to Man. This is essentially the great chain of being – a concept developed in ancient times however Young's analysis provides a deeper insight into the dynamics of involution-evolution with emphasis on degrees of freedom and constraint. Spirit first descends into materiality acquiring greater degrees of constraint at each stage and then rebounds back, removing that very constraint a degree at a time on the way up, revealing the implied intentionality behind human evolution.

Young model of involution-evolution

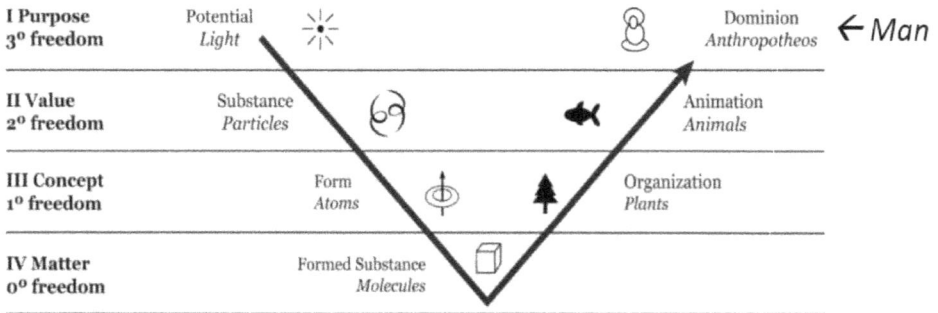

DUALITY REALITY

The involution-evolution dynamic offers a clue to the origin of purpose and intention. Humans are more than just machines interacting with physical surroundings, reacting on survival instincts. We're higher consciousness beings with self-awareness and the ability to make value judgments – to consider moral value in decisions with inner soulful guidance. I suspect the spirit that enfolded into the lowest physical domain yearns to return – to its origin. During one's lifetime journey of processing experiences anchored in the physical realm, the soul persistently strives to reconnect with higher levels; the original source. It's no accident that humans spend a third of their lives sleeping, enabling necessary dream channeling to rebalance the inner with outer, lower with higher.

Duality dynamics are connected to human consciousness. We're the fulcrum that amplifies duality. The moment we self-identify and embrace "I" we separate ourselves from everything else. It's now me *in here* and The World *out there* – literally a man centered reality. And while I'm in here I unwittingly put God and all else out there. In this manner the entire lower material plane becomes a realm of separated things with each of us occupying a self-created center. Ironically, our descriptions of The World out there say more about us and less about it. When you point your index finger at something the other three are pointing right back at you.

Man is a natural dual-reality filter. Every moment of our experience involves a micro event analogous to collapsing the wave function. Out of many possibilities we fix in place one actuality. From the higher-level infinite we create/experience a particular life, moment by chosen moment. But we do have free choice, expressed in decisions made every second of every day. Our conscious thoughts, desires and attachments become instant destiny. Infinity is filtered through us into a finite, singular reality. You may have access to hundreds of cable channels but only one is chosen to be viewed. Many possibilities, one actuality. This is how man the dual filter creates a distinct domain of self-imposed constraint, for better or worse.

Human duality-reality is largely an illusion of material primacy. Our senses obscure the higher primary realm which precedes the lower, secondary physical plane. Things and objects are patterns of spirt in temporary structured form. Your body is perceived as a static whole – a stable, fixed identity. Day by day you look in the mirror and see the same person, repeat the same habits and follow a regular routine. Yet from a higher, long-term perspective this is merely a snapshot illusion of temporal bias. We are all changing continuously, day by day, year by year along a lifetime path. Our true essence is an abstract pattern flowing over time and not limited by any short-term snapshot. We experience a dyad patten of time living in the here and now juxtaposed by an ongoing life journey timeline of many years. The present is transcended by past and future timelines.

Child-youth-teenager-young adult-middle age-senior is one spirit, one soul witnessed and filtered through a series of lenses at various ages and stages. One essence projected into multiple perceptions creates multiple realities. Human spirit endures while daily experience is temporary and constrained – tugging, pulling, twisting and shaping our souls. Life is really just the ongoing dynamic interaction between our inner being and the external constant turmoil of being in *The World*.

Ironically *The World* out there is really just a 3D projection of what's in here. Consciousness creates reality from the inside out. Man is like a prism-valve-transformer-lens all in one. Like the infinite waves contained in light we resonate to particulars, similar to separated color frequencies. Attachments and desires act like antennas focusing external reality from infinite possibilities, and it all originates from inside. Our natural law of "As above so below" is more accurately "As within so without". Human consciousness localizes the infinite plenum, sets particulars and creates an external reality of separate subjects and objects.

Everything in the material realm is subject to human perception which separates all, producing division, polarity and thus duality. In our man-centered reality we're always right in the middle so naturally we divide and categorize. Our privileged relationship to everything introduces new attributes to previously unified wholes. Size becomes big and small. Speed becomes fast and slow. Neither of these concepts are absolute, just values imposed by individual perception. What's big to you might seem small to others. It's all relative. But what's key is how every phenomenon becomes polarized by human perception. At the highest level there is no polarity – all is one. Good and evil, far and near, many and few…all are just human concepts that become meaningless beyond the physical realm.

Moments of synchronicity also reveal the desire to reconnect with the higher realm. This inner unconscious intention first emerges into lower life forms as the drive to simply survive and later incentivizes higher conscious beings to pursue progressive, moral ends. Much of it is masked by the daily grind and burdensome requirements of living in *The World yet* always present just beneath the surface. The descent of spirit down into the lower, constrained physical realm is like a stretched rubber band that is always pulling back to reassume its original state. That's the inner pull that mystically surfaces as higher intention and purpose. It's built into the living universe and directs all life, including the little acorn that becomes a fully grown oak tree.

Physical Realm	Non-Physical Realm
Objects, bodies, things, structures	Ideas, intension, creative energy
Science, conventional truth	Metaphysical – intuitive, higher truth, wisdom
Intellect of knowledge & experience	Intuition open to limitless domain, 6th sense
Laws, Codes, Mathematics	Metaphors, Analogies, Allegories
Conscious	Unconscious

Awareness of reality-duality is only a first step. False perceptions must be overcome, including the tendency to think in terms of two distinct, separate realms. This is a partial truth. More accurately, the higher realm transcends AND includes the lower realm. Both are blended, linked and inseparable. Like two sides of the same coin whose essence is linked. Like the matter/energy pair where energy is simply just a higher octave of the two – both different states of one essence.

What we perceive as permanent structures are merely temporary forms, like our own very limited vessels we call bodies. Everything we perceive and experience in the physical realm are illusions produced by our conscious identity, separating our ego-based self *in here* from all that's *out there*. Each of us becomes a spirit filter, projecting higher reality into the world. Man in the middle is the filter between two realms. Collectively, man creates The World out there in an identical manner. Same with quantum physics where the local observer creates physical reality.

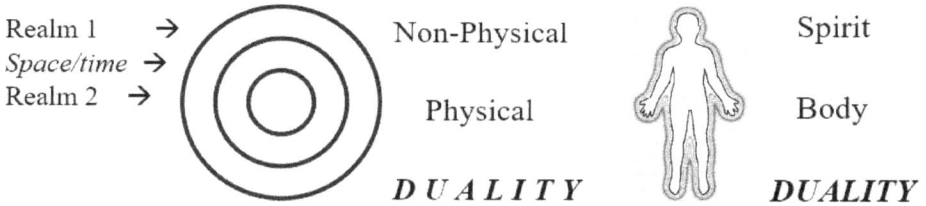

Realm 1 →
Space/time →
Realm 2 →

Non-Physical

Physical

D U A L I T Y

Spirit

Body

DUALITY

Realm 1 transcends & includes Everything, including Realm 2 + space/time membrane.

Our physical existence is tethered to a state of Dual reality. We naturally default to the lower, dense realm of things and structures separated into time-bound constraints. Human consciousness and limited senses imprison us within – out of practical necessity. Ego, mental attachments and desires divide us from higher reality yet they're necessary for our survival navigating ever-present challenges in Nature and *The World*. It's a push-pull dynamic where competing aspects of reality complicate our ability to distinguish between truth and illusion.

Newborn babies perceive their local world as a timeless, spaceless realm, where things just appear. It's a magical experience of imagination without constraint. Kids playing games experience timeless moments of being in the now, with little concern about outcomes. It's pure creative expression. Games are fun, art is fun, exploring is fun. Gradually over time these moments of pure being transition into structured experiences where ego emerges as a divider and separator. Games become rule-based competitions, creativity gets stifled by outcome-based goals and exploration is diminished by familiarity. Entering adulthood is a transition away from realm 1 freedom to realm 2 constraint.

Both realms interact through an experience and the experiencer – event and observer. Both are inseparable, complimentary aspects of one reality. Conscious beings born into the world become individualized identities, attached to senses, feelings and thoughts, embracing the I-me *in here* and everything else *out there*. Thus, the illusion of separation becomes the fundamental experience of daily life.

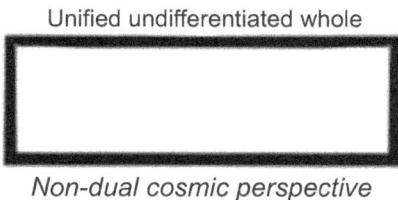

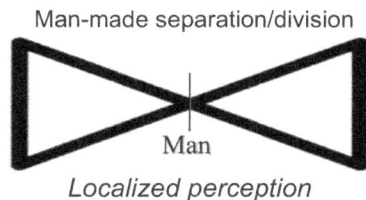

Unified undifferentiated whole   Man-made separation/division

Man

Non-dual cosmic perspective   *Localized perception*

Our daily reality largely consists of duality dynamics; operating in the physical and non-physical realms simultaneously, diminished by limited perception, biased intention and creative action. The higher realm is the timeless eternal now we sometimes experience briefly in between daily distractions and fragmentations. Our perception of past and future are mental constructs that have no meaning in the timeless higher realm.

Living within two realms necessitates us to possess elements of each built into our being. This is reflected in our body-mind-soul-spirit composition. Each of these subparts come from the same original essence; consciousness, differing by degree on a continuous spectrum. Our physical, mental, emotional and spiritual aspects resonate at different vibrational levels, lower to higher. These are like nodal points on the spectrum where new qualities emerge. This pattern is repeated in other essences including water, with its own nodal forms of solid, liquid, and gas (ice-water-steam). Lower levels contain the potential of higher levels (ice is just dense steam). All physical reality is similar to this pattern; same stuff differentiated by degrees and states. The transition from the higher non-physical realm to the lower, physical, constrained realm is no different; changes in degree along a continuous spectrum producing novel emergent forms.

Dual Realm Experiential Differences — *Man must reconcile living with both*

DUALITY – Illusions of separation	UNITY
Past and Future.	*Here and Now*
Relative truth.	*Absolute Reality*
Reflections, refraction, absorption, filtered images	*Pure White Light*
Noise, words, beats, vibration	*Silence – Pure Stillness*
Time & Space bound: dimensions, relationships	*Timeless, Spaceless*
Doing-Becoming	*Being*
Physical	*Non-Physical*
Constraint	*Freedom*
Blockage	*Flow*
Local	*Universal*
Manifest materiality	*Pure Spirit, unmanifest*
I Think *"I think therefore I am"* (*Descartes*)	*I Am* " *I am therefore I think"*
Both Objective and Subjective	*Non-Dual*
Polarities	*Wholes*

◄————————— *Jung's Dual Pathways* —————————►

Worldly	*Inner calling: myth's Hero journey*
Shadows, demons/serpents *overcome by →*	*Liberation, transcendence*

*Dynamics of unconscious: slaying the dragon is transformation of your own consciousness

As beings operating within two realms simultaneously, we express dual aspects to everything. In our own bodies we have the dichotomy of a realm 2 brain and a realm 1 mind. Emotions operate in realm 2 while the Soul emanates in realm 1. Emotions are realm 1 spirit oozing into your realm 2 body, like water seeping into earth. Imbalance creates problems; too much water mixed with earth makes mud, too little leaves dry, barren land. As conscious human beings we live and act in Realm 2, but our guiding purpose originates in Realm 1. The ultimate source of everything is realm 1; a boundless, formless, spaceless, timeless plenum, where we find limitless non-physical energy without shape, form or limitation of any kind. To tap into the unlimited power, we must stop seeing ourselves as separate from it.

Being present, attaining inner peace and achieving centeredness transcends the self-imposed limitations of duality.

Higher realm 1 is embedded in everything, including us. We are each at the center of the universe – an intersection between both realms. It's no different than religion where God is embedded in everything including us. As living beings we're expressions of spirit tethered to the physical world until death – Realm 1 induced into realm 2. Man perceives God as a great "being" from realm 2 but God is actually simply *being* in Realm 1 – less noun and more verb. Religious rituals aim to bridge the perceived gap between both realms, including prayer, yoga, Tai Chi, meditation, and channeling. Duality in The World is overcome through liberation from attachment traps that lead to unnecessary suffering and pain. Bliss is our natural state, obscured by distractions in The World pulling us from our center of unity. And just where do YOU end and the external world begins? There's really no defined boundary other than the physical membrane surrounding our bodies (skin), the rest is a self-induced illusion of separation.

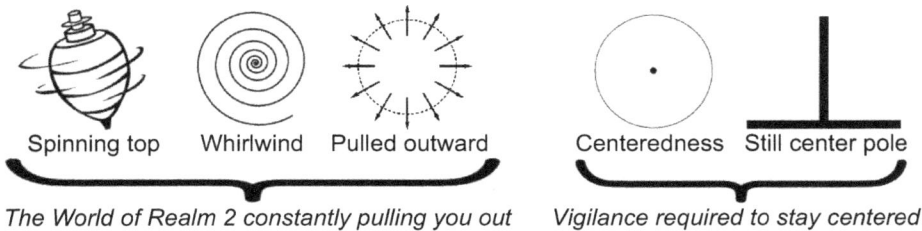

Spinning top   Whirlwind   Pulled outward   Centeredness   Still center pole

The World of Realm 2 constantly pulling you out   *Vigilance required to stay centered*

BRIDGING DUALITY

We all have a self-limiting tendency to default to a clinging embrace of the lower physical realm. It's really just a survival instinct as we navigate a stressful demanding world. Illusions on the material plane are difficult to see past and require ongoing vigilance to overcome, yet techniques and practices can be performed to overcome duality and achieve a more authentic real experience.

Transcendental meditation is the most popular and compelling method for bridging duality. Relax, be still, connect. Related forms and practices include chanting, prayer, drumming and dance. A variety of religious rituals offer similar pathways that enable one to reconnect with the sacred original source. The common denominator is suppressing the body and mind with measured action and opening awareness to deeper, higher levels. Achieving success requires repetitive effort and ongoing vigilance.

Bridging begins with changing your state of being. Like tuning a radio, our internal frequencies resonate to different states, with lower and higher vibrations in each one. Attitude is key. An open mind is essential but even more so, an open heart. Child-like curiosity and sense of adventure are conducive states. Romance, wonderment and humor equally resonate to a higher frequency. There's something mystically powerful within moments of excitement, surprise or a deep belly laugh.

Music is a simple, effective way to change your state of being. When music and dance are combined, even more so. There's a melody for every mood, a song for every occasion. Nothing can soothe the soul quicker than the harmony of great track, quickly arousing a more receptive state. Creative arts in all its forms can achieve similar results. Painting, sculpting, drawing, even simple doodling will open creative channels elevating one's state where receptivity is heightened. Artistic expression is a two-way process, inside and out, attuning to a higher realm. The body becomes a vessel – an activated conduit for spirit to flow. Achieving liberation from the physical realm via the arts, music or meditation is about suppressing the mind, allowing inner spirit to flow naturally without inhibition.

Dreaming is a bridging state where the subconscious and unconscious communicate using coded imagery in a timeless-spaceless domain. Symbols replace verbiage, feelings direct action and memories unfold into the present. It's an ambiguous experience where meaning is conveyed through a foreign language. Messages are exchanged at a very deep level, usually lost upon waking. Dreams are to individuals what myths are to societies; meaningful clues, feedback loops and reminders. It's a two-way flow between higher and lower states. Dreaming is a necessary, ongoing duality bridging process required to maintain a healthy psyche.

Religion is the primary vehicle civilizations have adopted for man to reconnect with the higher realm. Holy men served as interpreters of divine truth, translating the word of God and prescribing rituals for believers to follow the right path – opening potential gateways to the eternal. Eastern spiritualism focused on holism, connection with nature, polytheism and eternal life-death cycles of Karma. Western religion stressed one God above and separate from man and lower animals, good deeds and remission of sin, and the concept of a judgement day. In simple terms, Eastern philosophy connects God within us and everything else while the Western perspective projects God *out there*, above and separate from us *in here*, ironically reinforcing duality, not bridging it.

Bridging is challenging due to built-in language barriers. Metaphors and analogies are really the best way to describe attributes of the higher realm. Parables and storytelling are used extensively in holy books to convey higher truths. Mythology was a pre-modern pseudo-language using archetypes to convey meaning on a deeper universal level. Occult fields use esoteric language to capture the context of dual realms. Ancients relied on various gurus, prophets and spiritual oracles who bridged the language gap and performed as middlemen – spirit guides for seekers of higher wisdom.

Science is the antithesis of bridging duality. It's objective, reductionist, and anchored in the material realm. It reinforces separation from the world: microscopes below it, telescopes above it. Physics representing hard, objective science pushed limits ever further until it hit the dimensional seam where our two realms meet. Quantum physics produced quantum space, time and energy units –

the very boundaries of physical reality. Collapse of the wave function revealed the quantum moment – where a higher realm penetrates into our lower realm. Relativity shed insight on the mystical qualities of light – a phenomenon that has no mass, no charge and no time. Matter can never reach the speed of light and light can never be at rest. Like Lady Hawk never the twain shall meet. Light resembles spirit in pseudo-physical form, operating at the seam connecting both realms. It's the ultimate metaphorical bridge of our dual reality, physical and non-physical.

Mathematics reveals duality reflected in relationships between attributes of both realms. Squaring the circle is futile as squares represent our finite physical realm (straight lines and structure) while circles represent the eternal (curves and flow). π is an infinite transcendental number because like light it resides at the seam between 2 realms. The square root of 2 is likewise an irrational number as it represents the very symbol of the dyad and duality. Geometrically the circle conveys unity and wholeness from a higher domain.

To be incarnate or living in the flesh is to be anchored in the physical realm. Nobody and *No Body* can cross over. However, your spirit can certainly connect higher without the baggage of a body. We are beings present in both realms, like the symbolism of man having his feet on the ground and head in the clouds. Only our non-physical essence can traverse realms – bodies and objects, like squares and straight lines, can't.

Bridging duality is fundamentally an exercise in reconnecting with the original, hidden source. The higher realm is always present, always an integral part of the physical reality we're grounded in. Like a starry night portending a higher mystical heaven temporarily visible only after a long day fades into evening, or glimpses of sunlight shining through the trees in a deep forest, clues and reminders persist. Access to the higher realm cannot be forced in the same way a string cannot be pushed. Your state of being must be attuned with authentic intention along with congruent behavior and ongoing diligence. You'll know you're on the right path when you attract synchronous support from unexpected sources in a sort of cosmic feedback loop that rewards right action. Confirmation will follow.

"*A human being is part of the Whole...He experiences himself, his thoughts and feelings, as something separated from the rest...a kind of optical delusion of his consciousness*" – Albert Einstein.

When conscious and unconscious domains intersect, we experience *synchronicity*. It creates experiences of absurdity where both realms temporarily overlap, producing all sorts of odd nuances. The higher realm is always with us but masked, well beyond our ordinary awareness distracted by continual daily challenges of coping in The World. Occasionally, fleeting moments surface where portions of the curtain are lifted, exposing elements of the unfamiliar that were always there, like nighttime stars concealed by daylight. They awakening us to see personal dramas unfolding in real time. Increasing frequency of synchronicity confirms you're following the right path. Unrelated things suddenly connect, along with "coincidences", which are often accidents by design. People you've thought about appear out of nowhere. It starts to make more sense once you realize human reality starts on the inside, then creates the outside, not the other way around. Our experience of The World is a projection, both individually and collectively. Everything we need is out there but we won't see it or find it until we're ready on the inside. We change The World by changing ourselves. When we change the way we look at things, those things begin to actually change. Once your consciousness changes, you begin to see with new eyes. You experience more Aha moments: during a daydreaming vision, a meaningful coincidence, chance encounters, traveling in new territory, or a sudden premonition. It's just your hidden higher self speaking to you in code.

Sync events occur when the outside reveals what's going on inside. It can be a dream-like feeling while awake, an accidental meeting that turns out to be fateful, or a range of things that can't be explained by science or reduced to physical cause and effect. It's a non-linear, non-local quality with hints of the magical. A Trickster may come out of nowhere to reveal hidden things lurking in the shadows. It could be an odd experience of something previously felt in a dream state. Once you're centered you begin to follow the right path, where opportunities unexpectedly present themselves. Focused purposeful intention pulls you closer to your higher self, producing synchronicity that opens doors and attracts what you need.

Our natural tendency to explain reality from our physical realm perspective results in partial truths – only half of the whole picture. It unnecessarily fosters mysteries, enigmas, riddles and a variety of paradoxes. Confusion arises when properties of one realm are misapplied to the other, or language from our domain is used to describe phenomena from a higher realm. It's like the distortion of 2D flat maps depicting a spherical world. Modern science is guilty of this, examining wholes from a narrow, partial perspective. Its counterpart, religion, engages in its own share of misapplied absurdities by trying to explain things that should be left to science. Both should stick to their particular domains and avoid the other. It doesn't help when our natural egocentric localized bias creates a reality of partial truths and polarities of opposites. Once you embrace the existence of dual realms with dual attributes and their interconnected dynamics, everything makes much more sense. Even the particle/wave duality of light which still befuddles physicists to this day becomes instantly explainable. Paradoxes suddenly become less puzzling – just particular artifacts of duality. It becomes easier to see the folly of squaring the circle, mental versus physical time travel, unity versus infinity, time versus timeless, and the connection between inside *You* and outside *World*.

Bridging duality really comes down to detachment from The World. It's sounds simple but it's never easy. When you consider how anchored we are to all our material possessions, official titles, assigned roles and required responsibilities, it's a daunting task...but not impossible. Sure, we're pulled a little off center every day but every subsequent day is an opportunity to recenter. It just takes ongoing vigilance. Our reality is co-created; an interactive dynamic between You and The World, and all that's needed is more You and less World.

Dual Realms Explain Man-Made Paradoxes – *It's never either-or…it's both!*

Freedom vs Determinism
Randomness vs Order
Separation vs Unity
Science vs Religion
Body-mind duality
Time travel in either direction
Psychic, clairvoyance
Paranormal experiences
Infinity and eternity
Synchronicity
Polarities

Dual Realm Duality

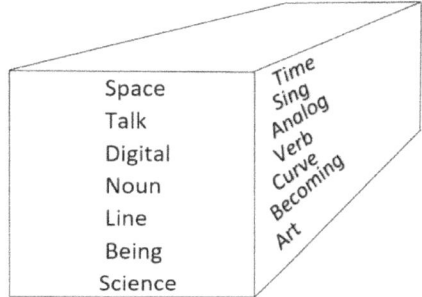

Space
Talk
Digital
Noun
Line
Being
Science

Time
Sing
Analog
Verb
Curve
Becoming
Art

OMNIPRESENT DUALITY DYAD PATTERN.

The dynamic interface between dual realms imbeds a dyad pattern into everything. This twoness template takes the form of structure and motion, form and process, fixed and fluid. Every level of physical reality contains this complimentary pattern. Every subject, thing or field reveals it. It's nature's way of combining attributes of two realms into everything. The closer you look at anything the more you'll see the pattern was present all along. It goes beyond simple basic polarities which are just opposites sides of one unity – it's actually two transcendent attributes that are complimentary. Black and white are basic polarities while colors transcend them. Dual aspects are related but different in kind. Consider the difference between science and art, quantity and quality.

The ubiquitous dyad pattern takes the form of *structure* and *flow* everywhere we look. It's natural since all things are simultaneously part of two realms; one free, one constrained. In such context how could anything not contain dual features? This is consistent everywhere, built into the foundational sub-structure of physical reality. Accordingly, physics reflects it at every level including relativity and quantum mechanics. The most fundamental essential ingredients in the physical domain all exhibit a dyad pattern of structure and flow; one essence, two forms. In every case there is a stock aspect and a flow aspect. This universal dyad template represents the dual pattern of substance and process, fixed and variable, structure and motion. Since both realms of reality intersect and apply to everything within, it must follow that all phenomena will express aspects of both. The dyad pattern is what makes our matrix map a complete whole. It integrates the substance and structure aspects of the lower material realm with the process and transcendent aspects emerging out of the higher realm.

Particle	Space	Matter	Magnetic	**Structure**
				Fields
				Currents
Wave	*Time*	*Energy*	*Electric*	**Flow**

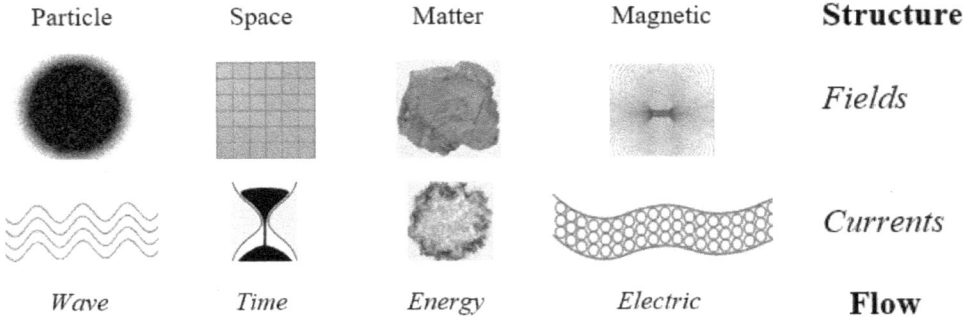

Man at the center of the universe is a living dyad pattern. As living beings we exhibit all sorts of them: brain-mind, glands-emotions, thoughts-intuition, soul-spirit, heart-circulatory system, left brain-right brain, rest-exercise, intelligence-wisdom, being-doing. The choices we make are a combination of fixed constraints (determinism) and flexible options (freewill). Our knowledge is a combination of memory and learning. When we engage in problem solving, we apply both critical and creative thinking (art and science). As human Dyads we express dynamics between inner states and external expression. Emotional states can be confined by dwelling, depression and apathy or uplifting flows of enthusiastic eager inspiration and creativity. All these dual aspects emerge from a built-in universal dyad pattern.

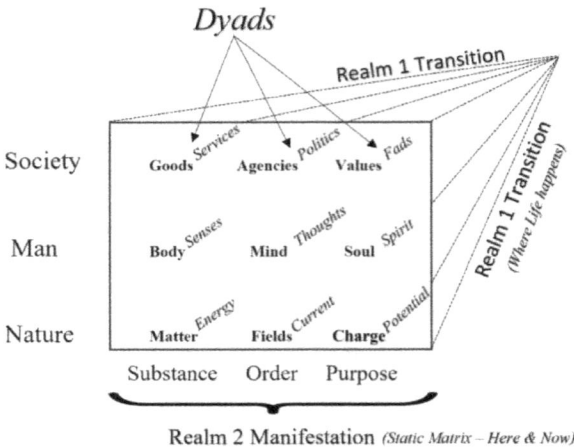

Realm 2 Manifestation *(Static Matrix – Here & Now)*

Duality is omnipresent at every level. It's highlighted at each of the 3 primary levels in our TOE model of everything as well as every point within, up and down, left and right. We covered the examples in physics and can just as easily add chemistry where the dyad shows up as substances and reactions, and the biological level with body and spirit. All living organisms at this level exhibit the dyad pattern as wholistic systems and living processes. At the social level the pattern emerges in economics, politics and culture as goods/services, laws/legislative process, and norms/fads respectively. Businesses have fixed and variable costs, produce using either batch or process approach, and keep track of finances with both a balance sheet (snapshot) and an income statement (process timeline). Advertising is either temporary product focused or enduring brand based. Businesses persist long term as a fixed structure while management and employees come and go continuously.

Universal codes reflect the duality pattern as a mirror of reality. Mathematics describes all manner of phenomena using equations with fixed and variable elements. Algebra and calculus capture the dual nature of things having a combination of stability and variation. A fixed position that suddenly moves is a dyad pattern. Same with something moving at a constant rate that suddenly accelerates. A bell curve capturing hundreds of individual actions in a single stable pattern is also a dyad pattern. DNA coding directs organic processes using fixed and variable mechanisms that maximize survival value by mirroring the natural environment's dual conditions of stable niches with unpredictable change events. Language contains narratives that range from objective, concise plain verbiage to artistic, poetic prose speaking to the divine. Consider the contrast between talking and singing, or at the most basic level of language, the dyad of nouns and verbs.

Ouroboros duality   Mobius strip: dual & non-dual   Passive-Active duality

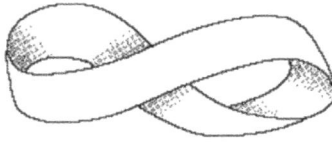

Duality and the Dyad pattern have been subtly referenced in each of the previous chapters. We introduced it with duality of the primary essences, and then displayed it in a TOE model highlighted by the process dimension. We observed The World with its indifferent threats and opportunities only to be transcended by the secret cypher of human intention. Our deep dive into human reality revealed multiple layers of duality expressed in body-mind-soul-spirit, states of being, man as spirit filter, and personality as a composite output of intersecting dual realms. We then uncovered duality in time, with a higher timeless realm and a lower domain constrained by time; man caught in the middle. Next, we examined the duality of hidden reality where man in the middle operates in between everything, above and below, larger and smaller, way before and way after. Our comprehensive look at cumulative knowledge revealed a dual spectrum of hard and soft science, rational fields and religion, critical and creative thinking, and the duality of art and science. Universal codes were full of dyad patterns, including math, music, language, DNA, pattern codes and divine codes. Our look into the nature of life concluded that the living beings we associate as "life" are merely temporary structures left in the wake of an eternal flowing spirit of progress, where life is more process than form.

Looking back there were plenty of clues in each chapter revealing a ubiquitous presence of the dyad pattern in literally everything. Dual realms produce duality. Freedom and constraint were built-in right from the beginning. The stickiness of our physical domain set the stage for structures to emerge out of continual flow, flux and fluidity from an unconstrained higher realm merging into the lower one. It also serves as a fixed anchoring force counterbalancing the perpetual motion of free-flowing energy, resulting in omnipresent vibrational dynamics everywhere we look. Man becomes the ultimate fulcrum between both realms – an intersecting duality filter: human being = *human* (form) *being* (formless). Every experience we have is processed through a dyad filter. It's a universal pattern you can find everywhere once you know what to look for. It's the secret pattern of our dual reality that's been hiding in plain sight.

Omnipresent Dyad Patterns

Space/Time	Matter/Energy	Particle/Wave	Gravity/Radiation	Field/Current
Magnetic/Electric	Position/Motion	Line/Curve	Light/Shadow	Noun/Verb
Body/Soul	Goods/Services	Climate/Weather	Talking/Singing	Concrete/Abstract
Thought/Intuition	Destination/Journey	Material/Intangible	Western/Eastern	Black-White/Color
Hierarchy/Network	Physics/Metaphysics	Intellect/Wisdom	Objective/Subjective	Brain/Mind
Being/Becoming	Form/Emptiness	Parts/Whole	Digital/Analog	Mundane/Beautiful
Science/Art	Words/Images	Order/Chaos	Think/Know	Quantity/Quality
Batch/Process	Sound/Silence	Now/Past-Future	Glands/Feeling	Wake/Dream

INFRASTRUCTURE OF REALITY CONCLUSION. The basic template of all physical reality in simple terms is a duality dynamic – an arrangement of transcendent physical and non-physical domains that overlap and interact. *Everything* is subject to this dual structure, especially man anchored in the middle like the center of a circle. Consciousness is the immaterial stuff embedded in all, comprising *everything* in different degrees, forms and states. At the highest level is the absolute – a formless, timeless, emptiness preceding both realms. Pure spirit descends from this original nothing to creatively become something; a mystical process transposing freedom for order. Descent into the physical realm is a transformational process of increasing constraint that manifests into differentiated *everything*. And the further spirit descends downward and outward, the more it builds reverse tension, yearning for a return to its original state. It's a subtle force built into the very substructure of physical matter pressing an evolutionary urge to ascend; to reconnect with the original absolute. What results is emergent forms of higher order where complexity increases along the great chain of being elevating man's consciousness along the return journey.

The transition from non-physical into physical is a mystical transformation revealed in pseudo-physical interactions near the seam between both domains. Space-Time is a pseudo-physical boundary; the barrier where *nothing* transitions, transforms and emerges into *something*. Light originates in this seam, exhibiting qualities of spirit in action. Quantum mechanics emerge here as well; infinite realities collapse into singularity, creating actual something from potential nothing.

The infrastructure of physical reality is not just about the ingredients and relationship of dual realms but the interactive dynamics driving all the action, expressed in processes and transformation of energy into materiality. A perpetual motion machine churning out temporary forms that appear solid in a grand illusion of resonant mechanics where subtle fields generate patterns of apparent structure. Energy transforms into matter one level at a time, from fundamental particles of matter and force (quarks and bosons) to increasingly complex forms of materiality (electron, neutron, proton, atoms, molecules, compounds, etc.). Parts form into larger wholes, then proceed further into even larger whole-parts (holons). The physical transformation between energy, force and matter brings with it an emergent mental attribute expressed in the selection process as particles interact and socialize with one another. The interaction is spirit driven from an original higher source. It comes from the same place as all other creative processes that turn nothing into something; where the non-physical manifests into the physical, one quantum moment at a time in a perpetual state of creative becoming.

Everything manifested into physical reality is merely a derivative product, tracing back to a form originating in a higher realm. Spirit resonates from a subtle, non-physical form, descending into physical energy patterns at lower levels. Matter is the lowest level, consisting of patterned, structured energy. All perceived things are actually nothing (no-thing) since they're primarily empty space possessing knotted energy patterns forming subtle templates of materiality. Atoms are the key energy structures of physical reality yet they're largely empty, mirroring the emptiness of the cosmos at large. Living beings are similarly "empty" energy structures but with higher states of complexity and resonant energy, elevated by emergent mental and intuitive attributes. And all of these "empty" things are actually interconnected fields in a far-reaching web of dynamic energy patterns, continually changing, flowing, and transforming into new manifested resonant temporary patterns.

Consciousness ascends, bringing with it increasing complexity and self-awareness. Human consciousness operates like a filter where all reality is man-centered – interpreted, internalized and projected outward. Man's consciousness localizes higher reality, creating a self-induced relational context of limited here and now perception; choosing particulars out of infinite possibles, much like the quantum wave collapse. When applied collectively, human consciousness creates *The World*, through experiences, attachments, desires and responses. Though anchored in the physical plane, man always remains connected to the higher spiritual domain. This duality of living in two realms simultaneously creates the central ongoing challenge of human life with competing demands of each realm, expressed in awkward situations, misguided pursuits and comedic predicaments. Duality reality presents a need to balance the competing pulls from higher and lower, driven by an unconscious desire to reconnect with the divine where man reaches ahead, progresses, evolves, pursues higher purposes and ultimately raises the collective consciousness of *The World*. Like a plant spiraling upward drawn to the sun, it's the urge to remerge with the source, essentially a resonant expression of our yearning to re-embrace the divine. And in this process where sentient life collectively advances in consciousness with man at the privileged tip of the spear, the universe as a grand complex whole is divinely becoming aware of itself.

Spectrum of Physical and Non-Physical Infrastructure

NOTHING (*No Thing*) **0**

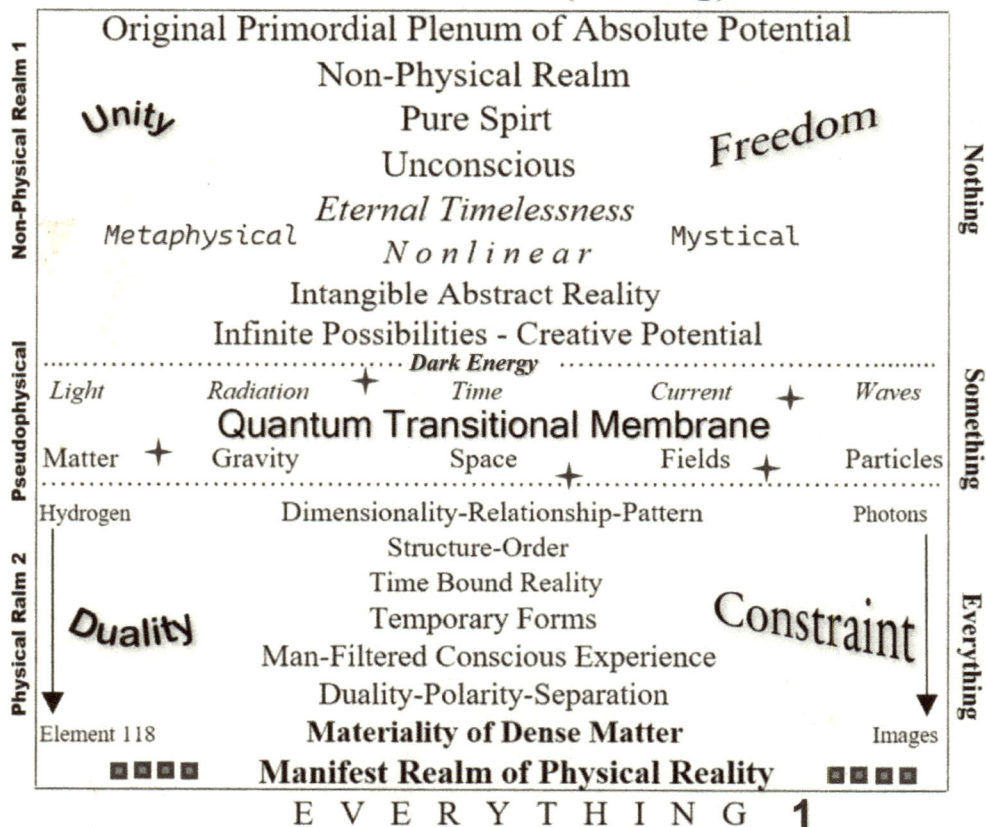

Original Primordial Plenum of Absolute Potential

Non-Physical Realm

Unity

Pure Spirt

Freedom

Unconscious

Eternal Timelessness

Metaphysical   *N o n l i n e a r*   Mystical

Intangible Abstract Reality

Infinite Possibilities - Creative Potential

································ *Dark Energy* ································

| Light | Radiation | Time | Current | Waves |

Quantum Transitional Membrane

| Matter | Gravity | Space | Fields | Particles |

| Hydrogen | Dimensionality-Relationship-Pattern | | | Photons |

Structure-Order

Time Bound Reality

Duality   Temporary Forms   Constraint

Man-Filtered Conscious Experience

Duality-Polarity-Separation

| Element 118 | **Materiality of Dense Matter** | | | Images |

Manifest Realm of Physical Reality

(left margin, top to bottom): Non-Physical Realm 1   Pseudophysical   Physical Ralm 2

(right margin, top to bottom): Nothing   Something   Everything

E V E R Y T H I N G   **1**

Realm 2 emerges from Realm 1, producing the universal Dyad pattern in everything. It produces universal dynamics of vibration: resonant forms embedded in Realm 2, and the experience of time, moment by quantum moment. Realm 1 isn't "above" 2, it permeates it.

12   THE UNIFYING CONNECTEDNESS
OF EVERYTHING

With experience and wisdom comes the inevitable conclusion that so many things are interrelated at various degrees and levels. We don't notice it much early in life as we learn by trial and error, acquire knowledge by formal education, look at subjects in isolation, and compete in a society that incentivizes specialization. Despite these structural limitations, we eventually begin to notice the subtle connections between things in unrelated areas of interest. As your awareness expands there's a subtle revelation that more and more of everything is interrelated and interconnected…a shift in awareness sensing the bigger picture, connecting the dots, tracing each path and linked chain back to an original unified source.

Some connections are more obvious than others. While it's easier to notice physical connections in plain sight it's a little more difficult to see the hidden, subtle, derivative relationships beneath the surface. Scientists and artists are more adept at finding hidden connections; between things physical and non-physical, seen and unseen. With a little bit of detective work, imagination and curiosity you can reveal masked forms of relationships, associations, linkages, derivatives, and interdependent connections of seemingly random parts.

PHYSICAL CONNECTIONS

Quantum physics reveals a built-in connection between every part or particle in physical reality. Quantum entanglement confirms every particle is connected to other particles regardless of position or distance from its counterpart. Likewise, gravity connects every particle in the universe as well, extending infinitely in every direction, though weaker and weaker with distance. Matter and energy everywhere remain forever connected, sharing a common link back to their big bang origin.

All material substances consist of recycle parts, present in everything including us. Lots of the atoms and molecules in your body were likely used by someone who lived centuries before you were born. Food and drink are consumed and discarded over and over and over again, in a long chain of recycled stuff. The air you breathe follows the same pattern of recycled substance used by everyone else on the planet. We're all connected by the ongoing sharing of used material substances, getting continually transformed and recycled.

Every moving object produces ripples in space projected outward in every direction, interacting with all other ripples generated by everyone and everything. The distortion of space produced by planets and large bodies takes place all around us, all the time, just on a very undetectable scale. We're all connected by a unified web of spatial dynamics that link each person and object to every other one.

Light traveling through deep space links its source to everything it illuminates. Spectroscopy reveals the unique signature produced by each substance light originates from. Doppler shifts in light's color provide information on the direction objects are moving, regardless of how far away they are. Despite great distances in the cosmos, the light emissions of stars produce a defining spectral finger print, revealing their exact composition and velocity, billions of miles away.

DNA links every organic life form backward through time millions of years, generation after generation. DNA also links species and their unique branches of development. Despite vast differences on the surface, humans and chimpanzees are genetically 99% identical. Every cell in your body links back to 1 single fertilized egg. Every living organism links back to a single original ancestor. DNA connects ALL organic life, regardless of species, type or extinction. Humans alone have produced well over 12,000 generations, each one participating in an ongoing passing of the baton, physically handing over a set of genes to the next set of contestants competing for survival in The World. The evolutionary tree of life is a biological blueprint confirming the absolute interconnectedness of living forms.

A wide range of fields and specialties engage in detective work to uncover physical connections. Science and its diverse set of branches examine, probe, explore and research related phenomenon, but it's equally matched by non-science fields as well; economists, philosophers, historians, linguists, lawyers, anthropologists, etc. All engage in connecting dots in time and space, essentially a reverse engineering process. Sherlock Holmes and later, modern day Forensic detectives (CSI) exemplify the task at hand; examining physical evidence at every level and scale, piecing together elements of a larger investigative puzzle, tracing backwards the sequence of events, and formulating a wholistic, big picture conclusion connecting what only appears disconnected. Sometimes all you need is means, motive and opportunity.

TEMPORAL LINKAGE

Non-physical connections include indirect relationships, abstract patterns and temporal reality. Everything is linked in time, both individually and collectively. Where spatial relationships link you with other things, timelines link you with you. Your life path is a single connected timeline with sequential stages resembling a linked chain. One long continuous unbroken path navigated by you with various exterior facades unified by your unchanging inner essence. It's no different than a series of back-to-back trips where the scenery may change and your state of being may fluctuate but the driver remains the same. Life is a long continuous series of stages and events connected by YOU; Many parts, one unbroken whole.

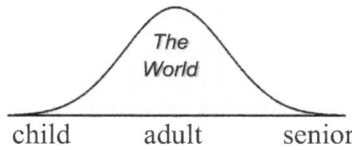

Beginning and end symmetrical
Dependency-isolation-sleep
Growth up – decline down

The World

child   adult   senior

One soul, many faces. Each life as spirit in an aging vehicle whose driver never changes

History connects the tapestry of collective human culture forming a world-wide web with multiple branches, all moving through time together. Political, cultural and economic shifts move in recurring waves, causing corresponding cycles in every aspect of societal activity. As history repeats so does every experience and event associated with its particular phases, including economic boom and bust cycles, and their corresponding social unrest, political turmoil and outbreaks of war. Every present *now* contains remnants of past *nows*. This time and every time are inescapably linked to every moment that preceded it. Clues to this linkage include relics, residue, ruins, and countless artifacts left behind. YOU are a living artifact connected to a birth long ago and further connected to even longer sequences of parents who preceded you. Past-Present-Future are one, artificially separated by our limited perception. Past is prologue. Every tomorrow is promised by every yesterday. It seems the more things change the more they stay the same.

Connections with our past are hidden by the passage of time, masking numerous links hiding in plain sight. Think of all the outdated artifacts right in your own home, useful just a few years ago but no longer serve any useful purpose. Telephone jacks, milk delivery doors, coal chutes, water cisterns, dumb waiters, boot scrapers, are all lingering vestigial remains linking a forgotten past to today. Strange words and common everyday phases are used by everyone routinely without a clue to their origins. By and large, resting on laurels, turning a blind eye, bury the hatchet, hands down, etc. are spoken frequently without regard to where they came from. No matter how far back we look, our past is always very much a part of today whether we see it or not.

ANCIENT TIMES	MODERN TIMES
Gladiators – Coliseum	NFL Stadium
Chariot races	NASCAR
Brothels/Orgies	Online Porn
Barbarians	Terrorists
Peasants	Rednecks
Senate Politics	Senate Politics

INTERCONNECTING PATTERNS

Universal laws and codes confirm the connectedness of everything. Intelligence is directly related to our ability to recognize patterns – to connect the dots. To break the code and see the forest through the trees. Math is the ideal tool to accomplish this; the fundamental language of connectivity, both direct and indirect, physical and non-physical. Over 3000 categories of mathematics have emerged, reflecting the diverse expressions of connected relationships in physical reality. These patterns are expressed graphically though Venn diagrams, matrices, graphs, equations, sets, etc. Universal constants, principles and laws of nature reveal the interdependent link between all things. The interconnectedness of everything is what allows us to solve for X. Algebra is simply 3 aspects: a known, an unknown, and a universal law connecting them. Problem solving in general deals with these same 3 aspects, whether it's a CSI detective at work or kids on a scavenger hunt.

Connections are about relationships, with patterns of spatial geometry, temporal sequences, cycles, general associations and degrees of connectedness. Relationships have a minimum of 3 elements: a thing, another thing and the aspect between. With people it starts with a pair or couple. It could be friends, lovers, a business partnership or just two people with something in common. This is the dyad minimum. You could apply this minimum pattern to an individual also: The outer you, the inner you, and the way they relate as a single personality. As more elements or people come into play there are group dynamics that emerge leading to complex relationships. Heck, even just 3 people in a love triangle can be a complex relationship. Large, connected groups (hundreds, thousands, millions) can evolve into grand networks that take on completely new characteristics even though the parts are generally similar, just lots more of them.

Relationships reflect us like a mirror. While we identify with our localized self, relationships provide feedback to us with an unbiased, non-localized view. Objects and things out there do the same as the universe mirrors back at us; *You-there* reflecting back to *Me-here*. We all have 3 very different perspectives of ourselves: How you see yourself, how others see you, and how you actually are. The greater the difference between these perspectives the less aware or honest you are.

General Relationship Aspects		*Perspectives*		
1 Form	interrelation of parts	1st person	*I*	individual values
2 Position	from Me to it	2nd person	*We*	collective values, societal standards
3 Scale	distance to it	3rd person	*It*	projected values by others
4 Orientation	reference point to it.			

Reality is context dependent. Nothing has a 100% independent existence. Interpretation is context based; nothing happens in isolation. In relationships each element becomes the context for the other elements. Reality is also observer based. Subject and object are inseparable. You and the world *out there* become one in a relationship of perspective. Perception is perspective, subjective and context based.

*A hole in a piece of wood can't be described without referring to the surrounding wood

In an interconnected universe everything must be a derivative of the original source – the physical realm is derived from the non-physical realm. A derivative universe is revealed through code breaking and reverse engineering, tracing the connections back to the source. Science is one approach, non-science the other – both equally valid. Much of our reality is experienced in a derivative manner, individually and collectively. Language is derived from the culture it serves, mirroring its unique values, customs, and attributes. Similarly, body language mirrors personality. You can tell a lot about a person by how they behave, dress, speak and facially express without hearing a single spoken word. Non-verbal language often provides more information than the actual verbiage. The way you walk, do a handshake, exhibit nervous tics, or simple handwriting are open clues to who you are. Reading a room is the subtle art of sensing others' body language, positioning, word choices, emotional state, tone, and demeanor. Intentions are revealed in behaviors where actions speak louder than words, exposing the difference between what's said and what's done.

All scientific progress is derivative, building on prior work where today's genius stands on the shoulders of earlier pioneers. Newton built upon the work of Kepler and Galileo, Einstein went further building upon works of Maxwell and Faraday, followed by a hoard of fellow physicists building upon all of them. Each successive paradigm raises the bar higher in series of ongoing derivative stages. The material universe they all examined was likewise derivative, with substances derived from smaller and smaller sub-particles, dividing molecules into atoms into protons/neutrons, into quarks, each going a little further down a derivative rabbit hole. Their comrades in biology followed a similar path where minerals, plants, animals, and man are similarly just derivatives of earth components (hydrogen, carbon, nitrogen, oxygen), which are further derivatives of the sun.

All art is derivative representation of something else, dependent on feelings, thoughts, perceptions, and intuitions. Like all human expression, the external is derivative of the internal. Same with body language. YOU in the middle are simply a filter between derivative realms, inside and out, including subconscious and unconscious currents. Inside is always connected to the outside. Blaming others or projecting is denial of the inner-outer connection. The "evil" we confront on the outside are projections of our repressed fears on the inside. The universe is a collective projection of us, our collective inner thoughts and states of being. Creations become extensions we project into The World. Everything we create is an extension of us; children, crafts, machines, vehicles, structurers. Everyone we associate with around us is a further extension representing aspects of what's inside. Every relationship reflects a part of us projected onto someone else.

"All things extend to all things, from all things, and through all things" – Walter Russell

Reverse engineering and derivative analysis have one primary goal; to reveal origins. Historians trace the ruins, relics, and legacies left behind history's wake. Linguists track words, dialects, and branches of languages spoken all over the world, such as the path from Latin to Greco Roman to European to American. Common names of locations (streets-towns, cities, states) buildings, bases, schools, hospitals, events, all have a backstory, often going back centuries. Modern holidays bear less and less resemblance to the original purpose and meaning of their origins, getting lost in translation as culture changes over time.

On a purely *physical* level light is the ultimate origin of everything. All matter and energy come from light. It was God's very first act "let there be light". Monet noted "the subject of every painting is light". All physical energy is essentially atomic, made from atoms and their interactions; nuclear fission and solar fusion (sun powered), electricity from electrons, charged particles, and electromagnetic radiation. The sun is the original source of matter and energy on earth. In a real sense everything is solar powered, directly or derivatively. And in perfect symmetry it applies on a purely *non-physical* level where one could say God or spirit is the absolute origin and ultimate connector of all there is or ever was.

INTERDEPENDENCE

Nothing really ever happens in complete isolation, especially where human beings are present. For us objective reality is more illusion than absolute. Everything we engage in is observer based, participatory and thus subjective. We don't just observe The World, we create it. This applies to every field and discipline. Interdependence is an inescapable aspect built into the physical infrastructure of reality and we're an integral part of it at its very core. Culture, economics, politics, geography, resources, climate and technology are all interconnected. A change in one area can affect all other areas.

Codependent relationships pervade our interconnected universe where everything interacts with everything else to varying degrees. Codependency includes man and all his objects of interest: warrior and weapon, artist and brush, athlete and ball. Elements in direct relationship with other elements become unified wholes. Joined pairs dynamically interact with each half influencing the other and vice versa. Like Yin/Yang each side contains the potential seed of the other. It's not unusual for roles reversals to occur. In protracted war the participants gradually begin to mirror their opponent (see cold war Russia and the US).

All symbiotic relationships are codependent: predator/prey, abuser/victim, leader/follower, performer/audience. Parasites and hosts interact in sync as each affects the other, common through the ecological biosphere. Over half of all species are parasitic, like codependent plants and insects (pollination partnership). The food chain in general consists of organisms feeding off other organisms.

You are much less independent than you think you are. Much of your life is influenced by those around you on a daily basis. Family, friends, coworkers and of course your spouse, engage in a continual push-pull, give and take, action and response dynamic that shapes who you are in any given moment. You are somewhat a byproduct of those you spend the most time with, positive or negative, better or worse. And each of us in turn are codependent cogs in every large group we're a part of, including city, state and country. Collective progress influences and is influenced by every part of the whole. Nothing evolves in isolation. Every improvement we make in ourselves makes the world a better place.

Complimentarity is the interdependence between two "separate" elements which are never really separated. They're connected relationships where each influences the other in mutual response: Predator-Prey, Spear-Shield, Virus-Antibody, Crime-Police, Offense-Defense, Parent-Child, etc. In some cases it's an evolutionary arms race where both elements temporarily get the upper hand until the other catches up or surpasses it. This happens with every sports team rivalry. It's an embedded dynamic in every 1 on 1 competition regardless of the arena. Strategy and tactics are always being revised as new techniques and technologies upset the temporary balance. Sports analytics emerged as mathematics revealed opportunities to gain temporary advantages over the competition: baseball stats favor batting solely for homers instead of steady base hits, basketball stats favor possible 3-point shots vs. probable 2 pointers, football stats favor more 4th down attempts vs automatic punting. Game theory applies complimentarity by calculating the best options based on likely responses of others. Every one-on-one competition becomes a unified connected single dynamic. It's like a chess game of continual back and forth, move-countermove, tit for tat, action-reaction, where both players respond to each other and to new developments as they arise, until the final move. Complimentarity leads to merging with your opposition; becoming a part of each other. Even physical fighting transitions into a dance between two partners. One moves, the other countermoves in lockstep. The difference between wrestling and sex is physicality. And it isn't limited to just individuals. Consider athletes merging with the object: ball, bar, racket, bat, club, disc, etc. becoming one with it. See the ball, be the ball.

The universe is dynamic at every level, so of course everything within it interacts in a dynamic manner. Every part moves, changes and influences other parts around it in a constant flux of competing forces producing temporary forms. Things and events emerge like nodal points in a sea of eddies and currents, each affected by all others in an interlinked web of fluid action. Dynamics connect forces interacting in mutual relation. Laws of association reflect this principle in strategy and tactics, economics, engineering, politics and group dynamics. Change is generally a process of dialectics (*thesis-antithesis-synthesis*) where an initial force is met by a counterforce that interacts to form a new resultant force.

Systems theory embraces the interdependent nature of the universe with its holistic orientation. It rises above the disconnecting reductionist tendencies of science by focusing on integration and mutual interaction. No independent parts but rather connected sub-parts of a greater whole. Every part is a linked element in a local system of many parts that influence each other, produce feedback and adjust together in reciprocal synchronization. You are no different than any other part in a universe of parts, influencing and being influenced by every part around you.

The universe is in a state of constant change where the path of forward progress is shared by every element within it. Nothing changes in isolation. There's no such thing as independent evolution but rather interdependent coevolution. Parts and wholes working together, changing together, evolving together. Changes in you affect me and everyone else around you. Changes in an organization affect everyone within the organization, some more than others.

Our interdependent reality is further revealed by physical forms sculpted by parsimony as a natural process where all creation, construction, and production of forms are restrained or shaped by external forces in dynamic interaction. It's no coincidence that buildings get pushed higher and higher in limited, expensive real estate zones of a city. Organic life is a comprehensive showcase of living forms molded by external forces. All processes, growth, expansion, and performance are constrained by equal and opposite regulating limitations. The shapes and forms of things and beings are simply products of equilibrium, the balance point between opposing forces. You are the exact shape, size and weight our environment allows modified by what went into you…and as you change the inputs, it too will respond with a specific counterbalancing limit.

CAUSE AND EFFECT. As everything changes in a coevolutionary manner, forward progress unfolds in a series of step-by-step stages of action-reaction, situation-response. The past creates the present; both linked in an unbroken temporal chain. Every moment proceeds in like manner, each dependent on and linked to its predecessor. Events are nodal points along an ongoing temporal conduit, each tethered in time, one after another. Where space connects everything horizontally, vertically and diagonally, time connects everything sequentially.

Change is constant but occurs at different rates, influenced by various catalysts, inhibitors, and mechanisms influencing the process. Gradual change gives way to rapid change, then transformative all-at-once change. Sometimes conditions are more sensitive, creating a butterfly effect where small differences lead to dramatic change. It's one reason good intentions often lead to unintended consequences.

Causality means no act is ever done in isolation since everything we do influences and limits what can come next. This is exemplified by the Heisenberg uncertainty principle where the simple act of observing something influences the outcome. It's not just quantum physics but life in general. Reality TV isn't really "reality" since the presence of cameras and crew influence behavior. Even celebrities who are used to cameras aren't completely natural given they're aware of being watched and filmed.

Law of Causality Universally Expressed

You reap what you sew
You get what you give
No pain no gain
Garbage in garbage out
What goes up must come down
As above so below
The circle of Karma
For every action there's an equal and opposite reaction – *Newton's third law*

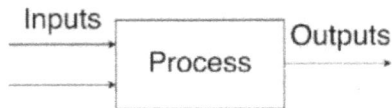

* Aristotle's 4 types of causes: material (substance), formal (design), efficient (action) final (purpose)

Cause and effects are mirrored by inputs and outputs linking parts sequentially in time. What goes in is generally what goes out. Systems are basically wholes where parts go in, get modified, and eventually go out. Quantity and quality of inputs directly correlate to outputs. You are a system of inputs and outputs expressed in a wide variety of aspects; health, knowledge, skill, strength, etc. The harder you work the more you can achieve. No pain no gain. 95% of success is tied to unseen advance preparation. Elite level performance generally requires 10,000 hours of work in a given specialty.

INTERCONNECTEDNESS IN EVERY MAJOR FIELD OR INTEREST

Meteorology – Climate and weather are a boundless interconnected system of ongoing change in rhythmic unison. Land, air, sea and space interact in a dynamic dance of tension-release, pressure-dissipation, hydration-evaporation, and countless reciprocal exchanges of energy in motion. Weather influences our health, activity, and social life. Our bodies continually adjust to daily weather patterns, which impact allergies, moods, headaches, arthritis, and general attitude. A sunny day boosts vitamin D, reduces high blood pressure and kills bacterial. Overcast skies increase depression, heat and humidity marginalize stamina, and extreme cold escalates illness. Lots of moving parts, each influencing the others.

Ecology – We think of the environment in a local sense buts we overlook the interconnected nature of the greater whole. All the parts interact with their neighbors, from microorganisms to mammoth beasts, exchanging energy and information. Soil below and atmosphere above connect us and every other element around us. Life is continuous, boundaries are illusions. The air we breathe and the water we drink is the recycled stuff our ancestors breathed and drank. Millions of atoms in our bodies are just recycled parts, most are billions of years old. All organic life is symbiotic with the earth, interacting in a codependent relationship. The earth itself is a complex living organism; host to a vast diversity of parasites, including man. Our ties with nature were captured in the organic architecture of Frank Lloyd Wright, and further by the socially interactive architecture of Christopher Alexander, connecting nature to the public square of human activity.

Health – Wholeness and balance are the key to wellness in any living being. Understanding the dynamics of the human body requires knowing the parts (cells, tissues, organs, systems) and how they interact with each other. Health is always a multifaceted matter involving nutrition, diet, exercise, stress, attitude, breathing, etc. Each area affects the others. Changes in one correlate to adjustments in others. You are what you eat. Health = wholeness and connected unity, illness = disorder, imbalance, disconnection, and fragmentation. Body language provides overt clues to one's physical and mental health. Posture does the same, revealing problems in specific parts of the body. Footprints reveal posture issues. Pain is a universal feedback mechanism directly connected to health. Other feedback clues include discoloration, voice, stiffness, leaning, facial expression, sweating, breathing pattern, and gate, all expressions of a comprehensively interconnected human system of systems.

Gestalt Psychology – Perception and personality are multifaceted; a holistic pattern of diverse integrated parts. Perception is influenced by interconnected relationships between elements in proximity, similarity, continuity and general context to each other. Personalities are similarly holistic with a complex mix of interconnected aspects; the whole being greater than the sum of the parts.

Economics – Organic bodies and national economies are similar integrated systems of inputs, outputs and interconnected dynamics where each element affects the others. Goods and services depend on supply and demand. Costs and benefits are affected by interest rates and inflation. Economic growth is restrained by unemployment and productivity. Government policy always has tradeoffs where incentives improve one but diminish another. Like health, the key is balance and moderation. Too much growth leads to inflation, not enough leads to recession. Every action has an equal and opposite reaction. Where have we heard that before? Economics is not limited to a single nation but extends along a continuous spectrum from local town to entire global markets, each level interconnected and impacted by interest rates, exchange rates, taxes, tariffs, fees, supply/demand, and government regulations.

Accounting – Tracking assets and liabilities is a zero-sum game. Ledgers must balance out so inputs equal outputs. Income statements and balance sheets follow the same corresponding equivalence of inputs and outputs. Both systems capture equilibrium, one as a snapshot, the other as a timeline.

Governing – Introducing new laws, rules and regulations is usually done with the best of intentions leading to unintended consequences. Human nature is easily overlooked, especially when the few control the many. People don't respond well when forced to do things they don't like. Imposing restrictions leads to a creative pursuit of loopholes. Adding costs or benefits to things in isolation incentivizes behaviors that take advantage of it, sometimes in ways that defeat the whole purpose. It's Newton's 3rd Law again – every action has an equal and opposite reaction. Effective governing requires wisdom of human nature and understanding the interconnectedness of policy incentivizing corresponding changes in behavior.

Culture – Fads come and go in sync with the changing zeitgeist shaped by the collective mood of a population. Changes in music tastes influence the arts and vice-versa. Same with fashion, entertainment, literature, celebrity types and news cycles. Generational cohorts interconnect with every cultural development in a codependent dynamic. Innovations in one area eventually impact all other areas. Culture is the composite expression of individuals collectively linked as one whole.

Geopolitics – Actions and reactions at the national level involve a variety of complex variables where changes in one area may result in numerous, out of proportion effects in other areas. International entities are intertwined with neighbors in a tangled complex web. Move and countermove often lead to unpredictable responses. Economics, culture, demographics, religion and geography all play a part in the interconnected dynamics making optimal policy choices a difficult task, often skewed by parochial bias.

Strategy and Tactics – Gaining relational advantage over a competitor requires a good grasp of the elements within an arena. Variable factors include the disposition of each side, terrain, weather, resources, intelligence accuracy, goals, readiness, etc. Every move induces a countermove, followed by successive countermoves. Optimum strategy is derived through analysis of opponents' most likely moves in a self-regulating system of action. Game Theory is based on the simple dynamic of how opponents' behavior influences each other's moves. Military offensives are directly shaped by terrain, similar to flowing water taking the path of least resistance, and also shaped by weather, which either incentivizes action or limits it.

Communication – Transferring information between people is a connecting process where sender and receiver interact, both influenced by the medium and the message. In ordinary conversation each person begins subconscious micro movements mimicking the other in a synchronous reflexive pattern. Feedback is an essential communication mechanism, letting you know something is right or wrong; brakes squeal, audience applauds, engine sounds weird, touching the stove hurts. Laughing, crying, protesting, growing, shrinking, garbage, graffiti and even silence are all forms of feedback.

There's no end to the fields of study that examine the interconnected nature of elements within them. Hard science, soft science, humanities, parascience, pseudoscience and even religion all correlate the dynamics of interrelated things, relationships and mechanics of what goes on within each area. The great pioneers of each field stand out not because of their knowledge of the parts but because of their wisdom and insight into how the parts interact and influence each other in a holistic, integrated manner.

We're all part of a system of systems, interacting and interdependent with every other one. Man in the middle is a connecting nodal point, a central hub in a comprehensive web linking communities, networks, spectrums and everything above, below, within, without, before and after, here and there. Every experience in our lives affects all other areas of our lives, individually and collectively. Isolation is a localized illusion. At the highest level everything is integrated and interrelated, just difficult to see at our limited individual perspective.

CONNECTING SPECTRUMS

Hierarchies and Holons are omnipresent in the universe as everything is part of a larger whole. At the lowest level parts become specialized and differentiated from other parts, only to be then integrated into something more. It's a universal pattern in physical reality where everything is both a whole itself and part of a greater whole above it. Hierarchies are limited versions of holons, but the relationship is basically the same. Though new attributes emerge as higher levels are formed, each element remains linked with every other one, tethered by a shared essence, related by a continuous vertical conduit.

Traditional Hierarchies (by title)

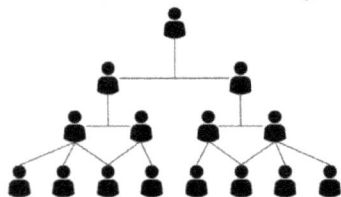

Military: Private, Corporal, Sergeant E5-9, Lieutenant (2), Captain, Major, Colonels (2), Generals (5)
Nobility: Knight, Lord, Baron, Viscount, Count, Marquis, Earl, Duke, Prince, King
Church: Laity, Deacon, Priest, Biship, Archbishop, Cardinal, Pope, God
Corporation: Worker, Team Leader, Supervisor, Manager, Director, Vice President, President/CEO

Holon chains are like nested dolls, each higher level extending beyond but including the lower. At the physical level we see the hierarchical pattern of whole-parts manifested as particles, atoms, molecules, substance and solid things of greater and greater scale. At the level of living beings, we see the holon pattern of organic life following a pattern of increasing complexity. The food chain is holonic with a vertical parasitic relationship of higher-level creatures feeding off lower-level organisms and often vice versa. As creatures grow in size their populations proportionally diminish. Miniscule microorganisms far outnumber their giant hosts, both symbiotically coexisting and sharing the planet's surface.

Traditional Hierarchy   *Holon Patterns*

Galaxy
System
Star
Planet
Moon
Asteroid

At the societal level the holon pattern emerges as evolutionary progress builds up step by step, each a little higher than the preceding one. Technology rachets up as each major advance lifts the collective capability higher and higher. But as new advances are made, previous lesser versions tend to linger on. The rapid progress of leading-edge civilizations hasn't eliminated the presence of 2nd and 3rd world states. Similarly, people advance at different rates of consciousness, creating a tiered segmented society of multi-level status.

Holons consist of one continuous, hierarchical, vertical pattern, in both space *and time*. Each new moment contains the residue of its predecessor. Even more so, the future contains every now and all of the past. YOU are a holon, here and now, containing every you of the past, and will soon be just a part of the greater future you. The whole of your life can only be experienced as individual parts, due to localized temporal perception. Your life journey is a single connected whole that you only get to experience part by continuous part. Consider an ocean of infinite waves which you can only experience a tiny segment at a time; one unitary whole perceived as many individual parts. Notice how the same wave may seem large to a surfer but tiny to a ship; they are just local perceptions of a single, continuous unity, experienced uniquely by different observers. This is true everywhere we look as the universe is one giant holon pattern of patterns.

Army	World	Universe	Organism
Division	Region	Supercluster	Organ
Brigade	Nation	Galaxy	Tissue
Battalion	State	Solar System	Cell
Company	County	Star	Molecule
Platoon	City	Planet	Atom
Squad	Town	Moon	Proton
Soldier	'Hood'	Asteroid	Quark

New properties emerge at each higher level/stage. Military units add complexity going higher: company adds a staff, Battalion adds sections (logistics, operations, admin), Brigade adds engineers, artillery etc.

Spectrums and Scales. Where hierarchies and holons are vertical concepts with depth, spectrums and scales are horizontal qualities of breadth. Hierarchies change in kind as you go up; emergent qualities appear as higher levels transcend lower ones. Conversely, spectrums and scales remain the same as you go across, changing in degree. Scales reflect raw measurements while spectrums have a more qualitative flavor. Spectrums and scales unify every element within. The spectrum itself is the whole while each variation along it is a corresponding part.

Spectrums – *Qualitative*: Colors (ROYGBIV), political ideology, emotions, knowledge
Scales   – *Quantitative*: Speed, size, weight, numbers, quantities, time, distance, etc.

Spectrums are sometimes applied to hierarchies and holons where qualities change in kind
Anything on a spectrum is connected to everything else on it, a unifying concept, similar to holons.

Connecting Spectrums and Scales

Organic Life Spectrum	All living things connected by DNA chain
Matter Spectrum	All material substance connected by shared atomic particles
Man-made objects	All things made on earth come from the same recycled materials
History Spectrum	All events are connected by shared timelines
Light/Energy Spectrum	All radiation has the same wave form and electric/magnetic parts
Space Spectrum	Gravity connects all objects along an infinite scale of attraction.
Consciousness Spectrum	Fundamental unifying fabric of reality within everything

Consciousness varies beyond measure or concise description.

Matrices and Networks. Combining multiple scales creates a cross-connected tapestry of unity. Matrices serve this purpose by graphically portraying related items on a grid. These checkerboard patterns are simply dual scales, connecting things on a plane. We could also add a third scale to create a cube pattern of connected elements. Networks join elements in a similar manner without the rigid structure of a squared array. Here the community of parts are connected and linked by nodal points called hubs. Together they form a unified whole with emergent attributes transcending individual parts. Networks can become highly sensitive interlinked systems synchronizing to resonate in an unpredictable manner. Hubs are key sources of influence, some more, some less. Networks are rarely static; dynamic conditions can suddenly emerge with unpredictable results.

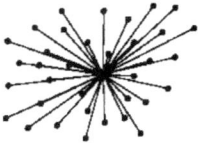

Centralized Network

Decentralized Networks

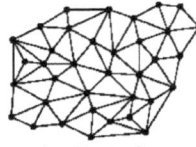

Distributed Networks

Tabular Matrix

Hierarchies, holons, spectrums, scales, matrices and networks are derivatives of the relational essence, serving as unifying templates of physical reality that express the interconnectedness of everything. The concept of 6 degrees of separation reveals the built-in connectedness of parts even when a network is enormous in size or scale. Nodal points integrate the many into the few. They act as bridging shortcuts that bring connections closer to increase the responsiveness of the whole.

All life on earth is connected in a coevolving biosphere where nothing happens in complete isolation. Every organism, species, habitat, or ecosystem interconnects as one dynamic interacting whole, culminating with the earth itself operating as a complex living entity. Man's collective intelligence forms a vast interconnected web referred to as the *Noosphere*, where aggregate mental activity coalesces into an integrated holistic global brain. Each mind is a linked part, connected to a broad network of other mind-parts where a greater collective whole emerges. Its tentacles reach out in every direction via old school word of mouth supplemented by electronic media, mass communication mediums and the world-wide web internet.

Theory of Everything (TOE) Model

We've applied combined concepts to construct our big picture model of how everything interconnects. It's a universal 3D Matrix of connecting spectrums vertically, horizontally and diagonally. The vertical axis captures increasing evolutionary complexity from the physical to biological to societal (nature, life, culture). The horizontal scale connects primary essences of physical reality, from *substance* to *order* to *purpose*. The diagonal scale shows the flowing, *process* essence. The whole matrix itself expresses the *relational* essence. Each of these spectrums unifies, capturing the interrelatedness-interconnectedness of Everything. Anything you can imagine falls somewhere within this matrix, like the coordinate points of a graph. The TOE matrix is *the* spectrum of all spectrums. It highlights 3 primary nodal scales while relating every other scale between and around them. It's an oversimplified, practical map with 3x3 nodal points and levels serving as a framework for every other possible spectrum/scale of anything and everything.

Everything captured in the 3D model is produced by process; the flowing essence represented by the diagonal axis (dotted line). The grid structure represents the physical here and now (horizontal and vertical axis solid line). Our natural tendency is to focus on the fixed grid pattern as primary rather than the flowing process axis that is truly fundamental. The grid is the physical realm; a derivative by-product from the non-physical realm. It's produced moment by moment from the creative process advancing forward along the diagonal axis, much like sands passing through an hourglass…or quantum mechanics where only 1 actual reality emerges instantly from many possibilities, moment by quantum moment. The front grid structure may be highlighted for convenience but realize this; the dotted line in the background representing process is fundamental, not the other way around.

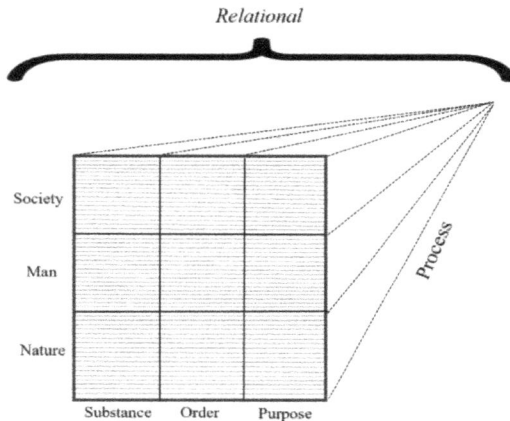

*3x3 Matrix plus transcendent essences: Process and Relation
**Everything falls along a scale or spectrum somewhere within

Complexity			
Societal	Above	Collective organization, ideas, language, civilization
Biological	*Man*	Organic molecules, tissues, organs, whole beings
Physical	Below	Matter-Energy dynamics, physics, chemistry

Evolution

ILLUSION OF SEPARATION

Despite the interconnectedness of everything we still perceive most of reality as being fragmented, separated and disconnected. It's a natural confusion reinforced by our isolated, localized experience of daily life. We're comfortable as individuals here looking at everything out there. We create artificial boundaries that produced the illusion of separation. It starts with your ego creating a false separation between "I" and everything else. It's the origin of the *us vs. them* perception, xenophobia and the catalyst of many bloody wars. Drawing boundaries provides temporary, necessary order through illusions of separation.

Our perception first separates us from objects and things. Soon it expands to people and relationships, the us versus them experience. These perceptions of separation create distinct points of view, reinforced over time:

1st person: *I*   Subjectivity where we place our own *individual values* on things and experiences.
2nd person: *We*   Relationships; we interact with *others' values* on things, shared experiences.
3rd person: *It*   Objective reality; science removes human element from things and experience.

Separation is an illusion of local perspective. Everything connected at the highest level merely appears disconnected and separate at our lower level. Time is likewise a function of local perception. Einstein's relativity demonstrates there's no absolute time anywhere; it's always dependent on the observer, so it's a man-made construct. And man's cognitive nature is to analyze, divide and categorize things. The mind's primary purpose is to establish order in a chaotic environment, and it does this by separating everything into neat, structured compartments. Separation is further reinforced by the ego, which for practical reasons clings to me versus you, us versus them. It's a coping mechanism exaggerating illusions of separation. Both of these influences can be overcome by higher awareness and love, transcending separation by unifying. Ironically, the very smart phones that connect us to all of humanity end up separating us from those closest to us.

Separation vs. Connection = *Inverse relationship*
It's always a degree of difference between 1 and 0

Western separation has a long history beginning with Plato, who saw the outer world as a by-product of our perception. Descartes' body-mind split laid the foundation of dualism, further separating man from the world. Newton reduced the world to a mechanical, lifeless machine. Darwin diminished life to just a struggle for survival; each of us engaged in selfish pursuit to gain over his fellow man. We steadily distanced ourselves from nature and diminished the Divine to a mere afterthought. Human reality was now founded on a split separated relationship between everyone and everything; between each of us *in here* and all *out there*.

Our experience and perception of separation is expressed on multiple levels: polarity, duality and transcendence. Some are differences in degree while others are changes in kind. *Polarity* emerges the moment man in the middle becomes consciously aware of something out there, establishing a physical, mental or emotional relationship with it. Objects become big or small solely based on our physical or mental relation towards it. An object is neutral until we make a conscious judgement about it. Nothing is good or bad in and of itself until someone subjectively associates one of those qualities with it. Polarity often manifests as polar opposites; things that are similar in kind but reversed.

Duality is a self-relation pattern. Wholes that differentiate are still connected. A piece of paper folded multiple times or crumpled up mercilessly may be differentiated but it's still whole. Much of what appears to be separated is just whole forms in differentiated states. Duality is a dilemma, meaning something having two premises. It emerges when something takes on another quality, transitioning from a single unitary essence into a thing with dual attributes.

Transcendent relationships go above and beyond the original form or state. Though a higher-level essence it still includes its former state. Transcendent change is a difference in kind. It's the maximum aspect of separation, represented by a 90-degree relationship. Like dimensions that split off at right angles, their difference is sharp yet they're still connected at a point.

Polarity-Duality-Transcendence begins with man's biased subjectivity of The World out there creating polarities and illusions of separation and division. All things become subject to self-comparison with YOU in the middle. Sizes, quantities, qualities, all become defined in the context of YOU perceiving them. Subjective judgements create dichotomies of good and bad, valuable and worthless, etc. Man's polarizing nature turns unities into polarities; single essences with opposite values. Despite this splitting process, opposites retain a quality of self-similarity, such as two sides of the same coin. It's ironic how similar opposites can be: pain-pleasure, hate-love, black-white, thrill-dread, birth-death. Contrary to popular opinion, *both* love and hate unify while fear or apathy separates. Polarity is a single essence relating to itself – a unified spectrum. Every relationship begins with self-relation which then extends beyond into complimentary relation. Nothing can escape self-relation or its complimentary connection with everything else.

* 0 and ∞ (*nothing and infinity*) are polar, dual and transcendent.

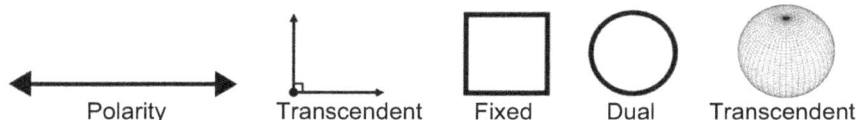

Polarity	Transcendent	Fixed	Dual	Transcendent

Transcendent is going beyond: Metaphysical is beyond the physical. Supernatural is beyond nature.

Transcendent levels of Reality

Spirit	Mysticism
Soul	Theology
Mind	Psychology
Body	Biology
Matter	Physics

Transcendent Dyad Pattern

FLOW	STATIC
Life	Death
Analog	Digital
Fluid	Fixed
Curved line	Flat line

Demands of The World pull us outward from our natural peaceful center place like a spinning spiral. The center is our link to a transcendent realm, always present in the here and now though masked by continual distractions of daily life. Western religion associates the spinning disc with separation from God via sinful life on earth. Its goal is to bring man back to God in the center. Eastern religion sees the center pole as God always being with us and in us. Western medicine reduces man to parts that need fixing, Eastern health is holistic. Western science separates man from nature, Eastern philosophy embraces both as complimentary connected parts of a unified whole. Milton's *Paradise Lost* covers this universal theme of man's self-separating move away from nature and the sacred. Man's superior mind may have led to great progress enriching civilization but its propensity for dividing, reducing and compartmentalizing condemns us to self-imposed separation, including losing the divine. The manifest world expresses everything in patterns of separation via polarities, dualities and opposing forces seeking reunification. Unity is genderless; the opposite of differentiated polarized sex. The whole purpose of sex is reunification – to achieve unity. God is whole, complete and the ultimate unity…and therefore genderless, despite the continual male reference.

Love bridges duality as subject and object merge via reunification. When you love something, you become one with it, unified. Healthy partnerships supplement you, each filling the gaps of the other, making the separate two into one whole. In the ideal love relationship, the other fulfills and completes you.

Levels of Connectedness:

1 – Unity. A single whole (Coin)
2 – Polarity. A whole perceived with equal, opposite aspects (Coin with heads and tails)
3 – Duality. A whole expressed with different aspects (Coin's flat surfaces and edge)
4 – Transcendence. A whole transformed on a different level (Coin represents money)

Polarity = *Opposites*: black-white, big-small, love-hate, good-bad, fast-slow, near-far,
Duality = *Inverses*: Key-lock, pos/neg space, love-apathy, plug-outlet, male/female genitals

*Yin/Yang is both polarity and duality

Degrees of Connectedness. Our perception of separation can be equated to positional value on a spectrum or fractional placement on a scale. Partial ratios, decimals and fractions all represent parts of a greater whole. Ironically, the very dimensions we use to describe separation, space and time, are the things that reveal connectedness. In our coin example, does the middle separate or connect the opposite sides? Does the internet cable between your computer and mine separate our systems or connect them? Nothing is ever totally separated – they're just less connected. Space connects you to everything else out there. Time connects you to every other you, past or present. The differences in either are just degrees of connectedness, more or less. Our entire concept of separation is essentially a way of perceiving differences; again, just degrees on a particular type of unified scale.

Separation as Degrees of a Circle　　　　　*Separation Expressed in Motion*

0	"0"	degrees	Identical, no difference	0 Stationary
1st	1-179	degrees	Simple variations by degree	1 Moving at constant speed
2nd	180	degrees	Polarity – exact opposite	2 Constant acceleration
3rd	90	degrees	Transcendent – greatest difference	3 Accelerating acceleration
*	360	degrees	Separation and reunification	**stationary to moving is transcendent*

"Everything is connected but some things are more connected than others."– Howard Pattee

Statistics measure the connectedness of things mathematically using a correlation coefficient, expressed as a degree between 100% and zero, or 1 and 0. 100% connectedness equals unity and perfect symmetry. The other end of the scale increases randomness and disorder. Randomness is a relative experience created by narrow, localized short-term perception. A single dice roll is unpredictable but multiple rolls become more and more predictable until eventually it's a virtual certainty. Individual actions are likewise unpredictable while collectively they become predictable in the aggregate (the basis of every insurance business). Similarly, the inputs that form a bell curve are unpredictable individually but inevitably produce a consistent, predictable pattern with time. Perceived separation is actually just lower degrees of connectedness. Randomness involves degrees of order approaching zero without ever quite getting there.

High degrees of randomness mask hidden order. Complexity, ambiguity and abstraction increase uncertainty, making it difficult to see the order. Expanded awareness and longer time frames can counteract this limitation. In the long run, there are no coincidences. Locally we can be fooled by short term events and the differences between linear and nonlinear, noise and information. Quantum entanglement reveals all particles remain connected regardless of any distance apart. Even noise has degrees of order, with white noise being featureless and unpredictable, black noise having a higher degree of order and brown noise somewhere in the middle. Complexity produces the appearance of randomness, but it's a false illusion. Fractal patterns seem magical and mysterious, yet they consist of repeated self-similar patterns generated from simple equations. The sensitivity to small changes in the butterfly effect also produce the appearance of randomness, but it's just another form of complexity.

Chaos theory deals with low degrees of order, non-linear dynamics and non-symmetrical patterns that appear completely chaotic, however they eventually transition into complex self-organized patterns. Chaos is essentially hidden, cryptic order. During your life journey there were certainly plenty of seemingly random events and developments but looking back and seeing them from the context of your entire life, they begin to resemble recurring themes that fit into a larger, somewhat orderly pattern. Chaotic systems appear more random than they actually are due to higher sensitivity to initial conditions, but they still follow deterministic laws. Complex systems appear random due to the sheer depth of interacting parts, but whether something has a few elements or a million, the dynamics involved do not become random simply because they're more complex.

Randomness vs order is identical to the question of free will vs determinism; it's always a combination of the two. Nothing is ever 100% random, just a very low level of order. In a universe with built-in purpose and intentionality, how can anything really be completely random?

Equal and Equivalent are relational concepts that describe degrees of connectedness. In math or science there isn't much grey area; it's either equal or it isn't. However, in other areas beyond hard science, connectedness has a quality of equivalence; similar, like, comparable, related, analogous, congruent, etc. The Arts, religion, humanities, pseudoscience, and other fields are full of things that exhibit linkages without exactitude, but definite congruence. Where math uses equal signs these other areas use metaphors, analogies, similes, and allegories to connect equivalent concepts. Art portrays connectedness with graphical metaphors, religion does it with parables, and dreams use symbolic archetypes that link realms.

Universal Equation. The *Theory of Everything* in physics is a quest to unify physical reality in a single framework, ultimately captured in a simple, elegant equation. Einstein pursued it without success, however he did provide us with

E=MC². Today the challenge remains; unifying the 4 fundamental forces of nature: strong and weak nuclear force, electromagnetism, and gravity. No one's done it yet but they seem to get ever so closer. If or when they do it still comes down to just physics, not *everything*. Hard science is not an end all be all.

The folly of a TOE equation is that math can never connect science and non-science, physical and spiritual. Duality reality requires something transcending the language of science. Connecting hard science, soft science and non-science is beyond the simple language we're accustomed to. The only possible option is using pattern language, the highest-level code of reality. Pattern language connects everything through relationship, symmetry, congruence, and correlation. It's the highest-level language connecting everything, the only code that transcends the limitations of hard science, mathematics and logic.

Hypothetical Equation of Everything
$$0 \rightarrow 1 \rightarrow \infty \cong 0$$

Nothing begets something/onething, which begets everything, equivalent to the original nothing

UNIVERSAL CONNECTING PATTERNS

Reality is a comprehensive pattern of interconnected relationships. Once you know what to look for, you'll see them everywhere. There are all sorts of hints that things are connected on a micro level. Most of it is subtle, hiding in plain sight all along…until a sudden aha moment when you notice something right in front of you that seems oddly familiar. Little nuances, patterns of similarity between different things in different settings. It takes elevated awareness and heightened powers of observation to notice it, the same way artists see "ordinary" things differently, or comedians making funny connections between random typical situations. For the vast majority who see The World with the same set of eyes, very little stands out. But for those who see with new eyes, everything changes, including yourself.

Simple Observations - Subtle Veiled Connections

Shape of continent coastlines mirror each other like puzzle pieces
US states get larger and straighter going West.
Lightbulbs were such a good idea they became the symbol for good ideas
Olympic and professional athletes have exceptionally strong thigh and hip muscles
Higher income people dress nicer and speak clearer
People are getting fatter every generation
Out of 7 continents, half of the world population lives in half of 1 of them
You have a near identical twin somewhere in your country, and others around the world
Sun and moon are perfectly symmetrical in size, distance and apparent motion.
Every city uses the same general street names.
Firetrucks are actually water trucks
Individuals have fewer freedoms and greater isolation than ever, but more safety
80% of the US population lives in the East half.

*Some of the most revealing subtle observations are made by comedians: "did ya ever notice...", "here's something I know you all hate...", "you're never gonna believe this but..."

Clues to the connectedness of everything are lurking everywhere, including your place of work, home, play, travel, entertainment, mundane activities, relationships and events. They're present in everyday objects, structures, trash, accidents, "coincidences" and idle conversation. They're revealed in recurring themes where common qualities and attributes are applied to a variety of different things and situations. Behind every subtle pattern is a simple explanation, an interesting story or counterintuitive reason.

Universal Archetypal Themes

Human nature doesn't change much as demonstrated by so many recurring activities and events over time. History continually repeats itself; same stuff, different names and places. Greek athletic competitions and Roman gladiator spectacles have been replaced by modern sports arenas. Ancient spear and shield fighting carries on to this day, only with modern missiles and steel tanks. Public square rhetoric and debate is now seen on television talk shows or heard on audio podcasts. Yesterday's freakshow is today's reality TV. The platform, tools and venues may have changed but it's still essentially the same stuff.

Archetypal themes remain the same as well, since the nature of man and the dynamics of physical law don't change. Thousands of years ago our predecessors observed and appreciated unity, harmony and proportion; they understood when something feels right, seems natural and looks beautiful. They knew then as we do now that symmetry in your face and body shape is attractive. Things that are disconnected, fragmented, or distorted are still perceived as ugly and repulsive. Monsters, demons and even clowns still scare people with their distorted physical features. Rhythmic, resonant, harmonic sound is considered musical, sounding as good today as it did way back when. It's opposite, dissonant sound, is uncomfortable and perceived as noise. Visual and auditory balance equals order and beauty, distortion and fragmentation express chaos and ugliness.

Surface Change, Essence Remains. A universal theme in life is the difference between outer and inner change. Much of our experience is focused on outer surface activity while overlooking the deeper, truer reality. Think of chemical reactions where it's a zero-sum game of parts rearranged with energy exchanged, yet the original components never change. Despite big differences in attributes produced by rearranging substances, the parts are always the same in the end. This mirrors the conservation of energy law, stating energy can never be created or destroyed, just transformed from one type to another. Material substances are no different. Life itself is an ongoing process of transformation where temporary forms (living organisms) grow and die while inner evolutionary forces press forward. Birth-death-renewal is a universal cyclic process of life where beings come and go, children replace adults, parasites feed off hosts, empires rise and crumble – all while the steady progressing life-force remains constant.

Light-Energy-Life-Consciousness. Light is the essential ingredient of physical reality – filtered, reflected, colored and expressed in a variety of ways that subtly connect everything. Dawn emerges from dusk; darkness disappears as sunlight bursts across the landscape illuminating everything in its path. Life suddenly awakens as creatures spring into action, hustling and bustling, interacting, burning up stored solar fuel until evening rest. Gradually darkness sets in, revealing a sky filled with billions of radiating stars, many much larger than our very average size sun. Looking down on earth, a night map of any country reveals patterns of concentrated light, nodal points metaphorically indicating areas that are rich in life; resonant levels of collective consciousness. Observing car traffic at night reveals a flowing pattern of approaching white light, and receding red light, mirroring the body's pattern of flowing blood, progressing red in arteries, receding blue in veins.

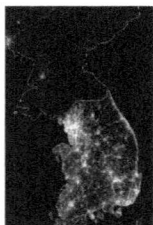      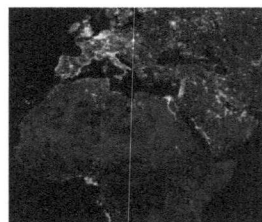

North vs South Korea   Western vs Eastern USA   Europe vs Africa

Ancient man perceived the light of the sun as derivative of divine energy – God's brilliance. Most religions equate light with spiritual relevance. Achieving wisdom is to be enlightened. The unity of pure divine light filters down into infinite variations: colors, intensities, wavelengths, amplitudes. It mixes with itself and everything else: reflections, refraction, absorption, interference. Sun light energizes and sustains all organic life; It's the spirit that sustains man. Light is life.

The sun is a super-bright, super-high energy source you can't stare at, exactly mirroring the original source of the higher realm; God-like radiance is so powerful that lowly human beings are humbled by sheer brilliance. The only sunlight we can handle is either reflected or absorbed. Similarly, and in perfect symmetry, the higher realm source of everything is transduced into physical reality, lowered in frequency, filtered and molded into dense forms (matter and material).

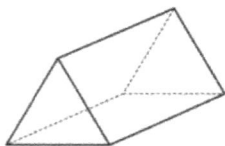   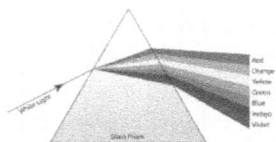   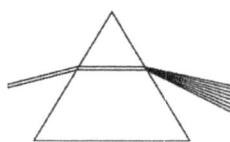

The World

UNITY FILTER   **PURE LIGHT**   **COLORS**   **SPIRIT** **MAN**   **MAN***ifestation*

Recurring Archetypal Patterns are imbedded within every culture, carrying perennial themes spanning the history of man. Universal symbols and images are ubiquitous in movies, television and literature: Hero, seeker, sage, adventurer, Trickster, wise old man, mother and father, mentor and apprentice, ordeal and triumph, failure and redemption. Each of us learns lessons in life which are no different than those experienced by people thousands of years ago. Most of us assume roles and play parts without realizing we are repeating universal plots carried out by others millions of times before. Whether it's work or play, conflict or romance, our lives are linked by universal actions connecting all of humanity.

Nature's Devine Archetypal Symmetry Patterns   *Realm 1* → Realm 2

Sun	Fire	Energy	Spirit	Intuition	Spiritual	Being	*Spirit*	
Sky	Air	Gas	Mind	Thoughts	Mental	Identity	*Light*	
Sea	Water	Fluid	Heart	Feelings	Emotional	Passion		Particles
Land	Earth	Solid	Body	Senses	Physical	Reaction		Materiality

Life Sustaining Elements in descending order   *Metaphorical equivalent*

Fire – You can't survive without physical energy.   Spiritual Realm
  Air – You can survive without air for a few minutes.   Consciousness
    Water – You can survive without water for a few days.   Organic Life
      Earth – You can survive without food for a few weeks.   Physical World

Cosmic-Heaven Metaphor. Outer space is a metaphor for heaven *above* and *out there*. The universe is composed of a vast number of stars; enormous spheres of high energy output and the source of everything physical. We could consider a single star, or the collective whole of stars, as the ultimate metaphor of God – the source of everything, bright white light, heavenly, possessing perfect spherical symmetry. Mythology and religion connect God and the sun as male energy, earth and moon as female counterparts (productive and receptive, giving and receiving)

Mothering Metaphors. The mother's womb mirrors the oceanic cosmic plenum from which everything emerges out of. It's the original source of unity, peace, bliss, union. It's metaphorically expressed as Goddess earth, and the modern concept of Gaia earth. Out of the mothering concept there is birth; the unfoldment into the physical plane, separation from the source, emergence into The World as a mailable form to be molded by external forces. As living beings birth creates a passenger entering a life-long journey of pain, pleasure, lessons and expressions. We are the needy children of mother earth and the creative expressions of father sun. Flowers reach for the sun, twisting and wiggling to face it and embrace it.

Separation and Reunion are an archetypal theme built into our collective psyche. Not only is it a recurring principle in religion's story of man's fall from grace (biting the apple, falling into sin) but it's part of the birth process, where a developing fetus forms in an unconscious state within a dark, whole, connected, comforting womb until unseen forces draw it out into The World. It travels through the birth canal and then there is light. Next there is the painful spanking from the physical world and cutting of the umbilical cord; completing a transformation from unity to separation, from conjoined whole to individuated identity.

Creation Narratives are universal stories and myths describing the physical origin of the universe, life and man. They incorporate the metaphorical fall of man into sinful life, and the split into heaven/hell, which parallels the descent of spirit into matter. They capture the departure from a timeless-spaceless eternity into the constraint of a physical world. Light again plays an essential role starting with *Involution* – the descending process where light transforms into matter, eventually followed by *Evolution* – the ascending process where material complexity emerges into higher levels of consciousness culminating with man. Genesis begins with "let there be light" and ends with creation of The World. "In the beginning" (time) "let there be light" (energy), God created Heaven (spirit) and Earth (matter). The creation of *Heaven and Earth* establishes duality, the basis of 2 realms.

Duality Metaphors. Everything originates in a higher realm of pure potential. Anything we discover or "invent" *down here* was already promised *up there*. The higher realm is a limitless plenum containing all original forms reflected, mirrored and manifested into the lower-level physical world. It's a simple yet profound process that looks like this: Nothing → Something → Everything, or (Realm 1 → Realm 2). This abstract process is reflected in religions' creation narratives, mythological stories of birth, death, and rebirth, and physics' Big Bang theory. Separation from a divine source is a universal, recurring archetypal theme.

Numerology's symbolic creation story is summarized as 0 → 1 → 2 → many. It's physically expressed in the periodic table of elements. A photon with "0" mass represents unmanifest energy, which on a scale of matter represents "0". Hydrogen is the first material element with 1 Proton and atomic number 1. Next, we get Helium with 2 protons and atomic number 2. After that we get many; a wide variety of different elements with new emergent properties that make up everything in our physical reality. This is how nothing becomes something and then everything.

Connecting Spectrum of Life. All organic life forms are connected by a common ancestor. All living things, materials, and environments are eternally linked in an ongoing coevolving process. Literally everything is alive, falling somewhere along a scale by degrees of aliveness. What we call life is a continual process of wholes separating into parts only to seek reunion. We're born single but desire partnership, attracted to the opposite sex to reunify as a complete whole. Religions emphasize man's separation from the holy spirit and then offer the path towards reunion with the divine. Creation began with the Big Bang, the greatest act of physical separation in history, immediately followed by evolution and its progressive force of reunifying and reconsolidation of the fragmented parts.

Recurring Themes Exhibit Symmetry at Every Level

Physical   Periodic Table 92 base elements, →molecules → compounds → complex forms
Biological Organic compounds → thousands of species → plants, mammals, primates, man
Societal   Human groups →social structures → businesses, agencies → complex societies

> * *Physical world sustained by energy and forces*
> * *Biological life sustained by water and food*
> * *Social order sustained by money and power*

UNIVERSAL TEMPLATES

The built-in interconnectedness of everything is expressed in universal laws, codes, and recurring relational patterns. Common, original forms generate a multitude of extending patterns in diverse, seemingly unrelated areas. These recurring patterns are produced by templates imbedded within the infrastructure of physical reality; products of dual realm dynamics where non-physical original forms emerge into universal mechanisms, processes and relational patterns that influence the way things manifest everywhere. They're revealed in mathematics, science, arts and metaphysics, reinforcing connectedness everywhere we look.

Archetypal Forms*.* Basic geometric shapes, structures and patterns mirror their original forms from which everything is created. Elemental forms derive their elegance from natural symmetry and simplicity. Numbers are man-made symbols mirroring quantitative attributes found in relationships between things. Numbers become pseudo templates themselves expressing qualities of oneness, twoness, threeness, etc. Nature's geometric forms and proportional structures are physical byproducts of templates from a higher non-physical realm. Terrain features, waterways, channels, hills and valleys are just the residue of orderly processes that shape and mold material substance through templates of geological dynamics.

Territorial borders are byproducts of dynamically balanced external forces; a form of parsimony or equilibrium where static form emerges from active pressures. Sometimes it's neighboring competitors, other times it's nature's terrain features that form physical barriers. State and national borders often trace rivers and lakes, mountain ranges, and impassible surfaces. The shape and composition of houses change in different geographical regions to accommodate local forces of nature and climate. Homes in hurricane zones are built with sturdier materials and flatter roofs while those in flood zones are built on stilts. Where materials are scarce, resourcefulness comes into play; igloos emerge in frigid regions while straw huts pop up in hot regions.

Maps of battles are really templates of terrain. Positions of combatants and avenues of attack are dictated by terrains constraints and natural pathways. Opposing enemies create a dynamic force filtered through surrounding shapes and obstacles of the land. Each side adjusts to maximize relative advantages, pushing back against counterforces while pressing forward where opportunities open up. It's parsimony again as constraints and niches incentivize actions and reactions –

navigating through and around obstacles, pathways, high ground, concealed areas and rough terrain. It's no surprise that battles taking place in the same geographic area hundreds of years apart follow the same general patterns of orientation.

Form Shaped by Physical Constraint (Realm 2)				*Genesis Describes Physical Reality*
Level 1	Point	0 Constraint	*Spirit* (potential)	1. Let there be Light - *ENERGY*
Level 2	Line	1 Degree of constraint	*Time* (flow)	2. Separated day from night – *TIME*
Level 3	Plane	2 Degrees of constraint	*Space* (form)	3. Divided the waters from the sky – *SPACE*
Level 4	Cube	3 Degrees of constraint	*Matter* (stuff)	4. Dry land appears, called Earth – *MATTER*

Images are fundamental template forms. They're non-physical patterns that can take the shape of anything in the physical realm. Image is the root word of imagination, which reflects its limitless potential of representing everything. Consider a blueprint which is the image form of an actual physical building. In time, the temporary structure of a building will disintegrate, however the blueprint will live on forever. And this gets to the essential dynamic of how things are manifested; they always originate in the higher, non-physical realm which contains the potential for everything. Images and pattern are the non-physical substance permeating the higher realm. They serve as abstract templates from which everything is manifested into physical form.

Processes are fluid templates directing how things and events unfold in time. They include sequences, catalysts, chain reactions, synergy, incremental development, tipping points, cycles, transformation, differentiation, bifurcation and integration, growth and decay. Change dynamics are the dual counterpart to relational images; one flowing in time, the other structured in space. Process emerges through quantum moments where potential becomes real, where many possibles become one actual. Evolution is the progressive advancement of change, moving steadily forward through thresholds of emergent innovation, creatively advancing as parts differentiate, specialize and reintegrate. Forward progress moves in incremental steps and stages as new forms unfold with derivative effects from preceding causes. It's an expression of creative spirit, channeled from intentional purpose to produce new things – similar to previous stuff yet different, like a spiral coming back to a higher original point. Evolutionary dynamics applies to organic life, social development (towns, cities, states), business products, personal growth and development, technology innovation.

Change Dynamics are active everywhere, influencing everything simultaneously. All things co-evolve, changing in both sequence and in parallel at all levels. Natural landscapes are shaped through geological processes on and within the earth's surface through seismic and volcanic activity, plate tectonics, erosion and build up, etc. Meteorological dynamics drive weather and climate, oceanographical currents, waves, tides, coastline reshaping, etc. Chemical reactions produce compounds, mixtures, alloys, solid-liquid-gas forms, plasma, etc. Biological dynamics direct cell mechanics: division, replication, separation, specialization, and integration into organic systems. Societal dynamics drive changing cultural values, economic boom and bust waves, governmental tensions between freedom and control, and cycles of conflict between internal domestic interest groups and external foreign adversaries.

Change dynamics are revealed in the universal process of evolution where everything goes through long gradual transitions, interrupted periodically by dramatic revolutionary events. Organic life progressed through long eons of excruciatingly slow change followed by major upward moves caused by unexpected punctual change events. Society likewise moves along gradually, steadily until interrupted by black swan events. Collective knowledge advances forward slowly until inevitable paradigm shifts completely changes entire fields. Businesses gradually improve routine operations until a competitor-disrupter introduces a dramatic innovation. YOU live your life in a series of enduring stages until life-altering events shake you out of your comfort zone and force transition.

The most fundamental change dynamic is the creative act itself, built right into the fabric of physical reality. It's embedded into every micro-moment as the universe is recreated continually, rhythmically, moment by moment at every level. At the physical level it's manifested in quantum mechanics where multiple possibilities are reduced to one actuality. Like grains of sand in an hourglass passing through the narrow middle, one grain at a time, suddenly appearing in the bottom half, the universe is recreated and sustained in a similar manner. Biological life follows this same pattern at a much slower pace, with heart beats and breaths that cycle in and out 1 second at a time. The World is slower yet, following a daily regeneration period of 24 hours; day followed by night, in an eternal cycle of fresh starts and culminations. Society's regeneration cycles are annual with tax collections and replenished budgets, new election campaigns, new school graduating classes and holiday renewals of collective spirit.

Regeneration Cycles at Every Level

Entire Universe	1 quantum moment
Biological life	1 second heart beats, breaths
The World	1 day (work/sleep/wake)
Society	1 year annual renewals: budget, taxes, holidays, graduations

Mechanics Dynamics are interacting parts, sequences, inputs and outputs, and systematic applications, including an assembly line, digestion, computer transistor gate system, DNA gene mixing and filtering, social politics (persuasion, coercion, cooperation, competition) and scientific trial and error. Mechanics dynamics are a key aspect of evolution with incremental building, progressing in sequential steps, and generating a holonic pattern of upward development. Science likewise progresses incrementally through pioneers expanding on work of predecessors. Knowledge expands in similar fashion. Much of physical reality consists of the dynamic interplay between parts and wholes, fragments and unities, driven by mechanics of change. A major part of that action involves cyclic action; growth and decay, struggle and triumph, fall and redemption, separation and reunification.

Micro-Macro Level Dynamics reveal how things operate differently between higher and lower levels. Collective aggregates produce changes in scale leading to emergent properties, including unpredictable nodal points. Small groups of parts behave very differently from large clusters of the same, even when those parts consist of people. Macroeconomics deals with wholes while microeconomics analyzes the parts; the former tracking GDP, interest rates, and inflation while the latter covers individual behavior and individual markets. Macro physics is subject to relativity while micro physics deals with quantum mechanics. The same dichotomy between higher and lower levels applies to most all fields of science and non-science.

Emergent Aggregate Properties — Concepts without meaning at the lowest level

Temperature	1 Molecule has no temperature
Economy	1 Consumer has no economic value
Music	1 Beat has no musical quality
Elements	1 Atom has no properties: metallic, density, ductile, hardness, soluble, tensile
Gravity	1 Atomic particle has no gravity
Pattern	1 Point has no relational quality

*All these properties only emerge in collective quantities – "more is different"

Resonance patterns produce structure and form, visually seen in mandala patterns or filings shaped by a magnet. Vibration radiates outward while resonance synchronizes locally, generating stable forms. Geometric shapes are resonant patterns. The same applies to energy fields. Vibrating particle fields produces resonant radiation, which then propagates in all directions in waves that carry information throughout the physical environment, connecting everything it interacts with. Fields are patterns of force with polarities that attract or repel. Antennas resonate by tuning into specific radiation wavelengths.

When we engage in mental imaging, visualization, affirmations or obsessive thoughts, we're creating resonance; synchronized patterns of energy on the inside that influence the outside. Intention becomes attention, which then leads to attainment. The spoken word is a resonant form; a specific tone produced by shaped vibrational sound. "AUM" is the sound resonating with the universal source containing all sound. AUM contains all vowel sounds. Consonants are mere fragments or fractions in between vowels. Before AUM there is total silence, mirroring the original plenum source of nothingness containing the potential for everything. Out of that silence a spoken word emerges, producing something from nothing. In a similar manner light emerges out of darkness, radiates, integrates and transforms into a rich variety of resonant patterns in the physical world.

Laser light is coherent, in-sync, super resonant, and capable of much more than ordinary light. Music is resonant sound; consonant vs. dissonant, harmony over discord. Moving melodies draw from coherent patterns of whole notes, richer than infinite fragments of fractional noise. Archetypes are resonant patterns of the collective psyche. All of us resonate to certain people, places and circumstances.

Memory is a resonant pattern. The universe stores everything that has and will happen. Memory is a built-in attribute of physical reality. Every activity, motion and event ripples interactively like waves in an ocean, exchanging and transferring bits and bytes everywhere. It's all connected and stored in patterns of information and energy, serving as an abstract record of the universe. Man's increasing consciousness translates to a higher awareness of what was already out there. Ruins, relics, and residue are physical markers no different than digital information stored on a hard drive. Every object around you is a coded marker; bits and bytes in the universal program operating physical reality. *The Matrix* portrays everything as a series of 1's and 0's, set in particular spaces in particular times, interrelated, full of stored information, accessible the same way we draw on our own memory.

States of Being are temporary resonant patterns of energy in living beings and environments they occupy. They're reflected in places that thrive for limited periods until eventual diminishment (towns, businesses, local hangouts, empires). With individual people it's expressed in levels of consciousness, moods, tastes, personalities and general dispositions. Attraction to individuals or places changes in direct correlation with state changes. States can be altered by outside forces or individual resonant acts; meditation, prayer, chanting, drumming, dancing, hypnosis, etc. It's a focused movement toward stable patterns of form.

   States of being are present at every level in every set of circumstances – Individual and collective, living and non-living, here, there and everywhere. Everything is differentiated by its particular condition or disposition. There are different states of readiness, wholeness, maintenance, stability, energy and aliveness. They're expressed in the conditions of organizations, individuals, teams, special interest groups, businesses, systems and environments. In people they're manifested in levels of excitement, physical and emotional energy, health, sensitivity, attitude, level of consciousness and general disposition – all expressed in body language and directed actions. Each state is associated with a particular energy/vibration. Higher consciousness = higher vibration. In each of us it's a combination of physical, emotional, mental and spiritual states of being.

Microcosms are ubiquitous throughout the universe. Man is the central example; mirroring earth's water ratio, chemical components, Vitruvian body ratios, mental attributes, operating as a self-sustaining system. A human body is a machine, vehicle, process center and super system of systems, encased in a skin membrane creating the illusion of separation between the mini universe within and the macro universe without. As Protagoras observed in 1500 BC "Man is the measure of all things". Man is both IN The World and collectively IS The World. Man is a mini universe while connected to everything else in the universe and consciously becomes *the* center of the universe. Everything in the universe is in us – component parts, processes, dynamics, symmetry, integrated and interconnected within and without, coevolving with everything around us. We're a living, breathing time capsule, carrying with us ties to a 14-billion-year-old universe.

We're right in the middle of it all; an integrating node in a comprehensive network, a central hub connected to everything, everywhere, everywhen.

TEMPLATES OF RELATIONAL PATTERNS

Recurring themes manifest everywhere we look as the same universal forces of physical reality apply to everything. Similar patterns will repeatedly reappear regardless of the situation. Pre-set forces limit what can happen and how it will happen, so despite infinite possibilities there are really very limited actualities. Like the principle of parsimony limiting the size and shape of organisms, all man-made things are subject to the same limiting forces in the physical environment. I suppose you could make a boat or plane square shaped, but it wouldn't operate very effectively. Of all the infinite shapes you could make a boat or plane, very few are chosen, and for good reason. The same thing applies with your own body.

Universal Templates Produce Recurring Themes

Enclosed center – Atom nucleus, seed, cell, egg, cocoon, womb, brain, headquarters, capsule
Membrane – Skin, hull, cocoon, sheath, placenta, pelt, casing, shell, package, husk, clothes, bark
Homes – Nest, cave, lair, hive, hut, nest, enclave, tent, nook, crib, igloo, domicile, lodge, den, base
Vehicles – Car-boat-plane-train-bike-submarine-spacecraft, motorcycle, subway, raft, snowmobile
Integrated systems – Organism, factory, computer, machine, colony, assembly line, city-state
Code systems – Languages, alphabets, numbers, genetic codon, I-Ching, Binary code, sheet music
Leadership roles – Coach, director, maestro, commander, governor, godfather, captain, CEO.
Images – Construct, form, geometry, design, picture, figure, graph, frame, outline, pattern
Circular systems – Atom, gyro plants, spiral galaxy, comet, Ferris wheel, flower, solar system
Fractal patterns – Branches, roots, tributaries, lightening, veins, broccoli, leaves, dendrite, crystals
Regulators – Resistor, lever, pump, processor, amplifier, transformer, filter, switch, valve, gate,
            screen, brake, feedback loop, governor, counterweight, funnel
Webs – network, lattice, meshwork, wiring, nexus, skeleton, fiber, channels, labyrinth
Conduits – arteries, veins, branches, capillaries, tributaries, roots, ant colony, circuit board, city grid
Hubs – shipping port, train junction, bus depot, airport, spine, command post, switch board, router

Man Centered Reality; You're a special hub connecting everything; literally in the middle

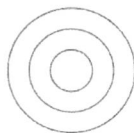

The World
Man
Human Reality

Man 2nd ring
Inner/Outer

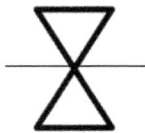

Big/Fast/High
Man
Small/Slow/Low

Man in Middle
Above/Below

Past-Present-Future

Man in Middle
Before/After

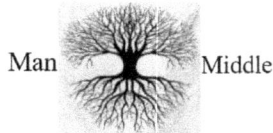

As Above

Man   Middle

So Below

Template of Life's Recurring System Patterns

Man	Car	Circuit Board	Home	City
Arteries	Cables	Wires	Pipes	Roads
Coiled Intestine	Solenoid	Inductor coil	Radiator	Treatment Plant
Stomach	Gas Tank	Capacitor	Pantry	Reservoir
Lung	Vent duct	Board Channels	HVAC ducts	Airport
Anus	Exhaust pipe	Outlet Plug	Exhaust fan	Sewage
Brain	Driver	Microprocessor	Owner	Mayor
Skin	Frame	Casing	Siding	Boundaries
Heart valve	Fuel Valves	Transistor	Doors	Gates

Cell	Nucleus	Atom	Physics	Atom	Mostly empty space
Man	Brain	Molecule	Chemistry	Man	Mostly empty space
Society	Government	Cell	Biology	Cosmos	Mostly empty space

Template of Materiality – Periodic Table of Elements

1st Row	(H, He)	99% of mass in the universe
2nd Row	(C,O,N)	Organic life key elements (carbohydrates, fats, proteins)
3rd-6th Rows		Metals and Metalloids
7th Row		Radioactive elements

Stages of Complexity – Atomic properties emerge in the aggregate

Metals – Bonding leaves electrons free to move
Salts – Ionic bonds join elements with extra and lacking electrons (attraction +1/-1)
Methanes – Covalent bonds join elements sharing electrons (no attraction)
Polymers – long chains of repeating parts (mirrors plant stems)
Proteins – Long chains with side chains (mirrors animal limbs)
DNA – Coded pattern of parts

The Periodic Table is a coded pattern of qualities, including harmonic musical octaves. Each element has a distinct shape and frequency (H-hydrogen: key of E). Isotopes are variations of elements mirroring sharps and flats musically. Inert gases are the octave nodes. Matter is essentially energy differentiated by patterns of shape (spatial) and frequency (time). All substance and materiality in general consist of one original undifferentiated pure energy expressed in a kaleidoscope of various patterns in space and time.

TEMPLATES OF GEOMETRIC PATTERN

A *Line* symbolizes first level unity; oneness. It represents direction, action, movement, force, and produces polarity, transitioning into twoness and the illusion of separation. A line reflects unity of motion as either a *constant* speed or *constant* acceleration. A curved line transcends the limits of a straight line and represents acceleration. Linear equations are easy to solve with fixed ratios between parts and the whole. Non-linear equations involve variable change and complexity.

A *Circle* is the fundamental pattern of higher-level unity and wholeness, derived from the original source of everything. Rotating around the circle expresses action in the here and now, while the center is the still, non-moving original source. A moving circle represents the quantum creative act, depicted as a sinewave. It's a universal cycle of growth, peak, decay and return.

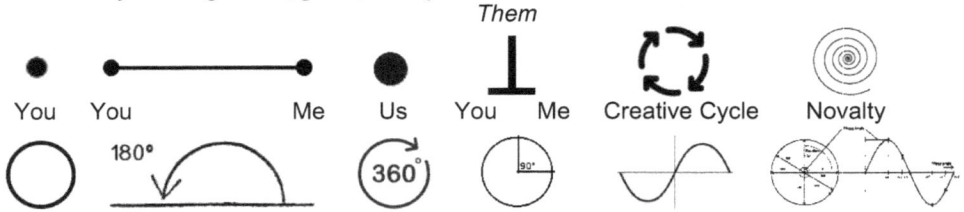

A Circle contains every polygon within it – a geometric holon and perfect symbol of unity and wholeness. Every shape, form and partial segment in between are within a circle. Primary shapes are resonant forms; nodal points between an infinite variety of possible shapes and partial fragments.

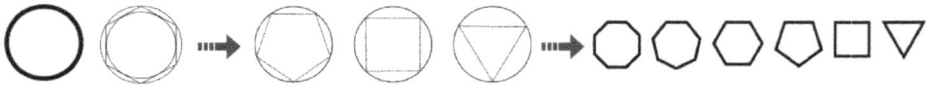

A Dot = *unity of spiritual realm* 1, a Circle is the dot manifested as *unity in realm* 2
   "0" = *unity of spiritual realm* 1, "1" = *unity of physical realm* 2
   "0" = *unity of emptiness*, "1" = *unity of fullness/completeness*
Everything in Physical Realm 2 is stuff between 0 - 1: fragments, fractions, partials.
Symmetrical polygons above are *Nodal points* on a spectrum between 0 and 1

A Circle represents unity as a complete whole. "Universe" in Latin means "1 turn" (a completed circle). Plato described it as the whole of wholes. After unity (1) comes duality (2). This is represented by a simple line, which separates two points. When a third point emerges apart from the line, we now have a pattern; (3). So, from unity to duality we get a trinity, meaning Tri-unity, or 3 as 1. The trinity pattern is a new level of wholeness – a cycle of completion with beginning, middle and end (birth-life-death, past-present-future).

A Circle is the universal relational pattern of integrated wholeness and duality. Sun, moon, planets, orbits, particles – all circles or *spheres* (3D circle). Circles are cycles representing beginning, ending and beginning again – life, death, and rebirth. Circles are both eternal and ever present.

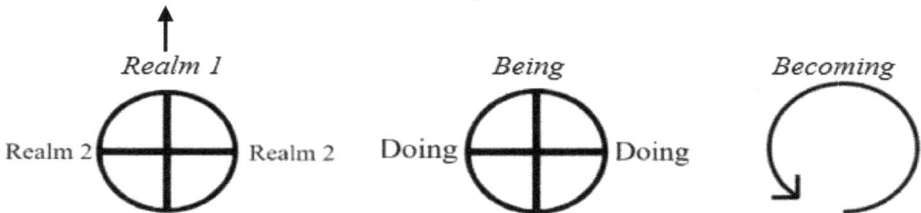

Spheres containing infinite circles appear everywhere as omnipresent patterns of wholeness, manifested as a nucleus, seed, egg, home or vessel. In the beginning there was nothing, represented by an empty circle or hole containing the potential for everything. 0 is the seed begetting every number after it $(1, 2, 3, \infty)$. A seed contains the pattern for that which will unfold in time. Think of the tiny acorn (sphere) that becomes a giant fully developed tree, or the single fertilized egg (sphere) that becomes a complex fully grown human being. And mother earth; the spherical egg that begat millions of life forms that unfolded through eons in time.

Spirals represent cyclical progress, moving forward and higher, attaining new beginnings that mirror the original starting point; a ubiquitous pattern found everywhere: shell, galaxy, sunflower, staircase, horns, tornado, DNA, whirlpool, pinecone, snakeskin, cabbage, fingerprints, vine tendrils, hair whorls, etc. Spiral polarity reflects the expansion-contraction template imbedded in all living processes where growth is followed by decline. It's the centripetal-centrifugal balance expressed in the ongoing rebalancing of forces in nature that build temporary structures which over time crumble and disintegrate. It mirrors the charge and discharge process of life; a perpetual cycle of beginnings and endings like the start and end of a song or the first breath and final exhale experienced by everyone of us temporary living beings.

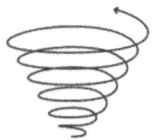  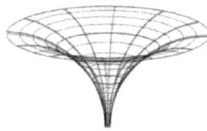  

| Cycle in Time | Centrifugal | Spiral-Funnel Transform | 90⁰ Dimensional Shift |

Cycle in Time   Centrifugal   Spiral-Funnel Transform   90⁰ Dimensional Shift

All spiral motion expresses 1 of 2 possible states: Expansion/contraction, Charge/discharge, Pressure/release

TEMPLATE OF TRANSCENDENT RELATIONSHIP

The earth is a flat, horizontal plane; a playground arena for living organisms to survive and thrive. While plants are stuck to the ground, animals move around on all fours, with heads not much higher than the rest of the body. Evolved *Man* stands erect, oriented vertically towards the heavens, perpendicular to the horizontal earth plane. A spinal cord is a vertical shaft pointing upwards, culminating with the brain at the body's very top. When humans are ill or killed, they collapse back to the ground, assume a horizontal posture, and eventually are buried in the same horizontal orientation returning back to the flat earth plane.

Man-made structures are generally square with straight lines meeting at 90⁰ corners. Homes, buildings, boxes, pallets, shipping containers, and boundary lines are generally squared off constructs. All the TVs, smart phones and computer screens we stare at all day are squared constructs. Ironically, there are no straight lines in nature, just curved surfaces and nonlinear shapes: spirals, vortices, circles, spheres, and free flowing motion.

Spirit
△
MAN – in the center
▽
Matter

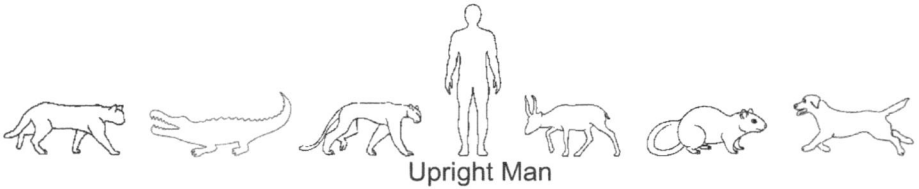

Heaven

Heaven

Earth

Upright Man

Earth Plane

Perpendicular Template

Physical dimensions are 90° apart
Rotational angular momentum is 90°, Torque leverage is maximized at 90°
Light Electromagnetic radiation: electricity and magnetic waves are synchronized 90° apart
Centripetal/centrifugal force; moving objects always go on a 90° tangent from the radius
Right hand rule – currents always flows 90° away from magnetic fields
Earth's precession force is 90°, Plants shoot up 90° from earth surface
Punch has maximum force when arm is bent 90°, Legs work hardest when bent 90°
The World plane and heaven above are 90°
Pitcher throws ball 90° from rotational arc, batter hits the ball 90° off the bat
Funnel Pattern transforms horizontal opening to vertical bottom hole 90°
*90-degree *transcendent relation is greater difference than 180-degree polarity of opposites*

Light wave is triple 90°
90° from source
90° Electric axis
90° Magnetic axis

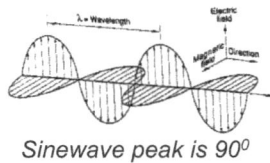

λ = Wavelength

Electric field

Magnetic field   Direction

Sinewave peak is 90°   *Quad 90°*   *Quad 90°*   *Triple 90°*

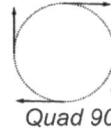

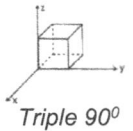

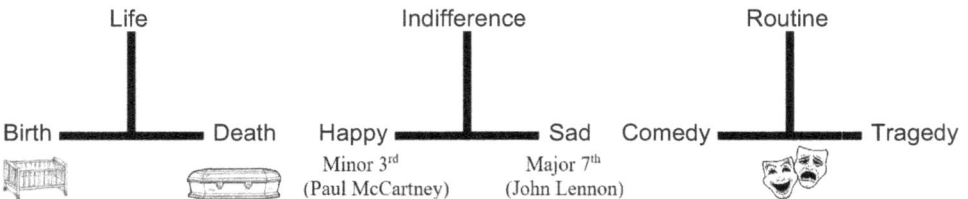

Life

Indifference

Routine

Birth ———————— Death   Happy ———————— Sad   Comedy ———————— Tragedy

Minor 3rd
(Paul McCartney)

Major 7th
(John Lennon)

Transcendent Process of Evolution

0 – Mineral: passive inactive element (point)
1 – Plant: expands: roots down, stem up (line), flower around (plane), fruit (sphere) → seeds (point)
2 – Animal: transcends fixed immobile plants, gains freedom to navigate environment (plane)
3 – Man: transcends animals limited by instincts & environment; man controls and directs it (sphere)

Hierarchies and Holons are universal and embedded everywhere. Languages (letters-words-sentence-paragraph), Matter (particle, nucleus, atom, molecule), Organic life (organic molecule, cell, tissue, organ), Music (note, chord, phrase, melody), Military (Soldier, squad, platoon, company, battalion), Government (town, city, state, federal). Each higher level supported by previous ones, extends further as a greater whole while maintaining the original lower parts.

Dyad Pattern is embedded everywhere. It's the fundamental pattern of duality originating from the interaction between dual realms. Space/time, Matter/energy, Gravity/radiation, Field/current, Particle/wave, Body/soul, Structure/flow, Left Brain/Right Brain, Form/process, Talk/sing, Wake/dream, Digital/analog, etc.

Complimentary Duals.   Art/science, noun/verb, object/subject, etc. Patterns of duality are manifested everywhere as reciprocal relationships. Transcendent dualities create the illusion of separation.

Complimentary Duality Pairs

Body	Mind	Math	Music	Object	Subject
Quantity	Quality	Nature	Nurture	Being	Doing
Word	Image	Think	Know	Static	Flux
Inner	Outer	Order	Chaos	Parts	Whole

Sun and moon express connected duality: heavenly circles of equal size moving in harmonic cycles in the skies above. Each connected to the other, to earth and us: connected through gravitational motions and dynamics of light. Sun representing spirit – the source of living, conscious energy, and the moon representing the soul, an unconscious basin of duality, each interacting as ambassadors of two realms.

Template of Binary Duality – Single Essences Polarized

All scales represent 1 single essence with opposite expressions on the ends; two partial truths. A single coin is one whole with 2 sides. The entire physical realm is polarized by man in the middle creating binary aspects to everything experienced, expressed in every field of interest:

Philosophy	Materialism vs Idealism, Empiricism vs Realism, modern vs post modern
Politics	Progressive vs Conservative, Liberal vs Traditional
Religion	Modern vs Orthodox, Old Testament vs New Testament
Art	Impressionism vs Expressionism, Formalism vs Realism
Psychology	Gestalt vs Structuralism, Psychoanalysis vs Humanism
Economics	Classical vs Keynesian, Capitalism vs Socialism

Every perspective is countered by an opposing one, each going in/out of favor over time

Binary Duality Expresses Difference Between West-East Perspective:

WEST	EAST
Space as empty void	Space as full plenum of potential
Life is linear, sequential	Life is cyclic, non-linear
Flat paper for writing symbols	Paper folded into forms (origami)

Western landscape is off in the distance, viewed from a perspective of separation.
Eastern landscape is connected to you; you're viewing it from inside as a part of it.

Binary Duality mirrors dual realms. Binary code consists of 0 and 1, expressed in many forms including on-off, yes-no, plus-minus. 0 represents nothing as an original whole, 1 represents unity as a manifested whole. It could also be portrayed as a dot and a circle. Nonphysical realm unity = 0, Physical realm unity = 1.

Fundamental Source Patterns built into the infrastructure of physical reality

Point .	Position	0D	2^0	Seed	Mineral	Fire	Sun	Plasma	Spirit
Line —	Distance	1D	2^1	Stem	Plant	Air	Sky	Gas	Mind
Plane ▱	Area	2D	2^2	Leaf	Fish	Water	Ocean	Liquid	Heart
Cube ▤	Volume	3D	2^3	Fruit	Animal	Earth	Land	Solid	Body

1 Unity	o
2 Relation	o---o
3 Pattern	o---o---o
4 Structure	o---o
	⁞ ⁞
	o---o

- • Potential
- — Emergence
- ☐ Form
- ▱ Substance

1 Point	Unity
2 Line	Duality
3 Triangle	Mind
4 Pentagon	Life
6 Hexagon	Man
7 Diamond	Divine

Dimensional Constraint: **0D** (Point)   **1D** (*Line*)   **2D** (*Flat Plane*)   **3D** (*Volume*)

	0D (Point)	1D (Line)	2D (Flat Plane)	3D (Volume)
Vehicles		Capsule	Train/Car	Boat/Buggy	Plane/rocket
Organic Life		Seed	Plant stem	Flower	Fruit

Universal Forms and Shapes in Nature

Seeds	Trees	Stone	Honeycomb	Snail	Snake	DNA	Veins
Fruit	Mountains	Land	Snowflake	Flower	River	Plants	Thunder
Stars	Petals	Rock Salt	Turtle Shell	Galaxy	Intestine	Pinecone	Corals

Physics unites everything at the most fundamental level. The theory of relativity connects energy, matter, space and time. Energy cannot be created or destroyed, only transformed. Matter is just a structured form of energy. Everything is connected as patterns of the same energy, expressed in different temporary forms. Objects consisting of matter produce gravitational fields that warp space. Objects traveling through space must follow that warped path. Each influences the other in perfect complimentary symmetry. All matter and all space are infinitely connected.

Matter and energy, space and time, are simply two sides of one coin. Since all matter/energy and space/time are connected, *everything* is physically connected.

Unities in Physics		Structure – Flow (Duality)	
Matter	Energy	Particle	Wave
Space	Time	Static	Flow
Gravity	Acceleration	Field	Radiation
Magnetism	Electricity	Charge	Current

The Universal Dyad Pattern is built into the physical infrastructure, revealed above

Physical Laws of Connectedness

Law of Thermodynamics – energy can't be created or destroyed, only transformed.
Law of Gravity – every *thing* attracts every other *thing*, as diminishing force that never ends.
Law of Quantum Nonlocality – particles are always connected regardless of distance apart.
Law of Special Relativity – All things warp space, which in turn directs their very own path.

Newton's Law of Gravity and Coulomb's Law of Electric charges are identical equations:

$$F = G \, m^1 m^2 / r^2 \qquad F = k \, q^1 q^2 / r^2$$

Symmetrical Flow Systems

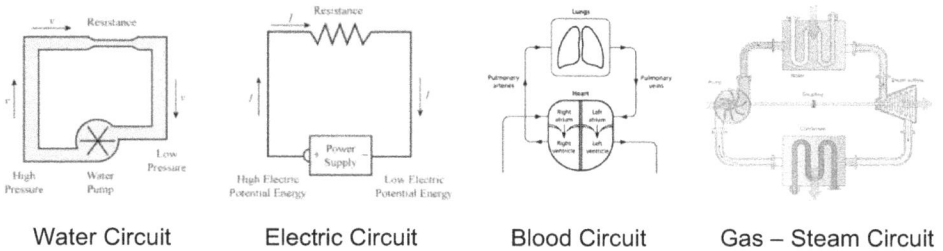

| Water Circuit | Electric Circuit | Blood Circuit | Gas – Steam Circuit |

As above so below

| Lakota | Celtic cross | Tree of Life | Shatkona | Hermetic | Duality |

Cosmic Habits as Recurring Universal Patterns
Dyad pattern, dualities, polarities, hierarchies, holons

Symmetry – Natures forms, polarity, geometric shapes, vibrations, equations, crystals
Circularity – Music, moods, timing, tastes, poetry, weather, metabolism, habits, cycles
Repeating Patterns – Hierarchies/Holons, cycles, resonant forms, fractals (scale invariant)
Dependency relationships – Mother-child, master-apprentice, host-parasite, earth-mankind
Systems – Connected integrated parts working together, transferring energy, materials, info
Sustaining substances – Energy, fuel, oil, blood, food, money, water, air.

Conflict dynamics – Games, sports, fights, court, debate, diplomacy, combat, politics.
States of Being – Mood, health, conscious, communal spirit, readiness, stability, resonance
Modes of constraint – Laws, rules, regulations, warnings, boundaries, barriers, punishment
Wholeness – Health, balance, unity, equilibrium, symmetry, reunion, "0" – "1", completion

The Golden Mean: Wholeness = Holiness

Balance and Excess: *Universal Perpetual Pendulum Cycles*

Every thing, situation or condition is always in 1 of 2 possible states; Equilibrium or imbalance. Stability is always a temporary state, always the result of 2 equal and opposite forces held in check; a product of dynamic equilibrium. Even a motionless rock on the ground is in a state of dynamic balance – force of gravity pulling down countered by the earth's surface pushing back. Even YOU just standing there are working subtly to keep from falling down as bones and muscles flex, maintaining a tension structure countering gravity's constant pull.

Achieving balance is a universal theme in human reality dating back to the ancient Greeks who championed the concept of a Golden Mean, emphasizing the value of moderation in all things. Avoiding extremes is a life lesson applied at every level; individual, groups and nations. We get in trouble when taking a good thing too far, pushing beyond the "normal" range and into the realm of excess. Too much eating, drinking, drugs, gambling, lust, and obsessions of all types will get you in trouble. Going too fast, working too hard, spending too much or even partying too long just turns a good thing into a bad thing. Nations do the same at the collective level where governments spend too much or reach to far, markets boom and bust, and political parties go too far left or right. The same principle applies in nature where life sustaining sunlight, water and air can become life threatening in excess: boiling desert sun, crushing tidal waves, or hurricane winds. It's the Goldilocks principle where life is just right somewhere in the middle.

UNIVERSAL ESSENCES

As we analyze connecting threads tying everything together, we must include anything considered an essence; the foundational qualities present in all things. While primary essences of *Substance, Order, Purpose, Process, Relation* are fundamental to everything we experience in reality, there are other derivative essences, also omnipresent in the physical universe. Essences are the key ingredients of not just life but all reality, and built right into man: *Substance*-body, *Order*-mind, *Purpose*-soul, *Process*-spirit, *Relation*-Identity. *Relational includes internal vs. external, self vs. others, and who you are vs. who you were.

Essence	Attribute	*Expressed in Man*		
Substance	Physical parts,	Body	Cellular tissue	Organs
Order	Form, control	Mind	Brain	Thoughts
Purpose	Directed action,	Soul	Desire	Emotions
Process	Change dynamics	Spirit	Metabolism	Being → becoming
Relation	Contextual pattern	Identity	Personality	Relationships

SUPPLEMENTAL ASPECTS OF THE PRIMARY ESSENCES

Substance – both the material quality of stuff and parts they comprise. Trillions of cells in your body are the stuff of you, yet they are themselves composed of various organic raw materials. Similarly, atomic particles are the stuff of atoms while atoms are the stuff of molecules and greater wholes of material substance.

Order – a bit confusing with its dual aspects expressed as both form and flow, structure and direction. Order can manifest as static control seen in rigid forms and variations of constraint, or guided movement toward a goal. A cop can either restrain you or guide you into a squad car. Order has further derivative aspects: *attraction* (gravity, magnetism, pressure, desire, static charge, passion, etc.) and *direction* (influence, force, persuasion, manipulation, politics, guidance, etc.).

Purpose – built into the infrastructure of reality, it's the evolutionary progressive force driving everything towards higher levels of complexity and consciousness. What we see as life is merely the temporary structures and organic forms that come and go, left behind in evolution's ongoing wake, simple vehicles used by spirit while pressing forward without end. It's an intentional force producing greater complexities, higher hierarchies and deeper holons (whole-parts).

Process – dynamics and mechanics of change. The ongoing dance between balance and imbalance of forces, equilibrium and disequilibrium. Mechanics are the micro level of change including cycles and waves of motion, difference, expansion-contraction, ups and downs, departures and returns. Process is expressed in repetitive cycles of life, your life path, The Hero's journey, growth and decay, and what goes up must come down.

Relation – is a *Pattern Essence* – The highest-level language of reality. Patterns transcend both parts and wholes. Patterns are the ultimate connectors of all there is, and more permanent than any material object or structure. Most things consist of recycled parts, including us, leaving the only stable aspect of *you* as simply an overall pattern itself. Our bodies are made from used elements, continually recycled on a daily basis. Emotions are fluid, cyclic patterns, used over and over again. Thoughts come from shared information that gets regurgitated for further reuse. It's all transient and temporary. What makes us and our surrounding environment seem orderly are the enduring patterns we perceive and identify with.

DERIVATIVE ESSENCES

Nonphysical forms are the quintessential ethereal substance preceding physical objects. Everything has an associate form built into its makeup otherwise there would be just unorganized raw materials. Archetypes and images are a type of nonphysical form preceding their physical counterparts. These pseudo-blueprints shape everything we experience; the original source of all substance and order.

Spectrums, holons and networks are ubiquitous derivatives of the relational essence. Laws and codes represent derivative expressions of relationship. When we add the essence of process we get vibration, cycles, and temporal patterns of relationship. Whenever those patterns become repetitious, they lead to resonance and synchronization, which are attributes of the order essence. This is an example of how the primary essences manifest derivative variants. Harmonic rhythmic vibrations are derivative forms of process, order and relational essences combined.

Intelligence is a derivative attribute of order, built into the universe at every level. Human intelligence and sentient beings are just localized expressions of it. Even an atom possesses a degree of intelligence, choosing which particles to join with and which states to adopt. Memory likewise is built in, present everywhere in various forms. The entire landscape of earth is a physical archive, memory in structured form. Layers of sediment and tree rings are visual records. Genes and heredity are physical memory. Information is the quintessence of both intelligence and memory, serving as an abstract, immaterial cellular substance of both. Everything is connected by patterns of information, and when combined with energy and purpose, it leads to complex forms and organic life.

Interdisciplinary Symmetry of Essences

Chemistry	*Physics*	*Biology*	*Economics*
Substance	Matter	Tissue	Supply
Charges	Forces	Muscles	Demand
Reactions (ions)	Energy transfer	Digestion	Exchange (money)
Catalysts	Leverage	Instinct	Incentives
Enzymes	Fields	Hormones	Regulations

ELUSIVE ESSENCES – *Difficult to Define*

Life – is a universal essence permeating everything at different degrees, whether animate or inanimate. Life's essence is not confined to physical forms or temporary organisms we associate it with but rather, the spirit that manifests those things; the ongoing progressive force through which everything emerges into physical reality. We see it's attributes in ordered energy and patterns of material structure continually evolving into greater forms of complexity. A machine and a city are both very much *alive* despite being inorganic systems. Popular definitions of life are way too limited by parochial bias. Even official descriptions of what qualifies as organic life vary greatly, as they include inconsistent elements, and fail to address all the ambiguities of where life begins or ends. Intuitively we know if something seems alive or has a degree of life in it, whether it looks like us or not.

Love – Despite the ongoing obsession with love in popular culture very few actually define it clearly. Music, books and movies are rife with examples of misguided expressions of love, often just cases of selfish, possessive projection. Love connects and unifies along a spectrum of levels and types. The duality created by man in the middle produces expressions of higher and lower aspects. Ego separates, heart attaches, mind rationalizes – all play a part as we filter pure love into *peaks* of compassion through *valleys* of lust and possessiveness. Love is an integrative resonant multilevel state of being and expression. Love unifies, non-love separates (fear, apathy, depression, anxiety). Love is a transcendent state of being that can't be quantified by science or rational analysis.

Energy – is a pseudo-physical substance present in everything with a vibrational living quality. All things are connected through patterns of energy in continuous states of transformation, leading to the creative emergence of new forms. Energy itself has many faces and attributes; thermal, radiant, electrical, chemical, nuclear, gravitational, mechanical. Its normal state is continual motion, especially in light which only moves at the fastest speed possible, however it's just as well bottled up inside structured patterns of matter where knotted energy produces the illusion of solid objects. Hidden beneath the surface is a dynamic web of tension where millions of atoms and particles vibrate at super high speed, producing strong local forces that keep a largely empty object together as a stable material substance.

Light is the universal energy conduit – the ultimate source connecting everything. It's the life sustaining force that literally feeds us the energy we use on a daily basis, whether directly from the sky or indirectly though photosynthesis. Its heat catalyzes the biosphere and agitates the atmosphere, dynamically influencing both weather and climate. Light connects us to everything else through reflection, refraction, absorption and interference, illuminating objects and images everywhere. We're even connected to stars billions of miles away as their light rays touch our eyeballs, linking each of us to the vast cosmos we falsely perceive as separate. Light is the preeminent foundation of the physical universe, a fundamental constant possessing no mass or charge. At the speed of light there is no space or time. There's no out there, only here. Light is synonymous with spirit and God, expressed in The World through the filter of man, manifested as both being and becoming. Walter Russell describes it aptly: "Light is the universal language in which the Divine Concept is plainly written"

Time – may be *the* most elusive essence yet, with its abstract immaterial serpentine quality that can never be seen, only experienced. It drives everything in an ongoing process of change as life dynamically emerges, develops and transforms along a traceable path, leaving a wake of residue we perceive as physical memory. The passage of time is always relative, never the same for any two observers, both physically and psychologically. Time seems to pass differently based on age, mood, activity, expectation and culture. Man's meager attempts to grasp and control time have resulted in clocks and chronometers that merely represent arbitrarily agreed upon units based solely on convenience. Time moves along both linearly and cyclically, moment by creative moment, each never quite the same as the preceding one, producing a continual successive link where man in the *middle now* perceives conscious separation between a divided *past* and *future*.

God – like love, is commonly referred to in popular culture in ways that are embarrassingly inadequate. How can an omnipresent, all powerful, invisible being or entity possibly be understood by those beneath it? We imagine and describe God using our limited language and experience, often assigning a male gender, which is a ridiculous, xenophobic projection. Of course, every religion and culture describe God differently, and *their* God is always the true, real deal. Even where there's agreement in overall description, there's still disagreement on whether God is an active participant in The World or just an uninvolved observing creator. Where is God? What is God? How does an all-powerful, all-omniscient God produce a world containing death, destruction, misery and evil? Any concept of God certainly has as many questions as it does answers.

Consciousness – poses one of the greatest mysteries of science. There's no consensus on what it is, where it comes from or how it operates. It's the underlying universal essence linking all the others. It's the fundamental basis of the non-physical realm yet its ground state is realm "0"; the indescribable nothingness from which everything originates. It precedes intelligence, which is merely a derivative emergent aspect of it. Consciousness precedes everything and then becomes everything. It's the pervasive substance of reality as both raw material and finished product. Analogous to water, it's shapeless, formless, flexible, ubiquitous, changes through states, takes on new attributes and has no standard form (steam, ice, liquid). Consciousness seems like a specific thing to us as we associate it with thinking and awareness, however it goes far beyond. It's every state and level of intelligence, and in every object and process. Perhaps the best way to grasp consciousness is to think of it as a refined aspect of *Spirit,* emerging as particular resonant states. Spirit and consciousness are what makes up everything; both being *transcendent* (above and beyond) and *immanent* (embedded within).

Omnipresent Patterns of Unifying Connectedness

Self-similarity patterns, proportional ratios, Fibonacci sequence
Universal themes and templates
Quantum entanglement, action at a distance
Interconnected biosphere integrating all organic life with the environment
Ubiquitous vibrational resonance patterns, complimentary color, harmonic music
Metaphors, analogies, similes, allegories, tropes, parables
Complimentarity, hermetic law, Karma, feedback, synchronicity
Polarity, Duality, Transcendence
Coevolution, heredity, lineage, tree of life, DNA
Universal laws and codes, language, cyphers, life lessons
Parsimony, equilibrium, micro principles
Hierarchies, holons, spectrums, scales, networks, matrices
Systems, microcosms, machines, mechanisms, operations
Omnipresent forms of life with extensions, layers, appendages, artifacts
Rhythms, cycles, recurring events, nodal points
Communication, information, feedback
Mathematic equivalents, invariants, inverse, sets, roots, derivative, function, logs, constants
Symmetry, fractals, poetry, repeating patterns, sacred geometry

Timelines, generational cohorts, ruins, relics, legacies, vestiges, memory, CSI, history
Codependency, Parasites, Food chain, Symbiosis, Predator-prey, Leader-follower
Quantum entanglement, General relativity, Special relativity
Consciousness, spirit, life, universal essences
Universal linking equations ($e^{i\pi} + 1 = 0$, $G = m^1m^2/r^2$, $E = MC^2$, $\nabla \times \mathbf{E} = -\partial\mathbf{B}/\partial t$)

IT'S ALL HERE NOW

Everything is already here and now. It's not obvious on the surface of our familiar physical realm but it's all here, concealed by our very own limited sense of perception and awareness. Reverse engineering reality reveals it all, lifting the veil to expose every masked link and connecting conduit that ties YOU to everything else, in here, out there, everywhere and everywhen.

We exist in a *Holographic Universe* where every whole contains the parts and *every part* contains the whole! Holons are wholes transcending parts while holograms are parts that transcend the whole! Every part is connected to its whole, forming a super-holon relationship. Just like every cell in your body contains the DNA blueprint of the whole. Just like a fractal pattern where the same form is repeated at every level infinitely. The universe is a comprehensive field of energy and information where each enfolded part reflects the whole. Our comprehensive holographic template of reality produces a hierarchy of microcosms where universal patterns are repeated at each level. YOU are a microcosm of the universe; a whole within a whole. Every individual action reverberates with the collective, adding to the mix in an integrated web of resonating patterns.

Hologram forms are connected by the conduit of light, which is reflected, refracted, filtered and absorbed everywhere. The holographic universe is created moment by moment as the unseen, original cosmic plenum of potential becomes manifested into the physical reality we experience. Moment by moment possibilities become actualities. Potentials become formed things, whose essence was already there. Like formless energy becoming structured matter through concentration, similar to blood clotting or water freezing. Everything is here already in various temporary forms, popping in and out of perception, emerging, manifesting, taking shape…and then transforming, receding, unraveling back into something else. Hindu's call it Maya, the original one with many faces; illusionary forms experienced in the moment, all temporary, transformative and essentially something else. Our reality is a partial truth, limited by the veil of local experience.

In a holographic universe what happens inside is mirrored outside. What happens inside man is expressed outside into The World. Man is a localized expression of the universe, and simultaneously, the universe is the non-local expression of man! In a holograph, every part contains the whole and vice versa. Which inescapably implies everything is already here and now. Man's consciousness is what draws the potentials of realm 1 into realm 2. It's the catalyst converting the *enfolded* potential into the *unfolded* manifest reality.

Quantum mechanics requires an all-here-now reality as it represents infinite possibilities prior to each actuality. It's like hundreds of cable channels playing everything imaginable simultaneously, but you can only watch one at a time. Think of the electromagnetic waves saturating our environment with everything from tv signals, to radio signals to microwaves and every possible band width and frequency out there. It's all right there, hidden from our awareness yet accessible any time with the right antenna tuning into it. It's a perfect metaphor of reality in general. Realm 1 is like this, always with us but beyond perception.

Universal Patterns of Original Unity

Ocean – metaphor of a consciousness plenum; an infinite whole with localized parts (waves, currents, whirlpools, droplets)

Light – metaphor of flowing spirit, pure white unity filtered locally into infinite patterns (colors, reflections, absorption, refraction, images, shadows, silhouettes)

Energy – metaphor of spirit-source essence, transformed into varied pseudo-physical forms and ultimately matter

Sound – metaphor of directed, creative expression of energy; vast sea of noise channeled into meaningful auditory patterns

Silence – metaphor of nothingness; original void, plenum of potential – vast emptiness transformed into somethingness with a single note, then a pattern of notes

Ancient sages intuitively understood the deep connections linking everything above and below, here and now. Cosmic consciousness is achieving a state of expanded awareness, seeing beyond the limitations of localized experience, witnessing the ever-present one without attachment or self-made barriers. You may very well attain a temporary glimpse during a peak experience as physical realm dimensions merge into unity; the true higher reality masked by our reinforced illusions of separation and fragmentation. That blank piece of paper already contains everything before the artist puts pen to paper, "creating" something from nothing. Creation is a process of manifesting; an act of extraction. But in order to extract, it already had to exist. Nothing is ever really created but rather unveiled. Inventions are really just discoveries, of something that was always there waiting to be manifested. Physical and non-physical realms are ever connected, interrelated and integrated; a marriage of being and becoming to create everything.

DUAL REALM IMMANENCE – Everything is Always Present

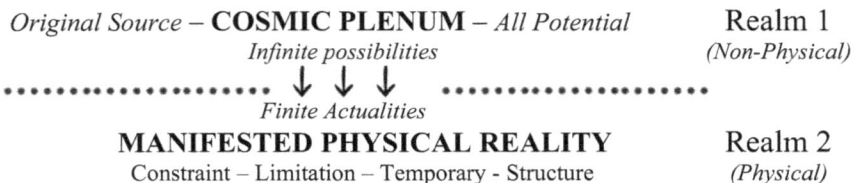

Original Source – **COSMIC PLENUM** *– All Potential*   Realm 1
Infinite possibilities   *(Non-Physical)*
••••••••••••••••••• ↓ ↓ ↓ •••••••••••••••••••
Finite Actualities
MANIFESTED PHYSICAL REALITY   Realm 2
Constraint – Limitation – Temporary - Structure   *(Physical)*

* Both realms are always present – everything is already here and now.
  The *infinity* of space and *eternity* of time are always present here and now.

"Human beings and all living things are a coalescence of energy in a field of energy connected to every other thing in the world" – Lynne McTaggart (*"The Field"*)

Realm 1 is the higher reality permeating everything in realm 2, which merely manifests parts of it, moment by moment. It represents all, and it's everywhere, here and now. We reveal it little by little as our thoughts lead to actions and creations, perceived by each individual as something "new". Everything we experience in the physical realm is really an unveiling process where what was already available finally emerges. The process of evolution is simply the slow, gradual unveiling of everything possible into its actual physical form. In a comprehensively integrated fully connected universe, change anywhere becomes change everywhere. Likewise, there really is no *us vs. them*, it was always *we*.

"*The universe is represented in every one of its particles. Everything is made up of one hidden stuff*" – Ralph Waldo Emerson

Tracing the interconnectedness of everything comes down to comprehensively *Reverse engineering physical reality.* Our TOE Model does this starting in the center and working outward in every direction. That center is man in middle, the central hub, the intersection of everything. The Model expands further as a tapestry of scales-spectrums-hierarchies-holons, with 3 nodal levels and 3 primary axes.

It's All Here Now, filtered by man into parts and localized fragments, where the *one* is perceived and experienced as *many*. Consider a DVD whose digital content is frozen in a timeless realm as a complete whole, until the disc is played, coming to life, experienced part by part, only to recede back to timelessness upon completion. It's the same with a record needle that intersects the finite with the infinite, bringing life to music in the eternal now. Similarly, our personal experience creates the Past-Present-Future relationship out of a timeless unity. Your life journey follows this same artificial splitting dynamic. Human reality is nothing but filtering unity. Nothing is invented, only discovered or revealed. "There's nothing new under the sun" – just derivatives and variations of what's already here. "Originality is undetected plagiarism" (a quote I'm stealing from someone else ☺)

In our man-centered reality *everything* is an extension of You: Parent → Child, Individual → Collective, Artist → Art, Inventor → Invention, Actor → Act, etc. Extensions of You include clothing, tools, vehicles, home, family, town-city-state. Likewise, You are an extension of the entire universe; a mirror of it and all of its attributes…You are a perfect inverse of the universe, like a lock and key making each other whole…a perfect fit, magically connecting, affecting and completing an interlinked relationship with it. You are a localized center of that Universe, and like the particle from a quantum wave collapse, you're an *actual* chosen from infinite *possibles*. Every relationship you have with something out there is a mirror of You in the center. Reality is actually a holographic universe with You at the local center. Universe = Youniverse…and *everything* is connected to You, everywhere and everywhen.

"The universe is a circle without a circumference, and every one of us is the center of the universe"
— D.T. Suzuk

13   IT ALL FITS

The Theory of Everything presented here consisted of two basic segments; first, identify what everything actually covers and then show how it all interconnects. Part one captured the essential ingredients present in anything and everything, and how they're expressed at every level. This was depicted in a 3-D matrix linking 3 primary essences at 3 primary levels. The middle level represents You right at the center of it all. Then we revealed what was above and below you, comprised by The World *outside*; nature below and society above. Next, we looked inside, exploring human reality itself. We also looked at temporal reality, which covers everything before, during and after You, and then a deep dive into the nature of time itself, as well as a broad look at history, which is simply the societal portion of temporal reality. Finally, we revealed overlooked aspects of hidden reality to acknowledge what we don't know and things we think we do but actually don't.

Part two consisted of unifying everything through a variety of different lenses and aspects of reality. We showed how the extensive accumulated knowledge of civilized man makes up a comprehensive kaleidoscope of different categories connected along a unified spectrum of science and non-science. Then we revealed laws and codes that unify everything in every field, activity and special interest. Next, we captured the omnipresence of life as a progressive force nested everywhere, linking everything in a continual coevolving process. We conducted a deep dive into the physical infrastructure of reality and revealed unities in physics and the interconnecting unity of dual realms – physical and nonphysical. Finally, we captured the interconnectedness of everything revealed in unifying spectrums, unifying patterns and universal templates present everywhere. So, part 1 and part 2 can be summarized as *Everything*, *Fits*.

SUMMARY OF UNIFIERS

In our pursuit of unifying everything into a single comprehensive model we revealed a variety of different components and mechanisms built into the fabric of reality that link and connect everything. Part two, *Fits*, provided unifying spectrums in each chapter that connected the elements of broad areas and categories. Looking at *The World* out there and everywhere we identified an inescapable spectrum of opportunities and threats everyone faces. Our direct connection with The World both constrains us to its unpredictable whims *and* provides us with the ability to influence our relationship to it, thus altering our own destiny. Human reality involves numerous unifying spectrums including states of consciousness, personality types, emotional spectrums and an assortment of bell curves capturing multiple ranges of human capacity. The human urge to reunify with the divine produces a unique spectrum of experience and love, expressed in a variety of modes ranging from friendship to the romantic. Our examination of hidden reality produced a spectrum of the unknown, including misknowns (mistaken facts), known unknowns and unknown unknowns. We took a deep dive into the nature and dynamics of time, identifying its comprehensive, vast connecting spectrum. We also identified the tapestry of history with its multiple levels, scales, scope and depth, as well as recurring cycles in history and unifying repeated themes.

Universal laws unify everything at every level, including physical principles, life lessons, societal issues and general laws of nature. Universal codes are equally pervasive at every level, including physical world math, biological DNA, societal language, and the highest-level code; pattern language. Our extensive analysis of knowledge produced a unifying spectrum that began with hard science, then transitioned into soft science, the humanities, pseudoscience, parascience and nonscience. The spectrum of knowledge is unity filtered through a prism of multiple perspectives. Life itself is a great unifier, present in everything to varying degrees. All organic life is connected to every other life form in a comprehensive interrelated super-ecosphere where each and every organism is unified and connected by an unbroken ancient chain of DNA. The physical infrastructure of the universe likewise has multiple unified components, including matter-energy, space-time, magnetic-electric, particle unities and dual connected realms. Our universe itself resembles a giant hologram where every part contains the whole. The interconnectedness of everything produces universal recurring patterns and templates, along with a wide variety of spectrums, scales, hierarchies, holons and networks that link anything and everything, everywhere.

Symmetry is an omnipresent unifier manifesting as sameness, wholeness, self-similarity and synchronization. The universe exhibits ubiquitous supersymmetry, revealed in the residue of interconnectedness with patterns of harmony and proportion where parts and components look alike, balance out and mirror each other. Symmetry is embedded unity built into the universe, leaving its fingerprint everywhere we look. Nature's bountiful organic forms exude this quality though a wide variety of plants, animals and miscellaneous creatures possessing radial, bilateral and spherical symmetry, expressed as living geometry. Mathematic ratios and numeric proportions are likewise embedded everywhere in nature; at all levels, within every niche and throughout every domain.

Temporal symmetry is present everywhere as well, conveyed in recurring cycles, waves, rhythmic change and repeating themes. Music is geometry in time, mirroring the same divine balance and harmony of symmetrical shapes and proportional relationships, captured mathematically in Pythagorean octaves. It's the very same essence present in literary prose, comedic timing, poetry rhymes, and a captivating passionate speech. History demonstrates these qualities as well, moving in rhythmic cycles of its own, repeating universal themes and collective life lessons.

Symmetry is just a fancy word for sameness. It's present in every math equation that contains an equal sign. It's revealed in physics' unities, chemistry's Periodic table, biology's tree of life, and cosmology's holographic Universe. It's implied in universal themes of religion, mythology and pseudo-sciences, through similar symbols and rituals, subconscious meanings, and divine insights translated into human relevance. But we need only look at ourselves to see evidence of ubiquitous symmetry and the harmonious connection between man and everything within, above and beyond – a microcosmic mirror of the cosmos itself.

The signature of symmetry can be seen in manifestations of self-similarity. It surfaces as universal constants, harmonic motion, invariants, crystalline alignment, love, vibration, exponents (powers of 10), universal essences, resonance and fractal patterns. Self-similarity in its purest form is a circle, the geometric shape of unity, along with its higher transcendent self; the sphere.

CONNECTING THE DOTS

The interconnected nature of everything is subtle, illusive and largely hidden, masked by our tendency to focus on just what's in front of us. Linking the vast aggregate of *all* things from our entire reality is a comprehensive task revealed though a combination of spectrums, scales, networks, hierarchies, holons, maps, models, matrices, tapestries, recurring themes and relational patterns. Each of these were used throughout the chapters to capture the interconnectedness of everything as it applies to various areas of expression:

Ch 2 Identified universal essences in everything, expressed at every level, including their dual aspects
Ch 3 Produced a TOE model connecting everything along scales, levels, hierarchies, and tapestries
Ch 4 Revealed how *The World* connects nature, man and society in a comprehensive interacting matrix of opportunities and threats; a collective inside-out projection of man
Ch 5 Connected Man (YOU) to the universe as a microcosmic hub; a mirror image of it, a dynamic system of systems at the center of it all
Ch 6 Presented a spectrum of Time connecting everything in a vast tapestry of timelines linking all things to all other things and back to YOU
Ch 7 Uncovered scales of omnipresent unity hidden by limitations (human perception and awareness)
Ch 8 Unveiled a spectrum of unified Knowledge connecting all fields, disciplines and perspectives
Ch 9 Identified universal laws and codes connecting everything via abstract language, and patterns
Ch 10 Defined the essence of Life on a unified spectrum connecting all beings and things in an ongoing co-evolution of increasing complexity and progressing consciousness
Ch 11 Revealed the omnipresent Dyad pattern present in all things, and the holographic matrix connecting everything as whole-parts
Ch 12 Presented comprehensive connecting scales, levels, hierarchies, holons, networks and patterns

Mathematics is a fundamental connector of all things, equating and relating practically every phenomenon in physical reality. Math applies to nature, man and virtually everything in The World, capturing both sameness and degrees of difference wherever they appear. It even connects dimensional differences that are transcendent via 90^0 angles, conveys irrational relationships between physical and non-physical realms: infinites, irrational numbers (pie, Sq Root of 2, etc.). Moreover, it captures unity of self-similarity (X^2, $1/x$, ratios, logarithms...).

Laws and codes are equally important connectors of things, linking all through universal principles, associations, and correlations. Nothing, whether physical object or living being, can escape the laws of nature. The same applies to subtle regulating forces in human relationships and society as a whole. Veiled rules and codes direct how The World works, maintaining the system and *everyone* subject to it; no one is exempt. Society itself is subject to nature, connected to the ecosphere, ecology webs, weather and climate, geological disturbances and extraterrestrial impacts. The earth as a whole resembles a single living superorganism (Gaia), connecting every living sup-part in a vast interrelated, interacting network of coevolving life.

Connecting the dots involves reverse engineering reality. It's really just breaking down the process where man's subjective perception separates and fragments unity, putting the pieces back together and reconnecting it all. However, it does require acknowledging hidden reality produced by man's limited perspective wedged in the middle of everything. We accomplish this through a transcendent perspective of higher awareness to see the bigger picture; going beyond our limited timeframes, sizeframes and mindframes. It's what we experience watching an accelerated movie or time lapse. It's how we can capture history's tapestries and depth, omnipresent vibrating energy fields, radiation present everywhere, universal patterns recurring at every level, and the temporary nature of all things in the here and now. Suddenly the subtle becomes obvious, with connectors appearing everywhere: metaphors, divine codes, DNA, parsimony, complex networks, resonance, analogies, languages, hierarchies, cycles, symmetry, and repeating themes.

Our TOE model is a comprehensive connector, linking all things, essences, levels and dynamic processes. It's an oversimplified map highlighting the primary-nodal levels (physical, biological, societal), but it contains everything within it. Each level is a whole-part; whole at its own level and part of the next higher one. Physical matter at the lower level is part of man in the middle, who is also part of society at the higher level. This holonic pattern is a hierarchy of connectedness where complexity and degrees of consciousness increase vertically. Consciousness itself is a connector, just as spirit and the life essence are, omnipresent everywhere, linking all things on a great spectrum of varying degree. Consciousness is a form of resonant spirit, emerging as light at the physical level (spirit manifest), water in organic life, and the electric force driving everything in modern civilization.

When we connect the dots of anything we get one step closer to confirming that everything is, was, and always remains connected, though some less than others. Separation is merely lower degrees of connectedness. Our localized, subjective perception and experience of separation are both an optical illusion and a sensory delusion; a limitation both necessary and practical to function and survive in The World we created and willingly participate in. While unity is the true reality, it's filtered into polarized spectrums by man. Absolute reality *is* unity, and it can be reacquired or reunified using the ultimate connector of everything; pattern language. The fifth primary essence, relation, is the very definition of pattern language. It connects, links, correlates and unifies everything in both physical and nonphysical reality. If there is a universal equation that can connect everything together, it won't be written in math – it will be conveyed in pattern language.

UNIVERSAL PATTERNS

Everywhere we look we see forms, objects, shapes, images, arrangements…and within each thing there is both an internal pattern and external relationship with everything else. In a random universe most of it would be dissimilar however the closer we look at each the more we find common types, recurring themes, cosmic habits, repetitive schemes, reappearing templates, and resonant forms. Most of these ubiquitous patterns are scale invariant, meaning they're similar at every level we look, like a fractal image. They're ordered by universal principles, laws and processes that influence and shape everything in physical reality.

Universal patterns appear and reappear everywhere as recurring forms and templates permeating the entire landscape. We see it in basic geometry where a simple dot joins other dots to become lines, then multiple lines become polygons and 3 dimensional forms. Some are more fundamental than others, including the circle/sphere, spiral/helix, radial/spokes, and other derivative patterns such as particle-cell-seed-nest-womb-tomb, or cylindric forms including channels, tunnels, tubes, conduits, etc. Nature draws from these patterns and incorporates them within life's forms as organs, embryos, skeletons, and entire living systems. Those organisms copy other universal forms like the branching pattern of arteries, veins, lungs, nerves, etc. mirroring nature's design expressed in leaves, lightening, tributaries, roots and tree branches. Other borrowed forms include membranes-shells-scales-bark-sheath-skin-clothes, etc. Man-made objects and advance constructions all incorporate these same fundamental patterns, despite each designer believing they are creating new forms when in reality they are simply borrowing and recycling from age old original universal templates.

Patterns of process are as ubiquitous as their 3D object counterparts, but they're form has an abstract temporal nature. Change is serpentine; a fluid quality progressing through time in a variety of modes insuring nothing ever quite stays the same. Reality is a process of periods and events subject to transitions, phases, increments, thresholds, sequences, spurts, emergence, accelerations, growth and decline, transformation, reversals, and changing tempos. These patterns are fundamentally abstract, difficult to track in the limited here and now. They become more apparent as time passes, providing context in hindsight gained after the fact. Process patterns are the essence of life; ongoing and interconnected everywhere.

Relationship Patterns include both spatial and temporal dimensions, establishing meaning and context to anything and everything. They relate things through a multitude of attributes, including proximity, association, divergence, position, affinity, order, tension, alignment, differences, linkage, attraction, correlation, harmony and discord, stability and chaos. Everything is related to everything else in unique ways where the type of relationship involved is more important than the thing itself. You by yourself are nothing without comparison to everything around you in both space and time. Parts derive their meaning by the wholes they support. Polarities involve opposites that derive meaning from their counterpart ends. What would cold be without hot? What would big be without small? Levels and scales likewise derive meaning from differences between each level and point on a scale.

What we perceive as solid, stable forms are merely resonant patterns. All material matter is nothing but vibrating fields held together by atomic forces. Atoms themselves are more like wave clouds of motion than they are stable particles. Everything we touch, feel, sense and see is nothing but millions of vibrating parts balanced in equilibrium, appearing as a stable structure from the outside. All things are temporary resonance patterns, including you, held together by the same proportional counterforces that keep wholes from breaking down into smaller parts. A living being is maintained in a delicate balance between static and dynamic forces operating in the sweet spot that allows minor changes without major disruptions; a teeter toter range that adjusts to external demands without going too far from the ground state. Everywhere around us we see a multitude of things and structures that are either in a state of balanced resonance or are transforming into other forms – post resonant state.

The dynamic nature of our ever-changing environment of universal patterns and changing forms is fundamental to the infrastructure of physical reality. And the original source of it all is our dual-realm modality (physical and non-physical), interacting, interpenetrating and influencing everything, everywhere, all at once. The most fundamental pattern emerging from dual-realm dynamics is the omnipresent Dyad. This fingerprint of duality is revealed in each chapter:

Ch 2 5 Primary essences each have a dual aspect; a static vs. flow dynamic
Ch 3 Structure and Process aspects of the TOE Model
Ch 4 Secret cypher of The World, societal products/services, laws/politics, norms/fads
Ch 5 Body/soul, conscious/subconscious, being/doing, left/right brain, stages/journey
Ch 6 Time bound physical/timeless nonphysical, here/now vs past/future, events/trends
Ch 7 Hidden reality itself, surface/shadow, spirit/manifest, middle view/everything beyond
Ch 8 Hard science/non-science: art-religion-mystical-magical, para & pseudo,
    critical/creative thinking, art and science, schools of thought and dual perspectives
Ch 9 Finite/infinite, sound/music, light/color, line/curve, constant/variable, digital/analog
Ch 10 Organic form/evolutionary process, identical DNA parts vs. emergent life qualities
Ch 11 Dual realm, Space/Time, Matter/Energy, Magnetic/Electric, Gravity/Radiation, Field/Current
Ch 12 Illusion of separation, duality dynamics, Nature-world metaphors, Holographic parts

The Dyad pattern is derivative residue produced by our dual realm reality. The nonphysical realm manifests into the physical realm, producing mirroring duality patterns. It creates a dynamic of transcendent relationships connected by 90-degree difference, including science vs. religion, left vs. right brain, being-doing, digital-analog, matter-spirit, space-time, Yin-Yang, and all the dualities embedded in mythology connecting nature, man and the divine, expressed in the language of metaphors and symbolism.

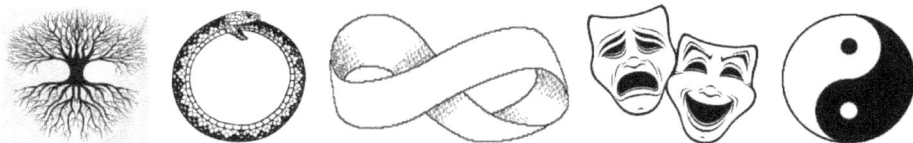

Pattern Language is the relational code connecting all things, processes, and derivative forms. Relationships between everything are expressed in a variety of subtle codes (math, music, symmetry, numbers, colors, geometry). Our TOE model is a pattern code, connecting everything in a language of relationships. Physical reality is a byproduct of the spirit realm manifesting into patterns of temporary forms, through filters, dynamic resonance, creative expression, and quantum processes. The stuff of life is primarily a composite of dynamic patterns of energy interacting, transforming, and recycling to create infinite forms in temporary states.

Pattern is fundamental. Out of nothing we create something, from something we get everything: numbers, elements, types, tones, images, events, etc. All matter consists of the very same stuff; 3 particles – protons, neutrons and electrons. Every type and form of matter has these same basic parts, yet every element is unique with its own special properties. The difference simply lies in their patterns. Materiality is secondary, pattern is primary.

Pattern is essentially about relationships. It's a code of relational information expressed in math shapes, music ratios, DNA sequences, color energies, written language, atomic forms, and even bar codes. "Solid" objects are an illusion of perception where their true form consists of resonant energy structures; knotted light. Patterns of resonant energy vibrations produce light and color, sound and music. Events are really just resonant convergence points, knots on a timeline. Resonance is a pattern of vibrational self-relation, like a rubber band's single circle twisting upon itself creating 2 circles (duality), then many; one unified connected whole with new emergent parts. All reality is created via this resonant pattern process: nothing becomes something, then everything (same as the TAO).

Our dynamic universe is a comprehensive pattern of patterns – the highest-level code linking everything. It's the highest language containing the most meaning, providing the most context. At every level there are patterns of relationship: spatial, temporal and resonant. There's omnipresent symmetry, universal templates and recurring forms. YOU possess all of these within. Man is a mirror pattern of the universe. Every pattern out there is right here within YOU.

Man the Microcosmic Template

Each of us occupies a position in the middle of everything; both the hub of a diverse spoke wheel and a center dot in the middle of a giant comprehensive sphere; the ultimate unifier and connector. While there's an enormous universe above and beyond our comprehension, there's also an incredibly tiny universe below and within. We're the central hub smack dab in the middle, connected to everything in every direction and scale. We're a central whole-part; a unit within an infinite holon spectrum, serving as a part to all above and a whole to all below. Each of us is like a cell containing both a sub-cell (nucleus) and DNA blueprint of the whole body above and beyond. It's a unified linked continuous spectrum where each rung connects with elements above and below.

Man is a byproduct of dual realms; physical and nonphysical. We emerge as a connecting bridge between spirit of the nonphysical realm and manifested physical reality of *The World*. So, man is composed of both: elements of both, experiences of both, and expresses both. Man is a spiritual being – a soul tethered to a physical body serving as a vehicle to navigate *The World* limited by physical constraint. It's an external world of threats and opportunities at every turn, with a variety of challenges lurking in arenas above (society), below (nature) and all around.

For practical reasons, man in middle is focused on the local here and now at the expense of everywhere else and everywhen else. This natural tradeoff produces hidden reality where man loses sense of anything way below or above, way before or after, or very deep within. Man's subjective perspective separates unity into partial truths – the mind separates, isolates and compartmentalizes. This local perspective creates the illusion of separation reinforced by ego's self-limiting perception of identity *in here* detached from everything *out there*. It's this self-imposed isolated experience that creates polarities out of unities; thus we get big/small, near/far, good/bad, etc. Ironically, this dynamic appears to transform the connecting fabric of space and time into disconnected dimensions of separation.

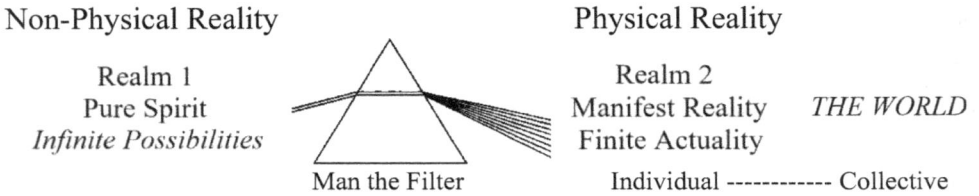

Non-Physical Reality   Physical Reality

Realm 1   Realm 2
Pure Spirit   Manifest Reality   *THE WORLD*
Infinite Possibilities   Finite Actuality
Man the Filter   Individual ------------ Collective

Man is a mirror of the universe, composed of the very same attributes present everywhere. Science confirms our material makeup traced back to the cosmos, and religion concurs proclaiming "God made man in his own image". The stuff and patterns of the universe are imbedded within each of us – a system of systems, the most complex machine ever created. We're all vibrating in a dynamic equilibrium state; a balance of form and flow, matter and spirit. Each of us is a living prism, a **spirit filter** expressing the universe locally; a process we experience through behavior dynamics, personality types, life lessons, developmental dynamics, and progressing states of being. We're all living time capsules of evolution as our DNA carries a record of biological history linked all the way back to the original single cell that started it all. Our linkage to every past generation that ever lived is no different than our connection to every past and future event that occurred or will occur, anywhere in the universe. We're *the* central hub with spokes extending in every direction connecting to us, to every now, before and after, and everything else along with it. Whatever your concept of God is, it's not something separate out there; it's everywhere, including within you, right here and now.

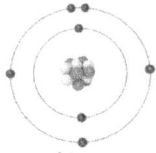 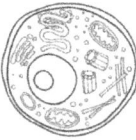  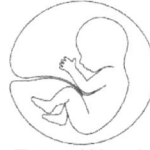 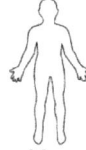 

| Atom | Cell | Seed | Fetus-Womb | Man | Earth |

Microcosms manifest at every level: a universal template of life

THEORY OF EVERYTHING SUMMARY

Part One set out to ID *Everything*, ID *universal essences*, set the universal model with Man at the center and work outwards, including outside/inside, above/below, before/after. Include every level: the outside *World* of nature and society, the inside world of *human reality*, the *temporal reality* of time, and *hidden realities*.

Part Two: Set out to Unify *Everything*, reveal unities in knowledge, laws and codes, life, physical infrastructure and the interconnectedness of it all.

Theory of Everything Requirements

Chapter one identified what a TOE must include and accomplish: reverse engineer reality, define and identify what *Everything* is, Identify required supporting elements and address fundamental concerns:

* Define and capture *everything* ✓
  A: Ch 1 Comprehensive categories of everything – page 5
* Identify the essential elements in *everything* ✓
  A: Ch 2 5 Universal Essences – page 16
* Reveal how the world works – mechanisms, processes, systems, etc. ✓
  A: pages 376, 390, 406, 422, 436, 449, 454, 459
* Identify universal truths, laws, and operating principles ✓
  A: pages 264, 270, 279, 288, 331, Ch 12 Whole chapter
* Reverse engineer reality – reveal its fundamental infrastructure ✓
  A: pages 190, 380, 385, 390, 397, 406, 411, 422, 436
* Demonstrate how seemingly random parts are actually unified wholes ✓
  A: Ch 12 Entire chapter
* Bridge gaps and expose the misleading appearance of separateness ✓
  A: pages 402, 426
* Produce a simple model relating how everything connects ✓
  A: Ch 3 Whole chapter pg. 28
* Reveal the deep source of connectedness and unmask hidden unity ✓
  A: pages 121, 149, 173, 247, 256, 263, 270, 426, All Ch 9 & 12
* Uncover universal patterns by connecting the dots ✓
  A: pages 301, 316, 318, 324, 331 345, 422, 429, 431, 434, 436, 441, 449, 454, 459

Answer Fundamental Questions:

What is the essence of life? A: pages 15, 279, 328, 386, 451, Ch 10 Whole chapter
What are the universal laws of life and reality? A: pages 264, 270
What are the hidden codes connecting everything? A: CH 9 Whole chapter
What is time and how does it actually work? A: CH 6 Whole chapter
What are the secret patterns in the universe? A: pages 301, 318, 320, 406, 445, 451
What is the foundational structure of physical reality? A: Ch 11 Whole chapter
How do human beings connect with everything? A: pages 28, 39, 80, 87,133, 156, 460
How do YOU connect with everything? A: pages 25, 31, 66, 121, 123, 151, 179, 476

Theory of Everything Model: (MOE – *Model of Everything*)

Chapter 3 introduced the TOE model as a 3-part matrix – a multidisciplinary, comprehensive format connecting everything along interrelated scales and spectrums. It reveals universal essences present at every level, along with a holonic spectrum of increasing complexity emerging from physical to biological to societal (nature, man, society). The model places man (YOU) at the center and branches outward. Most importantly, the TOE model incorporates the process nature of reality; the most fundamental element creating transcendent duality.

  Chapter 11 introduced the actual Theory of Everything which is the basis for a TOE model; the existence of dual realms, physical and nonphysical, connected, overlapping, interacting to produce duality dynamics in everything at every level.

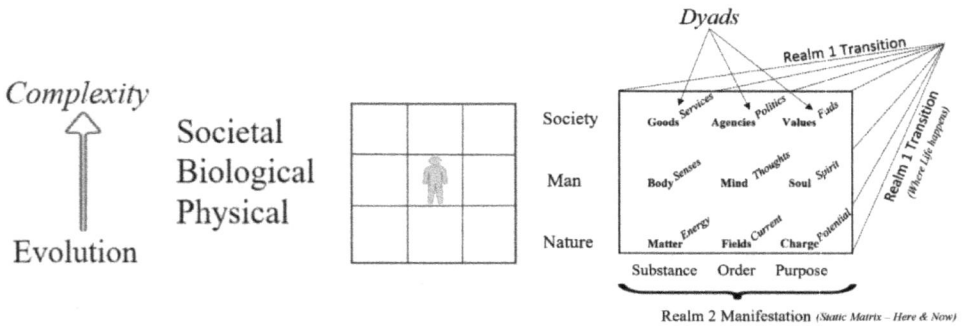

Realm 2 Manifestation *(Static Matrix – Here & Now)*

THEORY OF EVERYTHING KEY THEMES

Process is the creative origin of *Everything* – the fundamental basis of reality, represented by the diagonal line on our TOE model. It's expressed in quantum mechanics where moment by moment choices turn potentials into actuals. It produces an infinite series of micro moments that become enduring periods and distinct timelines. *Process* and *Time* are primary while the physical reality we perceive and experience are mere byproducts left in the wake. Process connects everything through dynamic interactions playing out in every situation and circumstance we experience in the various arenas of life. Processes are fundamental to physical reality at every level, expressed as dynamics of change via Time, Evolution, The World, Hidden reality, Laws of nature and Dual Realms. Dynamics involve subtle mechanisms built into the infrastructure of physical reality, including vibration, emergence, transformation, development, creativity, action-reaction and a universal constant state of flux, often hidden by balancing of counteracting forces in temporary states of equilibrium.

Temporal processes produce change dynamics present in everything, conveyed at each of the primary levels: natural, biological and societal. At the natural level, processes of time stretch over eons producing small, incremental change that adds up to great differences over long periods beyond human comprehension. These evolutionary dynamics can generate mysterious results due to unpredictable properties of emergence and complex branching of coevolutionary change.

Organic living systems are temporary forms coming and going through cycles of life and death, reproduction and continuation through regenerating progeny. Human beings pass the information torch through consecutive generations, leveraging language and media to capture individual experiences and transfer them collectively. Each life journey is a personal transformation leading to cumulative higher levels of awareness and collective consciousness.

Time progresses at the societal level through dynamics of history, expressed as a vast interconnected tapestry of linear webs with cycles, themes, tempos, and a wake of ruins, relics and legacies. The acceleration of technological progress further reveals the temporary nature of all things. Increasing acceleration of change in just the last century impacts culture where each generation no longer shares the values of its predecessor, reflected in recent distinct generational cohorts.

Physics exemplifies the dynamic interaction of forces through temporal processes. A variety of mechanisms and mathematic relationships pervade physical reality, serving as filters, levers, catalysts, inhibitors, and transformers that drive ubiquitous cause-and-effect interactions. Process dynamics are expressed along a spectrum of tempo, including slow, steady, moderate, accelerated and sudden (flat line, slight incline, curved incline). Change can be either small or dramatic, vary by degree or in kind – partial or transformational. Dynamic interactions are often abstract processes imperceptible at our level, like ocean waves that seem to move but are simply patterns of energy flowing across the surface while the water itself merely bobs up and down (90^0 off the force line).

Vibration is omnipresent in everything. Time itself is a form of vibration, connecting all things that come and go, live and die, grow and decline. Everything vibrates and interacts with every other vibrating thing. It all starts with elemental particles, atoms, molecules, fields and forces in nature. Light is a vibrating wave of electromagnetic energy. Alternating current driving everything in civilized society is vibrating energy. Predominant cycles present everywhere in nature, your body, economics and culture are all vibrational. Your daily repetitive routines are vibrational activities in time. Intercourse is simply vibration, not unlike a churning engine, a bobbing sailboat, a well pump, a tuning fork or a ringing bell.

Resonance is a special property of vibration where repetition creates form. All material things are just patterns of energy in resonant states. Atoms are essentially clouds of energy vibrating in a structured pattern. All "solid" objects are just aggregates of so many tiny vibrating parts, forming a larger resonant whole – a stable emergent form produced by repetitious patterns of vibration. A brick wall is mostly empty space held together by dynamic tension of atomic forces producing stability out of perpetual interlocking vibrations. Resonance is a template producing mechanism where forms emerge at nodal points, including primary shapes, living beings, special events and states of being. Resonance manifests as obsessive thoughts, persistent habits, repetitious action and recurring patterns.

States of Being are resonant temporary dispositions present in all things, places, organisms and time periods. They're conditions that distinguish the relative differences between how things are and how they could be. Human states include health, frame of mind, emotional mood, readiness, attitude, etc. Collectively, states of being apply to businesses, culture, government, city-state-nation, etc. and relates to organizational health, stability, vibrance, responsiveness, security, etc. States can also relate to places and time periods identifiable by specific unique attributes we resonate to. States of being are resonant energies that change over time like shifting frequencies, each characterized by unique temporary conditions.

Music is a special category of resonant vibration that comes to life in time, and dies in silence. It's an elusive quality bound by particular ratios of time – some harmonic, others discordant. Music is a temporal language that either feels right or doesn't, written in a code of beats, chords, tempos, crescendos, melodies, and harmonies. Music is a special vibration in time where rhythmic qualities emerge that can sync with your state of being, producing transformative power out of literally nothing. Music adds life to a movie in much the same way color adds life to black and white images, or the way passion gives life to an otherwise dull speech. Life, like music, is about creative participation, not the goal… the journey, not the destination. Music, dance, art, sports…all come to life in fleeting magical moments, sandwiched in between lots of dull space, not unlike your own mundane life highlighted by periodic special events.

Light is *the* fundamental vibration within the universe, the highest octave level of our realm, operating along the boundary seam of physical reality. Everything in our universe is derived from it, including every form of energy and matter. The objects we touch and perceive as solid are comprised of structured energy, which is simply constrained light. All matter is composed of vibrating particles, forming wave-like clouds; fields that are more etheric than material. It's all held together by an array of interacting forces vibrating at hyper speed. Every part and pattern of energy within is a derivative of light. And light itself is a derivative of spirit from the nonphysical realm, both being illusive and essential to everything. Light is never here and now; it's always moving away at incredibly high speed, far beyond comprehension. The moment you turn on a light it's gone, hundreds of thousands of miles away in less than a second. You never see light directly; only indirectly. Everything we see is either reflected, filtered, refracted or illuminated, always a derivative form. Even that lightbulb you're starring at is simply an illuminated bulb, you can't actually see the invisible light traveling between it and you!

Water mirrors consciousness much like spirit manifesting as light. It's serpentine and illusive, assuming any form and shape, solidifies into rigid structures or dissipates into free-flowing gas, and possesses cleansing properties. Water permeates the whole earth, flowing into every nook and cranny it can, weaving a snake like path, penetrating into everything, including saturating the air above and ground below. Like the earth's surface, our bodies are also composed of mostly water. Large population centers are always located next to major sources of it. Weather dynamics are driven by changing conditions of moisture in the atmosphere, mirroring the same emotional interactions expressed as moods in people and temperaments of societies. Water's accommodative properties make life possible, operating in a middle sweet spot between structure and flow, much like man in middle. Life operates in a middle degree of order; too little and you get chaos, too much and you get unresponsive fixity which equals death. Martial artist Bruce Lee described the ultimate combative style as no style, comparing it to water: *"Be formless, shapeless, like water. If you put water into a cup it becomes the cup. If you put water into a bottle it becomes the bottle. You put it into a teapot it becomes the teapot. Now water can flow or it can crash. Be water my friend"*.

Life is a dynamic purposeful process, mirroring water's omnipresent quality; expanding into every crack and crevice in nature, every niche and every possible environment. But where water settles into stable states, life relentlessly presses forward as a perpetual progressive force, stopping for nothing. Life is a transcendent essence powered by intention, pushing ahead, sweeping across, burrowing in and expanding in all directions, leaving behind temporary forms, used and discarded as it presses forward and upward. Its evolutionary forward thrust is powered by a dynamic process of energy interactions; a continual dance between degrees of order and chaos, changing tempos, and perpetual cycles of birth-death and rebirth. Life is omnipresent. Everything is alive at different degrees, expressed along a continuum including nature, organic molecules, biological beings, collective civilizations and the planetary biosphere. Life is a singular progressive process connecting everything in hierarchies of wholes and parts, systems and subparts, and persistent patterns of resonant forms. Its broad sweeping wake creates a vast interconnected environment where every part at every level coevolves in unison. It's no coincidence that societies and nations mirror living organisms. Life's incessant progression is a creative force with emergent properties producing synergistic whole systems. Life tinkers, tests, innovates, challenges, transforms, works overtime, and never quits. It's the most creative, purposeful, ruthless force in the universe.

Human Beings are just one of life's residual temporary products. We're more process than substance; a persistent pattern in time composed of basic organic elements presenting an exterior appearance of stable form while all the internal parts are constantly discarded and replaced. We're a patchwork of subsystems drawing off resources around us, processing them, incorporating essential elements, eliminating the rest. Every day billions of our cells are discarded and replaced. We're hyperactive machines working overtime to maintain a delicate balance of order among the chaos of competing internal and external forces. Even while we sleep our bodies are actively regulating temperature, pressure, energy levels, metabolism, air intake, cellular repair, and mental cleansing through dream activity. Human dynamics are far more than just physical processes; they include navigating states of being, personal development, and duality dynamics of physical/spiritual interaction (*body-mind-soul-spirit*).

Human beings occupy a convenient middle position at the center of the universe; a midpoint between everything above, below, inside and out – serving as the connecting link between dual realms, physical and non-physical... the exact intersection bridging the two, functioning as spirit filters where the higher realm above is manifested into the physical world below.

The World out there is driven by a variety of dynamic processes where competing external forces clash in a far-reaching domain of constantly changing conditions. First, weather events are fluid, dramatic and unpredictable as counterforces of air pressure, heat, moister, and terrain features interact in a perpetual action-reaction-counteraction process. Second, nature itself is full of opportunities and threats where various environments and conditions can be either helpful or harmful; sometimes providing you a meal while other times you are the meal. Third, our civilized social environment poses the very same mix of opportunities and threats, interconnected within a grand matrix of rules, regulations, prizes and punishments. The external world is a comprehensive arena of interacting forces regulated by competitors, roadblocks, traps, deception, loss, unpredictable curveballs and occasional treasure. The World serves as a collective super-arena of arenas, each with unique challenges, nuances, language codes, pain and pleasure, advantages and disadvantages, laws, rules and endless lessons.

Human experience in The World is governed by counterforces of *complimentarity dynamics* where 2 *"independent parts"* interact, balance out and operate as *connected wholes*. Examples include hunter and hunted, opposing competitors, parasite and host, leader and the lead. Complimentarity dynamics tether and unify everything at every level. At the human level it's expressed as inner thoughts-outer actions, conscious-unconscious, body language, romantic couples, 1 on 1 fighters, and Man-Universe microcosm/macrocosm. In the temporal realm there's the complimentary connection between past and future. Culturally we see the counterbalance between the West and East, science and spiritual, Yin and Yang. In organic life, complimentarity is conveyed through parsimony as counterbalancing forces produce living forms, whose natural symmetries mirror the hidden forces at work. The infrastructure of physical reality confirms complimentarity is a built-in fundamental attribute of the universe: matter-energy, space-time, electric-magnetic, gravity-radiation, particle-wave. It's captured in universal codes and laws as well, revealed in micro principals of science, life lessons of man, strategy and tactics of competitions, and macro issues shared by all societies – symmetrical connecting dynamics of life at every level.

Original Unities Split Through the Prism of Man's Local Perspective

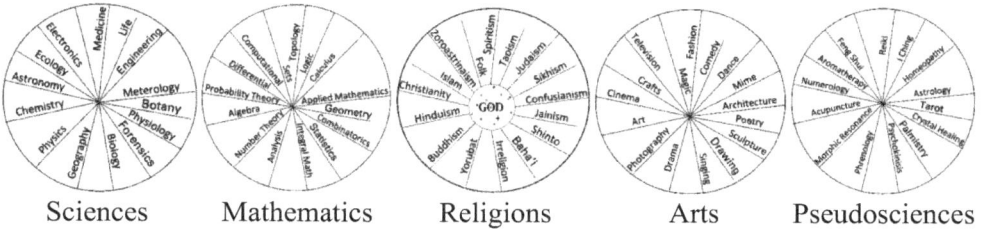

Sciences　　Mathematics　　Religions　　Arts　　Pseudosciences

Nations, states and societies exhibit the same process dynamics at a higher, collective level, including economic, political and cultural. Economics involve interconnected interactions between supply and demand counterforces, incentivized behavior (individuals and businesses), cycles of growth and decline, production and consumption, government intervention, and the linkage between interest rates, inflation, GDP, wages, goods and services, productivity and unemployment. Politics drive the dynamic competition for power among special interest groups, organized political parties, branches and levels of government, the entrenched establishment of bureaucrats and a hierarchy of local, state and national centers of leadership. Cultural dynamics link the changing tastes and moods of the public, interacting in an ever-shifting web of fads, emergent art styles, values, music types, fashion waves, and general cyclic renewal of *what's in*. Changing themes come, go and come back, often with a "Neo" prefix, expressed in revivals, movie remakes, song covers and periods of nostalgia. Man-made structures continually change as materials breakdown, fall apart and disintegrate, only to be re-stabilized by an ongoing cycle of maintenance and replacement, no different than organic beings replacing dying cells on a daily basis. Old buildings get knocked down and replaced, cars get junked and remade, streets get resurfaced and trash gets recycled, in an ongoing perpetual cycle of decay and renewal; a societal form of equilibrium.

UNIFYING EVERYTHING AT EVERY LEVEL AND SCOPE

KNOWLEDGE   LAWS & CODES   LIFE   PHYSICS   CONNECTEDNESS

DUALITY REALITY

Dual Realm Dynamics are *the* fundamental basis of our reality, responsible for all processes and interconnected interactions between the physical and nonphysical domains. It's the creative force present in literally everything, revealed by ubiquitous dyad patterns showing up everywhere in a variety of forms. The physical realm is a constrained version of its nonphysical counterpart, mirroring original abstract forms expressed in quantum fashion. The nonphysical spiritual realm manifests into the limited, lower, constrained physical domain, producing resonant structures that appear solid and permanent yet are neither. Impermanence, fluctuation, change and transformation are the truest reality, not static temporary forms. The dotted line in our TOE model highlights process dynamics which are fundamental, not the structured matrix itself. That dotted line applies to YOU as well because you're more process, flux and flow than static structure – more temporary and less permanent than you perceive.

Dual Realms are physical and nonphysical; one constrained, the other free from constraint. This produces the omnipresent **Dyad pattern** in everything, giving us polarities of structure/flow, fixed/variable, being/doing, temporary/eternal, event/trend, stable/chaos, digital/analog, right/left brain, etc. It's the source of experiential internal states of *being* in the here/now vs. the external action of *doing* through space and time. These differences can be either simple dualities or transcendent. Duality leads to all sorts of confusions: weird dream experiences, bizarre drug induced states, oddities of paranormal happenings and near-death experiences of transitioning realms. Duality creates paradoxes where properties of one realm are confused with the other. Examples include the folly of time travel, science vs religion, paranormal confusion, squaring the circle, and other man-made hypothetical brain teasers, puzzles and enigmas of ignorance. Duality is masked by our limited perceptions naturally defaulting to the physical realm. Man in the middle senses the clear, here and near easier than the abstract, far and away. The very tools allowing us to cope and survive within our local vantage point preclude grasping everything else beyond it. This reinforces illusions, distortions, deceiving perceptions and "misknowns".

Duality dynamics are the primary influence shaping man's experiential journey in and through The World. Navigating the challenges of daily life requires continual adjustment and adopting various coping mechanisms to survive and thrive. The soft spiritual half of man is continually pressured, pulled and tested by indifferent external physical demands coming from all directions. Staying centered and balanced requires ongoing vigilance but rarely lasts. Duality dynamics are omnipresent, impacting everything we do, built into the very fabric of the universe, applying at every level and field of interest, as covered in every preceding chapter

CONCLUSION

We've reverse engineered reality and broke the code to reveal the universal connectedness that was always right here hiding in plain sight. Everything was and is already here and now. Our universe is a self-reflecting mirror – an interrelated, interconnected dynamic system of systems, an omnipresent scale invariant pattern of patterns. The physical reality we experience is a fractal mirror pattern of the nonphysical realm, tracing back to the origin of everything, including You. Life is an ongoing dynamic fluid process of self-transformation, beginning and ending with You. There's no need to search the universe for the hidden patterns unifying it all…just look in the mirror; it's been right there in front of you the whole time. You are a perfect microcosm of the entire universe; a comprehensive, complex, connected, correlated machine. You are a local super system of systems. You are a local process within a universe of omnipresent processes. You are a whole-part; both a complete whole and subpart of a greater universe of whole parts, all connected from top to bottom. You are part of a grand hologram – an infinite fractal pattern linked to all other parts and patterns in perfect symmetry. You are the mean, median and mode simultaneously. You contain all the codes that connect everything; inside to outside, above and below, before and after.

You are the center of a vast universe connected at every level where embedded, symmetrical patterns link everything. Within this boundless plenum are ubiquitous repeating themes revealing universal qualities binding each part with every other part along scales, hierarchies, spectrums categories and dimensions. You naturally belong smack dab in the middle of it all, between a mini-universe within and a greater universe without – a comprehensive microcosm-macrocosm of interconnected parts and wholes. You are the greatest machine ever created, exemplified by a myriad of mechanisms, structures, programs and processes reflecting divine universal design. Your natural adaptability and resilience are complimented by unlimited potentials; a boundless yearning to grow, develop and transform into something greater than before. Every essence, attribute and special quality of the greater universe was already present, embedded deep inside you.

Reverse engineering reality is actually not complicated if you simply embrace your fundamental connectedness to everything. Just look at The World with your new set of eyes and see what's been hiding in plain sight. The universe out there was always right here in front of you. You're it! You are *the* center of everything; a magical Youniverse where it all connects directly back to You. The moment you realize this and embrace it…Aha…suddenly, wonderfully, everything fits!

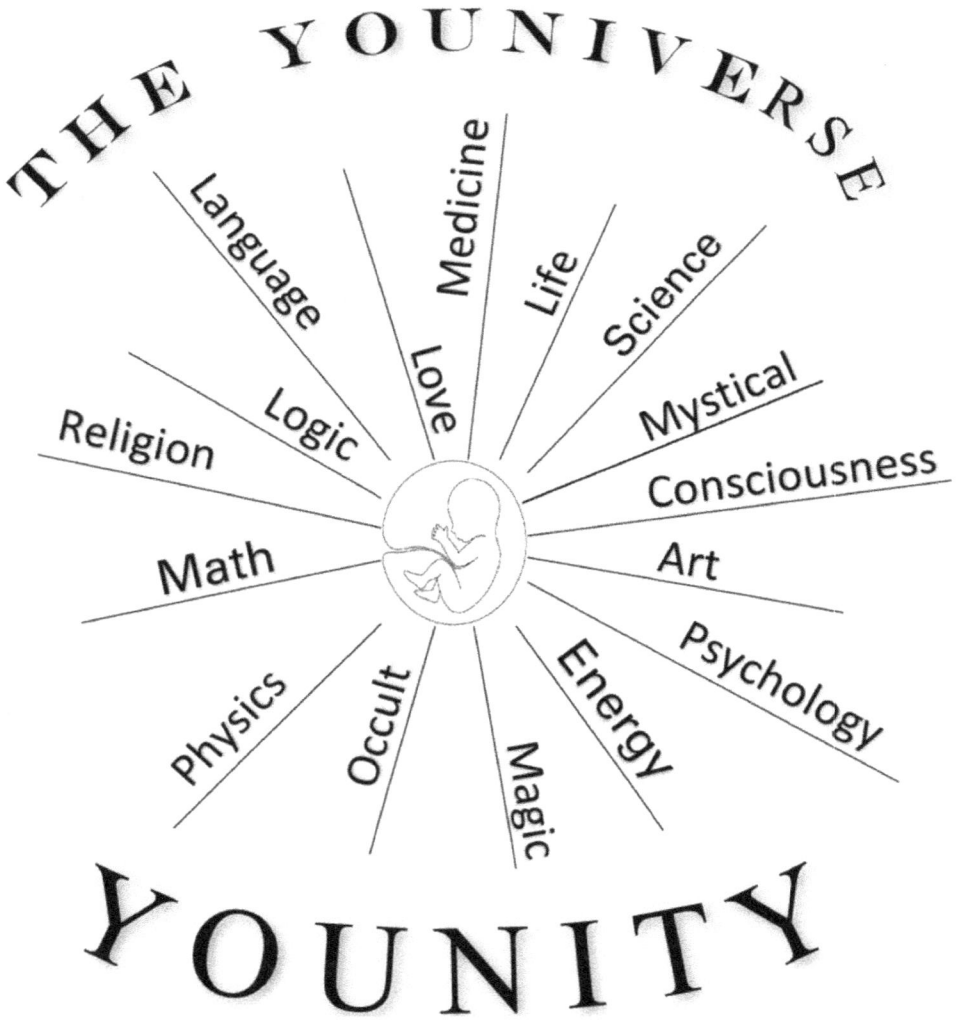

"We shall not cease from exploring…and the end of all our exploring will be to arrive where we started and know the place for the first time" – T.S. Elliot

Selected Bibliography

Armstrong, Karen. *A History of God*. (Ballentine Books. 1993)
Baker, Matt and Andrews, John. *Timeline of World History*. (Thunder Bay Press. 2020)
Barbour, Ian G. *Religion and Science*. (HarperCollins Publishers Inc. 1997)
Barrow, John D. *The Artful Universe*. (Back Bay Books. 1995)
Bradden, Gregg. *The Divine Matrix*. (Hay House Inc. 2007)
Calleman, Carl Johan. *The Purposeful Universe*. (Bear & Company. 2009)
Campbell, Joseph. *The Power of Myth*. (Anchor Books. 1988)
Capra, Fritjof. *The Tao of Physics*. (Shambhala Publications Inc. 1991)
Collins, Dianne. *Do You Quantum Think?* (Select Books, Inc. 2011)
Combs, Allan and Holland, Mark. *Synchronicity*. (Marlowe & Company. 1996)
Cook, Theodore Andrea. *The Curves of Life*. (Dover Publications. 1979)
Davies, Paul. *The Cosmic Blueprint*. (Simon and Shuster. 1988)
De Duve, Christian. *Vital Dust*. (Basic Books. 1995)
Dewey, Edward R. *Cycles*. (Manor Books Inc. 1973)
Diamond, Jared. *Guns, Germs, and Steel*. (W.W. Norton & Company. 1997)
Doczi, Gyorgy. *The Power of Limits*. (Shambhala Publications Inc. 1994)
Dyer, Wayne W. *The Power of Intention*. (Hay House Inc. 2004)
Falk, Geoffrey D. *The Science of the Soul*. (Blue Dolphin Publishing. 2004)
Field, Michael and Golubitsky, Martin. *Symmetry in Chaos*. (Oxford U Press. 1992)
Fontana, David. *The New Secret Language of Symbols*. (Watkins Media Limited. 2010)
Friedman, Norman. *Bridging Science and Spirit*. (Living Lake Books. 1990)
Garrant, Carl. *The Tao of the Circles*. (Humanics Publishing Group. 2000)
Ghyka, Matila. *The Geometry of Life*. (Dover Publications Inc. 1977)
Gladwell, Malcolm. *The Tipping Point*. (Back Bay Books. 2000)
Hall, George M. *The Ingenious Mind of Nature*. (Perseus Publishing. 1997)
Hawking, Stephen W. *A Brief History of Time*. (Bantam Books. 1988)
Hildebrandt, Stefan and Tromba, Anthony. *The Parsimonious Universe*. (Copernicus. 1996)
Howard, Lew. *Introducing Ken Wilber*. (AuthorHouse. 2005)
Huntley, H.E. *The Divine Proportion*. (Dover Publications Inc. 1970)
Jeans, Sir James. *Science and Music*. (Dover Publications, Inc. 1968)
Kauffman, Stuart A. *The Origins of Order*. (Oxford University Press. 1993)
Kelly, Kevin. *Out of Control*. (Addison-Wesley Publishing Company. 1994)
LaViolette, Paul A. *Genesis of the Cosmos*. (Bear & Company. 1995)
Marks, Robert W. *The Dymaxion World of Buckminster Fuller*. (Reinhold Publishing. 1960)
Murchie, Guy. *The Seven Mysteries of Life*. (Houghton Mifflin Company. 1978)
Noble, Denis. *The Music of Life*. (Oxford University Press. 2006)
Ouspensky, P.D. *A New Model of the Universe*. (Dover Publications Inc. 1997)
Pappas, Theoni. *The Magic of Mathematics*. (Wide World Publishing/Tetra. 1994)
Peterson, Ivars. *Fragments of Infinity*. (John Wiley and Sons, Inc. 2001)
Rensberger, Boyce. *Life Itself*. (Oxford University Press. 1996)
Russell, Walter. *The Secret of Light*. (The University of Science and Philosophy. 1974)
Schneider, Michael S. *A Beginner's Guide to Constructing the Universe*. (Harper P. 1995)
Schwartz, Gary and Russek, Linda G.S. *The Living Energy Universe*. (Hampton Roads. 1999)
Shlain, Leonard. *Arts & Physics*. (William Morrow and Company Inc. 1991)
Small, Jacquelyn. *Transformers*. (DeVorss & Company, Publisher. 1982)
Strogatz, Steven. *How Order Emerges from Chaos in Sync*. (Hyperion Books. 2003)
Talbot, Michael. *The Holographic Universe*. (HarperCollins Publishers Inc. 1991)
Taylor, John C. *Hidden Unity in Nature's Laws*. (Cambridge University Press. 2001)
Tolle, Eckhart. *The Power of Now*. (Namaste Publishing & New World Library. 1999)
Wade, David. *Crystal & Dragon*. (Destiny Books. 1991)
Watts, Duncan J. *Six Degrees*. W.W. Norton & Company. 2003)
Whitehead, Alfred North. *Process and Reality*. (The Free Press. 1978)
Wilson, Edward O. *Consilience*. (Alfred A Knopf, Inc. 1998)
Woodhouse, Mark B. *Paradigm Wars*. (Frog Ltd. Books. 1996)
Young, Arthur M. *The Reflexive Universe*. (Robert Briggs Associates. 1976)

About the Author

Before reaching the age of 10, Brendt showed early signs of a curious active mind, browsing the entire Time-Life book series with a special interest in rockets, submarines, physics, and science in general, then drawing up his own imaginative designs. As a teenager his interest shifted into contemporary subjects including economics, politics, pseudoscience and history…along with a healthy dose of science fiction. Following in his father's footsteps he joined an Olympic style weightlifting club and became a teenage champion at the state level, and placed 3rd in a national competition.

As a young adult Brendt owned and operated a videotaping business that he ran for 25 years. As a side hobby he produced documentary videos for cable television that won awards in Milwaukee, Wisconsin. During this same time period he practiced Astrology professionally for over 10 years.

Brendt graduated from the University of Wisconsin Whitewater, earning a degree in Economics and achieved the designation of Distinguished Military Graduate. He served in the Army Reserve for 32 years, is a graduate of the United States Army War College and retired at the rank of Colonel. Brendt is a licensed Florida Realtor and currently lives in Fort Myers Florida with his wife Debbie.

For more information and access to online videos visit
www.howeverythingfits.com

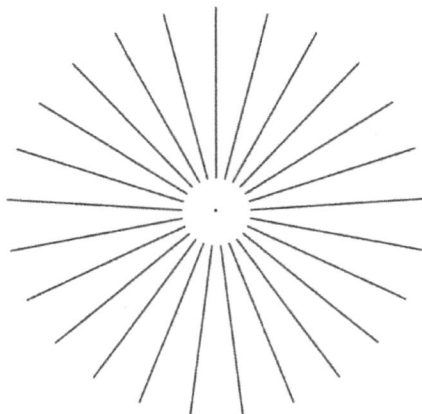

www.ingramcontent.com/pod-product-compliance
Lightning Source LLC
Chambersburg PA
CBHW021209090426
42740CB00006B/164